D1301117

Alkaloids

ALKALOIDS

Chemical and Biological Perspectives

Volume One

Edited by

S. WILLIAM PELLETIER
Institute for Natural Products Research
and
Department of Chemistry
University of Georgia, Athens

A Wiley-Interscience Publication

JOHN WILEY & SONS

New York Chichester Brisbane Toronto Singapore

Library of Congress Cataloging in Publication Data:

Pelletier, S. W., 1924–
Alkaloids: chemical and biological perspectives.

"A Wiley-Interscience publication."
Includes index.
1. Alkaloids. I. Title.

QD421.P39 1982 574.19'242 82-11071
ISBN 0-471-08811-0 (v. 1)

Printed in the United States of America

10 9 8 7 6 5 4 3 2 1

Dedicated to
Leona Jane, Bill, Dan, Rebecca,
Lucy, David, and Sarah

Contributors

M. H. Benn, Chemistry Department, University of Calgary, Alberta, Canada

Murray S. Blum, Department of Entomology, University of Georgia, Athens, Georgia

Manfred Hesse, Organisch-Chemisches Institut der Universität Zürich, Zürich, Switzerland

John M. Jacyno, Department of Pharmaceutical Chemistry, Medical College of Virginia, Virginia Commonwealth University, Richmond, Virginia

Tappey H. Jones, Department of Entomology, University of Georgia, Athens, Georgia

M. Volkan Kisakürek, Organisch-Chemisches Institut der Universität Zürich, Zürich, Switzerland

Edward Leete, Natural Products Laboratory, School of Chemistry, University of Minnesota, Minneapolis, Minnesota

Anthony J. M. Leeuwenberg, Department of Plant Taxonomy and Plant Geography, Agricultural University, Wageningen, The Netherlands

S. William Pelletier, Institute for Natural Products Research and Department of Chemistry, University of Georgia, Athens, Georgia

Preface

Today alkaloids are extremely important in chemistry, industry, and medicine. In the latter category, alkaloids find use for the following types of medicinal activity: analgesic potentiator (cocaine), antiamebic (emetine), anticholinergic (atropine, hyoscyamine, scopolamine), antihypertensive (reserpine, rescinnamine, deserpidine, protoveratrine A), antimalarial (quinine), antitumor (vinblastine, vincristine), antitussive (codeine, noscapine), cardiac depressant (quinidine), central stimulant (caffeine), diuretic (theobromine, theophylline), emetic (emetine), gout suppressant (colchicine), local anesthetic (cocaine), narcotic analgesic (codeine, morphine), opthalmic cholinergic (physostigmine, pilocarpine), oxytocic (ergonovine), skeletal muscle relaxant [(+)-tubocurarine], smooth muscle relaxant (papaverine, theophylline), sympathomimetic (ephedrine), and tranquilizer (reserpine, deserpidine). It is no exaggeration to say that, in many circumstances, alkaloids are indispensable.

The purpose of this series is to provide comprehensive and authoritative reviews of the chemistry and biological properties of the various classes of alkaloids. This series will include chapters on structure elucidation, synthesis, biogenesis, pharmacology, physiology, taxonomy, spectroscopy, and X-ray crystallography of alkaloids. Certain chapters will include treatment of several subjects, such as structure elucidation, synthesis, and pharmacology, whereas other chapters will treat a single aspect of alkaloids.

Because no current book or monograph provides the interdisciplinary treatment planned for this series, we hope these monographs will prove attractive as a source work for investigators working in such diverse fields as medicinal chemistry, natural products chemistry, pharmacology, pharmacognosy, biochemistry, phytochemistry, plant taxonomy, oncology, forensic science, and medicine. Publication of a volume is planned every 12–18 months, and each volume will contain 350–400 pages.

Chapter 1, Volume 1, provides a detailed treatment of the nature and definition of an alkaloid. The term *alkaloid* has been redefined along modern, functional lines. Chapter 2 is an excellent treatment of arthropod alkaloids, their distribution, functions and chemistry. Chapter 3 provides a detailed review of the biosynthesis and metabolism of the tobacco alkaloids, and Chapter 4 presents an excellent summary of the toxicology and pharmacology of diterpenoid alkaloids. Chapter 5 is a monumental chapter on a chemotaxonomic investigation of the plant families **Apocynaceae**, **Loganiaceae**, and **Rubiaceae** by their indole alkaloid content. Each chapter in this volume was reviewed by an expert in the field.

S. WILLIAM PELLETIER

Athens, Georgia
October 1982

Contents

Alkaloids

Chapter One

The Nature and Definition of an Alkaloid

S. William Pelletier
Institute for Natural Products Research
and
Department of Chemistry
The University of Georgia
Athens, Georgia 30602

CONTENTS

1. INTRODUCTION

What is an alkaloid? After discussing this question with several alkaloid chemists, a colleague suggested that "an alkaloid is like my wife. I can recognize her when I see her, but I can't define her." Indeed, defining what the term *alkaloid* means today is no easy task [1–13]. The reason is that over 5000 alkaloids of all structural types are known. No other class of natural products possesses such an enormous variety of structures. Steroids, for example, are all modeled on a few skeletal types. The same holds true for triterpenes, flavonoids, or polysaccharides. But alkaloids exhibit dozens of different skeletal types. This situation causes extraordinary difficulty in defining alkaloids so they may be readily recognized and differentiated from other classes of organic nitrogen-containing compounds.

This chapter treats the nature and definition of an *alkaloid* as this class of compounds is perceived by chemists and pharmacognosists. The term *alkaloid* was coined in 1819 by the pharmacist W. Meissner and meant simply, *alkalilike* (Middle English *alcaly*, from Medieval Latin *alcali*, from Arabic *alqaliy* = ashes of saltwort, from *qualey*, to fry). The first modern definition by Winterstein and Trier [14] described these substances in a broad sense as basic, nitrogen-containing compounds of either plant or animal origin. "True alkaloids" were defined as compounds meeting four additional qualifications:

1. The nitrogen atom is part of a heterocyclic system.
2. The compound has a complex molecular structure.
3. The compound manifests significant pharmacological activity.
4. The compound is restricted to the plant kingdom.

Alkaloids frequently occur as salts of plant acids such as malic, meconic, or quinic acid. Some alkaloids occur in plants combined with sugars (e.g., solanine from potato, *Solanum tuberosum* L., and tomato, *Lycopersicon esculentum* Mill.), whereas others are present as amides (e.g., piperine from black pepper, *Piper nigrum* L.) or as esters (e.g., cocaine from coca leaves, *Erythroxylon coca* Lam.). Still other alkaloids occur as quaternary salts or as tertiary amine oxides.

Compounds satisfying the definition of a "true alkaloid" are restricted to certain families and genera of the plant kingdom, rarely being distributed in larger groups of plants. Though about 40% of all plant families contain at least one alkaloid-bearing species, alkaloids have been reported in only 9% of over 10,000 plant genera. Among the angiosperms they occur abundantly in certain dicotyledons and particularly in the families Apocynaceae (dogbane, quebracho,

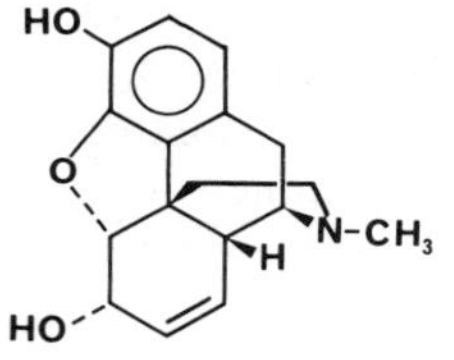

Morphine (1)

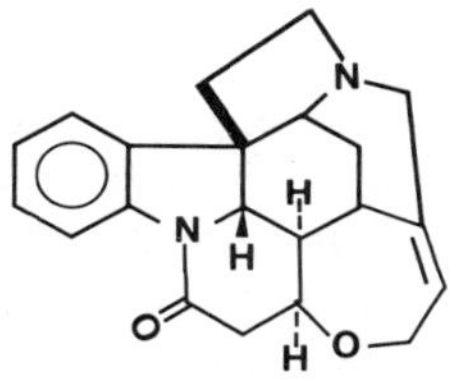

Strychnine (2)

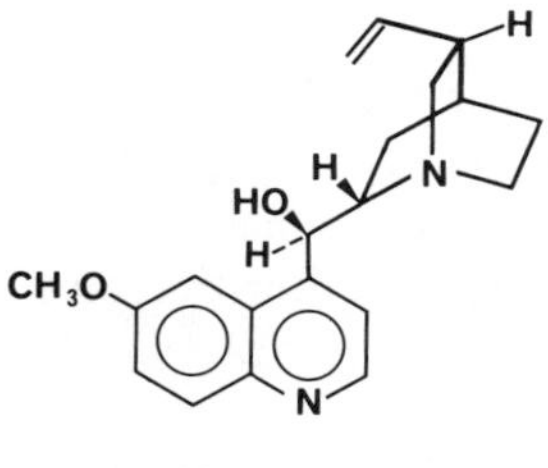

Quinine (3)

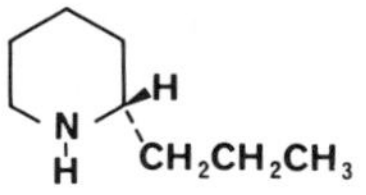

Coniine (4)

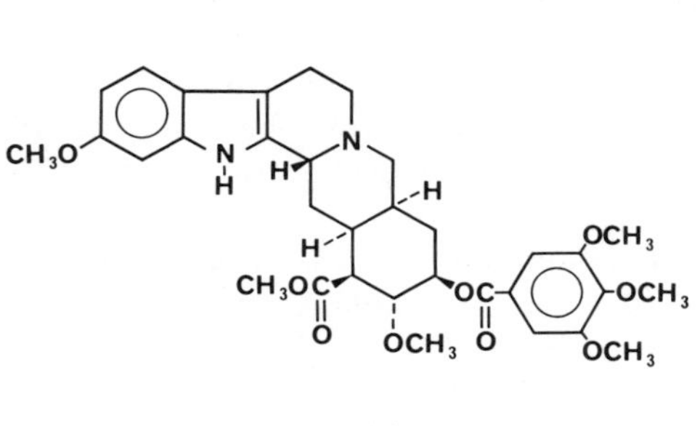

Reserpine (5)

pereira bark), Asteraceae = Compositae (groundsel, ragwort), Berberidaceae (European barberry), Fabaceae = Leguminosae (broom, gorse, laburnum, lupine, butterfly-shaped flowers), Lauraceae (rosewood tree), Loganiaceae (American jasmine, *Strychnos* species), Menispermaceae (moonseed), Papaveraceae (poppies, chelidonium), Ranunculaceae (aconite, delphinium, larkspur), Rubiaceae (cinchona bark, ipecac), Rutaceae (citrus, fagara), and Solanaceae (tobacco, deadly nightshade, tomato, potato, thorn apple). They are rarely found in cryptogamia, gymnosperms, or monocotyledons. Among the latter, however, the Amaryllidaceae (amarylis, narcissus) and Liliaceae (meadow saffron, veratrums) are important alkaloid-bearing families.

Examples of well-known compounds that satisfy the four-part definition above and that are universally accepted as alkaloids are morphine (**1**, opium poppy, *Papaver somniferum* L.), the first pure alkaloid isolated (Serturner in 1805); strychnine (**2**, *Strychnos nux-vomica* L. and *S. ignatii* Berg.); quinine (**3**, *Cinchona* bark, various *Cinchona* species); and coniine (**4**, poison hemlock, *Conium maculatum* L.). The latter three alkaloids were isolated by Pelletier and Caventou in 1817, 1820, and 1826, respectively. Coniine is of special historical interest in that it was the compound responsible for the death of Socrates in 400 B.C., when he drank a cup of tea made from poison hemlock, and it was the first alkaloid to be synthesized (Ladenburg, 1886). An example of an alkaloid of modern vintage is reserpine (**5**, *Rauvolfia serpentina* (L.) Benth. ex. Kurz.), a compound widely used as an antihypertensive agent and as a tranquilizer.

2. PLANT BASES SOMETIMES EXCLUDED

Unfortunately there is no sharp distinction between alkaloids and certain other naturally occurring plant bases. Most chemists do not regard such simple and widely distributed plant bases as methylamine, trimethylamine, and other straight-chain alkylamines as alkaloids. Many authorities also exclude the betaines (e.g., **6**), choline (**7**), and muscarine (**8**, fly agaric, *Amanita muscarea* L. ex. Fr.) [3]. Though these compounds are biosynthesized from amino acids and are basic, they are designated as *biological amines* [15], *biogenic amines* [5], or *protoalkaloids* [16] because their nitrogen is not involved in a heterocyclic system. Aliphatic diamino, triamino, and tetraamino compounds such as putrescine (**9**), spermidine (**10**), and spermine (**11**) are likewise excluded by most authorities, though macrocyclic derivatives of these polyamino bases are regarded as alkaloids (see Section 9.5). Some authorities even exclude the phenylalkylamines, such as β-phenylethylamine (**12**, mistletoe, *Viscum album* L.), hordenine (**13**, barley, *Hordeurn vulgare* L.), dopamine (**14**, banana, *Musa sapientum* L.), (−)-ephedrine (**15**, "Ma Huang," *Ephedra sinica* Staph.), mescaline [**16** "peyote," *Lophophora williamsii* (Lemaire) Coulter], and tryptamine (**17**, *Acacia* spp.) [3]. In certain cases, the distinction between what is an alkaloid and what is not is arbitrary. For example, thiamine (vitamin B_1, **18**), though a heterocyclic nitrogenous base with profound physiological activity, is

$^{\ominus}O_2CCH_2CH_2CH_2CH_2CH(CH_3)\text{-}\overset{\oplus}{N}\text{-}(CH_3)_3$

6

$HOCH_2CH_2\overset{\oplus}{N}\text{-}(CH_3)_3 \; ^{\ominus}OH$

Choline (7)

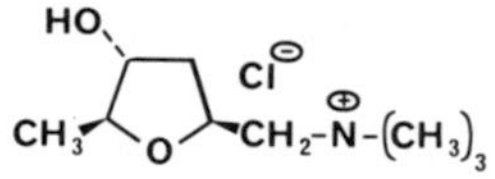

L-(+)-Muscarine Chloride (8)

$H_2N\text{-}(CH_2)_4\text{-}NH_2$

Putrescine (9)

$H_2N\text{-}(CH_2)_4\text{-}NH\text{-}(CH_2)_3\text{-}NH_2$

Spermidine (10)

$H_2N\text{-}(CH_2)_3\text{-}NH\text{-}(CH_2)_4\text{-}NH\text{-}(CH_2)_3\text{-}NH_2$

Spermine (11)

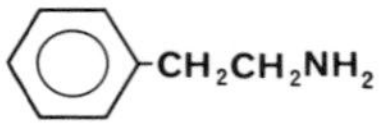

$C_6H_5\text{-}CH_2CH_2NH_2$

β-Phenylethylamine (12)

$HO\text{-}C_6H_4\text{-}CH_2CH_2N\text{-}(CH_3)_2$

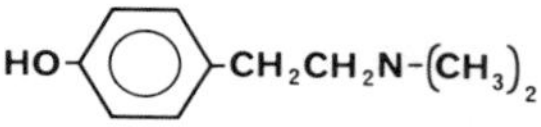

Hordenine (13)

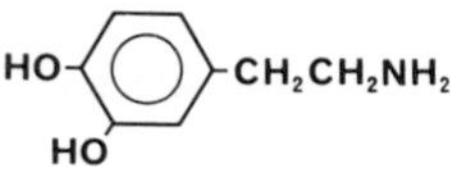

Dopamine (14)

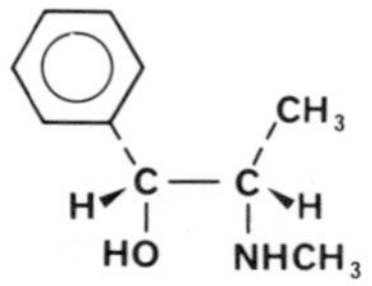

(−)-Ephedrine (15)

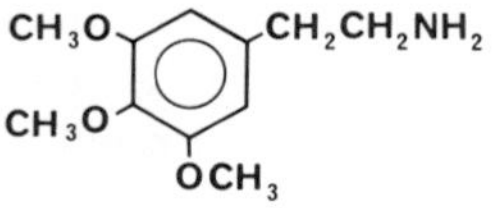

Mescaline (16)

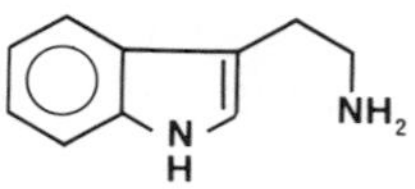

Tryptamine (17)

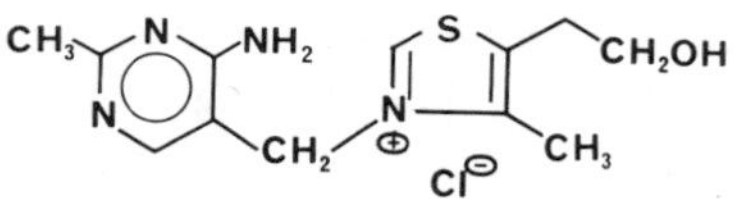

Thiamine Hydrochloride (18)

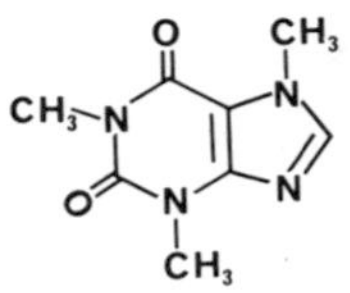

Caffeine (19) 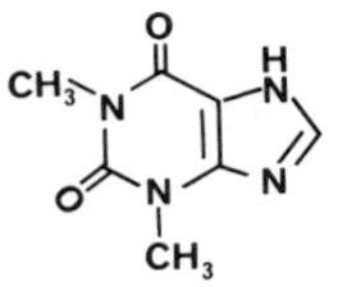 Theophilline (20) 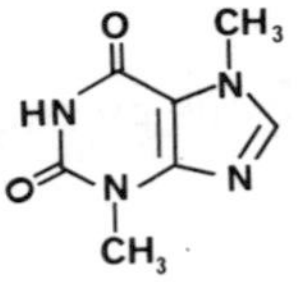 Theobromine (21)

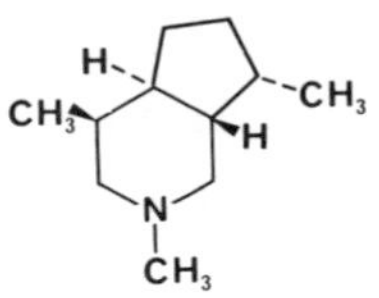

β-Skytanthine (22) 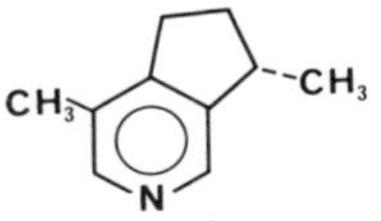 Actinidine (23) 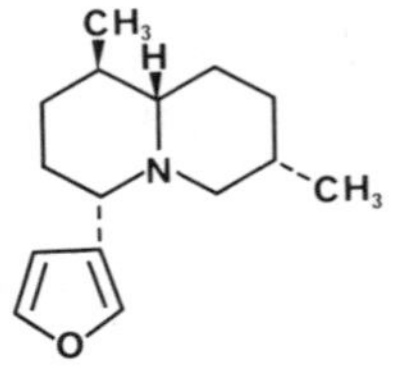 (−)-Deoxynupharidine (24)

not classed by most writers as an alkaloid because it is almost universally distributed in living matter.

Other examples of compounds sometimes excluded are the purine bases caffeine (**19**, coffee bean, tea leaves, cocoa), theophilline (**20**, tea leaves), and theobromine (**21**, cocoa beans, *Theobroma cacao* L.). These plant bases contain heterocyclic nitrogen and possess physiological activity, but because they are unrelated biosynthetically to the amino acids, certain writers do not consider them alkaloids [3]. Heterocyclic nitrogen-containing compounds modeled on monoterpenoid (e.g., β-skytanthine, **22**, *Skytantus acutus* Meyern; and actinidine, **23**, *Actinidia polygama* Franck), sesquiterpenoid (e.g., (−)-deoxynupharidine, **24**, *Nuphar japonicum* DC.), and diterpenoid skeleta (e.g., aconitine, **25**, *Aconitum* species) are regarded by most chemists as alkaloids, but are considered by some chemotaxonomists as *pseudoalkaloids* because these compounds are not related biogenetically to the amino acids [3]. Two further examples illustrating the extreme position that a few authors embrace are ricinine (**26**, castor beam, *Ricinus communis* L.) and gentianine (**27**) [12, 17]. Because ricinine is the *only* alkaloid in *R. communis*, and because the *only* alkaloid the majority of the species of Gentianacea contain is gentianine, these compounds are classed as pseudoalkaloids! The guiding principle invoked is that a *true* alkaloid-bearing plant must contain more than one alkaloid, the main component being accompanied by smaller amounts of a number of biogenetically related compounds [12, 17].

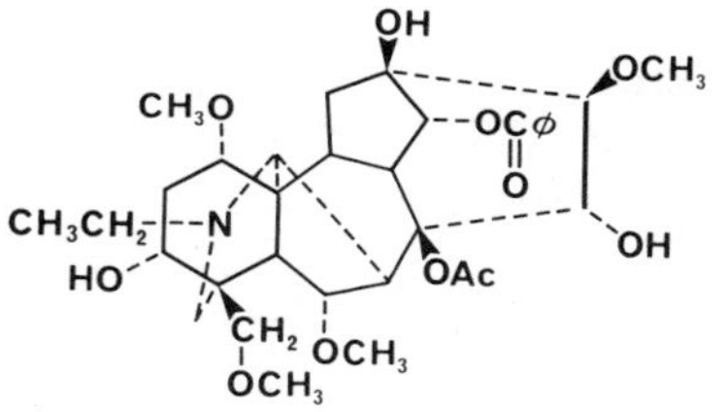

Aconitine (25)

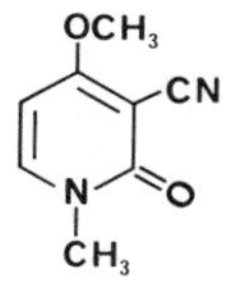

Ricinine (26)

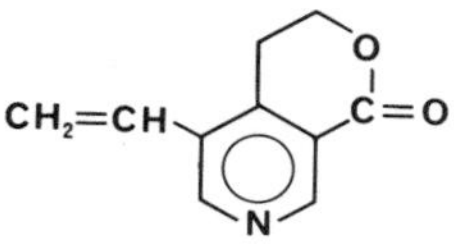

Gentianine (27)

3. NEUTRAL COMPOUNDS AS ALKALOIDS

Because the word *alkaloid* means *alkalilike*, basicity might be regarded as the hallmark of an alkaloid. Arbitrary conventions, however, have gained a secure foothold, for many neutral compounds now are classed as alkaloids. A prime example is colchicine [**28**, autumn crocus, *Colchicum autumnale* L. (Eur.)], an essentially neutral compound (pKa 12.35) in which the nitrogen is involved in an amide group. The only factors justifying inclusion as an alkaloid are its valuable medicinal properties and its restricted botanical distribution. Otherwise it does not fit the classical definition of an alkaloid. Another amide-type alkaloid is piperine (**29**) from black pepper (*Piper nigrum* L.). Other examples of neutral or weakly basic compounds that are usually classed by chemists as alkaloids are lactams, such as ricinine (**26**), mentioned above; certain *N*-oxides, such as the highly active antitumor agent indicine *N*-oxide (**30**, *Heliotropium indicum* L.), and the di-*N*-oxide trilupine (**31**, *Lupinus barbiger* S. Wats and *L. laxus* Rydb.); betaines [e.g., stachydrine (**32**, alfalfa, *Medicago stiva* L., and Chinese artichoke, *Stachys sieboldi* Miq.) and trigonelline (**33**, garden peas, oats, potatoes, coffee beans, hemp)]; and certain quaternary nitrogen salts such as the aporphine alkaloid laurifoline chloride (**34**, *Cocculus laurifolius* DC.); some authorities even accept certain nitro compounds, e.g., the aristolochic acids (**35** and **36**, *Aristolochia* species) [18–22].

4. TERRESTRIAL-ANIMAL-DERIVED ALKALOIDS

Alkaloids originally were considered basic compounds derived from plant sources. Certain purists still insist that alkaloids are found only in plants! Until

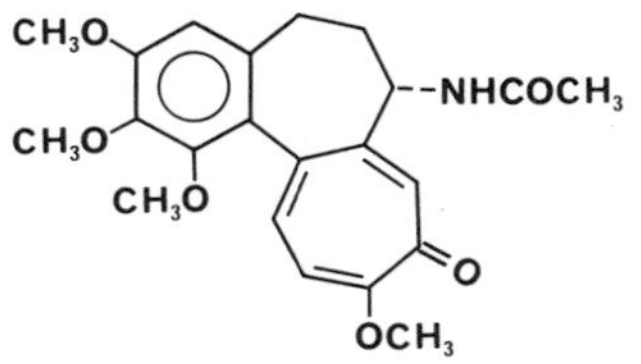

(–)-Colchicine (28)

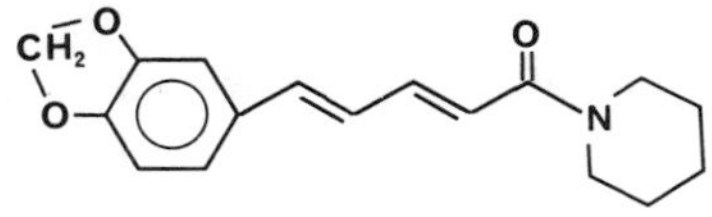

Piperine (29)

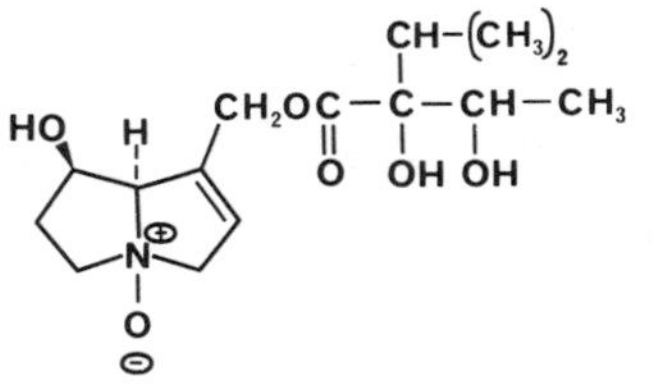

Indicine *N*-Oxide (30)

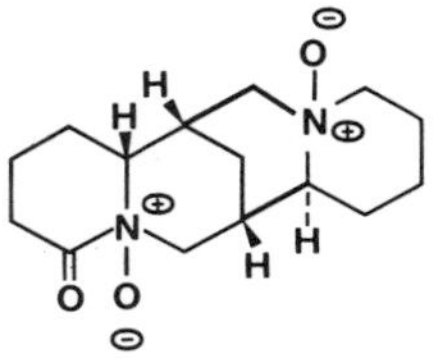

Trilupine (31)

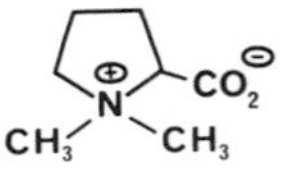

Stachydrine (32)

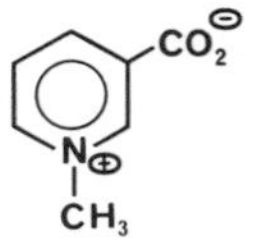

Trigonelline (33)

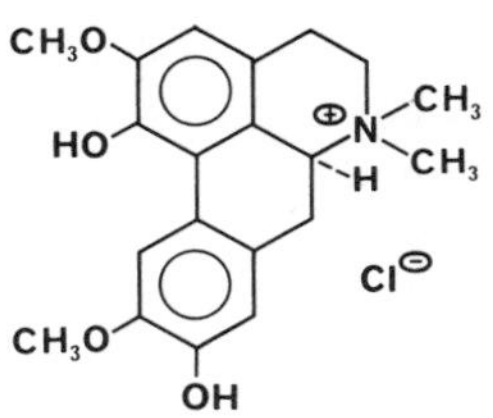

Laurifoline Chloride (34)

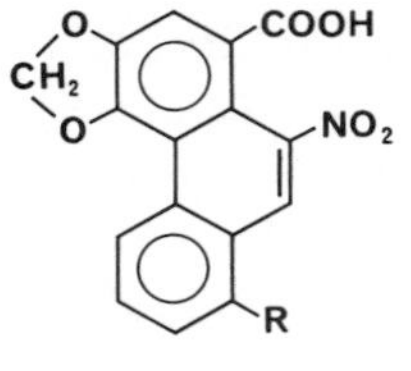

Aristolochic Acid-I (35) R=OCH_3

Aristolochic Acid-II (36) R=H

recently, compounds with alkaloid-type structures were found only rarely in animals. Today the situation has changed, with dozens of compounds with classical alkaloid-type structures having been isolated from animal sources. A brief survey of some of these compounds will illustrate the rich variety of alkaloidal structural types available from animals.

4.1. Salamander Alkaloids

The earliest authenticated example of the isolation of an alkaloid from animals is that of "samandarine" from the skin glands of the European fire salamander (*Salamandra maculosa* Laurenti) [23, 24]. "Samandarine" later turned out to be a mixture, from which the alkaloids samandarine (**37**) and samandarone (**38**) were isolated [25–27]. Subsequent work led to the isolation of other interesting structural variants: samandaridine (**39**), cycloneosamandaridine (**40**), and cyclo-

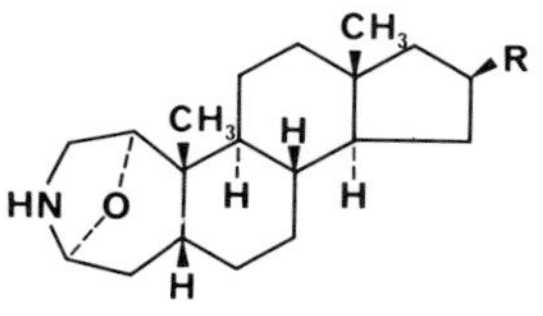

Samandarine (37) R=OH

Samandarone (38) R=O

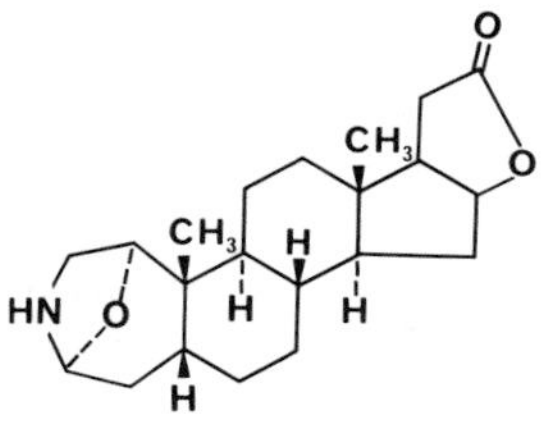

Samandaridine (39)

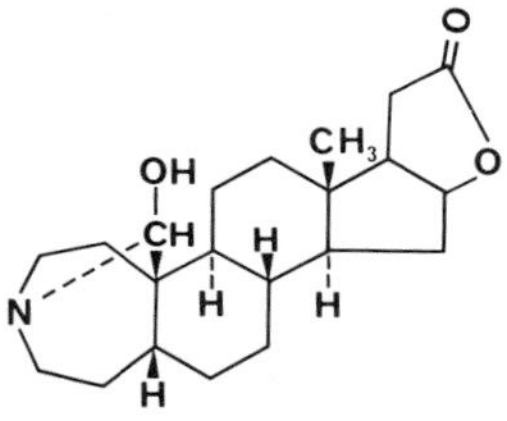

Cycloneosamandaridine (40)

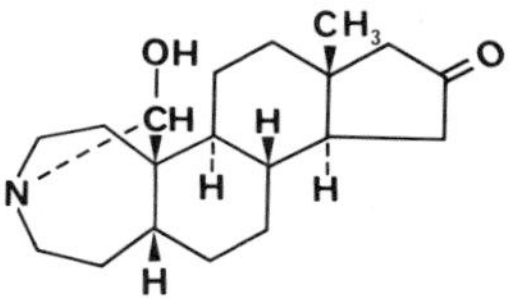

Cycloneosamandione (41)

neosamandione (**41**) [28–30]. These compounds are basic, form salts, and manifest the usual properties of alkaloids. To insist that because these compounds were not isolated from a plant they are not alkaloids is irrational.

4.2. Anuran Alkaloids

Toxic steroidal alkaloids occur not only in salamanders but also in certain anurans (frogs, toads, and tree frogs) [31, 32]. The skin of the brightly colored Columbian arrow-poison frog, *Phyllobates aurotaenia*, contains a highly lethal venum from which two major toxins, a cardiotoxin and a neurotoxin, both of which are steroidal alkaloids, have been isolated: batrachotoxin (**42**; LD_{50} 2 $\mu g/kg$, mouse) and homobatrachotoxin (**43**; LD_{50} 3 $\mu g/kg$) [33]. The skin extracts of another Columbian frog, *Dendrobates histrionicus* Berthold, yields several unusual acetylenic alkaloids, among which are histrionicotoxin (**44**), dihydroisohistrionicotoxin (**45**) and gephyrotoxin (**46**) [34, 35]. The *cis*-decahydroquinoline alkaloid pumiliotoxin C (**47**) has been isolated from the Panamanian frog *D. pumilio* [35, 36]. Subsequent studies on the companion alkaloids pumiliotoxin A and pumiliotoxin B have shown them to have the indolizidine structures **48** and **49**, respectively [37].

The parotid gland of the common European toad, *Bufo bufo bufo* (*B. vulgaris* Laur.), yields a number of tryptamine-type alkaloids, among which are bufotenin (**50**), *O*-methylbufotenin (**51**), and dehydrobufotenin (**52**). This toad also elaborates bufotoxin (**53**) a conjugate of a bufogenin with suberylarginine.

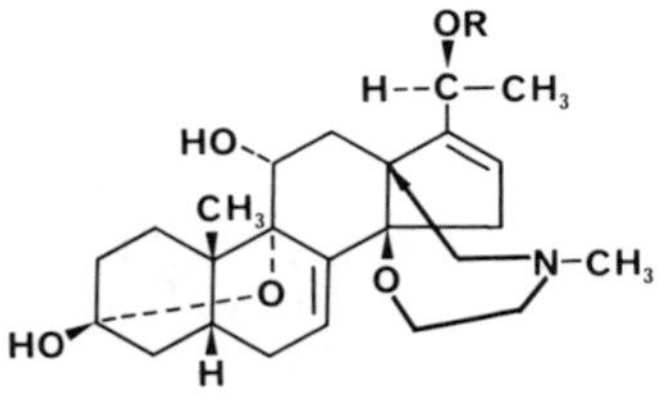

Batrachotoxin (42) R =

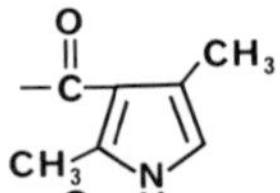

Homobatrachotoxin (43) R =

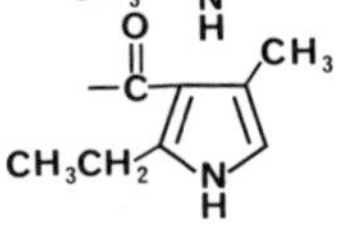

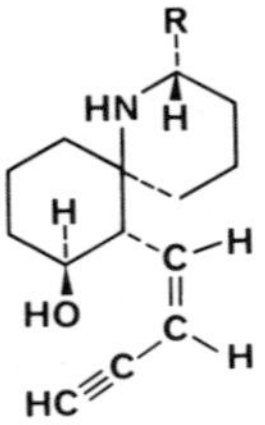

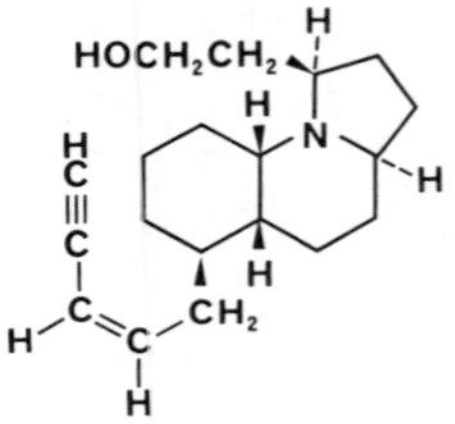

Histrionicotoxin (44) R = $-CH_2\overset{H}{C}=\overset{H}{C}-C\equiv CH$

(45) R = $-CH_2CH_2\underset{H}{C}=C=CH_2$

Gephyrotoxin (46)

Of interest is the occurrence of bufotenin (**50**) and *O*-methylbufotenin (**51**) in several plant species used by Indian tribes of South America and the Caribbean Islands as a ceremonial narcotic snuff called cohoba (e.g., in *Piptadenia peregrina* Benth, *P. colubrina*, *P. excelsa*, *P. marcocarpa*) [38–41], as well as in the fungus, *Amanita mappa* Batsch. [42, 43], and in human urine [44].

4.3. Mammalian Alkaloids

A few alkaloids have been isolated from mammals. Thus (−)-deoxynupharidine (**24**) and (−)-castoramine (**54**) have been obtained from the Canadian beaver (*Castor fiberi* L.) [45], and the interesting *ansa*-type pyridine structure muscopyridine (**55**) has been isolated from the scent gland of the musk deer (*Moschus moschiferus*) [46].

4.4. Arthropod Alkaloids [47]

4.4.1. Ant Alkaloids. The fire ant, *Solenopsis invicta* (= *saevissima*) Forel, secretes a powerful venom from which several 2,6-dialkylpiperidines (e.g., **56–**

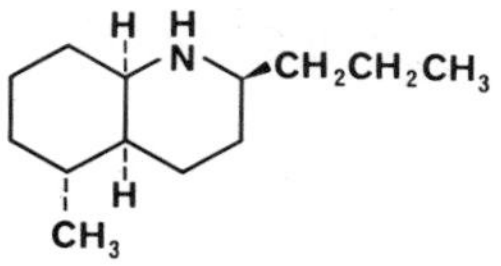

Pumiliotoxin C (47)

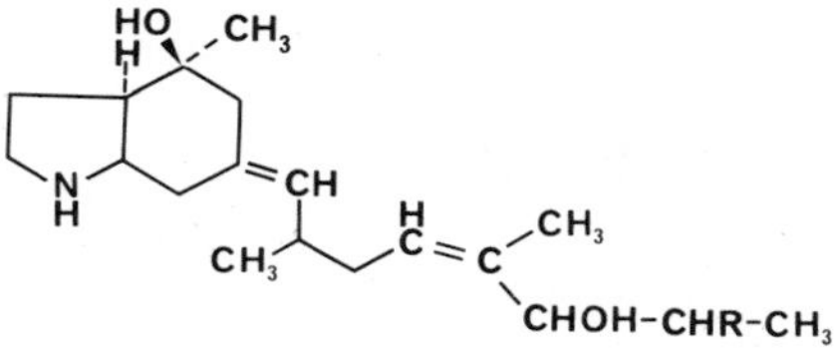

Pumiliotoxin A (48) R=H

Pumiliotoxin B (49) R=OH

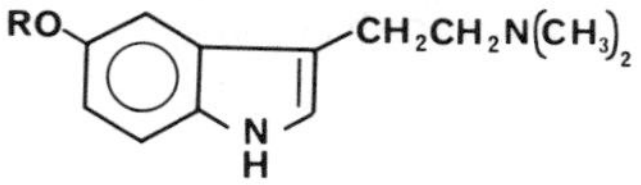

Bufotenin (50) R=H

O-Methylbufotenin (51) R=CH_3

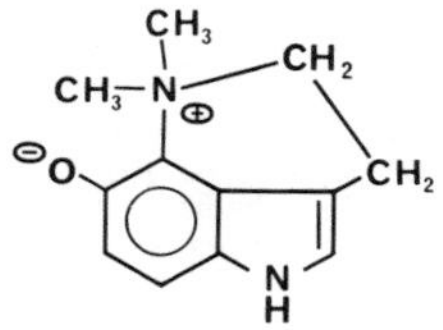

Dehydrobufotenin (52)

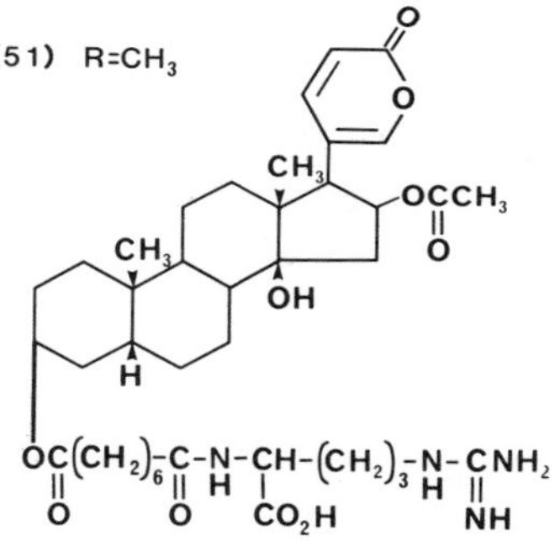

Bufotoxin (53)

60) have been isolated [48, 49]. These compounds are structurally related to cassine (**61**), an alkaloid isolated from the tropical shrub *Cassia excelsa* Shrad.

The large neotropical ant *Odontomachus hastatus* (Fabricius) discharges a chocolate-flavored secretion from which the dialkylpyrazine (**62**) has been isolated. The related species *O. brunneus* (Patton) furnishes the trialkylpyrazines **63–66** [50]. These pyrazines are potent stimulators of alarm behavior for

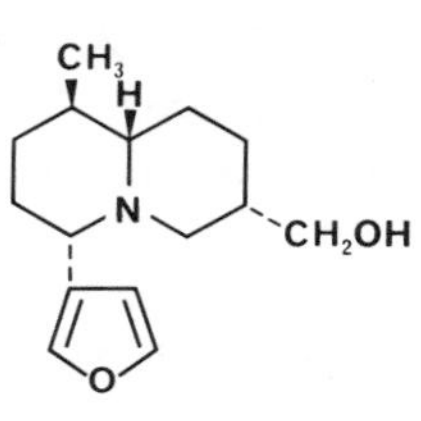

(-)-Castoramine (54)

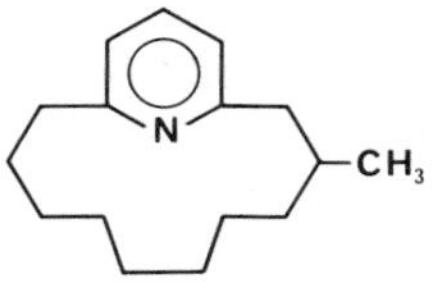

Muscopyridine (55)

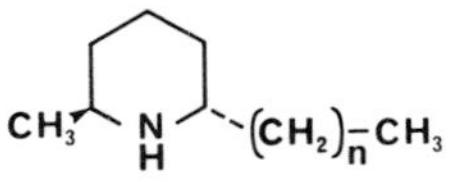

56 n=10

57 n=12

58 n=14

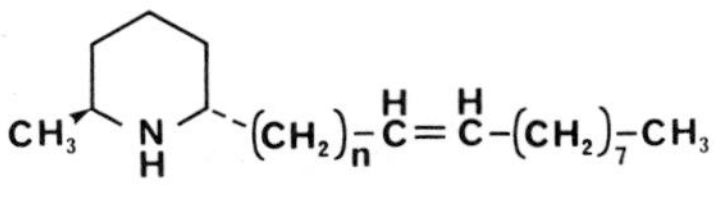

59 n=3

60 n=5

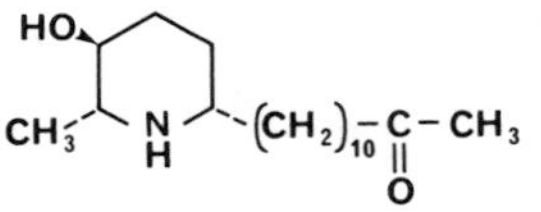

Cassine (61)

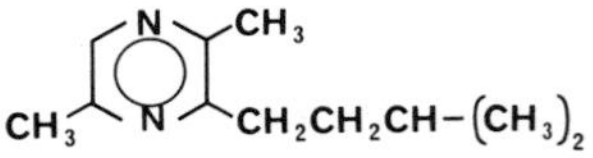

62

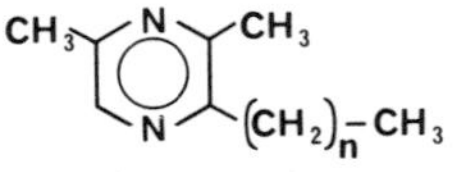

63 n=4

64 n=3

65 n=2

66 n=1

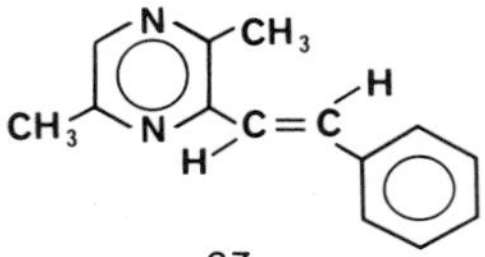

67

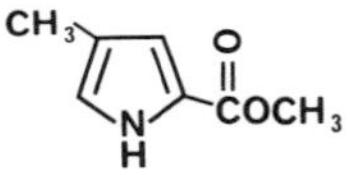

68

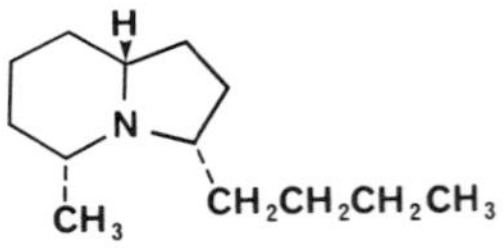

69

Odontomachus workers and probably are used also as defensive substances. The volatile fraction of the Argentine ant *Iridomyrmex humilis* (Mayr) contains minor constitutents that have been identified as 2,5-dimethyl-3-isopentylpyrazine (**62**), and (E)-2,5-dimethyl-3-styrylpyrazine (**67**) [51, 52].

The Texas leaf-cutting ant *Atta texana* elaborates a trail pheromone that has been shown to be methyl 4-methylpyrrole-2-carboxylate (**68**) [53]. This alkaloid is also the major trail-active compound of the ant *Atta cephalotes* as well as several other ant species of the tribe Attini.

Pharaoh's ant, *Monomorium pharaonis*, produces a trail pheromone from which the all-cis indolizidine alkaloid (**69**) has been isolated [54, 55].

4.4.2. Millipede Alkaloids (Diplopoda). Provocation of the European millipede *Glomeris marginata* causes discharge from the dorsal glands of a defensive

Glomerin (70) R=CH_3 Polyzonimine (72) Nitropolyzonamine (73)

Homoglomerin (71) R=CH_2CH_3

secretion containing the quinazolinones glomerin (**70**) and homoglomerin (**71**), compounds that the millipede synthesizes using anthranilic acid as a precursor [56–58]. The defensive secretion of the millipede *Polyzonium rosalbum* contains polyzonimine (**72**), as well as nitropolyzonamine (**73**) [59, 60].

4.4.3. Ladybug Alkaloids (Coccinellidae). Ladybugs (Coccinellidae), when molested, secrete hemolymph droplets of reflex bleeding at their joints. This

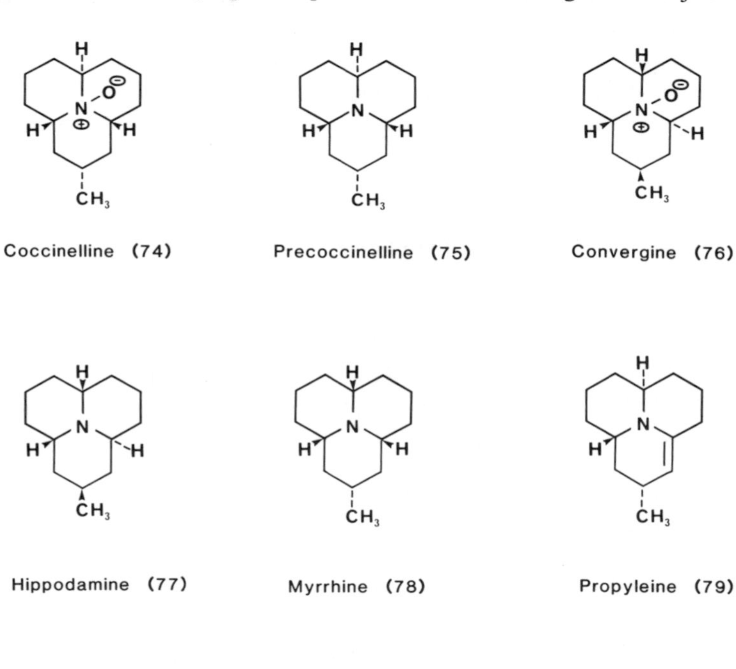

Coccinelline (74) Precoccinelline (75) Convergine (76)

Hippodamine (77) Myrrhine (78) Propyleine (79)

Adaline (80) Porantherine (81)

bitter-tasting secretion affords an efficient protection against predators. The hemolymph of the European ladybug (*Coccinella septempunctata*) has been shown to contain the *cis*, *trans*, *cis-N*-oxide coccinelline (**74**) and its free base, precoccinelline (**75**) [61, 62]. Azaphenalene alkaloids corresponding to each of the other two ring stereoisomers have been isolated from different species of ladybugs. Thus *Hippodamia convergens* affords the cis, cis, trans isomer, convergine (**76**), and its free base, hippodamine (**77**) [63], whereas *Myrrha octodecimguttata* contains the trans, trans, trans isomer, myrrhine (**78**) [64]. This alkaloid has also been isolated from the boll weevil, *Anthonomus grandis* [65].

The ladybug *Propylea quatuordecimpunctata* produces the alkaloid propyleine (**79**) [66] and *Adalia bipunctata* elaborates the homotropane alkaloid adaline (**80**) [67].

Since coccinelline (**74**) and its base (**75**) have been found in the eggs and larva of *C. semptempunctata* but not in the aphids upon which the ladybugs feed, these alkaloids most probably are synthesized by the insect itself [64, 68]. These azaphenalene alkaloids are not limited to insects. Thus porantherine (**81**) and closely related compounds have been isolated from the woody shrub *Poranthera corymbosa* Brogn. [69].

4.4.4. Water Beetle Alkaloids (Dytiscidae). The prothoracic defensive gland of the water beetle *Ilybius fenestratus* affords methyl 8-hydroxyquinoline-2-carboxylate (**82**) [70]. This alkaloid manifests powerful antiseptic activity and is apparently used by the beetle to prevent penetration by microorganisms.

4.4.5. Beetle Alkaloids (Staphylinidae). Several species of beetles of the genus *Paederus* elaborate a powerful cytotoxin and vesicant, pederin (**83**), which can be isolated from the hemolymph of the insects [71, 72]. The beetle *P. fuscipes* contains in addition to pederin the closely related compounds pseudopederin (**84**) and pederone (**85**) [73, 74]. Incorporation of (1-^{14}C)- and (2-^{14}C)-acetate into pederin, followed by appropriate degradation schemes, suggests that pederin and its analogs are biosynthesized via a polyketide pathway.

4.4.6. Butterfly Alkaloids (Lepidoptera). All the arthropod alkaloids mentioned above are of endogenous origin. Certain species of arthropods elaborate pheromones derived from exogenous pyrrolizidine alkaloid precursors. The male Danainae butterflies disseminate pyrrolizidine-derived pheromones during courtship. The pyrrolizidinone alkaloid **86** has been identified in the hair-pencil secretions of *Lycorea ceres ceres* and *Danaus gilippus berenice* [75, 76], where it elicits olfactory-receptor responses in female antennae and serves as the chemical messenger that induces mating behavior in the female. The related pyrrolizidine alkaloids **87**, **88**, and **89** have been detected in several other Danainae butterflies [77].

Some butterflies of the family Arctiidae are known to utilize pyrrolizidine-containing plants as larval host plants. Alkaloid **87** is present in the scent organ of *Utetheisa pulchelloides* and *U. lotrix*, and alkaloid **88** is also present in *U. lotrix* [78].

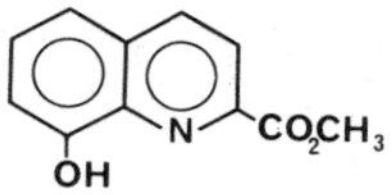

82

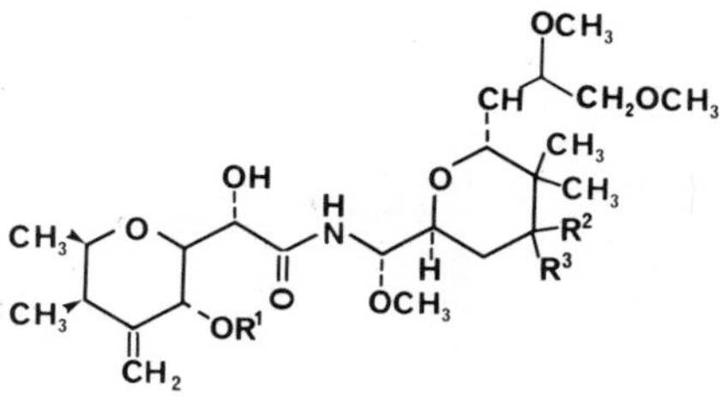

Pederin (83) R^1=CH_3, R^2=H, R^3=OH

Pseudopederin (84) R^1=H, R^2=H, R^3=OH

Pederone (85) R^1=CH_3, R^2, R^3=O

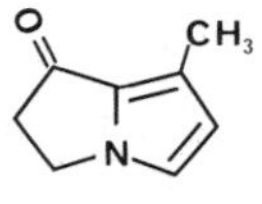

86

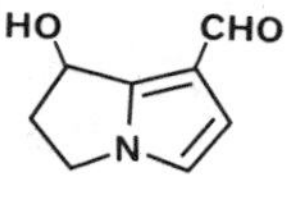

87

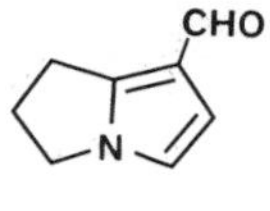

88

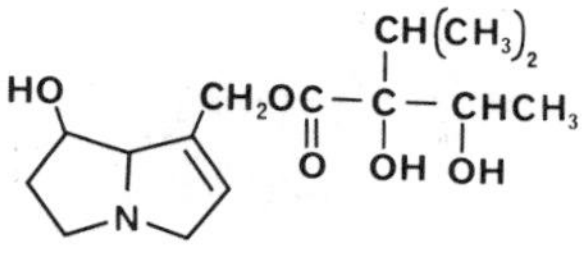

89

5. MARINE ALKALOIDS

Recently, alkaloidal structures have been isolated from several marine organisms, both plant and animal. A few of the compounds encountered to date will be summarized to illustrate the variety of alkaloidal structural types available from marine sources.

5.1. Dinoflagellate Alkaloids

The so-called red tides consist of high concentrations of toxic dinoflagellates, particularly *Gonyaulax tamarensis* in the north Atlantic and *G. catanella* in the south Atlantic. Filter feeders, such as shellfish, concentrate the toxic constituents of the red tide and when consumed by organisms higher on the food chain, such as man, can lead to severe and sometimes fatal poisoning. The toxic principles

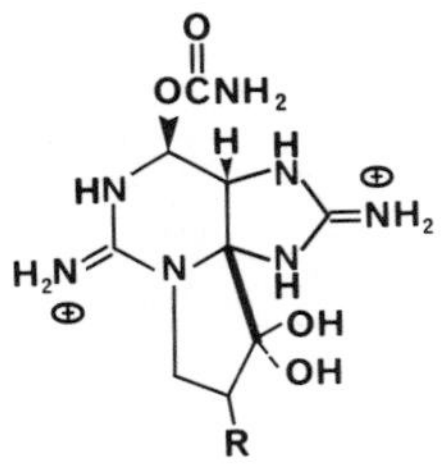

Saxitoxin (90) R=H

Gonyautoxin-II (91) R=--OH

Gonyautoxin-III (92) R=◄OH

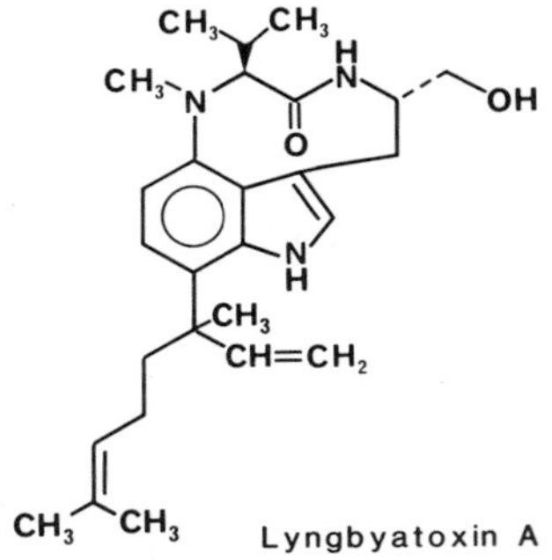

Lyngbyatoxin A (93)

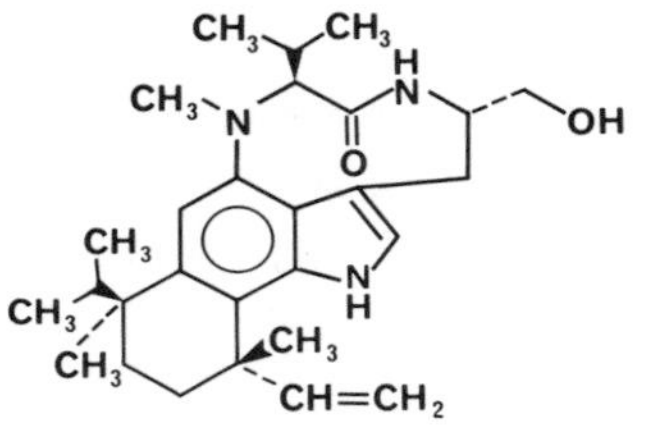

Teleocidin B (94)

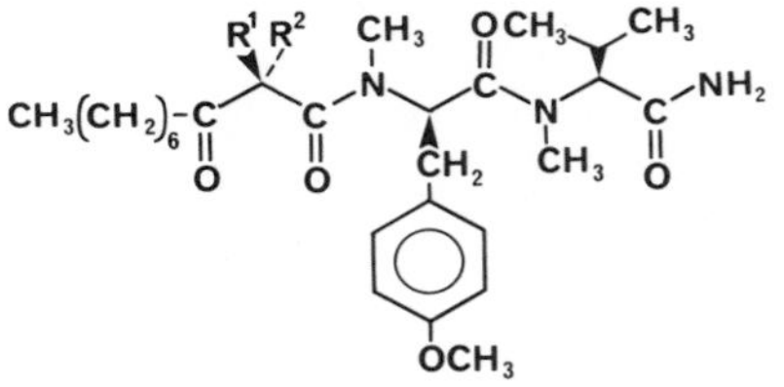

Majusculamide A (95) $R^1=CH_3$, $R^2=H$

Majusculamide B (96) $R^1=H$, $R^2=CH_3$

isolated include the alkaloids saxitoxin (**90**, LD_{50} 5–10 μg/kg in mouse), gonyautoxin-II (**91**), and gonyautoxin-III (**92**) [79–81]. Saxitoxin has also been isolated from Alaskan butler clams (*Saxidomus giganteus*), California mussels (*Mytilus californicus*), and softshell clams (*Mya arenaria*).

5.2. Blue-Green Alga Alkaloids

The shallow-water variety of *Lyngbya majuscula* Gomont. elaborates lyngbyatoxin A (**93**), a potent inflammatory and vesicatory agent [82]. Lyngbyatoxin A is the first indole alkaloid from a marine plant and has a structure similar to that of teleocidin B (**94**) produced by certain *Streptomyces* strains.

From Hawaiian *Lyngba majuscula* Gomont., Marner and co-workers [83] have isolated two amide alkaloids, majusculamides A (**95**) and B (**96**). From this same alga, Moore *et al.* have isolated the amide alkaloid **97** [84].

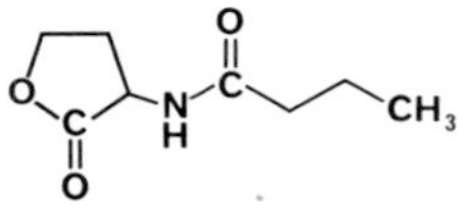

97

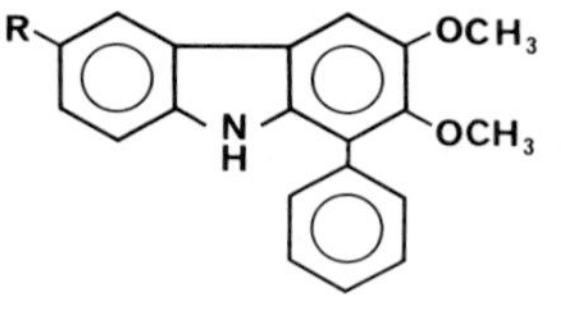

Hyellazole (98) R=H

6-Chlorohyellazole (99) R=Cl

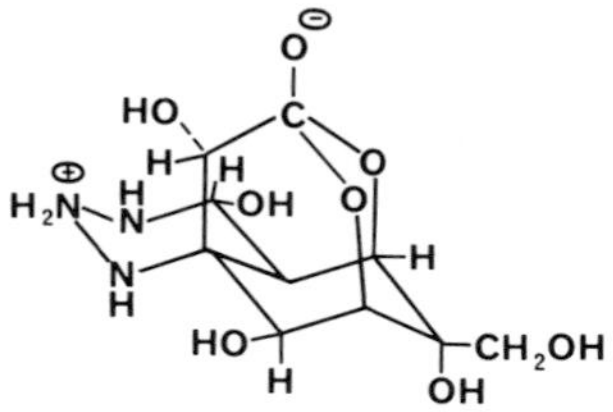

Tetrodotoxin (100)

A supralittoral variety of the blue-green alga *Hyella caespitosa* produces two unusual carbazole alkaloids, hyellazole (**98**) and 6-chlorohyellazole (**99**) [85]. These structures are different from carbazole alkaloids isolated from terrestrial plants.

5.3. Tetrodotoxin

The ovaries and liver of the puffer fish (swellfish; Japanese: *fugu*), *Spheroides rubripes*, *S. vermicularis*, etc., contain the alkaloid tetrodotoxin (**100**), one of the most toxic low-molecular-weight poisons known [86]. Poisoning from this toxin has long been a serious problem in Japan, where the puffer fish is highly prized as a food item. Tetrodotoxin has also been isolated from the goby fish, *Gobius criniger*, from Taiwan [87] and from the California newt *Taricha torosa* (formerly *Triturus torosus*), as well as several other newts such as *T. torosa sierrae* and *T. rivularis* [88]. It has also been isolated from the skins of frogs of the genus *Atelopus* from Costa Rica [89].

6. CLUB MOSS ALKALOIDS

Alkaloids are rarely found in cryptogamia. The club mosses, which are among the lowest forms of plant life, are an important exception, for the genus

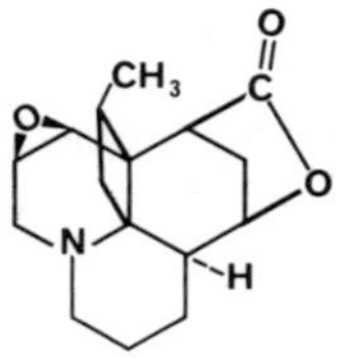

Annotinine (101)

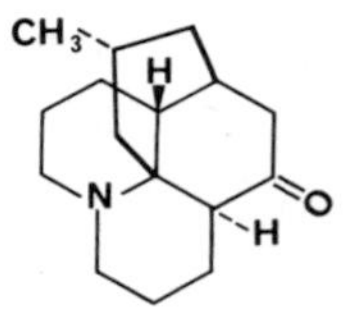

Lycopodine (102)

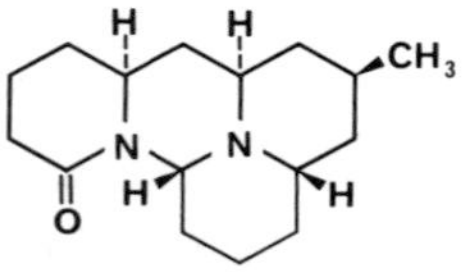

Cernuine (103)

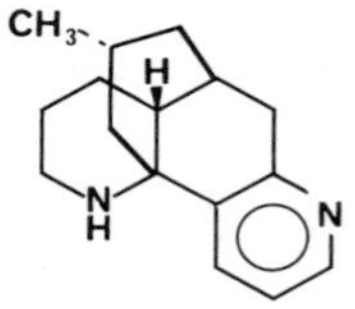

(−)-Lycodine (104)

Lycopodium is a prolific producer of alkaloids. With over 100 alkaloids now known, a few of the structural types are illustrated by annotinine (**101**, *L. annotinum* L.), lycopodine (**102**, *L. complanatum* L.), cernuine (**103**, *L. cernuum* L.) and lycodine (**104**, *L. annotinum* L.) [90–92].

7. FUNGAL ALKALOIDS

Alkaloids of a variety of structural types have been isolated from fungal sources. Perhaps best known are those produced by ergot, the dried sclerotium of the fungus *Claviceps purpurea* (Fries) Tulasne, which grows parasitically on rye and certain other gramineous crop plants.

Examples of a few of the structural types elaborated by *C. purpurea* include chanoclavine-I (**105**), agroclavine (**106**), ergonovine (ergometrine) (**107**), and the peptide alkaloid ergocristine (**108**) [93, 94]. *Aspergillus fumigatus* furnishes agroclavine and chanoclavine-I in addition to elymoclavine (**109**). *Penicillium concavorugulosum* produces alkaloids of a different structural type, rugulovasines A (**110**) and B (**111**). *Penicillium roqueforti* elaborates isofumigaclavine A (**112**), as well as the diketopiperazine alkaloid roquefortine (**113**) [95].

Several of these ergot alkaloids are also produced by higher plants of the family Convolvulaceae; the genera *Ipomoea*, *Argyreia*, *Rivea*, and *Stictocardia*

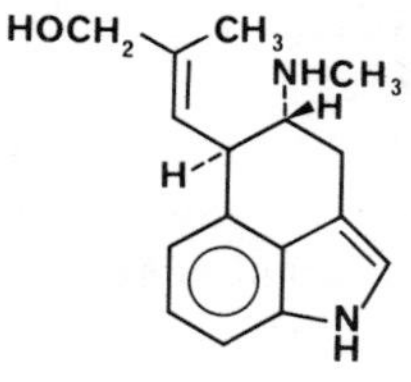

HOCH$_2$ CH$_3$ NHCH$_3$ H H N H

Chanoclavine-I (105)

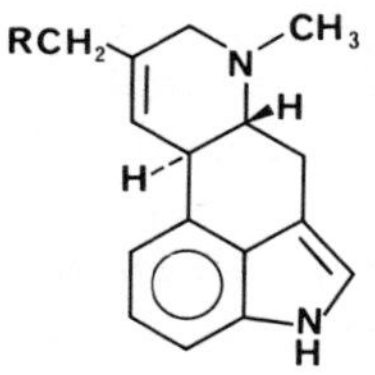

RCH$_2$ N CH$_3$ H H N H

Agroclavine (106) R=H

Elymoclavine (109) R=OH

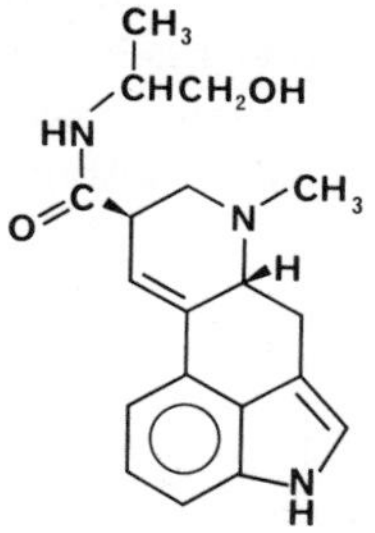

CH$_3$ CHCH$_2$OH HN C O N CH$_3$ H N H

Ergonovine (107)

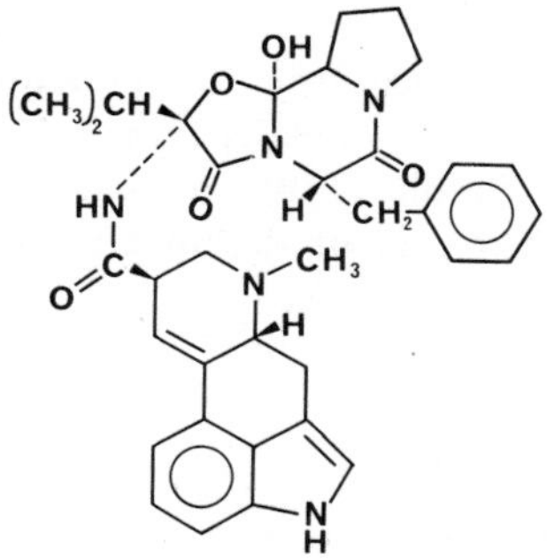

OH (CH$_3$)$_2$CH O N N O O H CH$_2$ HN C O N CH$_3$ H N H

Ergocristine (108)

O O NHCH$_3$ R CH$_3$ N H

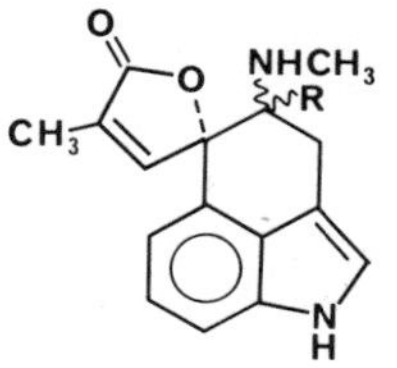

Rugulovasine A (110) R=β-H

Rugulovasine B (111) R=α-H

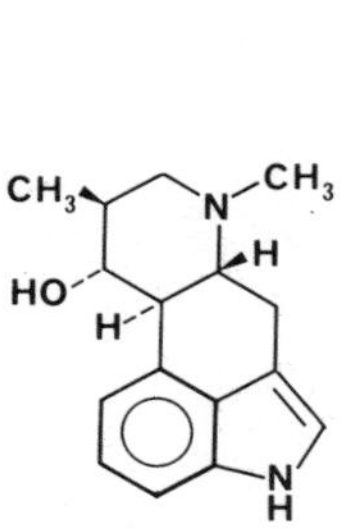

CH$_3$ N CH$_3$ H HO H N H

Isofumigaclavine A (112)

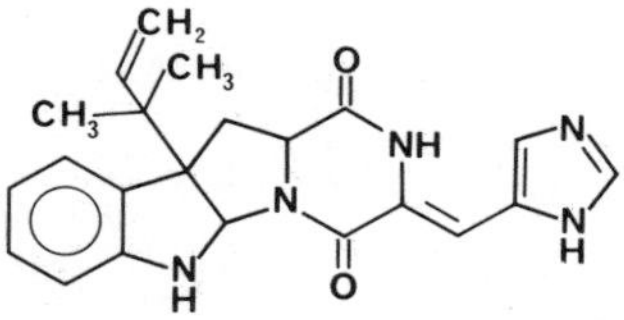

CH$_2$ CH$_3$ CH$_3$ O NH N N H O N N H

Roquefortine (113)

are particularly good sources. Thus chanoclavine-I (**105**), ergonovine (**107**), and elymoclavine (**109**) are produced by certain *Ipomoea* species.

The diketopiperazine alkaloid gliotoxin (**114**), which is an inhibitor of viral RNA synthesis, is elaborated by the fungi *Trichoderma viride*, *Penicillium terlikowskii*, and *Aspergillus fumigatus* [96]. Other related alkaloids are arantoin (**115**, *Arachniotus aureus*), echinulin (**116**, *Aspergillus echinulatus*), and sporidesmin A (**117**, *Pithomyces chartarum*) [97, 98]. A different type of diketopiperazine alkaloid is mycelianamide (**118**, *Penicillium griseofulvum*) [99].

Green peanuts furnish a fungus, *penicillium verruculosum* Peyronel, that elaborates a toxic metabolite, verruculotoxin (**119**). This alkaloid represents the first naturally occurring example of the octahydro-2*H*-pyrido[1,2-a]pyrazine system [100]. Recently the fungal strain *Penicillium herquei* Fg-372, has been shown to produce herquline (**120**), an alkaloid that inhibits platelet aggregation induced by adenosine diphosphate, but shows no antimicrobial activities [101].

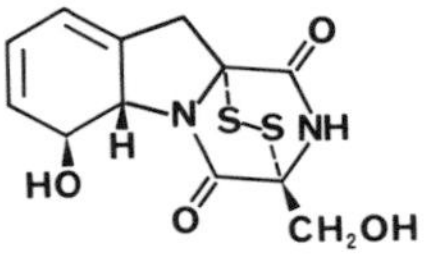

Gliotoxin (114)

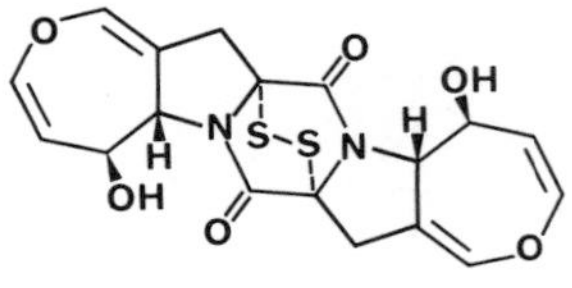

Arantoin (115)

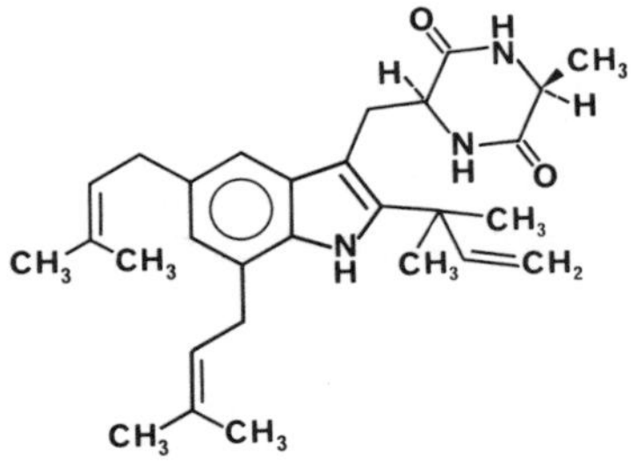

Echinulin (116)

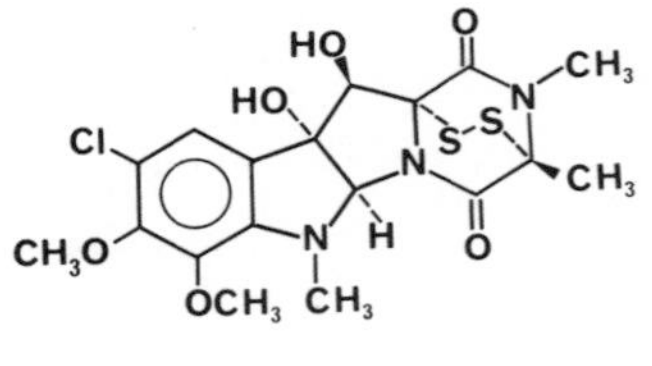

Sporidesmin A (117)

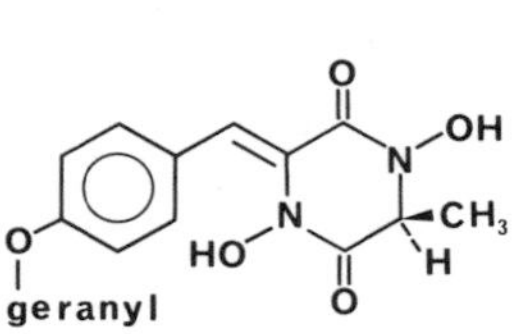

Mycelianamide (118)

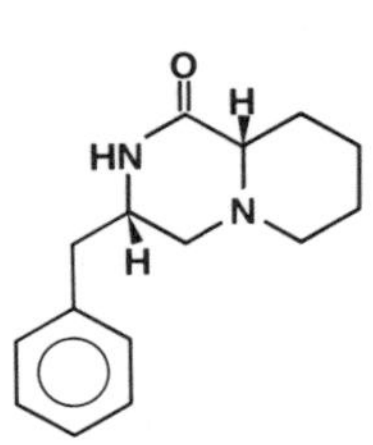

Verruculotoxin (119)

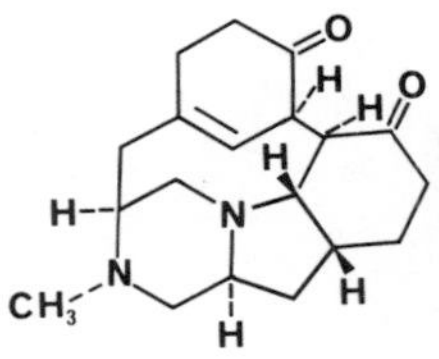

Herqulin (120)

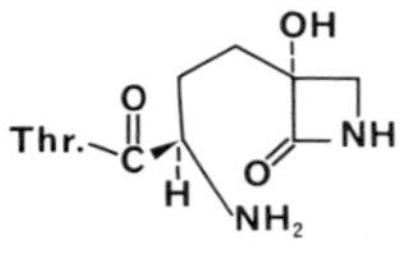

Tabtoxin (121)

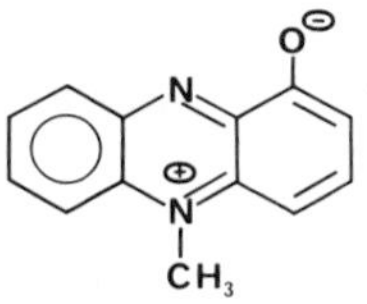

Pyocyanine (122)

8. BACTERIAL ALKALOIDS

Very few alkaloids have been reported from bacterial cultures. The chlorosin-inducing exotoxin tabtoxin (**121**) is produced by *Pseudomonas tabaci* and certain other phytopathogenic *Pseudomonas* species. It is toxic to tobacco, soybean, oat, and timothy [102].

The bacterium *Psuedomonas aeruginosa* elaborates a deep-blue-colored alkaloid, pyocyanine (**122**) [103].

9. FACTORS INFLUENCING CLASSIFICATION AS AN ALKALOID

9.1. Source

Does the source of a compound (i.e., plant, animal, fungus, or bacterium) have any bearing on its being classified as an alkaloid? I believe not. In certain cases the same compound has been isolated from both plants and animals. Thus actinidine (**23**) occurs in the plant *Actinidia polygama*, in two *Conomyrma* species of ants [104], and in two species of beetles, *Hesperus semirufus* and *Philonthus politus* [105]. (−)-Deoxynupharidine (**24**) occurs in the plant *Nuphar japonicum* and in the scent gland of the Canadian beaver. Similarly, bufotenin (**50**) occurs in several species of *Piptadenia*, in the toad *Bufo bufo bufo*, in the fungus *Amanita mappa*, and also in human urine. Is a sample of bufotenin from *Piptadenia* species an alkaloid, whereas a sample of the same compound from the toad is not an alkaloid? Obviously an alkaloid derived from a plant does not cease to be an alkaloid because it also occurs in an animal. As a consequence, the source of a compound is not significant in determining whether a particular compound is an alkaloid.

9.2. Biogenesis

Heterocyclic nitrogen-containing compounds modeled on monoterpenoid, sesquiterpenoid, and diterpenoid skeleta are not classed as alkaloids by some chemists and chemotaxonomists because these compounds are not related biogenetically to the amino acids. The concept of characterizing alkaloids by

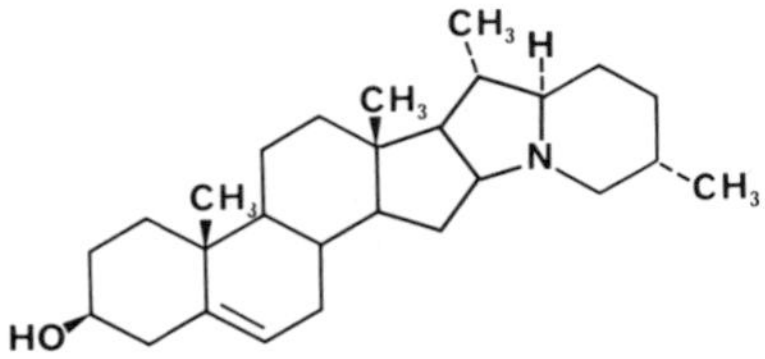

Solanidine (123) 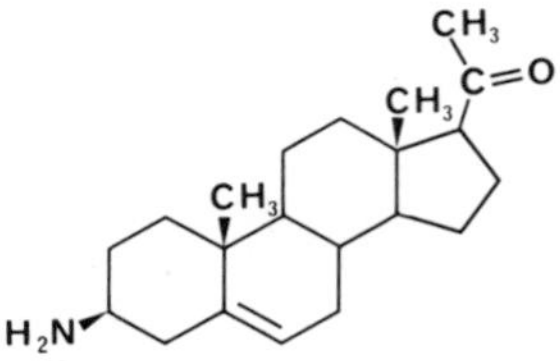 Holaphyllamine (124)

their mode of biogenesis led Hegnauer to propose the following definition for the purposes of plant taxonomy:

"Alkaloids are more or less toxic substances which act primarily on the central nervous system. They have a basic character, contain heterocyclic nitrogen, and are synthesized in plants from amino acids or their immediate derivatives. In most cases they are of limited distribution in the plant kingdom" [3].

In the arbitrary classification scheme followed by such writers, "true alkaloids" are those compounds in which decarboxylated amino acids are condensed with a non-nitrogenous structural moiety. These primary precursor amino acids are ornithine, lysine, anthranilic acid, phenylalanine/tyrosine, tryptophan, and histidine. The alkaloids formed from these precursors are regarded as by-products of protein metabolism that are methylated on either nitrogen or hydroxyl groups. In those cases in which "ammonia" is incorporated into isoprenoid or polyketide carbon skeleta, the resulting isoprenoid amines and polyketide amines are classed as "pseudoalkaloids" or "alcaloida imperfecta" [106]. The series of compounds relegated by some authors to this class includes not only those derived from monoterpenes, sesquiterpenes, and diterpenes (e.g., β-skytanthine, **22**, deoxynupharidine, **24**, and aconitine, **25**, respectively), but those from C_{27} and C_{21} steroids, (e.g., solanidine, **123**, and holaphyllamine, **124**, respectively), from purines (e.g., caffeine, **19**), and from nicotinic acid (e.g., ricinine, **26**, and trigonelline, **33**).

9.3. "True Alkaloids" Versus "Pseudoalkaloids"

This classification of nitrogenous compounds into "true alkaloids" and "pseudo-alkaloids" based on biogenesis is very arbitrary and needlessly complicates the already-difficult task of deciding what are and what are not alkaloids. This writer feels that the structure, and *not* the biogenetic pathway followed, should determine whether a compound is an alkaloid. To do otherwise means it is not possible to classify a compound of a new structural type as an alkaloid until one is reasonably certain of the biogenesis of the compound. Moreover, the term *alkaloid* serves functions for the chemist and pharmacognosist that are well beyond those necessary for the plant taxonomist. For these reasons I recommend that the term *alkaloid* be defined on the basis of structure and not on the basis of source or biogenesis. As a corollary, the term "pseudoalkaloid" should be abandoned, and appropriate nitrogenous compounds derived from terpenes, steroids, purines, and nicotinic acid should be classified as genuine alkaloids.

9.4. Antibiotics

Should antibiotics of appropriate structure be classified as alkaloids? Though some authors automatically exclude antibiotics, I believe this practice is not justified. The term *antibiotic* is generally defined as "a chemical substance produced by a microorganism that has the capacity, in low concentration, to inhibit selectively or even to destroy bacteria and other microorganisms through an antimetabolic mechanism" [107]. This definition has been expanded to include higher plants as the source and tumors as a site of action [107]. The term *antibiotic* thus has reference to a function the compound performs, not to its chemical structure. Therefore, following the concept that the structure of a compound determines classification as an alkaloid, one should accept antibiotics of appropriate structure into the alkaloid class. Examples of antibiotics that, on the basis of structure, can be classed as alkaloids are gliotoxin (**114**), pyocyanine (**122**), cycloserine (**125**), mitomycin C (**126**), and the macrocyclic maytansinoids such as maytansine (**127**).

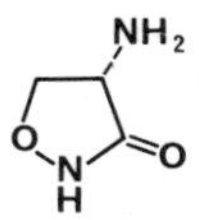

Cycloserine (125)

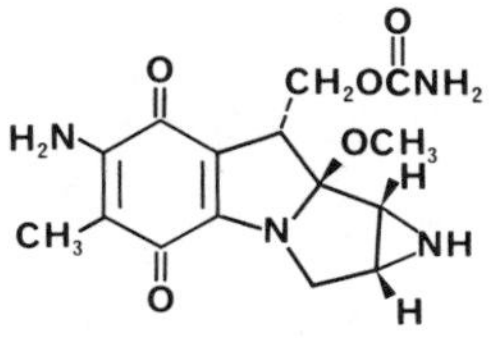

Mitomycin C (126)

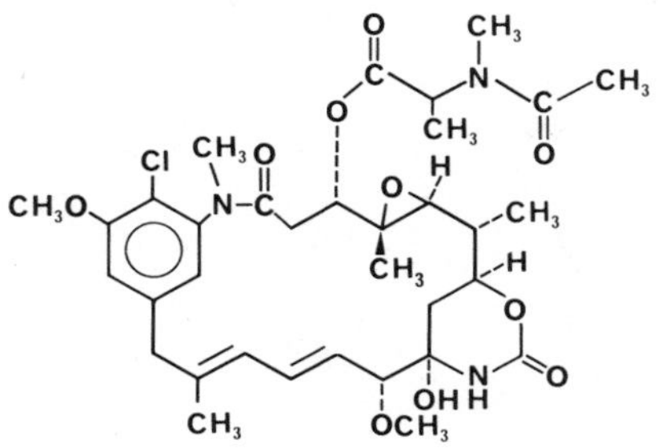

Maytansine (127)

9.5. Polyamines

The polyamines putrecine (**9**), spermidine (**10**), and spermine (**11**) are considered by most authorities as biogenic amines (i.e., alkaloid building blocks) rather than genuine alkaloids [19–22]. I concur in this view, for these compounds are simple, straight-chain amines that are widely distributed in animals, microorganisms, and higher plants. For example, among plants used for foods, they have been

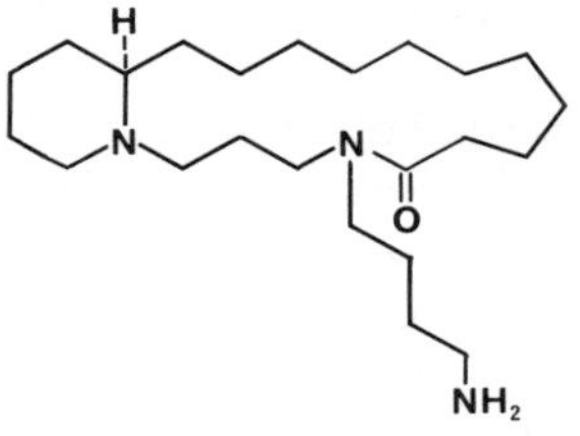

Oncinotine (128)

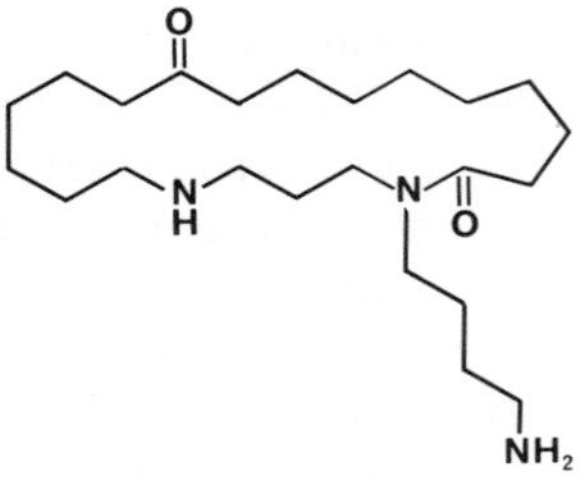

Inandenine-12-one (129)

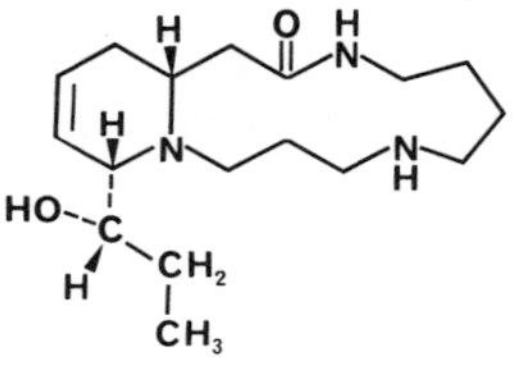

Palustrine (130)

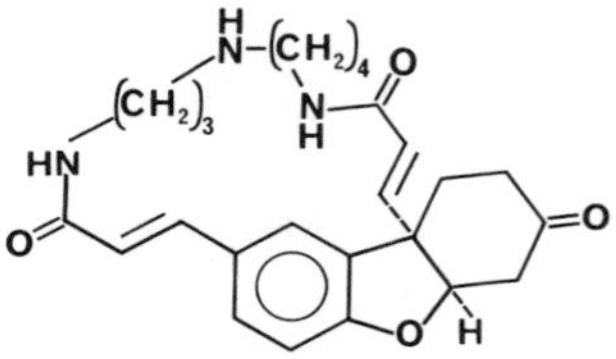

Lunarine (131)

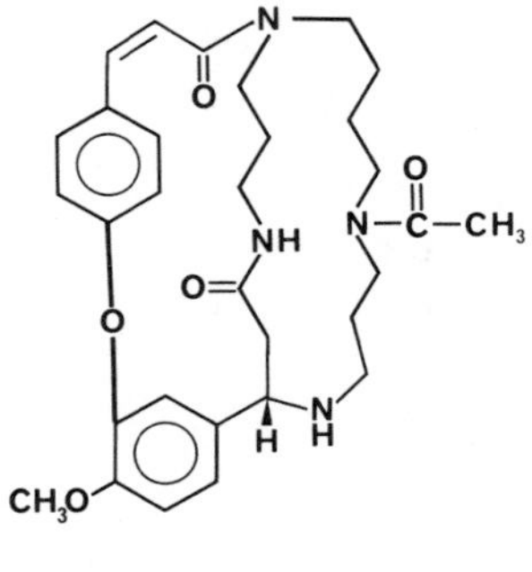

Chaenorhine (132)

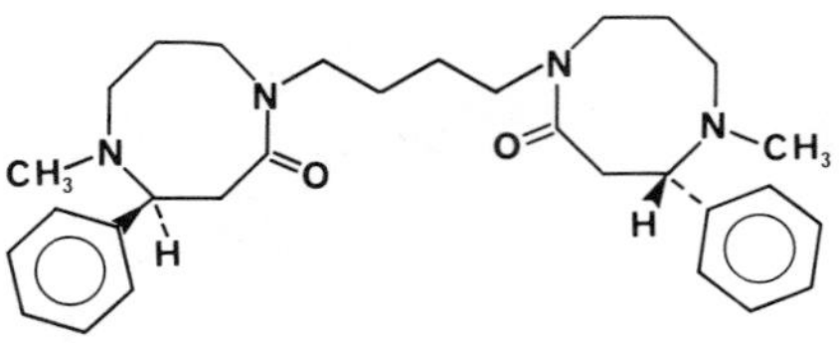

Homaline (133)

detected in apple, cabbage, spinach, tomato, and in the leaves of wheat, maize, pea, black currant, and tobacco. Spermidine and spermine are also found in substantial concentrations in human semen [108].

Recently a group of structurally complex derivatives of spermidine and spermine that are genuine alkaloids have been isolated. There are several dozen of these interesting macrocyclic alkaloids of which a few representative examples follow: oncinotine (**128**, *Oncinotis nitidia*), inandenine-12-one (**129**, *Oncinotis inandensis*), palustrine (**130**, *Equisetum* species), and lunarine (**131**, *Lunaria biennis*) are all derived from spermidine (**10**). The alkaloids chaenorhine (**132**, *Chaenorhinum origanifolium*) and homaline (**133**, *Homalium pronyense*) are derived from spermine (**11**) and two cinnamic acid units [109].

10. DEFINITION OF AN ALKALOID

That the classical definition of an alkaloid is no longer serviceable should be clear from the previous discussion. There are many examples of compounds that are universally accepted as alkaloids and that violate one or more of the cardinal requirements of definitions of the Winterstein-Trier [14] type. A brief summary of the inappropriateness of certain of these requirements in a modern definition follows.

10.1. Basicity

Though the term *alkaloid* originally meant *alkalilike*, we have seen that basicity can no longer be regarded as a necessary property of an alkaloid. Such glaring exceptions as colchicine (**28**), piperine (**29**), amine oxides (e.g., **30**), and important quaternary salts such as laurifoline chloride (**34**) require that basicity no longer be included in a definition.

10.2. Nitrogen as Part of a Heterocyclic System

Although most early classified alkaloids contained nitrogen in a heterocyclic system, there are now too many exceptions for this condition to be mandatory. If it is to be part of a definition, one must exclude colchicine (**28**) as well as all of the important β-phenylethylamines such as hordenine (**13**) and mescaline (**16**). Common sense dictates that rather than exclude these compounds, the definition must be modified.

10.3. Complexity of Molecular Structure

Complexity of structure is too vague a concept to include as a requirement. What is perceived as a complex molecule by one chemist may be considered simple by another. Thus this requirement has no place in a modern definition of an alkaloid.

10.4. Pharmacological Activity

Though most alkaloids manifest pharmacological activity, I believe it inappropriate to make such activity a requirement for classification as an alkaloid. Otherwise a newly isolated structural type must be submitted for pharmacological screening before classification. Also, just what is meant by *pharmacological activity*? Almost any compound if administered in large enough dosage will have a physiological effect on a living organism. A concentration level would have to be specified if pharmacological activity were to be made a part of the definition of an alkaloid.

10.5. Restriction to Plant Kingdom

We have already seen that dozens of compounds possessing classical alkaloid-type structures have been isolated from animal, fungal, and bacterial sources. One cannot on any rational basis exclude such compounds because they happen to occur in living tissue other than plants. A plant-derived requirement no longer has validity in a functional definition.

10.6. A Modern Definition

An appropriate definition must be workable in the sense of accommodating most, if not all, of the compounds regarded as alkaloids by most chemists and pharmacognosists. This means compounds such as colchicine (**28**), piperine (**29**), quaternary-salt alkaloids (**34**), certain amine oxides (**30** and **31**), and the β-phenylethylamines must be included. The various structural types of alkaloids discussed in this chapter may suggest that alkaloids should be defined simply as naturally occurring nitrogenous compounds. For practical considerations the definition must be more restrictive. By common consent most authorities would agree that the members of the following classes of compounds are *not* alkaloids: simple low-molecular-weight derivatives of ammonia (e.g., methylamine, triethylamine, and other acyclic alkylamines, choline (**7**), acyclic betaines such as **6**), amino acids, amino sugars, peptides (exception: peptide alkaloids such as ergocristine, **108**), proteins nucleic acids, nucleotides, pterins, porphyrins, and vitamins.

After much soul-searching and discussion with colleagues, I suggest the following simple definition for an alkaloid:

> *An alkaloid is a cyclic organic compound containing nitrogen in a negative oxidation state which is of limited distribution among living organisms.*

The requirement for a cyclic structure in some part of the molecule excludes simple low-molecular-weight derivatives of ammonia as listed above, as well as

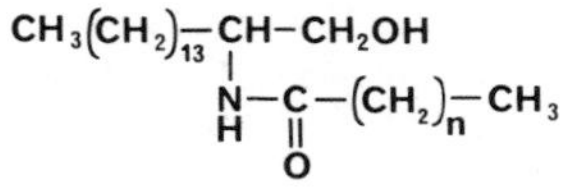

$$CH_3(CH_2)_{13}-CH(-N(H)-C(=O)-(CH_2)_n-CH_3)-CH_2OH$$

134 n = 23,24,25

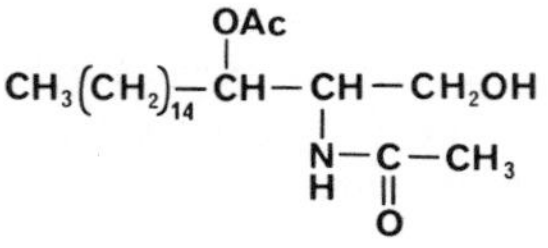

$$CH_3(CH_2)_{14}-CH(OAc)-CH(-N(H)-C(=O)-CH_3)-CH_2OH$$

135

acyclic polyamines such as putrecine (**9**), spermidine (**10**) and spermine (**11**). Also excluded are such acyclic amides as **134** and **135**, recently isolated from the green alga *Caulerpa racemosa* [110] and the red alga *Laurencia nidifica* [84], respectively, and incorrectly classified as alkaloids [111]. The requirement of nitrogen in a negative oxidation state includes amines (−3), amine oxides (−1), amides (−3), and quaternary ammonium salts (−3), but excludes nitro (+3) and nitroso (+1) compounds. Thus by this definition the aristolochic acids **35** and **36** are not classed as alkaloids, nor should they be! The requirement for a limited distribution among living organisms seems essential for practical considerations. Otherwise one must classify almost all naturally occurring nitrogenous compounds as alkaloids. By imposing the requirement of restricted occurrence among the various genera of plants, animals, and other living organisms, one excludes the almost ubiquitous compounds such as amino acids, amino sugars, peptides, proteins, nucleic acids, nucleotides, porphyrins, and vitamins. This definition has the advantage of including most of the compounds that are exceptions to the classical-type alkaloid definitions, viz., colchicine (**28**), piperine (**29**), the β-phenylethylamines, ricinine (**26**), gentianine (**27**), bufotoxin (**53**), and pederin (**83**). It also includes as alkaloids the purine bases caffeine (**19**), theophilline (**20**), and theobromine (**21**).

I welcome comments from others as to the suitability of this definition.

REFERENCES

1. K. W. Bentley, *The Alkaloids*, Interscience , New York, 1957, p. 1.
2. G. A. Cordell, *Introduction to Alkaloids. A Biogenetic Approach*, Wiley, New York, 1981, pp. 1–6.
3. R. Hegnauer, The Taxonomic Significance of Alkaloids, in *Chemical Plant Toxonomy*, T. Swain, Ed., Academic, New York, 1963, pp. 389–399.
4. R. Hegnauer, Comparative Phytochemistry of Alkaloids, in *Comparative Phytochemistry*, T. Swain, Ed., Academic, New York, 1966, pp. 211–212.
5. M. Hesse, *Alkaloid Chemistry*, Wiley, New York, 1978, p. 2–5.
6. R. Ikan, *Natural Products. A Laboratory Guide*, Israel Universities Press, Jerusalem, 1969, p. 178.

7. S. W. Pelletier, Alkaloids, in *The Encyclopedia of Chemistry*, 3rd ed., C.A. Hampel and G. G. Hawley, Eds., Van Nostrand Reinhold, New York, 1973, pp. 47–48.
8. S. W. Pelletier, Alkaloids, in *The Encyclopedia of Biochemistry*, R. J. Williams and E. M. Lansford, Jr., Eds., Reinhold, New York, 1967, pp. 28–32.
9. S. W. Pelletier, Introduction in *Chemistry of the Alkaloids*, S. W. Pelletier, Ed., Van Nostrand Reinhold, New York, 1970, pp. 1–2.
10. A. R. Pinder, Alkaloids: General Introduction, in *Chemistry of Carbon Compounds*, Vol. IV/C, E. H. Rodd, Ed., Elsevier, Amsterdam, 1960, pp. 1799–1800.
11. L. F. Small, Alkaloids, in *Organic Chemistry. An Advanced Treatise*, 2nd ed., Vol. 2, H. Gilman, Ed., New York, 1943, pp. 1166–1171.
12. G. A. Swan, *An Introduction to the Alkaloids*, Wiley, New York, 1967, pp. 1–2.
13. V. E. Tyler, L. R. Brady, and J. E. Robbers, Alkaloids, in *Pharmacognosy*, 8th ed., Lea & Febiger, Philasdelphia, 1981, pp. 195–196.
14. E. Winterstein and G. Trier, *Die Alkaloide, eine Monographie der natürlichen Basen*, Bornträger, Berlin, 1910.
15. M. Guggenheim, *Die Biogenen Amine*, S. Karger, Basel, Switzerland, 1951.
16. D. Ackerman, *Abh. Dtsch. Akad. Wiss. Berlin* **7,** 1 (1956).
17. R. Hegnauer, The Taxonomic Significance of Alkaloids, in *Chemical Plant Taxonomy*, T. Swain, Ed., Academic, New York, 1963, pp. 398–399.
18. G. A. Cordell, *Introduction to Alkaloids. A Biogenetic Approach*, Wiley, New York, 1981.
19. M. Hesse, *Alkaloid Chemistry*, Wiley, New York, 1978.
20. S. W. Pelletier, Ed., *Chemistry of the Alkaloids*, Van Nostrand Reinhold, New York, 1970.
21. G. A. Swan, *An Introduction to the Alkaloids*, Wiley, New York, 1967.
22. J. S. Glasby, *Encyclopedia of the Alkaloids*, Vols. 1–3, Plenum, New York and London, 1975, 1977.
23. G. Habermehl, The Steroid Alkaloids: The *Salamandra* Group, in *The Alkaloids*, R. H. F. Manske, Ed., Academic, New York, Vol. IX, 1967, pp. 427–439.
24. S. Zalesky, *Med. Chem. Unterss. Hoppe-Seyler* **1,** 85 (1866).
25. C. Schöpf and W. Braun, *Ann. Chem.* **514,** 69 (1934).
26. C. Schöpf and K. Koch, *Ann. Chem.* **522,** 37 (1942).
27. E. Wölfel, C. Schöpf, G. Weitz, and G. Habermehl, *Chem. Ber.* **94,** 2361 (1961).
28. G. Habermehl and G. Haaf, *Chem. Ber.* **98,** 3001 (1965).
29. C. Schöpf and K. Koch, *Ann. Chem.* **552,** 62 (1942).
30. G. Habermehl and S. Göttlicher, *Chem. Ber.* **98,** 1 (1965).
31. V. Deulofeu and E. A. Ruveda, The Basic Constituents of Toad Venoms, in *Venomous Animals and Their Venoms*, Vol. 2, W. Bücherl and E. E. Buckley, Eds., Academic, New York, 1971, pp. 475–495.
32. J. W. Daly and B. Witkop, Chemistry and Pharmacology of Frog Venoms, in *Venomous Animals and Their Venoms*, Vol. 2, W. Bücherl and E. E. Buckley, Eds., Academic, New York, 1971, pp. 497–519.
33. T. Tokuyama, J. Daly, and B. Witkop, *J. Am. Chem. Soc.* **91,** 3931 (1969).
34. J. W. Daly, I. Karle, C. W. Myers, T. Tokuyama, J. W. Waters, and B. Witkop, *Proc. Natl. Acad. Sci. USA* **68,** 1870 (1971).
35. J. W. Daly, B. Witkop, T. Tokuyama, T. Nishikawa, and I. L. Karle, *Helv. Chim. Acta* **60,** 1128 (1977); R. Fujimoto and Y. Kishi, *Tetrahedron Lett.* **22,** 4197 (1981).
36. J. W. Daly, T. Tokuyama, G. Habermehl, I. L. Karle, and B. Witkop, *Ann. Chem.* **729**, 198 (1969).
37. J. W. Daly, T. Tokuyama, T. Fujiwara, R. J. H. Highet, and I. L. Karle, *J. Am. Chem. Soc.* **102,** 830 (1980).

38. V. L. Stromberg, *J. Am. Chem. Soc.* **76,** 1707 (1954).
39. I. J. Pachter, D. E. Zacharius, and O. Ribeiro, *J. Org. Chem.* **24,** 1285 (1959).
40. G. A. Iacobucci and E. A. Ruveda, *Phytochemistry* **3,** 465 (1964).
41. M. S. Fish, N. M. Johnson, and E. C. Horning, *J. Am. Chem. Soc.* **77,** 5892 (1955).
42. T. Wieland and W. Motzel, *Ann. Chem.* **581,** 10 (1953).
43. V. E. Tyler, *Lloydia* **24,** 71 (1961).
44. G. A. Swan, *An Introduction to the Alkaloids*, Wiley, New York, 1967, p. 201.
45. B. Maurer and G. Ohloff, *Helv. Chim. Acta* **59,** 1169 (1976).
46. H. Schinz, L. Ruzicka, U. Geyer, and V. Prelog, *Helv. Chim. Acta* **29,** 1524 (1946).
47. B. Tursch, J. C. Braekman, and D. Daloze, *Experientia* **32,** 401 (1976).
48. J. G. MacConnell, M. S. Blum, and H. G. Fales, *Science* **168,** 840 (1970).
49. J. G. MacConnell, M. S. Blum, and H. G. Fales, *Tetrahedron* **26,** 1129 (1971).
50. J. W. Wheeler and M. S. Blum, *Science* **182,** 501 (1973).
51. G. W. K. Cavill and E. Houghton, *Austr. J. Chem.* **27,** 879 (1974).
52. G. W. K. Cavill and E. Houghton, *J. Insect Physiol.* **20,** 2049 (1974).
53. J. H. Tumlinson, R. M. Silverstein, J. C. Moser, R. G. Brownlee, and J. M. Ruth, *Nature* **234,** 348 (1971).
54. F. J. Ritter, I. E. M. Rotgans, E. Talman, P. E. Verwiel, and F. Stein, *Experientia* **29,** 530 (1973).
55. F. J. Ritter, I. E. M. Bruggeman-Rotgans, E. Verkuil, and C. J. Persoons, in *Proceedings of the Symposium on Pheromones and Defensive Secretions in Social Insects*, Ch. Noirot, P. E. Howse, and G. Le Masne, Eds., University of Dijon Press, 1975, pp. 99–103.
56. Y. C. Meinwald, J. Meinwald, and T. Eisner, *Science* **154,** 390 (1966).
57. H. Schildknecht and W. F. Wenneis, *Z. Naturforsch* **21b**, 522 (1966).
58. H. Schildknecht, U. Maschwitz, and W. F. Wenneis, *Naturwissenschaften* **54,** 196 (1967).
59. J. Smolanoff, A. F. Kluge, J. Meinwald, A. McPhail, R. W. Miller, K. Hicks, and T. Eisner, *Science* **188,** 734 (1975).
60. J. Weinwald, J. Smolanoff, A. T. McPhail, R. W. Miller, T. Eisner, and K. Hicks, *Tetrahedron Lett.*, 2367 (1975).
61. B. Tursch, D. Daloze, M. Dupont, C. Hootele, M. Kaisin, J. M. Pasteels, and D. Zimmermann, *Chimia* **25,** 307 (1971).
62. R. Karlsson and D. Losman, *J. Chem. Soc. Chem. Commun.*, 626 (1972).
63. B. Tursch, D. Daloze, J. C. Braekman, C. Hootele, A. Cravador, D. Losman and R. Karlsson, *Tetrahedron Lett.*, 409 (1974).
64. B. Tursch, D. Daloze, J. C. Braekman, C. Hootele, and J. M. Pasteels, *Tetrahedron* **31,** 1541 (1975).
65. P. A. Hedin, R. C. Gueldner, R. D. Henson, and A. C. Thompson, *J. Insect Physiol.* **20,** 2135 (1974).
66. B. Tursch, D. Daloze, and C. Hootele, *Chimia* **26,** 74 (1972).
67. B. Tursch, J. C. Braekman, D. Daloze, C. Hootele, D. Losman, R. Karlson, and J. M. Pasteels, *Tetrahedron Lett.*, 201 (1973).
68. B. Tursch, D. Daloze, M. Dupont, J. M. Pasteels, and M. C. Tricot, *Experientia* **27,** 1380 (1971).
69. W. A. Denne, S. R. Johns, J. A. Lamberton, and A. McL. Mattieson, *Tetrahedron Lett.*, 3107 (1971).
70. H. Schildknecht, H. Birringer, and D. Krauss, *Z. Naturforsch.* **24b,** 38 (1969).
71. A. Furusaki, I. Watanabe, T. Matsumoto, and M. Yanagiya, *Tetrahedron Lett.*, 6301 (1968).
72. A. Bonamartini Corradi, A. Mangia, M. Nardelli, and G. Pelizzi, *Gazz. Chim. Ital.* **101,** 591 (1971).

73. A. Quilico, C. Cardani, D. Ghiringhelli, and M. Pavan, *Chim. Ind. Milano* **43,** 1434 (1961).
74. C. Cardani, D. Ghiringhelli, A. Quilico, and A. Selva, *Tetrahedron Lett.*, 4023 (1967).
75. J. Meinwald, Y. C. Meinwald, J. W. Wheeler, T. Eisner, and L. P. Brower, *Science* **151,** 583 (1966); J. Meinwald and Y. C. Meinwald, *J. Am. Chem. Soc.* **88,** 1305 (1966).
76. J. Meinwald, Y. C. Meinwald, and P. H. Mazzocchi, *Science* **164,** 1174 (1969).
77. J. A. Edgar, C. C. J. Culvenor and L. W. Smith, *Experientia* **27,** 761 (1971); J. A. Edgar, C. C. J. Culvenor and G. S. Robinson, *J. Austr. Ent. Soc.* **12,** 144 (1973); J. A. Edgar and C. C. J. Culvenor, *Nature* **268,** 614 (1974).
78. C. C. J. Culvenor and J. A. Edgar, *Experientia* **28,** 627 (1972).
79. Y. Shimizu, M. Alam, Y. Oshima, L. J. Buckley, and W. E. Fallon, Chemistry and Distribution of Deleterious Dinoflagellate Toxins, in *Marine Natural Products Chemistry*, D. J. Faulkner and W. H. Fenical, Eds., Plenum, New York, 1977, pp. 261–269; Y. Shimizu, Dinoflagellate Toxins in *Marine Natural Products. Chemical and Biological Perspectives*, Vol. 1, P. J. Scheuer, Ed., Academic, New York, 1978, pp. 1–42.
80. E. J. Schantz, V. E. Ghazarossian, H. K. Schnoes, F. M. Strong, J. P. Springer, J. O. Pezzanite, and J. Clardy, *J. Am. Chem. Soc.* **97,** 1238 (1975).
81. J. Bordner, W. E. Thiessen, H. A. Bates, and H. Rapoport, *J. Am. Chem. Soc.* **97,** 6008 (1975).
82. J. H. Cardellina II, J-F. Marner, and R. E. Moore, *Science* **204,** 193 (1979).
83. F.-J. Marner, R. E. Moore, and J. Clardy, *J. Org. Chem.* **42,** 2815 (1977).
84. R. E. Moore, Algal Nonisoprenoids, in *Marine Natural Products. Chemical and Biological Perspectives*, Vol. 1, P. J. Scheuer, Ed., Academic, New York, 1978, p. 120.
85. J. H. Cardellina II, M. P. Kirkup, R. E. Moore, J. S. Mynderse, K. Seff, and C. J. Simmons, *Tetrahedron Lett.*, 4915 (1979).
86. R. B. Woodward and J. Z. Gougoutas, *J. Am. Chem. Soc.* **86,** 5030 (1964); T. Goto, Y. Kishi, S. Takahaski, and Y. Hirata, *Tetrahedron* **21,** 2059 (1965).
87. T. Noguchi, H. Kao, Y. Hashimoto, *Bull. Jpn. Soc. Sci. Fish.* **37,** 642 (1971); T. Noguchi and Y. Hashimoto, *Toxicon* **11,** 305 (1973).
88. H. S. Mosher, F. A. Fuhrman, H. D. Buchwald, and H. G. Fisher, *Science* **144,** 1100 (1964).
89. Y. H. Kim, G. B. Brown, H. S. Mosher, and F. A. Fuhrman, *Science* **189,** 151 (1975).
90. D. B. MacLean, The Lycopodium Alkaloids, in *Chemistry of the Alkaloids*, S. W. Pelletier, Ed., Van Nostrand Reinhold, New York, 1970. pp. 469–502.
91. D. B. MacLean, The Lycopodium Alkaloids, in *The Alkaloids. Chemistry and Physiology*, Vol. 14, R. H. F. Manske, Ed., Academic, New York, 1973, pp. 347–405.
92. G. A. Cordell, *Introduction to Alkaloids. A Biogenetic Approach*, Wiley, New York, 1981, pp. 170–178.
93. A. Stoll and A. Hofmann, The Ergot Alkaloids, in *The Alkaloids. Chemistry and Physiology*, Vol. 8, R. H. F. Manske, Ed., Academic, New York, 1965, pp. 725–783.
94. A. Stoll and A. Hofmann, The Chemistry of the Ergot Alkaloids, in *Chemistry of the Alkaloids*, S. W. Pelletier, Ed., Van Nostrand Reinhold, New York, 1970, pp. 267–300.
95. G. A. Cordell, *Introduction to Alkaloids. A Biogenetic Approach*, Wiley, New York, 1981, pp. 622–631.
96. M. R. Bell, J. R. Johnson, B. S. Wildi, and R. B. Woodward, *J. Am. Chem. Soc.* **80,** 1001 (1958).
97. A. Taylor, S. Kadis, A. Ciegler, and S. J. Ajl, Eds., *Microbial Toxins*, Academic, New York, 1971.
98. P. G. Sammes, *Fortschr. Org. Chem. Naturforsch.* **32,** 51 (1975).
99. A. J. Birch, R. A. Massey-Westropp, and R. W. Richards, *J. Chem. Soc.*, 3717 (1956); K. W. Blake and P. G. Sammes, *J. Chem. Soc. C*, 980 (1970).
100. J. G. Macmillan, J. P. Springer, J. Clardy, R. J. Cole, and J. W. Kirksey, *J. Am. Chem. Soc.* **98,** 246 (1976).

101. A. Furusaki and T. Matsumoto, *J. Chem. Soc. Chem. Commun.*, **1980**, 698.

102. D. L. Lee and H. Rapoport, *J. Org. Chem.* **40,** 3491 (1975).

103. F. Wrede and E. Strack, *Hoppe-Seyler's Z. Physiol. Chem.* **140,** 1, (1924); R. Schoental, *Br. J. Exp. Pathol.* **22,** 137 (1941); H. Hilleman, *Chem. Ber.* **71B**, 46 (1938).

104. J. W. Wheeler, T. Olagbemiro, A. Nash, and M. S. Blum, *J. Chem. Ecol.* **3,** 241 (1977).

105. T. E. Bellas, W. V. Brown, B. P. Moore, *J. Ins. Physiol.* **20,** 277 (1974).

106. R. Hegnauer, *Abh. Deut. Akad. Wiss. Berlin* **7,** 10 (1956); R. Hegnauer, *Planta Med.* **6,** 1 (1958).

107. V. E. Tyler, L. R. Brady, and J. E. Robbers, *Pharmacognosy*, 8th Ed., Lea & Febiger, Philadelphia, 1981, p. 326.

108. G. A. Cordell, *Introduction to Alkaloids. A Biogenetic Approach*, Wiley, New York, 1981, pp. 930–931.

109. M. Hesse and H. Scmid, Macrocyclic Spermidine and Spermine Alkaloids, in *MTP, International Review of Science*, Ser. 2, Vol. 9, K. Wiesner, Ed., Butterworths, London, 1977, pp. 265–307.

110. M. S. Doty and G. A. Santos, *Nature (London)* **211,** 990 (1966); G. A. Santos and M. S. Doty, Chemical Studies of Three Marine Algae Genus *Calerpa*, in *Food-Drugs Sea, Proc. Conf., 1st 1967*, H. D. Freudenthal, Ed., Marine Technology Society, Washington, D.C., 1968, p. 173.

111. R. E. Moore, Algal Nonisoprenoids, in *Marine Natural Products. Chemical and Biological Perspectives*, Vol. 1, P. J. Schuer, Ed., Wiley, New York, 1978, p. 119.

Chapter Two

Arthropod Alkaloids: Distribution, Functions, and Chemistry

Tappey H. Jones and **Murray S. Blum**
Department of Entomology
University of Georgia
Athens, Georgia 30602

CONTENTS

1. INTRODUCTION

Among animals, species in the phylum Arthropoda have achieved an incredible degree of success, accounting for more than 80% of animal species and containing from 1 to 2 million forms. These populous invertebrates, which include crustaceans, centipedes, millipedes, and insects, have achieved this success in spite of great predatory pressure from a variety of pathogens, invertebrates, and vertebrates that constitute omnipresent adversaries. The ability of arthropods to blunt the attacks of their predators is frequently correlated with the presence of potent defensive secretions containing compounds that are generally synthesized *de novo* in specialized exocrine glands. Frequently, the major defensive products in these glandular exudates are alkaloids, some of which constitute unique natural products. Indeed, among animals, arthropods constitute the alkaloidal biosynthesizers *par excellence*, and most of the alkaloids that are classified as animal natural products are derived solely from these invertebrates. Although the defensive chemistry of relatively few arthropods has been examined, the demonstration that these animals synthesize a wide variety of distinctive alkaloids strongly indicates that they will continue to be an outstanding source of nitrogenous compounds.

In general, complexity is not a characteristic of arthropod alkaloids, most of which possess molecular weights of less than 300. Since most of the alkaloids derived from arthropods are highly adaptive because they are fairly volatile, it is obvious that in general, compounds with excessively high molecular weights would not have been evolved by these animals. Many of these alkaloids function as pheromones that are capable of rapid information transfer between individuals, whereas others function as olfactory repellents for predators. These biological roles often require compounds with sufficiently high vapor pressures to insure that chemical signals are transmitted rapidly. In addition, many alkaloids are utilized as powerful olfactory or gustatory repellents that stimulate the chemoreceptors of potential predators. As we hope to emphasize, the alkaloids of arthropods function admirably in their roles as both pheromones and deterrents, demonstrating that these animals have adapted a large number of relatively small nitrogen heterocycles to subserve a variety of critical biological functions.

For the most part, arthropod alkaloids are biosynthesized in glands that secrete their products to the external environment. These organs, termed exocrine glands, are derived from the same ectodermal cells that are responsible for the synthesis of most of the compounds that make up the cuticle. However, since many of the compounds synthesized in exocrine glands are very cytotoxic, the final synthetic reactions usually occur in the glandular reservoir of the exocrine gland, a simple structure that generally consists of a sac that is lined with an impermeable chitinous membrane. This structural arrangement insures that highly toxic compounds such as alkaloids are isolated in the chitin-lined reservoir, effectively removed from the sensitive cellular systems elsewhere in the body.

In general, arthropod natural products are only produced in trace amounts, with submicrogram quantities of most defensive compounds being the rule. In addition, these compounds have rarely proved to be solids amenable to X-ray crystallography, and their identifications have generally required degradative research on submicro quantities in combination with a variety of spectral techniques. Ultimately, direct comparisons with standard compounds, the hallmark of quality research, have been utilized to identify conclusively the host of alkaloids detected as exocrine products of a variety of these invertebrates. To this end, synthesis becomes a *sine qua non* of the investigative process, especially when questions of stereochemistry or absolute configuration need to be answered.

Notwithstanding the quantitative limitations encountered in analyzing arthropod alkaloids, a large body of significant information vis-a-vis their roles are biological agents has been forthcoming. The functions of many of these glandular products are now unambiguously established, thus providing critical information with major correlates in chemical ecology and behavior. The elucidation of the functions of these compounds, which has often been undertaken with synthetic standards, has provided biologists with a means of comprehending the selective pressures that have resulted in their evolution. The ability to utilize animal behavior in order to study the *raison d'être* of these alkaloids presents zoologists with a luxury not generally available to those studying plant alkaloids.

In many cases, arthropod-derived alkaloids have been demonstrated to function as defensive compounds (allomones) against a variety of predators, frequently enabling their producers to escape unscathed from aggressive adversaries. These compounds may be delivered from external glands as an ooze or as an accurately discharged spray. In the case of ants, alkaloid-rich venoms are secreted through a sting that can result in subdermal administration of these pharmacologically active compounds. In some cases, alkaloids possess activity as chemical signaling agents (pheromones), but it is very likely that this function often accompanies the defensive roles that many of these nitrogen heterocycles are now known to possess. The parsimonious utilization of the same arthropod natural product for both communication and defense is a well-developed phenomenon [1] and further emphasizes the adaptiveness of these compounds in the biology of these animals. Some alkaloids, e.g., coccinelline, are not synthesized and stored in specific glands but rather are produced in internal organs and released into the body cavity, thus thoroughly fortifying the bodies of their producers with toxic and distasteful compounds [2].

Arthropod alkaloids have been identified as natural products of species in the classes Diplopoda (millipedes) and Insecta (insects). Among insects these compounds have been detected as glandular products of species in the orders Coleoptera (beetles), Hymenoptera (ants and wasps), Trichoptera (caddisflies), and Neuroptera (lacewings). Ants constitute the largest source of these nitrogen heterocycles; many species of ants biosynthesize a wide variety of alkaloids in their poison glands. In addition, many adult insects contain alkaloids that were sequestered from their host plants by their feeding larval stages. These compounds, which include a wide variety of pyrrolizidine alkaloids [3], will not be treated here since they clearly constitute plant natural products that are simply stored after ingestion, serving admirably to render their possessors distasteful, emetic, or toxic to vertebrate predators such as birds. On the other hand, adults of species in a few butterfly and moth families ingest pyrrolizidine alkaloids that are subsequently converted to compounds that are utilized as sex pheromones. These specialized cases will be discussed in Section 4.

Generally, alkaloids are defined as being basic, nitrogen-containing compounds, with the nitrogen incorporated into a heterocyclic ring. In the plant kingdom, the traditional exception to this rule is colchicine, whose nitrogen is neither basic nor is it incorporated into a ring. However, since it possesses important pharmacological activities and has a limited distribution in the plant kindgom [4], it is included in discussions of plant-derived alkaloids. Among arthropod natural products, pederin occupies a similar position, both because of its pronounced physiological properties and its restriction to a single genus of beetles [5]. In addition, this compound is characterized by one of the most complex nonproteinaceous structures elucidated from an arthropod source.

Except for one brief review [2], the subject of arthropod alkaloids has not been previously treated. In the present chapter, nitrogen heterocycles are organized according to their structures; the saturated monocyclic compounds are presented first, whereas the bicyclic aromatic compounds with more than one nitrogen, i.e., quinazolinones, are treated near the end of the chapter. A number of relatively simple heterocyclic amine compounds—the indoles—that are major exocrine products of a number of insects are treated separately, as is the complex acyclic amide pederin, the unusual nonexocrine compound only known to be produced by a few species of beetles in one arthropod taxon [5].

In this chapter we have endeavored to treat arthropod alkaloids with sufficient breadth to emphasize both their chemistry and their biology. In so doing, we hope that the biosynthetic versatility of the arthropods will be complemented by a consideration of the incredible functional diversity that characterizes the utilization of these compounds by a multitude of very successful animals. We hope this treatment will underscore our belief that for many arthropods, alkaloids make life possible in a hostile world.

2. PYRROLIDINES, PYRROLINES, AND PYRROLES

2.1. Distribution

A diversity of pyrrolidines and pyrrolines have been identified as venomous contituents of ant species in the genera *Solenopsis* and *Monomorium* [6], two taxa in the subfamily Myrmicinae. The *Solenopsis* species are members of the subgenus *Diplorhoptrum*, many species of which are known as thief ants because of their propensity for stealing brood from the nests of other ant species [7]. Although the venom of *S. punctaticeps* contains a complex mixture of both pyrrolidines and pyrrolines [8], those of other species generally contain a single pyrrolidine [9, 10]. In contrast, the venoms of *Monomorium* species are generally qualitatively rich in pyrrolidines and sometimes pyrrolines as well [6]. Pharaoh's ant, *M. pharaonis*, produces a variety of pyrrolidines in its venom [11], as do North American species in the same subgenus (*Monomorium*) [6]. An Asiatic species, *M. latinode*, is distinctive in producing a venom that is rich in both pyrrolidines and pyrrolines [6].

A unique pyrroline, polyzonimine (**4**), has been identified in the defensive secretion of the millipede *Polyzonium rosalbum* [12], a species in the family Polyzoniidae, order Polyzoniida. This compound is produced in paired exocrine glands that open on the margins of most of the body segments.

A pyrrole, methyl 4-methylpyrrole-2-carboxylate (**10**), has been identified as a trace constituent of the poison gland secretion of the ants *Atta texana* [13], *A. cephalotes* [14], *A. sexdens*, and *Acromyrmex octospinosus* [15], all species in the tribe Attini of the subfamily Myrmicinae.

2.2. Functions

The pyrrolidines and pyrrolines in the venoms of both *Monomorium* and *Solenopsis* species function as excellent repellents for a variety of ants. The venom of *S. fugax*, the European thief ant, repels foreign ants whose nests are normally raided by this species [16]. The repellency of the venom is due to the presence of *trans*-2-butyl-5-heptylpyrrolidine (**1c**, Table 1), the only alkaloid synthesized in the poison gland of this species [9]. Similarly, the pyrrolidine-rich venom of *M. pharaonis* is utilized to repel workers of other species in offensive contexts [16]. Field observations have demonstrated that a North American species of *Monomorium, M. minimum*, repels fire ant workers (*S. invicta*) [17, 18] with its pyrrolidine-rich venom [19]. Some of the pyrrolidines in the venom of *M. pharaonis* are reported to possess both trail-following and attractant activities as well [11].

Polyzonimine (**4**) functions as a defensive allomone for the millipede *P. rosalbum*, exhibiting considerable activity as a repellent for ants [12].

Methyl 4-methylpyrrole-2-carboxylate (**10**), the poison gland product of species in the genera *Atta* and *Acromyrmex*, is utilized as a trail pheromone for most of these species. *Atta texana* follows trails prepared with 2.7 pg/cm of the synthetic pyrrole [13], and *A. cephalotes* [14] is equally sensitive. *Acromyrmex octospinosus* also follows this pyrrole [15], but *Atta sexdens*, which produces this compound in its venom, utilizes a pyrazine for its trail pheromone. This pyrrole may be utilized by many related species in the tribe Attini for generating trails, since it is followed by species in the genera *Trachymyrmex*, *Cyphomyrmex*, and *Apterostigma* as well [20].

2.3. Chemistry

2.3.1. 2,5-Dialkylpyrrolidines. The carbon-nitrogen skeleton of 2,5-dialkylpyrrolidines is readily discernible using microanalytical techniques. Their mass spectra exhibit two intense peaks resulting from α-cleavage of the side chains as well as a weak molecular ion, all of which are unchanged by hydrogenation conditions. In addition, carbon-skeleton hydrogenolysis gives a mixture of the corresponding pyrroles, which show benzylic cleavage, and unbranched saturated hydrocarbons. Acetylation confirms the secondary-amine character of the natural alkaloids [8].

In the venom alkaloids from *Monomorium* species native to North America,

Table 1.

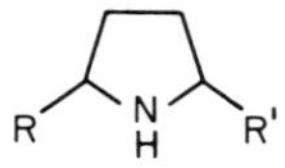

	R	R'	references
1a	C_5H_{11}	C_4H_9	8
1b	C_7H_{15}	C_2H_5	8
1c	C_7H_{15}	C_4H_9	8
1d	C_5H_{11}	C_6H_{13}	10
1e	C_9H_{19}	C_6H_{13}	19
1f	C_5H_{11}	$(CH_2)_4CH=CH_2$	79
1g	C_7H_{15}	$(CH_2)_4CH=CH_2$	79
1h	C_9H_{19}	$(CH_2)_4CH=CH_2$	19
1i	$(CH_2)_7CH=CH_2$	$(CH_2)_4CH=CH_2$	19
1j	C_5H_{11}	C_4H_9; $N-CH_3$	6
1k	C_7H_{15}	C_4H_9; $N-CH_3$	6
1l	C_9H_{19}	$(CH_2)_4CH=CH_2$; $N-CH_3$	19
1m	$(CH_2)_7CH=CH_2$	$(CH_2)_4CH=CH_2$; $N-CH_3$	19
2a	C_5H_{11}	C_2H_5	8
2b	C_7H_{15}	C_2H_5	8
2c	C_5H_{11}	C_4H_9	6
2d	C_7H_{15}	C_4H_9	6
2e	$(CH_2)_7CH=CH_2$	$(CH_2)_4CH=CH_2$	19

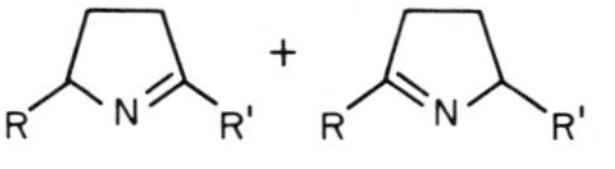

the terminal unsaturation in the side chains is assigned from methoxymercuration-demercuration on a microscale. This technique introduces a methoxy group only at the penultimate carbon of the side chain, as evidenced by the mass spectra of the resulting methoxy derivatives, which display only an intense ion ($m/z = 59$) for α-cleavage of the methoxy-substituted carbon from the rest of the side chain (see Table 1) [19].

The stereochemistry of the alkyl groups is derived from the fact that both *cis*-and *trans*-2,5-dialkylpyrrolidines have identical mass spectra, but different gas chromatographic behavior. A stereoselective synthesis of the cis isomer, by catalytic hydrogenation of the corresponding pyrrole, provides a mixture that can be compared directly to the natural material [10]. To this date, all of the 2,5-dialkylpyrrolidines reported from ants are of the trans configuration. A number of other synthetic routes to the 2,5-dialklpyrrolidines are available. The reductive amination of 1,4-diketones, which are also precursors to the aforementioned pyrroles and can be prepared by a convergent synthesis, permits the

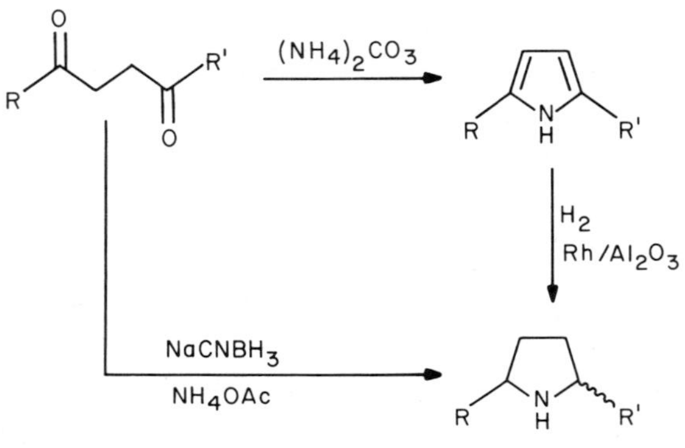

$(NH_4)_2CO_3$

H_2
Rh/Al_2O_3

$NaCNBH_3$
NH_4OAc

1

incorporation of the terminal double bonds [21]. However, this method as well as sequences based on the Hoffmann–Loffler reaction [8], the sequential alkylation of *N*-nitrosopyrrolidine [22], and the borohydride reduction of the corresponding 1-pyrrolines are not selective, all producing 50:50 mixtures of both the cis and trans configurations [23].

Sequential alkylation of 1-(methoxycarbonyl)-3-pyrroline (**3**) does form *trans*-2,5-dialkylpyrroline with high stereoselectivity; it can then be *N*-decarbomethoxylated and catalytically hydrogenated to yield only the trans dialkylpyrrolidine [24]. For example, *trans*-2-butyl-5-pentylpyrrolidine (**1a**), a component in the venoms of *Solenopsis punctaticeps* [8] and *Monomorium pharaonis* [11], can be prepared by this method in 38% yield [24].

2.3.2. Pyrrolines. The 2,5-dialkyl-1-pyrrolines, which occur as both possible isomers, can be reduced to the corresponding pyrrolidines with sodium borohydride or sodium borodeuteride. In the latter case, deuterium incorporation allows them to be distinguished from the pyrrolidines with which they occur and gives a rough idea of the relative amounts of each double bond isomer present. Their mass spectra can also be used to assign the location of the C=N bond toward the appropriate side chain by the formation of an odd electron ion resulting from a McLafferty rearrangement [8, 19]. Finally, 2,5-dialkyl-1-pyrrolines can be prepared essentially quantitatively from the corresponding pyrrolidines by N-chlorination followed by treatment with sodium hydroxide [10, 21].

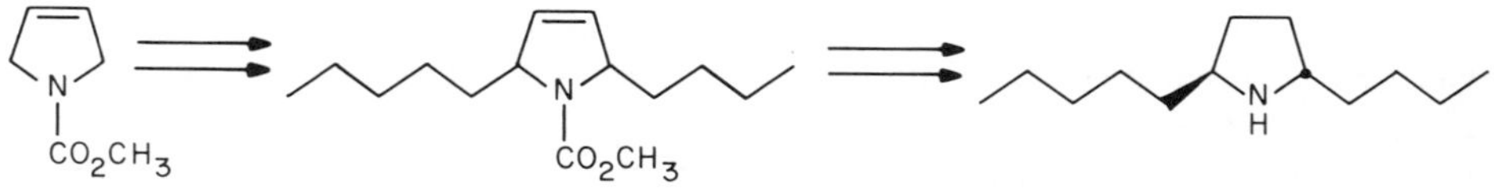

CO_2CH_3

3

CO_2CH_3

1a

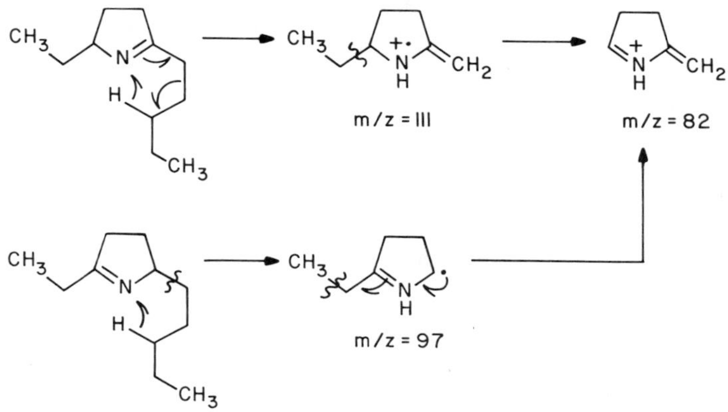

m/z = 111

m/z = 82

m/z = 97

2.3.3. Polyzonimine. The most volatile of the remarkable terpene alkaloids found in the defensive secretion of the millipede *Polyzonium rosalbum* is the spirocyclic 1-pyrroline polyzonimine (**4**) [12]. This alkaloid is characterized by its spectral data, which indicate a $C_{10}H_{17}N$ imine, substituted —HC=N—CH_2— and possessing two quarternary methyl groups. Although **4** is a liquid, X-ray crystallographic analysis of a related alkaloid from the defensive secretion suggests its structure, which can be confirmed by synthesis. The preparation of *dl*-**4** is based initially on the formation of 2,2-dimethylcyclopentane carboxaldehyde (**6**) by the lithium bromide assisted rearrangement of 3,3-dimethylcyclohexene oxide (**5**). The pyrroline ring can then be constructed by Michael addition of the morpholine enamine of **6** to nitroethylene and subsequent reduction of the nitro group to an amine in the presence of the protected aldehyde. Acid treatment during the workup gives **4** in 22% overall yield from the epoxide (**5**) [12].

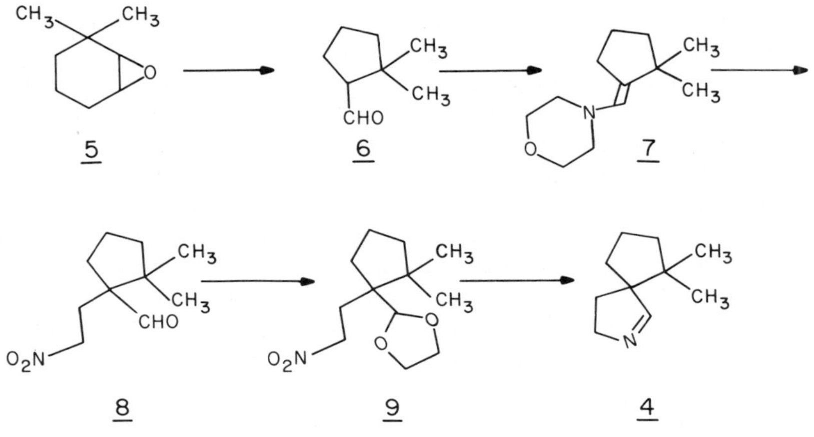

5 6 7

8 9 4

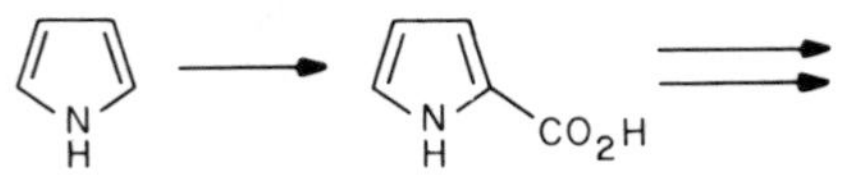

2.3.4. **Pyrrole.** The trail pheromone of the leaf-cutting ant *Atta texana* contains methyl 4-methylpyrrole-2-carboxylate (**10**) as its major active component [13]. This simple pyrrole ester is characterized by its spectrial data [25] and is available from the decarbomethoxylation of 4-methyl-3-carbomethoxypyrrole-2-carboxylic acid [26]. A more convenient synthesis is based on 2-pyrrolecarboxylic acid (**11**), formed directly from pyrrole with trifluoroacetic anhydride [27]. The methyl ester of **11** can be formylated selectively at the 4 position, and the resulting aldehyde can be hydrogenolyzed catalytically to give **10**. A number of synthetic analog of **10** also elicit trail-following behavior by *A. texana* and give some indication of the range of structural and functional group requirements for activation of trail-pheromone receptors [28].

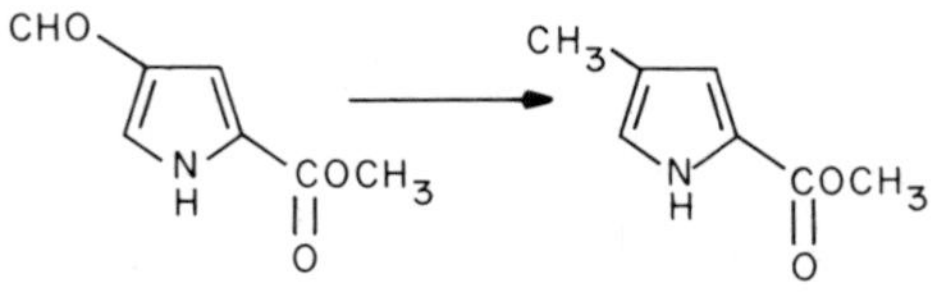

3. PIPERIDINES, PIPERIDEINES, AND PYRIDINES

3.1. Distribution

Dialkylpiperidines (Table 2) are particularly characteristic of fire ant workers and queens in the genus *Solenopsis*, a taxon in the subfamily Myrmicinae. These compounds, which are poison gland products, have been assigned the trivial name solenopsins [29] and have been identified in species in three different subgenera. All species analyzed in the subgenus *Solenopsis* produce these alkaloids [30, 31], which are often rather species specific. Queens usually produce piperidines that differ quantitatively and/or qualitatively from those of their workers, a fact that has been employed to construct an evolutionary phenocline for the genus *Solenopsis* [32]. In addition, the ratios of these compounds in the venoms of workers and soldiers are quite different [33]. Workers of a few species in the subgenus *Diplorhoptrum* also synthesize these alkaloids, and one species in the subgenus *Euophthalma* has been demonstrated to produce a piperidine [6]. Stenusin (**14**), the only piperidine not identified as a venomous constituent of ants, has been isolated from the larger paired pygidial glands of the beetle (Staphylinidae) *Stenus comma* [34].

Two piperideines have been identified as poison gland products of fire ants. 2-Methyl-6-undecyl-1-piperideine (**19**) is a minor constituent in the venom of *S. xyloni* [30], a species in the subgenus *Solenopsis*, and may serve as a precursor for the corresponding piperidines. An unidentified *Solenopsis* species in the subgenus *Diplorhoptrum* produces 2-(4-penten-1-yl)-1-piperideine (**21**) in its poison gland [6], an unexpected departure from the dialkyl theme that characterizes the nitrogen heterocycles of other species of fire ants.

Anabaseine (**26**) is one of two pyridines identified as exocrine products of insects. This compound is produced in the venom of the ants *Aphaenogaster fulva* and *A. tennesseensis* [35], two species in the subfamily Myrmicinae. Actinidine (**23**), which accompanies several iridoids in the anal (pygidial) gland secretions of two ant species in the genus *Conomyrma* (Dolichoderinae) [36], is also produced in the pygidial glands of two beetles, *Hesperus semirufus* and *Philonthus politus* [37], two members of the family Staphylinidae.

3.2. Functions

The dialkylpiperidines that fortify the venoms of fire ants are defensive compounds that exhibit an incredible diversity of pharmacological activities. In humans, ant stings result in pronounced necrosis of the epidermis with the formation of pruritic pustules [38] that are not characteristic of human reactions to the stings of other insects. The broad-spectrum activity of the venom is indicated by its pronounced antibacterial, antifungal, phytotoxic, and insecticidal properties [39]. These activities are directly attributable to the presence of the alkaloids in the venom, since synthetic piperidines possess both the necrotoxic [40] and antibiotic [41] properties exhibited by the venom. In addition, these alkaloids possess powerful lytic activity, completely hemolyzing rabbit erythrocytes in seconds [42].

The piperidines exhibit a wide range of additional pharmacological activities. At low concentrations they inhibit Na^+ and K^+ ATPases [43], an activity that is probably related to their ability to uncouple oxidative phosphorylation and reduce mitochondrial respiration [44]. They are also capable of blocking neuromuscular junctions [45] and releasing histamine from mast cells [46]. The latter activity probably contributes substantially to the potent algogenicity of fire ant venoms. It is also likely that the alkaloids with shorter 2-alkyl side chains ($n = 8$, 10) function as repellents for other ants, in much the same way as the dialkylpyrrolidines in other *Solenopsis* venoms do [9].

Stenusin (**14**), which is only slightly toxic to mammals [34], possesses a unique function in the biology of *Stenus comma*. This compound, when discharged from the tip of the abdomen, functions as a strong surface-active agent that propels the beetle rapidly across the water when it is disturbed. The rapid spreading velocity (32 cm/s) of **14**, which is the major glandular product secreted, provides this beetle with an elegant means of rapidly moving to shore when it is in an aquatic environment [34]. This phenomenon, termed *Entspannungsschwimmen*, is of rare occurrence in insects.

Actinidine (**23**), a glandular product of both ants and beetles, constitutes one of the defensive compounds utilized by these arthropods in antagonistic contexts. On the other hand, anabaseine (**26**) is reported to be an attractant for workers of *A. fulva* [36], but it will not prove surprising if this compound functions is a defensive allomone as well.

Both of the piperideines identified as venomous constituents of *Solenopsis* species are deterrents for other ant species [6] and probably serve as olfactory repellents against specific competitors.

3.3. Chemistry

3.3.1. 2-Alkyl-6-methylpiperidines. The known 2-alkyl-6-methylpiperidines occurring in *Solenopsis* venoms possess long *n*-alkyl groups that are either saturated or contain a carbon–carbon double bond at the ninth carbon from their terminal methyl groups (Table 2). Carbon-skeleton chromatography converts these compounds to a series of odd-numbered normal alkanes, whereas micro-ozonolysis produces only nonanal, and permanganate-periodate oxidation yields nonanoic acid [47]. The most obvious indication of the nitrogen-containing portion of these compounds is the base peak at $m/z = 98$ in their mass spectra, a fragment that is characteristic of 2-methylpiperidines and of 2,6-dimethylpiperidine itself.

The overall carbon–nitrogen skeleton of the 2-alkyl-6-methylpiperidines can be confirmed by synthesis, which also permits the assignment of the stereochemistry of the ring substituents. A generally applicable synthesis consists of alkylation of the lithium salt of 2,6-lutidine with the appropriate alkyl bromide

Table 2. 2-Alkyl-6-Methylpiperidines (**12**)

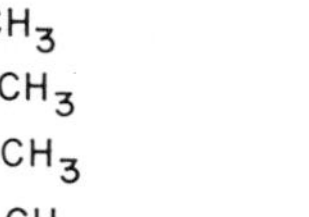

CH_3 N H R

12

	R	references
12a	$(CH_2)_8CH_3$	47
12b	$(CH_2)_{10}CH_3$	47
12c	$(CH_2)_{12}CH_3$	47
12d	$(CH_2)_{14}CH_3$	47
12e	$(CH_2)_3CH{=}CH{-}C_8H_{17}$	47
12f	$(CH_2)_5CH{=}CH{-}C_8H_{17}$	47
12g	$(CH_2)_7CH{=}CH{-}C_8H_{17}$	47
12h	$(CH_2)_8CH_3$; $N{-}CH_3$	6
12i	$(CH_2)_{10}CH_3$; $N{-}CH_3$	6

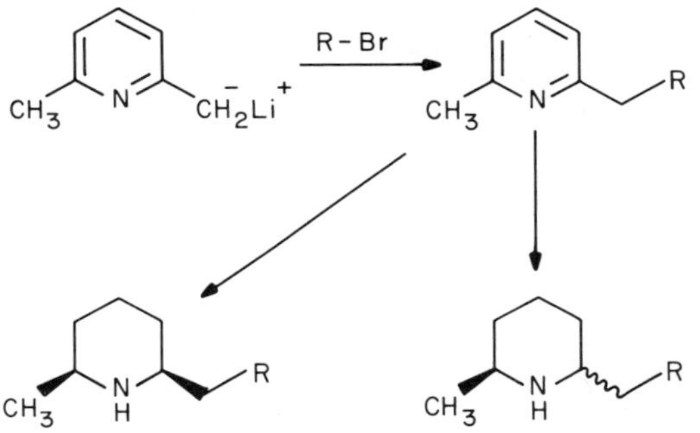

or tosylate, followed by reduction of the resulting 2-alkyl-6-methylpyridine [47]. Hydrogenation produces only the *cis*-2,6-disubstituted piperidine, whereas reduction with sodium in ethanol yields an 85:15 cis/trans mixture. Although the mass spectra of both configurations are almost identical, their gas chromatographic retention times are quite different, allowing assignment of the stereochemistry of the natural materials. Both configurationed isomers are present in a number of species of *Solenopsis* [31].

Since the natural 2-alkyl-6-methylpiperidines with side chain unsaturation display none of the usually strong infrared absorptions associated with trans olefins, their side chain double bonds are cis. This negative evidence is supported by synthesis. The appropriate alkenyl bromides with cis double bonds can be prepared using partial hydrogenation of an acetylene in the presence of Lindlar catalyst as the geometry-determining step, and carried through the 2,5-lutidine alkylation and reduction to give products that are chromatographically and spectroscopically identical to the natural alkaloids [47].

Aside from the alkylation of lutidine, the fire ant venom alkaloids can be prepared by a route utilizing the Mundy rearrangement and by a sequence based on aminomercuration. The rearrangement of *N*-acyllactams to give 2-substituted cyclic imines, "the Mundy rearrangement," in the case of 2-acylpiperidones gives 2-alkylpiperideines, which can be reduced to the corresponding piperidines [48]. Unfortunately, although the yields are good when the alkyl group is short, with long alkyl chains, the yields are poor. The yield of 2-methyl-6-undecylpiperidine (**12b**) is only 5% from the corresponding *N*-acylpiperidone [49]. On the other hand, aminomercuration–demercuration of 2-amino-6-heptadecene (**13**) produces **12b** in 72% yield with predominance of the trans-substituted isomer, although the preparation of **13** requires six steps from 6-hydroxy-2-hexanone [50].

The *N*-methyl piperidines (**12h**) and (**12i**) (Table 2) occur with the corresponding N-H piperidines. They show a characteristic peak in their mass spectra at $m/z = 112$ and can be prepared from the non-*N*-methylated piperidines by treatment with formaldehyde and formic acid [6].

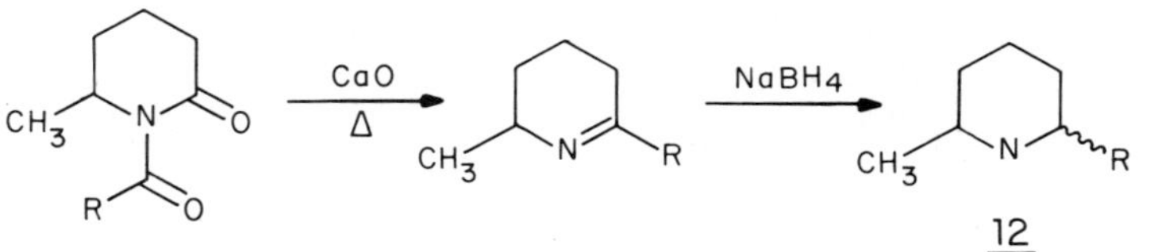

3.3.2. Stenusin. The structural elucidation of *N*-ethyl-3-(2-methylbutyl)piperidine, "stenusin" (**14**), from its spectral data and subsequent synthesis admirably demonstrates the importance of synthesis as a structural proof for natural products of limited availability. With its elemental composition established by high-resolution mass spectrometry, the IR and NMR spectra of **14** revealed the absence of double bonds and the presence of a tertiary *N*-ethylamine, perhaps a piperidine ring. A study of the metastable transitions in the mass spectrum showed that the important odd-numbered ion at $m/z = 113$ originates directly from the molecular ion by arrangement and loss of a pentyl group, which itself cannot loose a propyl group [34, 51].

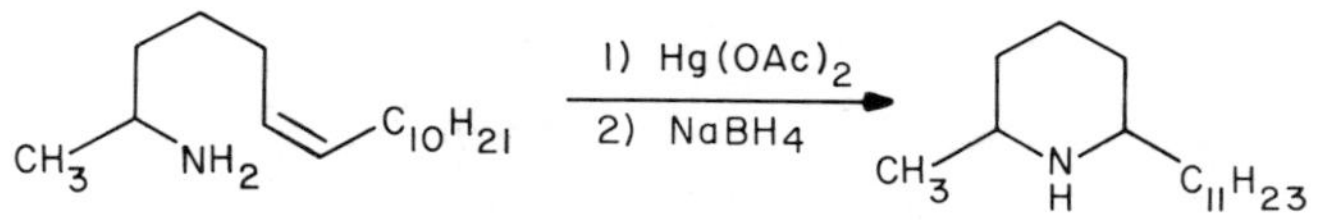

In the absence of X-ray crystallographic data, the above spectral information hardly constitutes a vigorous proof of structure, but it is confirmed by the relatively straightforward preparation of **14**. Thus the monoester of 2-methylbutylmalonic acid (**15**) can be converted by ethyl-2-(2-methylbutyl)acrylate (**16**) by a Mannich reaction. Treatment of **16** with ethylamine, and the resulting secondary amine with ethyl acrylate in a series of Michael additions gives the tertiary amine diester (**17**), which can be cyclized to *N*-ethyl-3-(2-methylbutyl)-4-piperidone (**18**). Wolff–Kishner reduction of **18** provides stenusin (**14**) whose IR, NMR, and mass spectra are practically identical to those of the natural product. Additionally, the incorporation of (−)-2-methyl-butan-1-ol in this synthesis results in a final product with a specific rotation of $[\alpha]^{20}_{356} + 5.40$, similar to that of the natural material ($[\alpha]^{20}_{356} = +5.8$) [34].

Since *Stennus comma* uses its pygidial gland secretion as a means of propulsion, a comparison of the spreading velocity over water of stenusin with other terpenoid component shows that stenusin spreads most rapidly and probably has a major role in driving the beetle [34].

3.3.3. Piperideines. The structures of the 1-piperideines occurring in *Solenopsis* venoms at submicrolevels can be determined from their reduction products and their mass spectral fragmentation patterns. 2-Methyl-6-undecyl-1-piperideine (**19**) is cleanly reduced to 2-methyl-6-undecylpiperidine (**12b**) by

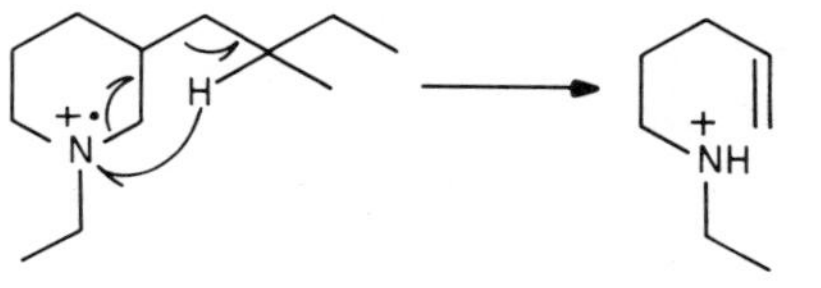

m/z = 183 m/z = 113

14

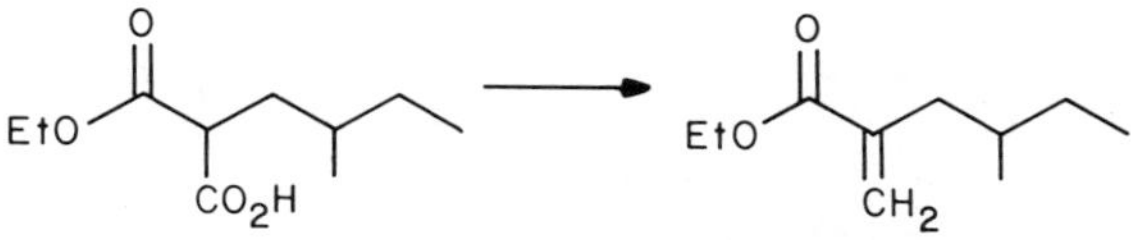

15 16

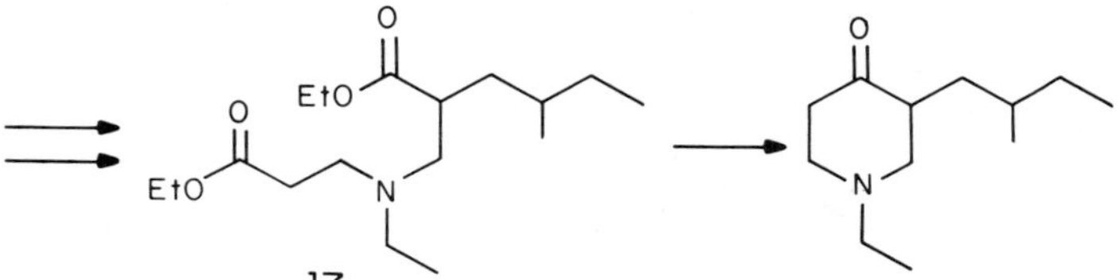

17 18

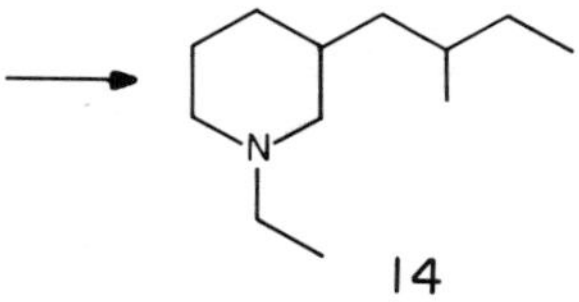

14

borohydride, establishing its carbon-nitrogen skeleton and the nature of its double bond. Treatment of piperidine **12b** with *t*-butylhypochlorite produces both **19** and its isomer 2-methyl-6-undecyl-6-piperideine (**20**), whose mass spectra are quite different. In the former, an intense odd-electron ion appears at $m/z = 97$, whereas in the latter, an intense odd-electron ion appears at $m/z = 111$. These diagnostic fragments both involve hydrogen transfer from the alkyl side chain. The natural material has an identical mass spectrum and gas chromatographic retention time as piperideine **19** [30, 51].

The mass spectrum of 2-(4-penten-1-yl)-1-piperideine (**21**) also displays an odd-electron ion at $m/z = 97$, and although borohydride reduction reveals one C=N bond, hydrogenation gives a mass spectrum identical to that of 2-pentylpiperidine (**22**). Thus, **21** is an unique monoalkylated piperideine whose structure is confirmed in a straightforward manner by the Mundy rearrangement of *N*-5-hexenoyl-2-piperidone [6].

CH_3 N H C_8H_{17} 19 → CH_3 $\overset{+}{N}$ H m/z = 97

CH_3 N H C_8H_{17} 20 → CH_3 $\overset{+\bullet}{N}$ H CH_2 m/z = 111

21 → 70 eV → $\overset{+\bullet}{N}$ H CH_2 m/z = 97

21 → H_2, Pd/C → N H CH_3 22

N O O CH_2 → CaO, Δ → N CH_2 21

3.3.4. Pyridines. The structure of actinidine (**23**) from arthropod sources is suggested by its mass spectrum (M^+, $m/z = 147$, $C_{10}H_{13}N$ with loss of methyl on the base peak) and by its UV spectrum, which is characteristic of the pyridine chromophore ($\lambda_{max} = 269{,}260$ nm), and is confirmed by comparison with synthetic material [36, 37]. Actinidine can be prepared from nepetalinic acid imide (**24**) by way of the 2,6-dichloropyridine [52], and from iridodial (**25**) by acid-catalyzed cyclization of its *cis*-2,4-dinitrophenylhydrazone or by direct ammoniation [53, 54]. The biosynthesis of **23** in plants has been shown to proceed through the mevalonate pathway [55]. Since the biosynthesis of other iridoid monoterpenes in insects has also been shown to follow the mevalonic acid route [56], it is likely that the biosynthesis of **23** in other species of arthropods also follows this pathway.

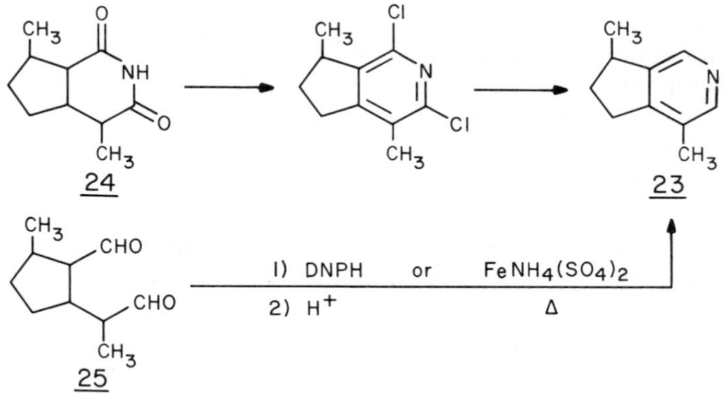

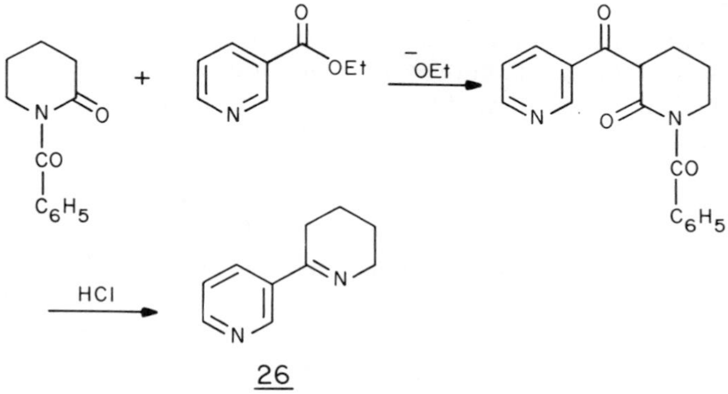

The proton magnetic resonance spectrum of anabaseine (**26**) shows the presence of a 3-substituted pyridine with eight methylene protons attached at the 3-position. The carbon-nitrogen skeleton of **26** is established by its conversion to 2,3-bipyridyl with chloranil [35]. Anabaseine can be prepared by condensation of *N*-benzoylpiperidone with ethyl nicotinate. Heating the resulting ketoimide with hydrochloric acid in a bomb-tube provides **26**, which can be purified as its dipicrate salt [57].

4. PYRROLIZIDINES

4.1. Distribution

Pyrrolizidine alkaloids (**27a, 27c, 27d**) have been identified as metabolites of adult male butterflies and moths in several unrelated families. Significantly, these insects visit and feed on injured or dead plants that contain these nitrogen heterocycles, and it now seems firmly established that, for the most part, many of these lepidopterans metabolize these compounds in specialized glands [58]. In a few cases, these alkaloids are reported to be taken up from the plant and utilized without being metabolically altered [59]. Brushlike scent organs that are extrusible [60] or, in some cases, present as wing tufts [61], are associated with these glands, thus permitting the metabolized pyrrolizidine alkaloids to be externalized. These insects can effectively disseminate these compounds by splaying the brushes or tufts, thus exposing the individual hairs with their great evaporative surface areas.

Males of many species in the family Danaidae, which includes the monarch butterfly and its relatives, feed on plants containing pyrrolizidine alkaloids that are metabolized and disseminated from paired abdominal brushes, the hair pencils [58]. Species in the danaid genera *Lycorea* [62], *Danaus* [63], *Euploea* [64], and *Amaurus* [65] have been demonstrated to contain these alkaloids in their hair pencils. Males of another group of butterflies, members of the family Nymphalidae, are known to produce metabolized pyrrolizidines and disseminate them from hairs of the costal fringe, a specialized organ on the hind wings. These nitrogen heterocycles have been identified as products of species in the genera *Hymenitis*, *Godyris*, *Pteronymia*, and *Ithomia* [59], all members of the subfamily Ithomiinae. Alkaloids have also been identified in the scent organs (brushes or coremata) of males in the moth genus *Utetheisa* [66], a taxon in the family Arctiidae (tiger moths). The males of many species in the family Ctenuchidae probably also contain alkaloids in their abdominal brushes, since a large number of species in this family demonstrate a male bias in visiting pyrrolizidine-containing plants [67].

A novel pyrrolizidine alkaloid, nitropolyzonamine (**30**), has been identified as a secretory product of the millipede *Polyzonium rosalbum* (Polyzoniidae: Polyzoniida) [68]. This compound is produced in paired glands occurring on the margins of most of the body segments.

The venom of a thief ant, *Solenopsis xenovenenum*, has been demonstrated to contain an unusual pyrrolizidine alkaloid [69]. Only pyrrolidines and pyrrolines have been identified in other species in this subgenus (*Diplorhoptrum*).

4.2. Functions

Males of many species of butterflies and moths have coevolved with plants producing pyrrolizidine alkaloids, and in many cases these insects convert these compounds to metabolites that possess critical reproductive functions. For example, removal of the hair pencils of the queen butterfly, *Danaus gilippus berenice*, substantially reduces the courtship success of males [70]. Blocking the chemoreceptors on the female's antennae achieves a similar result, and it has been demonstrated that specialized thin-walled sensilla basiconica on the antennae are the receptors for the alkaloids. Thus these insects have utilized plant-derived pyrrolizidine alkaloids as precursors for sex pheromones that are a *sine qua non* for successful mating. Preventing adult males from having access to alkaloid-rich plants reduces their ability to seduce females, but in some cases this handicap can be overcome by providing them with a synthetic sample of their sex pheromone [71]. The antennal chemoreceptors of both males and females of the queen butterfly have been shown to respond strongly to the hair-pencil constituents, or to the synthetic pyrrolizidine that has been identified as a male metabolite [72]. On the other hand, hair pencils of laboratory-reared butterflies, lacking these alkaloids, elicit very weak electroantennogram responses [73].

In some cases, the biosynthesis of the sex pheromone (danaidone, **27a**) requires specialized behavioral responses on the part of the male butterfly. For example, males of *Danaus chrysippus* possess, in addition to abdominal hair pencils, glandular (alar) pockets in the hind wings. The synthesis of the aphrodisiacal sex pheromone disseminated from the hair pencils of *D. chrysippus* does not occur if the male does not insert his hair pencils into the alar pockets [74]. On the other hand, males of *Lycorea ceres* also produce danaidone (**27a**) [62], but these insects lack alar pockets. Other danaids (e.g., *Danaus limniace petervirana*) produce the sex pheromone in the glandular alar pockets, and it is transferred to the hair pencils when the latter are inserted into the pockets [74].

Lycopsamine, a possible precursor of danaidone (**27a**), has been identified in hair pencils of two danaid species along with the ketonic pheromone [59]. Although plant-derived alkaloids do not appear to be frequently utilized directly as part of a pheromonal bouquet, some danaids are known to sequester these agents and store them in body tissues, and these hepatotoxic compounds can provide protection against vertebrate predators. For example, the monarch butterfly, *D. plexippus*, which does not contain pyrrolizidines in its hair pencils, and the African monarch, *D. chrysippus*, which does, store alkaloids such as seneciphylline and senecionine, which are believed to contribute to the unpalatability of these insects [75].

The compounds found in hair pencils or brushes do not appear to be produced with either quantitative or qualitative exactitude. Considerable variations in the ratios of alkaloids and other compounds have been detected in the hair pencils of an *Amauris* species [65], possibly reflecting dietary effects vis-à-vis exogenously obtained pyrrolizidines. In the case of two populations of an *Euploea* species, qualitative differences in alkaloidal constituents were present [64]. Similarly, qualitative variations in the pyrrolizidines present in the coremata of males of the arctiid moth *Utetheisa lotrix* demonstrated that two distinct forms were present [66]. Beyond these unexplained populational variations in sex-pheromone chemistry, the alkaloids have been demonstrated to be localized in different types of hairs in the hair pencils of *Amauris ochlea* [76]. The significance of the spatial distribution of these compounds remains to be determined.

The pyrrolizidines present in the secretions of the millipede *Polyzonium rosalbum* [68] and the ant *Solenopsis xenovenenum* [69] constitute defensive allomones.

4.3. Chemistry

4.3.1. Lepidopterous Pheromones. The structure of 2,3-dihydro-7-methyl-1*H*-pyrrolizin-1-one (**27a**) can be inferred from its spectral data. Its absorption spectra and mass spectral fragmentation pattern are analogous to those of 2,3-dihydro-1*H*-pyrrolizin-1-one (**27b**). The coupling constant of the pyrrole protons in the NMR spectrum of **27a** requires that they be α- and β-pyrrole protons and that the methyl group be at C(7) [62]. Synthetic can be obtained from 3-methylpyrrole by cyanoethylation with acrylonitrile followed by acid-catalyzed cyclization. The synthetic material is identical with the natural alkaloid in all respects [62].

Both **27a** and 2,3-dihydro-7-formyl-1*H*-pyrrolizine (**27d**) can be detected in the natural extracts by TLC comparison with authentic samples. In addition, the mass and NMR spectra of the natural alkaloids are identical with those of the synthetic compounds [63]. Heliotridine or retronecine (**28**) can be converted to **27c** by treatment with manganese dioxide, whereas supinidine (**29**) yields **27d** under similar conditions [63].

4.3.2. Nitropolyzonamine. Although rare in a natural product, the presence of the nitro group in nitropolyzonamine (**30**) is clearly indicated by its IR (1540 and 1375 cm^{-1}) and mass spectra—$m/z = 238$ ($M+$) and 192 ($M-46$). The remainder of the structure and stereochemistry of **30** are revealed by an X-ray crystallographic study of its perchlorate salt. The nitropyrrolizidine (**30**) can be prepared from its congener polyzonimine (**4**) directly by treatment with β-nitroiodopropane, followed by cyclization with pyridine [68].

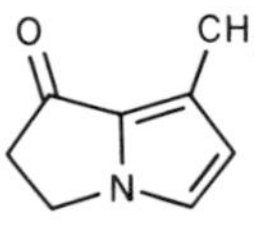

27a

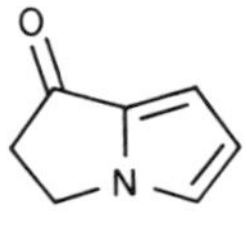

27b

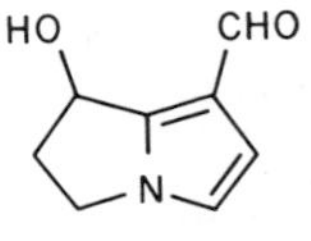

27c

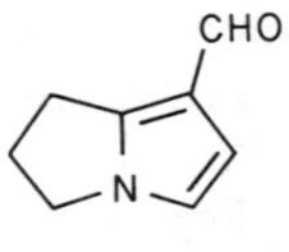

27d

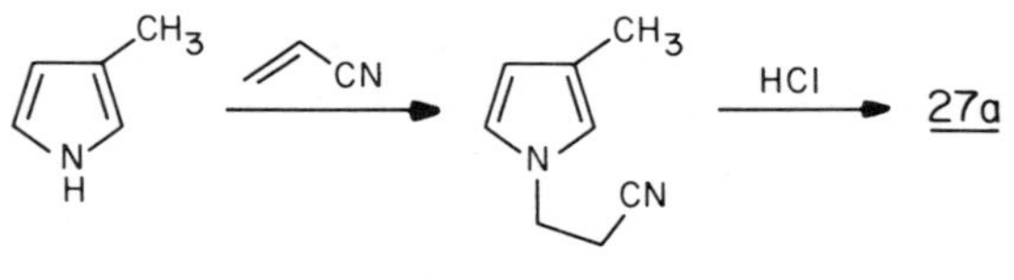

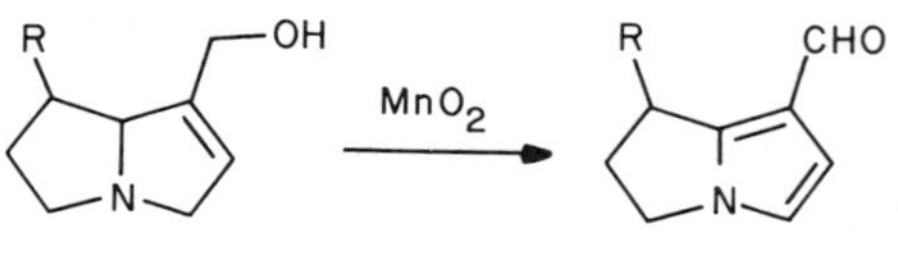

28: R = OH
29: R = H

27c: R = OH
27d: R = H

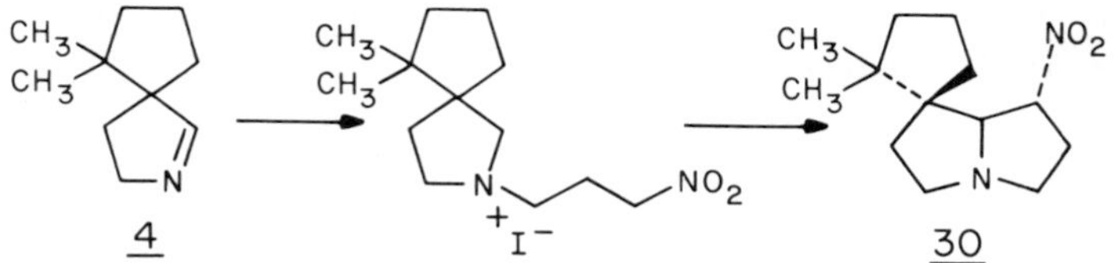

4.3.3. (5*Z*, 8*E*)-3-Heptyl-5-methylpyrrolizidine.

The structure of (5*Z*,8*E*)-3-heptyl-5-methylpyrrolizidine (**31**) is suggested by its mass spectrum, which shows intense ions for the loss of CH_3 and C_7H_{15} from carbons adjacent to nitrogen and a parent ion that indicates a possible molecular formula of $C_{15}H_{29}N$. A bicyclic structure is probable since this spectrum is unchanged by hydrogenation conditions. The required compound can be obtained by reductive amination of the known 2,5,8-pentadecatrione (**32**) [77], which provides all four possible stereoisomers. All four isomers have different gas-chromatographic retention times, and that of the natural product is identical with the retention time of the major isomer. The stereochemistry of each isomer can be assigned by ^1H and ^{13}C spectroscopy, which shows that the major isomer (70%) has a cis-fused ring junction and is the (5*Z*, 8*E*)-3,5-dialkyl configuration. Consideration of each ring of **31** as a separate trans-disubstituted pyrrolidine reveals a close structural relationship to 2-butyl-5-heptylpyrrolidine (**1c**), which also occurs naturally as the trans isomer [9].

5. INDOLIZINES

5.1. Distribution

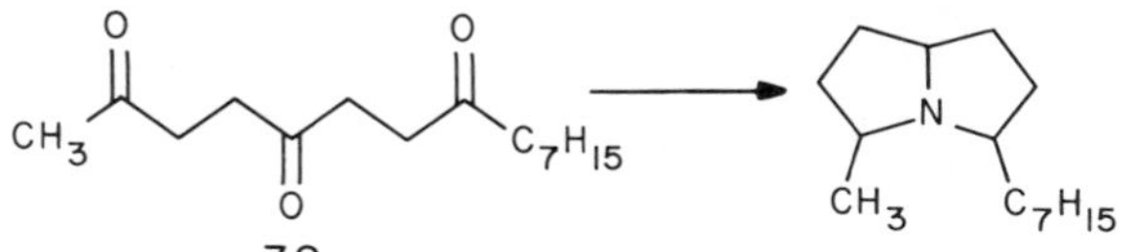

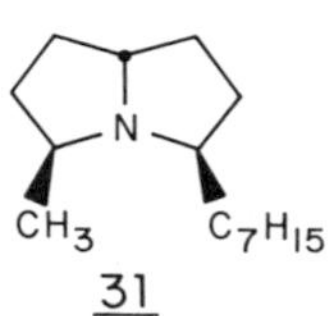

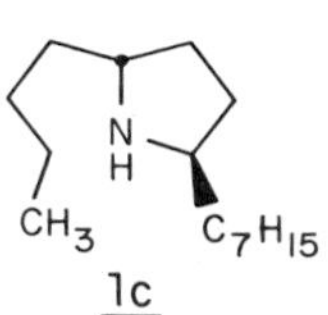

Indolizines are limited in their known arthropod distribution to Pharaoh's ant, *Monomorium pharaonis* [78, 79], a species in the subfamily Myrmicinae. These alkaloids, which are poison gland products of both workers and females, have not been detected in the venoms of other species of *Monomorium* that are in the same subgenus (*Monomorium*) [19].

5.2. Functions

5-Methyl-3-butyloctahydroindolizidine (**33**), termed monomorine-I, is attractive to ant workers and is present in their natural trails [78]. Workers produce considerably more monomorine-I than females (queens), and it has been suggested that this compound, along with other alkaloids, might have a different function in workers and queens [79]. This indolizine possesses slight trail-following activity, but it is clearly different from the natural trail pheromone [79]. Workers will follow trails containing 10 ng/cm of monomorine-I. A synergistic effect (attraction) resulting from combinations of **33** and a naturally occurring pyrrolidine has been observed, and concentrational effects eventuating in attraction, repellency, or alarm behavior may result. The possible role of the indolizines as natural repellents has not been evaluated, although Ritter and Stein have suggested that these compounds may function as repellents (allomones) for other insects [23].

5.3. Chemistry

The overall carbon-nitrogen skeleton of 3-butyl-5-methylindolizidine (**33**) is suggested by its NMR and mass spectra and confirmed by synthesis of a mixture of stereoisomers of **33**. Condensation of 4-aminooctanal diethyl acetal, diethyl-3-oxoglutarate, and ethanal forms 3-butyl-5-methyl-6,8-dicarbethoxy-7-oxoindolizidine (**34**), which can be saponified, decarboxylated, and reduced to give **33** as an isomeric mixture [78]. There are four possible isomers whose stereochemistry was established by unambiguous synthesis from aromatic precursors, utilizing the principle of cis hydrogenation of an aromatic ring to control their geometry [80, 81]. Thus the lithium salt of 2,6-lutidine was treated with 1-hexenoxide to give 1-(6-methyl-2-pyridyl)-3-heptanol (**35**). Cyclization was carried out by converting the alcohol to a secondary bromide and subsequent treatment with triethylamine. The resulting pyridinium salt in this case was hydrogenated to give only the all-cis isomer **33a**. On the other hand, when **35** was hydrogenated to the *cis*-2,6-disubstituted piperidine (**36**), the C_3 isomers **33a** and **33b** could be obtained by cyclization. In a similar manner, the trans isomer of **36**, obtained

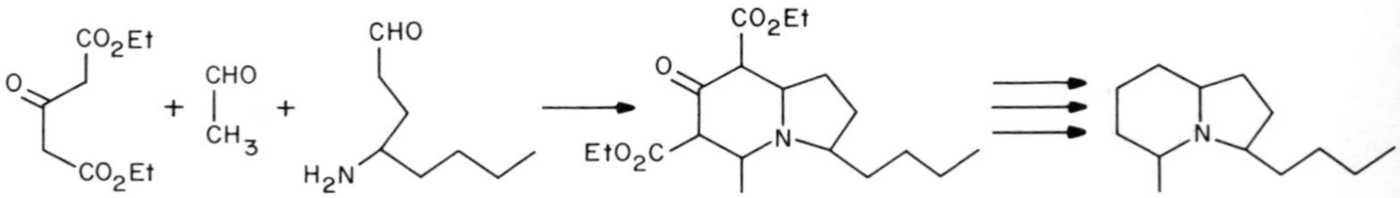

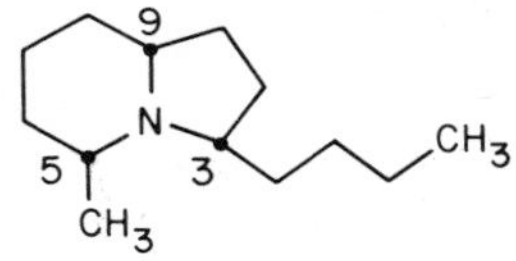

33a 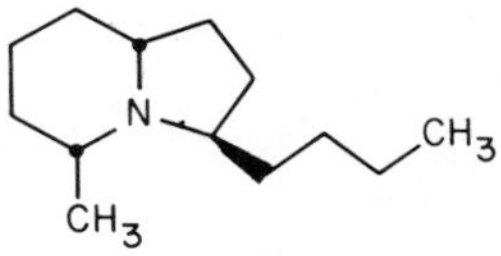 33b 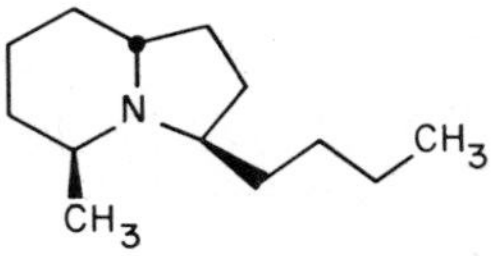 33c

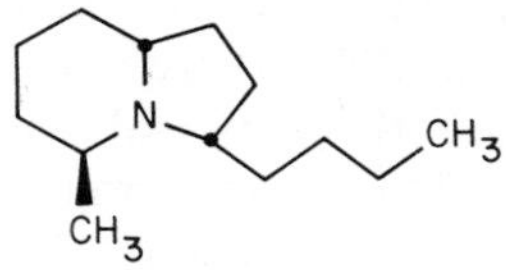

33d

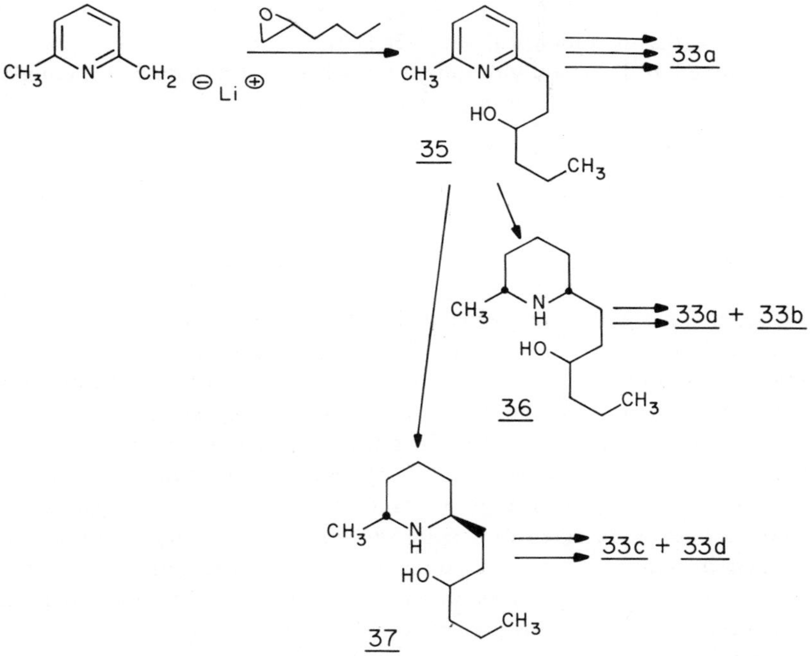

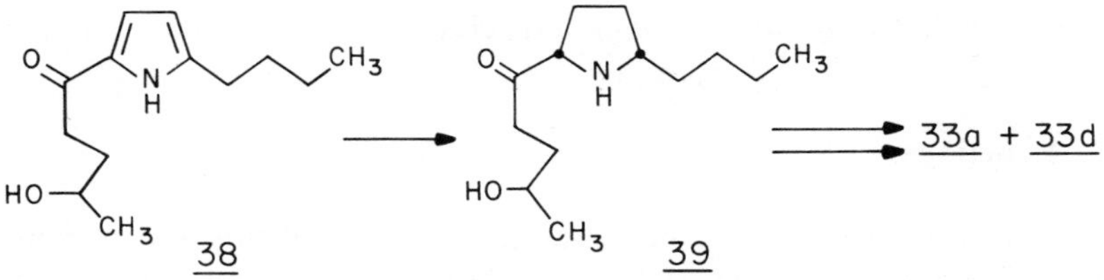

35 → 33a

36 → 33a + 33b

37 → 33c + 33d

38 → 39 → 33a + 33d

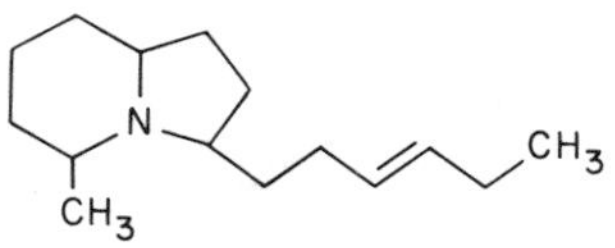

40

from the sodium and ethanol reduction of **35**, could be cyclized to produce the C_3 isomers **33c** and **33d**. These reactions established the C_5 and C_9 stereochemistry of all four isomers and the C_3 stereochemistry of **33a** and **33b**. Hydrogenation of 1-(5-butylpyrrol-2-yl)-4-hydroxy-1-pentanone (**38**) gives the *cis*-pyrrolidine (**39**), which can be cyclized to the C_5 isomers **33a** and **33d**, establishing the C_3 stereochemistry of the latter and therefore also of **33c**. These stereochemical assignments are confirmed by an analysis of the ^{13}C NMR spectra of each isomer [82]. Alkylation of the anion of 1-(methoxycarbonyl)-2-butyl-3-pyrroline with 1,4-dibromopentane followed by cyclization and reduction yields only stereoisomerically pure **33b** [24]. Comparison of the natural material with the isomers **33a–d** shows that it is identical with **33a** [79].

3-(3-Hexen-1-yl)-6-methylindolizidine (**40**) is also reported as a congener of **33a**, but little is known about its stereochemistry [11].

6. COCCINELLINES

6.1. Distribution

The coccinellines have been primarily identified as natural products of ladybird beetles (ladybugs), members of the family Coccinellidae. A large number of species in the subfamily Coccinellinae produce alkaloids, particularly those in the tribe Coccinellini [83]; alkaloids have also been detected in species in four other tribes in this subfamily. In addition, a species in the subfamily Epilachinae is also reported to produce alkaloids [83]. We wish to emphasize that alkaloids are produced only by warningly colored (asposematic) ladybugs, these compounds not being detected in cryptic species in this family. The coccinellines produced by ladybugs are not secreted from external glands but rather are present throughout the body as bloodborne constituents. On the other hand, a soldier beetle (Cantharidae) in the genus *Chauliognathus* has been demonstrated to secrete coccinellines from paired glands located on the prothorax and abdomen [84]. As in the case of the ladybugs, this soldier beetle is highly aposematic. A coccinelline has also been identified as a volatile constituent from the boll weevil, *Anthonomus grandis*, a species in the beetle family Curculionidae [85].

6.2. Functions

The coccinellines have been evolved to function as repellents or deterrents for a variety of vertebrate and invertebrate predators. In ladybugs these compounds are present in the blood, and although these beetles do not possess exocrine glands that discharge to the exterior, they are nevertheless capable of externalizing the alkaloids. Adult coccinellid beetles are capable of reflex bleeding at leg articulations [86], a development that enables them to discharge coccinelline-rich blood at weakened points in the cuticle. Reflex bleeding, or autohemorrhage, is highly adaptive and can be limited to legs that are proximate to the point of traumatic stimulation. The blood that is liberated reflexively clots instantly, and this phenomenon is not injurious to the beetle in any sense of the word. Some larval coccinellids also exhibit reflex bleeding [87], raising the possibility that they may be externalizing coccinellines in the same manner as the adults. Eggs and larvae of coccinellids have been demonstrated to contain these alkaloids [88].

Coccinellines are highly repellent to ants, and concentrations between 2×10^{-4} and 5×10^{-4} *M* will effectively deter ant workers from feeding [86]. These bitter-tasting alkaloids probably constitute the main line of defense of ladybugs against ants, insects that the beetles encounter frequently when they feed on the aphids tended by ants. In addition, European quail may reject these beetles as well, and the alkaloids have been determined to be responsible for this repellency [86]. Probably the coccinelline-rich secretions of soldier beetles, which also contain an acetylenic acid [84], are directed against avian predators that these free-flying beetles must frequently encounter.

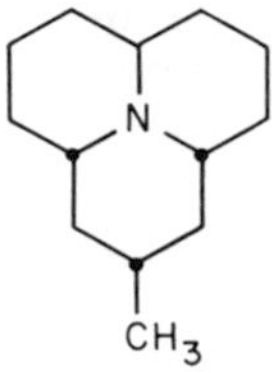

N
CH_3

41 Precoccinelline
42; N-oxide Coccinelline

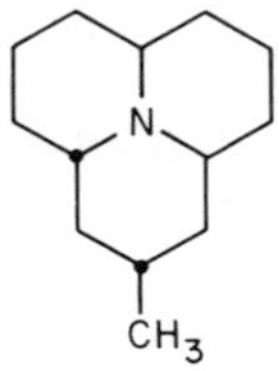

N
CH_3

43 Hippodamine
44; N-oxide Convergine

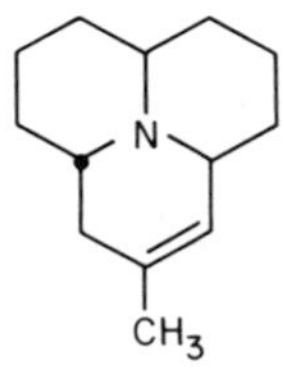

N
CH_3

45 Hippocasine
46; N-oxide Hippocasine oxide

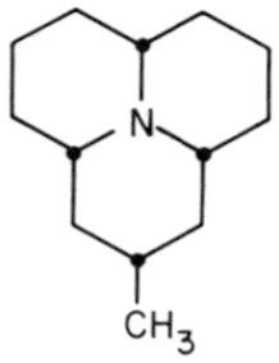

47 Myrrhine

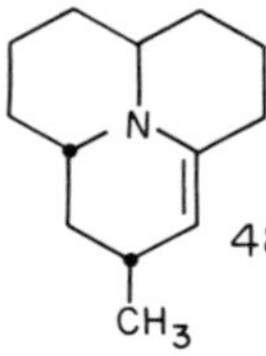

48 and 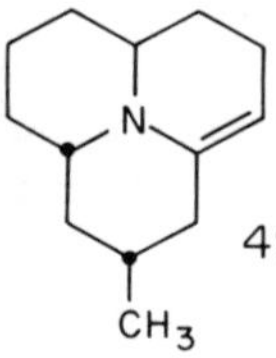 49 Propyleine

6.3. Chemistry

6.3.1. Structural Elucidation. Although the structure of coccinelline (**42**) is suggested by its mass spectrum, $m/z = 209$, ($C_{13}H_{23}NO$), by the lack of any spectral indication of double bonds [88], and by the symmetry apparent in its ^{1}H and ^{13}C magnetic resonance spectra [89], it is firmly established as the 2-methylperhydro-9*b*-azaphenylene (**42**) by X-ray crystallographic analysis [90]. Precoccinelline (**41**) is the free base of **42**, since it can be oxidized to **42**, which can in turn be reduced back to **41** [88, 89]. In a similar manner, the structures of hippodamine (**43**) and its *N*-oxide convergine (**44**) [91], as well as the structures of hippocasine (**45**) and hippocasine oxide (**46**) [92], can be assigned from their X-ray crystallographic data and their chemical interconversions. The structure of myrrhine (**47**), the all-trans fused perhydro-9*b*-azaphenylene isomer, can be presumed from the intense Bohlmann bands in its IR spectrum, indicating the trans antiperiplanar relationship between the three carbon α-hydrogens and the nonbonding electrons on nitrogen [94, 96, 97]. This can be confirmed by chemical correlation between **47** and **42** using the Polonouski reaction. Thus formation of the *N*-acetate of **42**, which immediately eliminates acetic acid to give an enamine, followed by catalytic hydrogenation provides only **47** and **41** in a 9 : 1 ratio [93]. The enamine alkaloid propyleine (**48** and **49**) is described as a single isomer from its spectra and its conversion to **42** [95]. On the other hand, preparation by total synthesis shown that it occurs as an isomeric mixture of **48** and **49** [96].

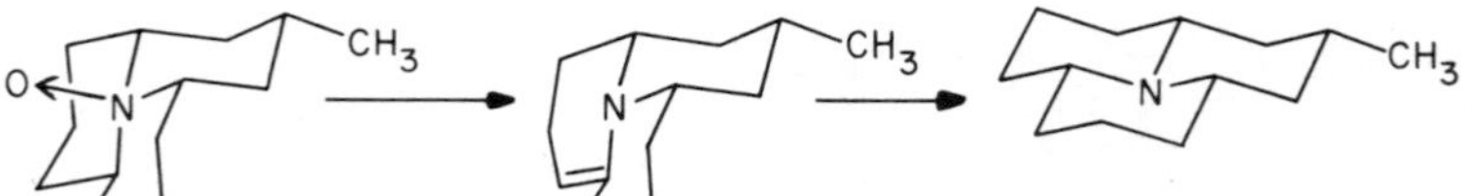

42 → → 47

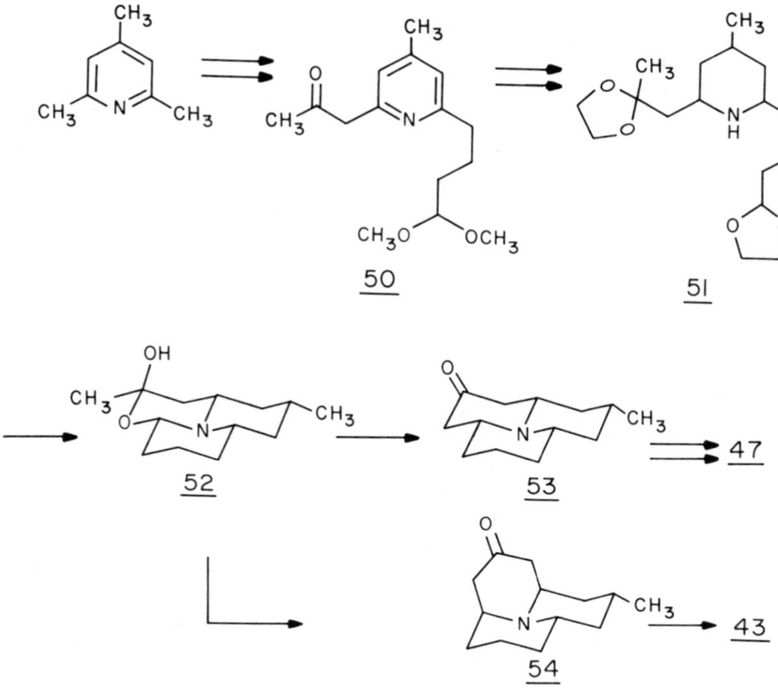

6.3.2. Synthesis. Two general routes to the coccinelline alkaloids are available. The first is based on sequential alkylation and acylation of 2-collidyllithium, first with bromopropionaldehyde dimethyl acetal, and following a second treatment with phenyllithium, with acetonitrile to give the ketoacetal **50**. Following conversion to the bis-ethylene acetal, the pyridine ring can be reduced with sodium in isoamyl alcohol to form the all-cis piperidine (**51**). The tricyclic hemiketal **52** is produced immediately upon removal of the protecting groups, and in the presence of acid, recyclizes to the ketoamine **53**, whose thioketal can be converted to myrrhine (**47**) by Raney nickel desulfurization. Alternatively, **52** can be cyclized with piperidine acetate to an isomeric ketoamine (**54**) whose thioketal provides hippodamine (**43**) upon desulfurization [97]. In a similar manner, *cis*-2,6-piperidine (**55**) can be prepared from 2,6-lutidine and then cyclized to ketoamine **56**. Treatment of **56** with methyllithium followed by dehydration and hydrogenation provides precoccinelline (**41**) [98].

A more elegant route to the coccinelline alkaloids is based on the known perhydroboraphenylene (**57**) [99], which is coverted to the enamine **58** by treatment with *N*-chloro-*O*-2,4-dinitrophenylhydroxylamine followed by Jones oxidation of the intermediate amino alcohol [100]. Hydroboration and subsequent oxidation of **58** gives the ketoamine (**59**). Since the usual methylation procedures are ineffective for α-methylation of **59**, the α-vinylogous amide of **59** is formed with Bredereck reagent [102] and reduced and hydrogenolyzed to the methyl ketone (**60**). Hippodamine (**43**) is then prepared by reduction of the thioketal of **60** and can be oxidized to convergine (**44**) [101]. Alternatively, the Bamford-Stevens reaction of **60** produces hippocasine (**45**) and with subsequent oxidation, hippocasine oxide (**46**) [101]. The methanesulfonate ester **61** can be prepared from **60** and subjected to elimination conditions to give propyleine as an equilibrium mixture of the two isomeric enamines **48** and **49**.

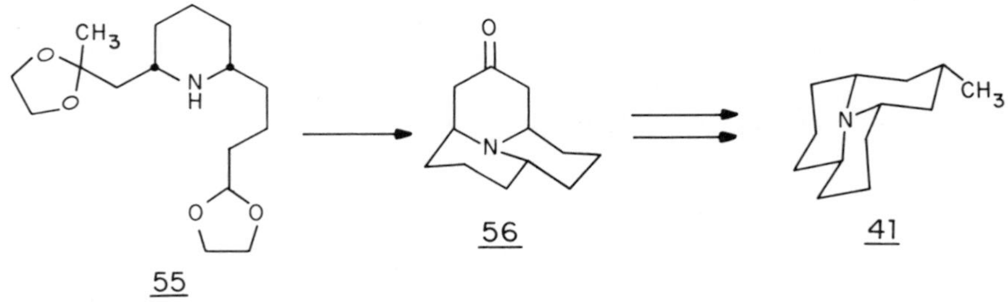

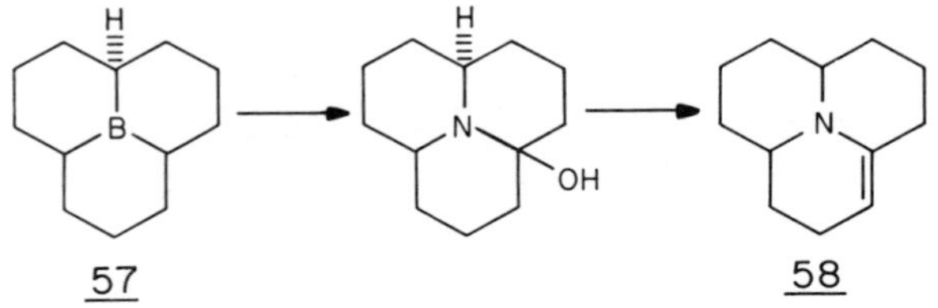

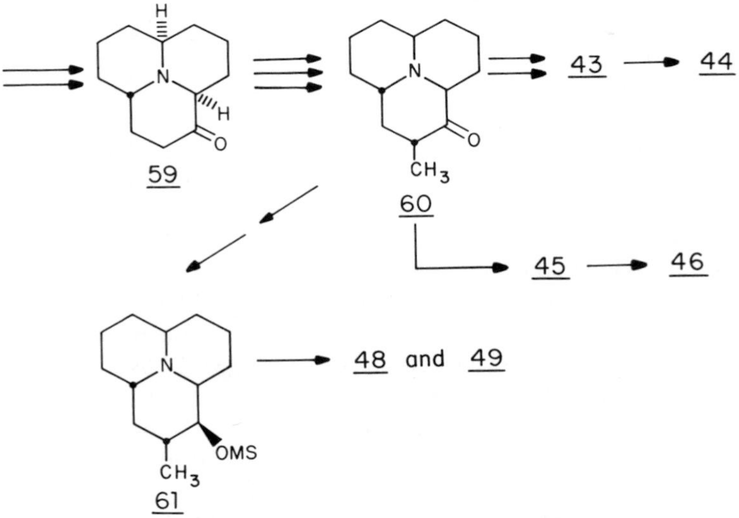

6.3.3. Biosynthesis. The coccinelline alkaloids are probably formed in insects via the polyketide pathway from a polyketo acid such as **62**. Feeding experiments

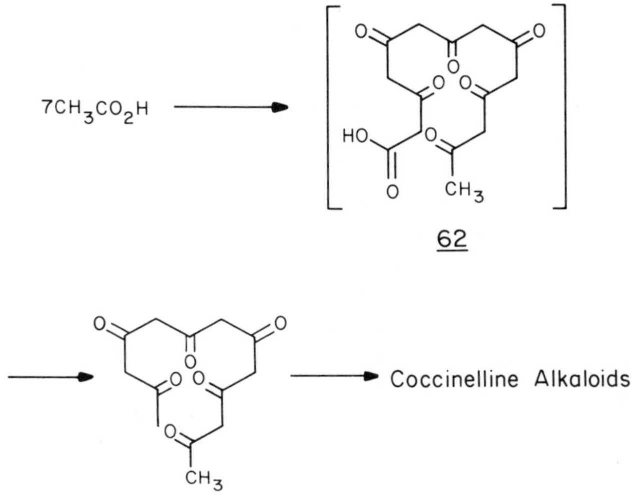

$7CH_3CO_2H$ → [62] → → Coccinelline Alkaloids

on *Coccinnella septempunctata* with $^{14}CH_3CO_2Na$ and $CH_3{}^{14}CO_2Na$ confirm with hypothesis. Degradation of the alkaloids produces acetic acid from the methyl group at C_2, and for both radiolabeled acetates the relative incorporation of the radiolabels at C_2 and C_{10} is approximately the expected 16% [103].

7. ADALINE

7.1. Distribution

Adaline (**63**) has been identified as a bloodborne product of ladybugs in the genus *Adalia*, a taxon in the subfamily Coccinellinae, family Coccinellidae. The alkaloid is present in *A. bipunctata* [103], and *A. decimpunctata* [104].

7.2. Functions

Reflexive loss of blood (autohemorrhage) from leg articulations was observed as part of the defensive behavior of *A. bipunctata* [105]. This reflex bleeding externalizes adaline and provides the beetle with a gustatory repellent for both ants (*Myrmica rubra*) and, in some cases, for avian predators such as European quail (*Coturnix coturnix*) [83].

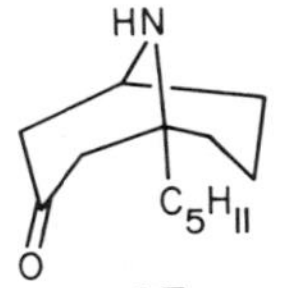

63

7.3. Chemistry

Adaline (**63**) is a unique, bicyclic ketoamine, whose complete structure was established by X-ray crystallographic studies [103]. The ORD curve of **63** exhibits a positive cotton effect that results in assignment of the *R*-configuration to C(1) of *l*-adaline (**63**) [104].

The traditional Robinson–Schopf reaction can be used to prepare *dl*-**63** from ketoaldehyde **64**, ketoglutamic acid, and ammonium chloride, although the yield is low [104]. A series of Grignard additions to nitrones followed by intramolecular 1,3-dipolar cycloaddition of the resulting olefinic nitrones provides *dl*-**63** in higher yield [106]. Thus 2-pentyl-*N*-hydroxypiperidine is oxidized to the nitrone mixture (**66**) and treated with alkyl Grignard reagent to yield a mixture of *N*-hydroxypiperidines (**67**). A final oxidation forms a second mixture of nitrones, two of which (**68**) are capable of intramolecular cyclization and upon heating, form the isoxazolidine (**70**). The formation of *dl*-adaline (**63**) is straightforward from **70** by hydrogenolysis and subsequent oxidation [106]. An elegant synthesis of *dl*-**63** is based on conjugate addition to cross-conjugated 2,7-cyclooctadienones. The copper-catalyzed 1,4 addition of pentylmagnesium bromide to 2,7-cyclooctadienone (**72**) followed by trapping and resulting enolate with phenylselenyl bromide gives the α-selenophenyl ketone (**73**) and upon oxidation the cross-conjugated dienone (**74**). The double conjugate addition of benzylamine to **74** followed by hydrogenolysis provides *dl*-**63** in 64–67% yield. Substitution of *R*-1-phenylethylamine in this sequence gives a separable mixture

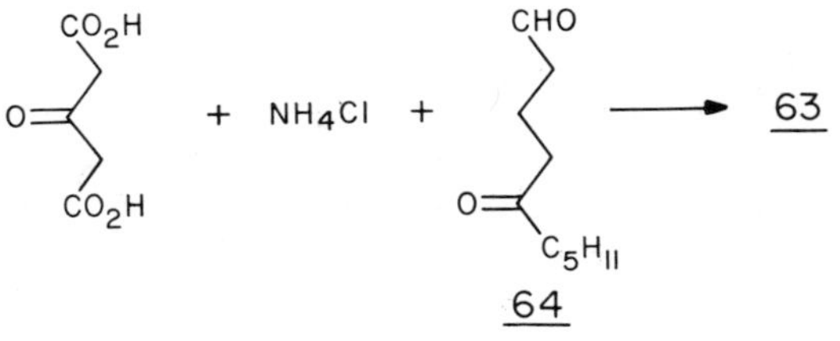

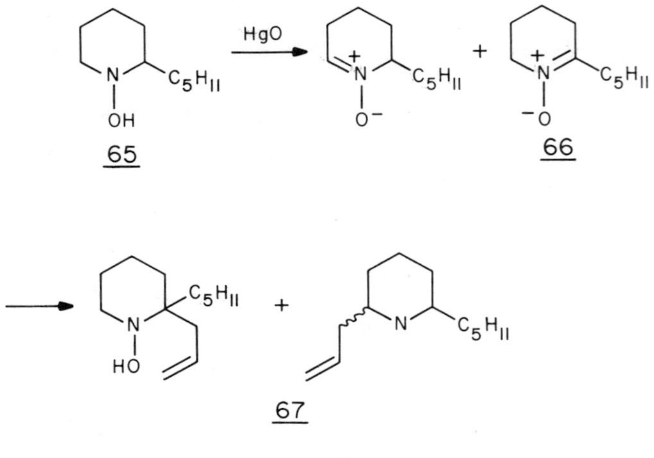

CO_2H, $O=$, CO_2H + NH_4Cl + CHO, $O=$, C_5H_{11} → 63

64

OH, N, C_5H_{11} —HgO→ N⁺, O^-, C_5H_{11} + N⁺, ^-O, C_5H_{11}

65 66

→ N, HO, C_5H_{11} + N, C_5H_{11}

67

HgO → 68 + 69

LXVIII →(Δ) 70 → 71 → 63

72 →(1) $C_5H_{11}MgBr$, 2) ØSeBr) 73 →(O) 74

→(ØCH$_2$NH$_2$) 75 →(H_2, Pd/C) dl 63

74 →((R) CH_3CH(NH$_2$)Ph) diastereomers →(H_2, Pd/C) (+) 63; →(H_2, Pd/C) (−) 63

of diastereomers, each of which can be hydrogenolysed to (+) or (−) adaline. The (−) adaline produced this way is identical with the natural alkaloid [107].

8. PYRAZINES

8.1. Distribution

Pyrazines have only been identified as exocrine products of species of insects in the order Hymenoptera. These compounds have a particularly widespread distribution in the family Formicidae, having been detected as glandular products of ant species in four subfamilies. Pyrazines appear to be especially characteristic of workers of species in the subfamily Ponerinae, having been identified as mandibular gland constituents of workers of species in the genera *Odontomachus* [108, 109], *Anochetus* [109], *Brachyponera* [109], *Hypoponera* [110], *Ponera* [110], and *Rhytidoponera* [111]. These compounds were also identified as mandibular gland products of males of a species of *Odontomachus*, *O. troglodytes* [109]. In the subfamily Myrmicinae, pyrazines have been identified as mandibular gland products of workers of *Wasmannia auropunctata* [112], *Aphaenogaster rudis* [113], and a poison gland constituent of workers of *Atta texana* [15]. Workers of the Argentine ant *Iridomyrmex humilis*, a species in

Table 3. Pyrazines

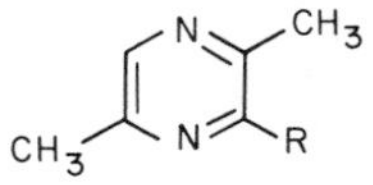

76

	R	
a	C_2H_5	[15]
b	C_3H_7	[118]
c	$CH_2CH(CH_3)_2$	[109]
d	$CH(CH_3)CH_2CH_3$	[109]
e	$n-C_5H_{11}$	[109]
f	$CH_2CH_2CH(CH_3)_2$	[108]
g	$CH=CH-C_6H_5$	[120]
h	$CH_2CH_2CH(CH_3)CH_2CH_2CH=C(CH_3)_2$	[111]

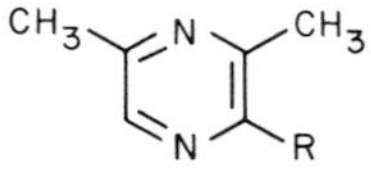

77

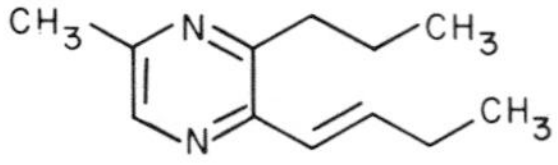

78 (113)

	R	
a	C_2H_5	[108]
b	$n-C_3H_7$	[108]
c	$n-C_4H_9$	[108]
d	$n-C_5H_{11}$	[108]
e	$CH_2CH(CH_3)_2$	[109]
f	$CH(CH_3)CH_2CH_3$	[109]
g	$n-C_6H_{13}$	[109]

the subfamily Dolichoderinae, have been demonstrated to produce pyrazines in an unidentified cephalic gland [114]. In the subfamily Formicinae, these nitrogen heterocycles constitute mandibular gland products of workers of a *Calomyrmex* species [115] and males of a *Camponotus* species [116].

Recently, pyrazines have been detected as mandibular gland products of a variety of male and female solitary wasps. These compounds have been identified in species in several genera in the family Eumenidae and appear to be characteristic natural products of this family. Species-specific pyrazines were detected in members of the genera *Ancistrocerus, Stenodynerus, Pseudodynerus*, and *Eumenes* [117]. Similarly, digger wasps in the family Sphecidae have been demonstrated to produce pyrazines in their capacious mandibular glands [118]. Males and females in the genera *Argogorytes* and *Philanthus* are a rich source of these nitrogen heterocycles, and males of a species of *Nysson*, a parasite of *Argogorytes* species, also secrete these compounds.

8.2. Functions

Alkylpyrazines are employed as pheromones more than any other class of alkaloids identified as arthropod natural products. In particular, a wide variety of ant species utilize these compounds as alarm pheromones that are secreted from their hypertrophied mandibular glands when these insects are subject to traumatic stimuli. Alarm pheromones generally function as attractants that accelerate the rate of movement of the workers while bringing them to the pheromonal emission source in a highly excited (aggressive) state. Ant workers frequently react to pyrazine-treated objects by biting and stinging [108], and the agitated workers may increase the ambient pyrazine concentration by secreting the contents of their own mandibular glands. These compounds thus function admirably to exploit the collective resources of the ant colony by rapidly recruiting aggressive workers to the source of the disturbance.

Odontomachus workers display such frenzied behavior when they are attracted to a pyrazine emission source that they sometimes kill their sister workers [108]. When the alerted (alarmed) ants approach the pyrazine-treated object, they may indiscriminately snap their poised mandibles on any moving insect while simultaneously stinging the "prey." Attraction of alerted workers to a pyrazine source has also been observed for species in the genera *Wasmannia* [112] and *Calomyrmex* [115], further demonstrating that these compounds are widely utilized as alarm pheromones by a variety of unrelated ant species. On the other hand, no demonstrable reaction to these compounds could be observed with workers of the dolichoderine *Iridomyrmex humilis* [114].

In some cases, ants may react to their liberated pyrazines by retreating from the emission source rather than being attracted to it. Such a response is probably of considerable value for species that do not form large colonies and/or are too small or fragile to defend themselves against larger aggressors. Diminutive ants in the ponerine genera *Hypoponera* and *Ponera* flee when they detect the pyrazine produced in their mandibular glands [110]. This response is highly adaptive, since these small, delicate ants would hardly be a match for any sizable predator. The same can be said for male ants, which possess weak mandibles and lack a sting. It has been demonstrated that males of *Odontomachus troglodytes* [109], in contrast to the workers, retreat from pyrazine sources and thus avoid potentially injurious conflicts for which they are ill suited. However, the males may secrete the pyrazines in their mandibular glands and thus augment the alarm signal to which the aggressive workers respond.

In contrast to the mandibular gland secretions of *Odontomachus* males, those of a carpenter ant, *Camponotus dumetorum*, are probably utilized for an entirely different purpose. Males of this species produce a pyrazine-rich mandibular gland secretion, in contrast to the workers, which do not produce detectable volatiles in these glands [116]. The pyrazines produced by males of *C. dumetorum* probably function as sex pheromones, possibly serving to draw females from the nest to participate in mating flights, as has been observed with other *Camponotus* species.

One species of myrmicine ant, *Atta sexdens*, utilizes a pyrazine, a trace constituent in the poison gland secretion, as a trail pheromone [15]. About 100 μg of pyrazine were isolated from 4.2 kg of ant workers, constituting about 2.5×10^{-8} of an ant's body weight. This is in contrast to another *Atta* species, *A. texana*, which utilizes methyl 4-methylpyrrole-2-carboxylate (**10**) as a trail pheromone. Although *A. sexdens* also produces the pyrrole, it preferentially utilizes the pyrazine to generate trails. Other related species of *Atta* and an *Acromyrmex* species exhibit weak responses to the pyrazine. Workers of one species, *Acromyrmex octospinosus*, were induced to leave the nest and mill around on the surface when the pyrazine was introduced as a vapor [15].

In common with many other pheromones, the pyrazine-based alarm pheromones of ants appear to serve a defensive function as well. Pyrazines produced by *Odontomachus* workers repelled workers of the fire ant, *Solenopsis invicta* [108], and 2,5-dimethyl-3-isopentylpyrazine, the alarm pheromone of the ant *Wasmannia auropunctata*, substantially reduced feeding by workers of the ant *Monomorium minimum* when applied to mealworms [112]. Probably these compounds also may act as gustatory repellents for predatory vertebrates as well. The mandibular gland secretion of a *Calomyrmex* species has been demonstrated to be a powerful feeding deterrent for a variety of insectivorous and carnivorous marsupials [119]. Since the mandibular secretion of this species contains three major pyrazines and several minor ones, these alkaloids may contribute substantially to the observed repellency of this exudate for vertebrate predators.

One would anticipate a priori that solitary wasps, unlike the highly social ants, would not utilize their pyrazines as alarm pheromones. In eumenid wasps these compounds are not released when the insects are molested, indicating they do not possess a defensive function. Hefetz and Batra have suggested that these compounds may play an important role in the wasps' presocial behavior, such as marking nocturnal roosting sites that serve as centers for aggregations of these invertebrates [117]. On the other hand, males of some sphecid wasps, e.g., *Philanthus triangulum*, mark stalks of grass or pine needles with their mandibular gland secretions, and these males orient to these marked spots when flying [118]. The exact roles of these pyrazine-rich secretions in sphecid biology have yet to be determined.

8.3. Chemistry

The simple, trisubstituted alkylpyrazines (Table 3) are generally identifiable from their mass spectra, although the mass spectra of the isomeric 2,5-dimethyl-3-alkylpyrazines (76) and 2,6-dimethyl-2-alkylpyrazines are usually almost identical [121]. These isomers have different gas chromatographic retention times however, and can be identified by direct comparison with authentic samples of each possible isomer. As a rule, the short-chain trialkylpyrazines show strong parent ions in their mass spectra, whereas the most intense ion in those cases having a three-carbon or longer side chain results from a McLafferty rearrangement.

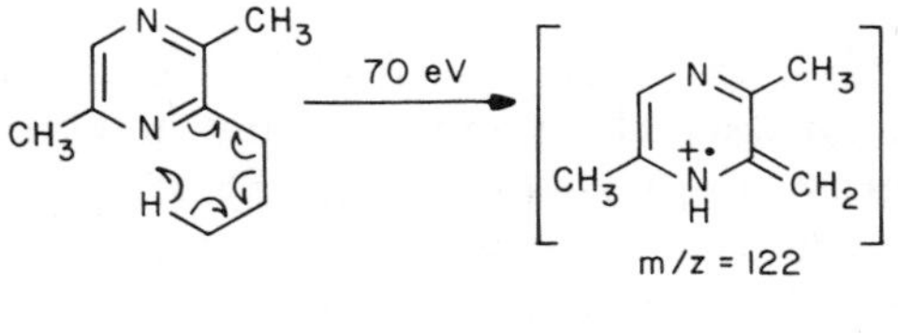

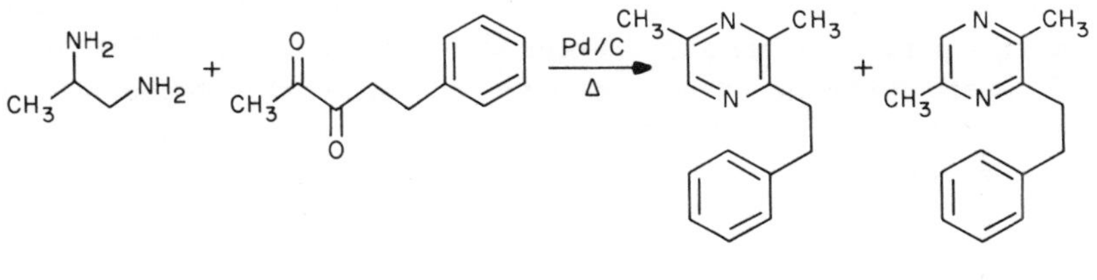

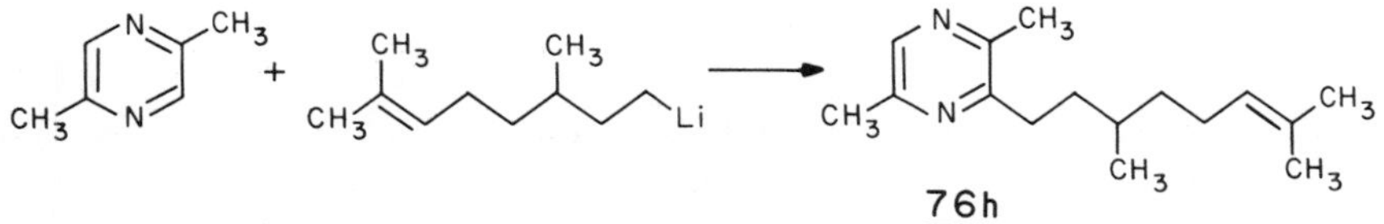

The structures of several of the more complex pyrazines deserve mention. The naturally occuring 2,5-dimethyl-3-styrylpyrazine (**76g**) can be hydrogenated to the corresponding 3-phenethylpyrazine. For comparison, both 2,5-and 2,6-dimethyl-3-phenethylpyrazine can be prepared by the condensation of 1,2-diaminopropane with 5-phenyl-2,3-pentanedione [120]. This reaction establishes the complete carbon-nitrogen skeleton of the natural material including the substitution pattern around the heterocyclic ring. The NMR spectra of **76g** indicates that the styryl double bond has the *Z* configuration, and this can be confirmed by acid-catalyzed isomerization to the *E* isomer and subsequent reisomerization to the *Z* isomer upon exposure to sunlight [114]. This sequence permits observation of the UV spectra of both isomers in direct analogy to the behavior of stilbene under these conditions and shows in another way that the longer-wavelength-absorbing *E* isomer is not present in the natural material [114].

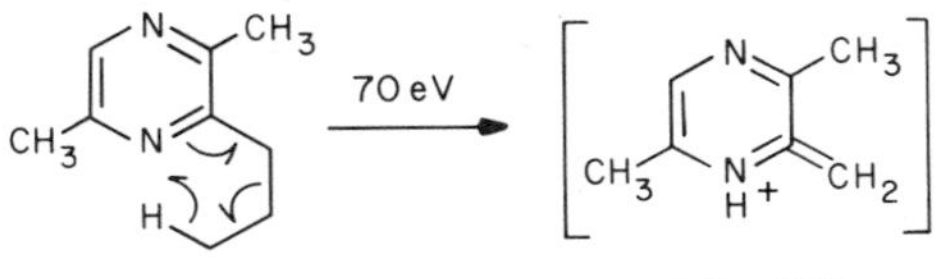

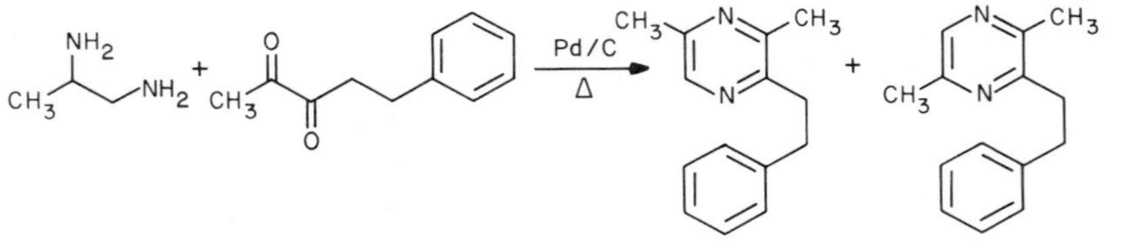

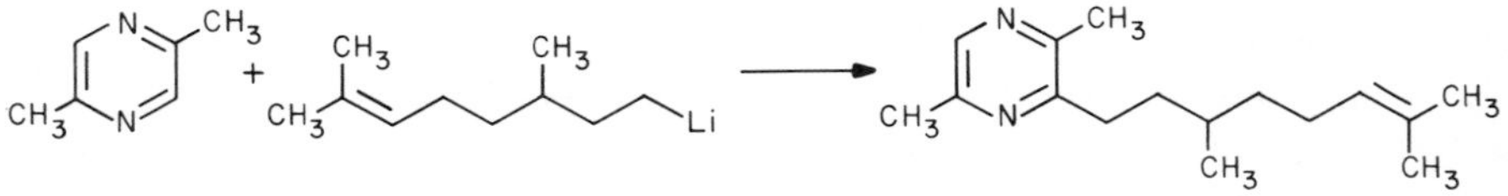

The mass spectrum of 2,5-dimethyl-3-citronellylypyrazine (**76h**) indicates the presence of a 10-carbon unsaturated side chain without α-branching [$m/z = 122$ (100%)]. Since the mass spectrum also shows loss of a methyl group but not an ethyl moiety, an isoprenoid side chain may be assumed. Treatment of 2,5-dimethylpyrazine with citronellyl lithium provides **76h** for direct comparison and the final proof of structure [111]. The direct alkylation of dimethylpyrazines with organolithium reagents is the most straightforward, general synthesis of arthropod trialkylpyrazines [120, 122]. For the biosynthesis of 2,5-dimethyl-3-isopentylpyrazine (**76f**) and 2,5-dimethyl-3-styrylpyrazine (**76g**), alanine and 2,5-dimethylpyrazine have been suggested as common precursors, with the isoprenoid or styryl moieties being introduced into the symmetrical pyrazine intermediate [120].

9. QUINAZOLINONES

9.1. Distribution

Two quinazolinones, glomerin and homoglomerin, collectively termed glomerins, have been identified in the defensive secretion of the pill millipede,

Glomeris marginata, a species in the family Glomeridae, order Glomerida [123, 124]. These compounds, in admixture with viscous proteinaceous constitents, are present in the single droplets discharged from eight pores on the dorsal midline of the millipede.

9.2. Functions

Glomerin and homoglomerin are reported to be very effective defensive compounds against both vertebrate and invertebrate predators. Several hours after ingesting *Glomeris*, mice became partially paralyzed, sluggish, and tremulous [125]. Two mice, which had ingested six or more *Glomeris* each, subsequently died. Birds that had consumed these millipedes exhibited retarded alarm reactions, and toads that had ingested *Glomeris* subsequently experienced emetic reactions. An invertebrate, a lycosid spider, after feeding on *G. marginata*, appeared to be normal for a sustained period of time. However, the spider eventually exhibited symptoms of motor impairment and in some cases became motionless, the catatonic state being evident for up to several hours [126]. The latter response to ingested quinazolinones is very maladaptive, since paralyzed individuals are then very susceptible to predation or desiccation during the heat of the day.

Quinazolinones, which are noted for their bitter taste, may act as very effective gustatory repellents for predators of *G. marginata*. When molested, these millipedes roll up into a ball, and only after continued molestation do they discharge their alkaloid-rich secretion. Predators may simply reject these arthropods after tasting the bitter alkaloids, thus providing the millipedes with a highly adaptive means of avoiding potentially injurious treatment. It is worth noting that the bitter qualities of glomerin can be readily detected by humans at a concentration as low as 6×10^{-4} *M* in aqueous solution [124].

9.3. Chemistry

Two alkylquinazolinones, 1-methyl-2-ethyl-4(3*H*)-quinazolinone (glomerin, **80**) and 1,2-dimethyl-4(3*H*)-quinazolinone (homoglomerin, **79**), are present in the defensive exudate of *G. marginata*. The structure elucidation of these compounds from their NMR, mass, IR, and UV spectral data has been reported by two groups [123, 124]. Fortuitously, these two heterocycles had already been synthesized in conjunction with a study of the plant alkaloid arborine (**81**) [125] by treatment of *N*-methylanthranilic amide with either acetic or propionic anhydride. Direct comparison of the synthetic material with the natural products from *G. marginata* confirms the structural assignments. In the light of the synthesis of these compounds from an anthranilic acid derivative, experiments show that feeding ^{14}C-labeled anthranilic acid to *G. marginata* results in incorporation of the radiolabel in **71** and **72** produced by the millipede [126].

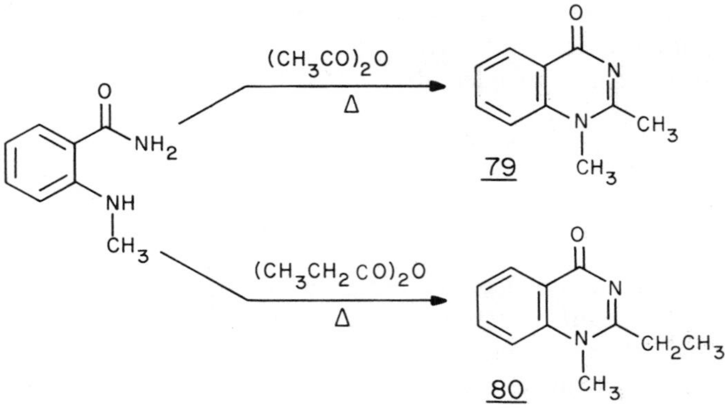

$(CH_3CO)_2O$, Δ → 79

$(CH_3CH_2CO)_2O$, Δ → 80

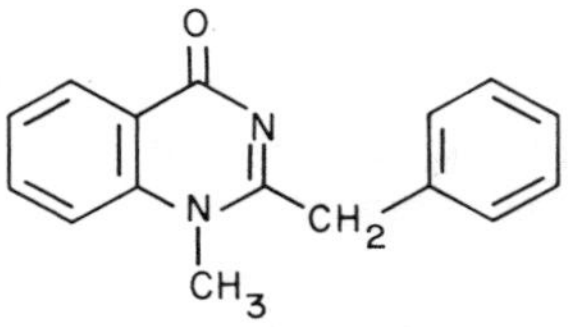

81

G. marginata

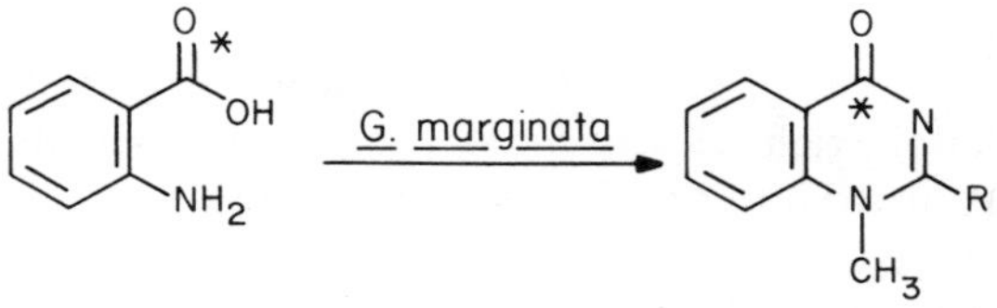

10. INDOLES

10.1. Distribution

In the insect orders, indoles have the widest distribution of any alkaloids identified as glandular products of arthropods. A variety of glands of species in four orders have been demonstrated to produce indoles, usually in admixture with other classes of compounds. Skatole (**82**) has been reported to be a major product of the hypertrophied poison gland of soldiers of the ant *Pheidole fallax* (Myrmicinae: Formicidae) [127] and the mandibular glands of workers of the army ant *Neivamyrmex nigrescens* (Dorylinae) [128]. This compound is also produced in the paired sternal glands on the fifth abdominal segment of males and females of the caddisfly *Pycnopsyche scabripennis* (Limnephilidae: Trichoptera), where it is accompanied by indole [129]. Paired prothoracic glands of adult green lacewings, *Chrysopa septempunctata* [130] and *C. oculata* (Chrysopidae: Neuroptera) [131] produce skatole (**82**) as a minor constituent of alkene-dominated secretions. A variety of diving beetles (Dytiscidae) generate pygidial gland secretions that contain indole-3-acetic acid (**83**) [132, 133]. This compound has been identified as a product of species in the genera *Hyphydrus, Hygrotus, Guignotus, Hydroporus, Graptodytes, Stictotarsus, Potamonectes,* and *Scarodytes* (Hydroporinae), as well as one genus (*Noterus*) in the subfamily Noterinae. This acid is also produced in the metapleural glands of ant species in the genera *Atta, Acromyrmex*, and *Myrmica* [134], members of the subfamily Myrmicinae.

10.2. Functions

The insect-derived indoles appear to be primarily utilized as defensive compounds that may be directed against pathogens and invertebrate and/or vertebrate predators. The poison gland product of soldiers of *P. fallax* [127] is almost certainly used in defensive contexts, as is the same compound (**82**) in the mandibular gland secretion of the ant *N. nigrescens* [128], which is both fungistatic and bacteriostatic [135]. This compound also is repellent to blind snakes that normally feed on *N. nigrescens* [128] and has been proposed to have a dual function in army ant biology.

The indolic secretions of both caddisflies [129] and lacewings [130] are discharged when these insects are disturbed, and these exudates have been demonstrated to repel both insect and some vertebrate predators. Ants (*Formica subsericea*) were effectively repelled by the secretion of *C. oculata*, and mice generally rejected these caddisflies after contacting them [131]. Similarly, the glandular exudate of the caddisfly *P. scabripennis* deterred fire ant workers (*Solenopsis invicta*) [129]. On the other hand, both gerbils and lizards consumed prey that had been treated with the latter secretion. In the case of both the lacewings and caddisflies, the indolic exudates appear to play a key role in protecting their producers from nocturnal flying predators such as bats. Compound **83** is reported to be a growth stimulant for the fungus cultivated by the ant *Atta sexdens* [136].

10.3. Chemistry

Indole has been identified by its gas chromatographic behavior and by its mass spectrum. Skatole (**82**) has been identified by its thin layer chromatographic behavior [127] and by its mass spectrum and gas chromatographic behavior [130,

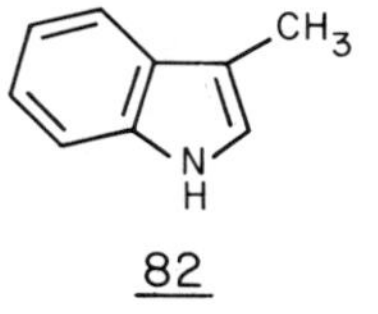

CH_3 N H

82

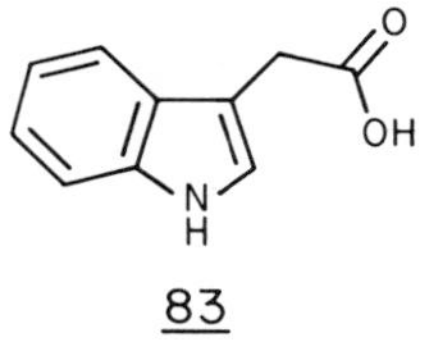

O OH N H

83

134]. Although **82** makes biosynthetic sense, care must be taken in assigning its structure from small quantities of natural material because its mass spectrum is nearly identical with those reported for other methyl indoles [137]. Indoleacetic acid (**83**) has also been identified from its thin layer chromatographic behavior along with its UV and mass spectra [132, 133].

11. QUINOLINES

11.1. Distribution

Only two quinolines have been identified as arthropod natural products, and in both cases they represent atypical compounds for the species that produce them. Methyl-8-hydroxy-2-quinoline carboxylate (**84**) is a prothoracic gland product of *Ilybius fenestratus* [138], a diving beetle in the family Dytiscidae. The prothoracic glands of these beetles normally produce steroids such as estradiol, estrone, and testosterone [139], and the occurrence of the quinoline in the secretion is truly anomalous. Adults of the beetle *Metriorrhynchus rhipidius*, an aposematic species in the family Lycidae, produce 1-methyl-2-quinoline (85), and this compound is present in the blood [140]. Each beetle contains about 1 mg. of the quinolone, which is accompanied by a second alkaloid. Lycid beetles typically produce acetylenic acids [141], and the occurrence of a quinolone in the blood of *M. rhipidius* appears to be very unusual.

11.2. Functions

Since the quinoline produced by *I. fenestratus* constitutes the major product in the secretion (350 μg) [138], its presence must be regarded as highly adaptive for the beetle. This compound is a powerful antiseptic, and it may prevent microorganisms from developing in the prothoracic defensive glands of the beetle [142]. Although the quinoline is not toxic to amphibians and fish, it does produce clonic spasms in mice [143]. The quinolone produced by *M. rhipidius* is very bitter to the human palate, and it is quite likely that this compound contributes to the distastefulness to potential predators of the beetle [140].

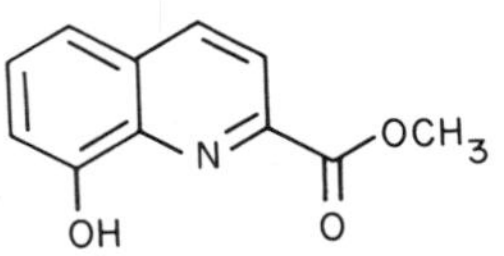

84

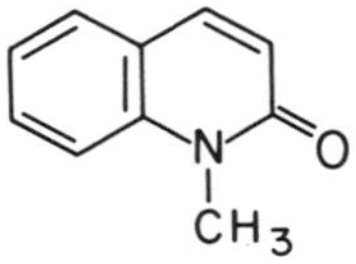

85

11.3. Chemistry

Methyl-8-hydroxy-2-quinoline carboxylic acid (**84**) was indicated from its UV, IR, NMR, and mass spectra and was confirmed by comparison with a synthetic sample [138]. The structure of 1-methyl-2-quinolone (**85**) was suggested by its mass spectrum and confirmed by comparison with an authentic sample [140].

12. PEDERIN

12.1. Distribution

Pederin (**86**) has been isolated from several species of beetles in the genus *Paederus*, a taxon in the family Staphylinidae. This compound, which is not secreted from an exocrine gland but rather is found in the blood and elsewhere throughout the adult's body, has been identified in *P. fuscipes* [145], *P. melanarus, P. litoralis, P. rubrothoracicus, P. rufocyaneus* [146], and *P. columbinus* [147]. Each beetle contains about 1 μg on the average, and females produce more than males, **86** constituting about 0.025% of the wet weight of the former.

12.2. Functions

Pederin is a nonexocrine defensive compound that possesses a great diversity of pharmacological activities [147]. When applied to human skin, this compound causes necrotization at acute doses and sustained desquamation at chronic doses [148]. However, low dosages of **86** have been demonstrated to cause dramatic healing in patients with chronic bedsores. In mice, topical application produces severe edema with a permanent loss of hair, even after healing has occurred. The great toxicity of **86** is indicated by the fact that the LD_{50} for the white mouse is 0.14 mg/kg, a dosage that corresponds to the amount of **86** in less than one beetle [148]. Both tumor cell lines and chicken heart fibroblasts are inhibited by concentrations of **86** as low as 1 ng/ml. Against plants the inhibitory effects of this amide are equally pronounced. Protein synthesis and growth of yeast cells are inhibited by **86**, and in higher plants, root growth is supressed and mitotic activity is blocked at the metaphasic stage [149]. Recently, the development of ascidian eggs was demonstrated to be inhibited by low dosages of **86** [150].

The *raison d'être* of the powerful cytotoxic effects of pederin (**86**) has been explored in several studies. *In vitro* studies with mammalian cells established that this compound causes an almost immediate block of DNA and protein synthesis without affecting RNA synthesis [151]. Studies with cell-free systems demonstrated that **86** has no effect on the DNA-polymerizing system, whereas protein synthesis is inhibited. These results are consistent with the conclusion that this amide acts primarily on the amino-acid-polymerizing system, the DNA effects being secondary. More recently, an *in vitro* study of the effects of **86** on enzymatic and nonenzymatic translocations by yeast polysomes demonstrated that this compound is a potent inhibitor [152]. These results strongly suggest that **86** acts on a specific ribosomal subunit in order to inhibit these important translocations.

12.3. Chemistry

Initially, on the basis of spectral data and degradation, the structure of pederin (**86**) was believed to be a shown, except that the ring hydroxyl group was placed vicinally to the tetrahydropyran ring oxygen [146]. Subsequent spectral and X-ray crystallographic studies provided the structure and stereochemistry shown for **86** [146, 153, 154].

Subsequent research efforts by two research groups have been directed at the synthesis of pederin. Both groups approached the synthesis of this complex molecule convergently by the preparation of pederamide (**87**) or a derivative and pederinal (**88**) or a derivative separately, and by joining them as a final step.

In one approach to **87** [155], malonate addition to *trans*-2,3-butene oxide provided *cis*-dimethylbutyrolactone (**89**), which could be converted to the dihydroxy alkynyl ester **90** by alkylation, reduction, and carbomethoxylation. This diol underwent base-catalyzed internal Michael addition to give the vinylidine furan **91**, which could be transformed to the bicyclic ketal **92** by epoxidation with *m*-chloroperbenzoic acid. From this point one, a series of functional-group modifications was carried out. The bicyclic ketal was opened with acidic methanol, and the resulting primary alcohol (**93**) was selectively selenated to give **94**. After oxidative elimination of the selenide to form the exocyclic double bond, the stereochemistry of the α-hydroxyl group was established by oxidation and stereoselective reduction, and the ester group was aminated to form **87**.

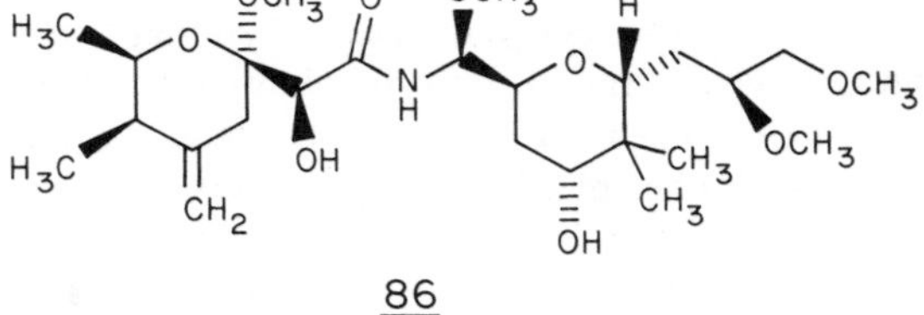

H_3C OCH_3 O OCH_3 H O N H O OCH_3 H_3C OH CH_3 OCH_3 CH_2 CH_3 OH

86

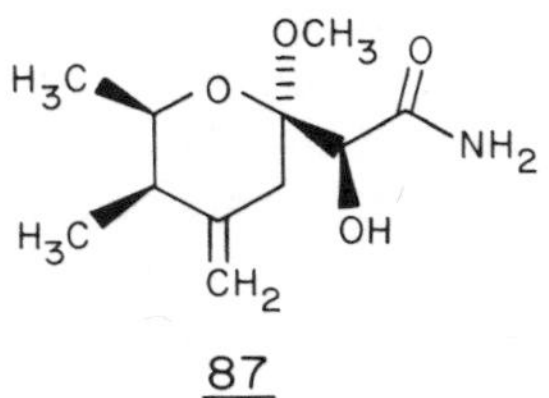

H_3C OCH_3 O O NH_2 H_3C OH CH_2

87

O HC H O OCH_3 OCH_3 CH_3 CH_3 OH

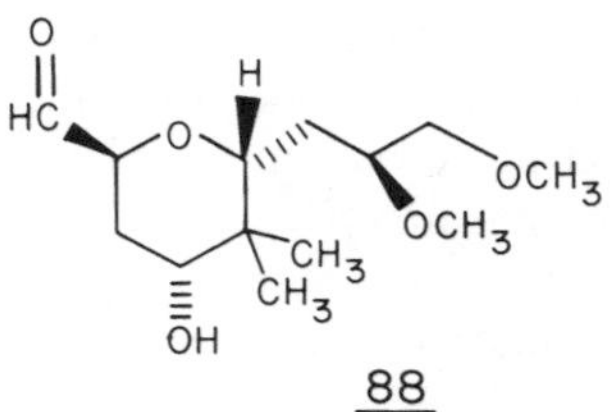

88

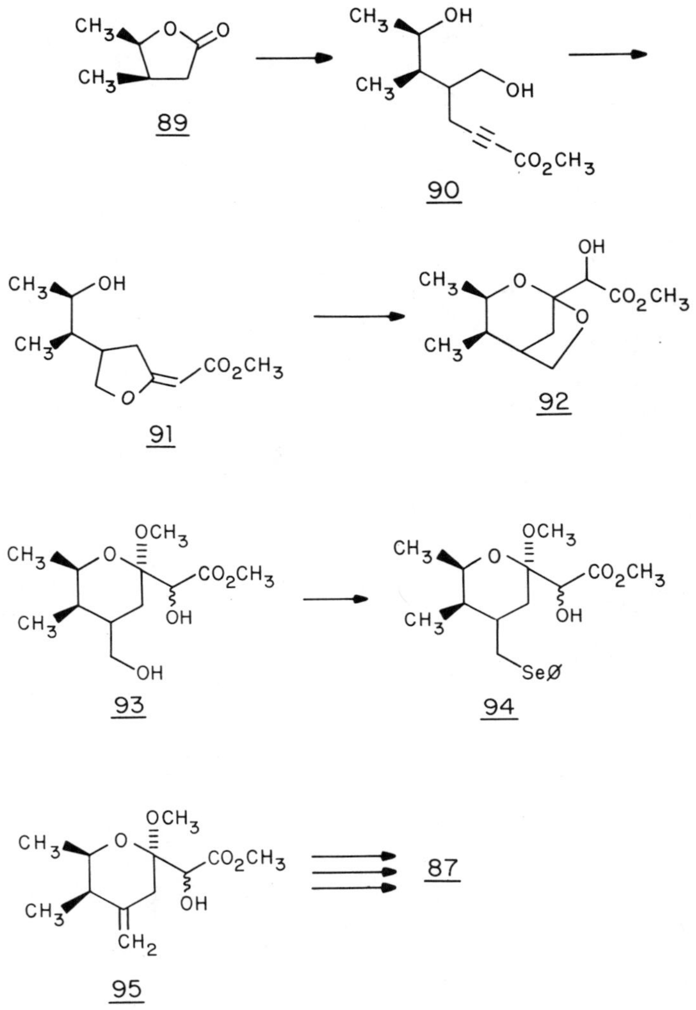

A more recent synthesis of **87** also began with *trans*-2,3-butene oxide, which was first treated with lithium acetylide to give the acetylenic alcohol (**96**) [156]. The unsaturated lactone **98** was formed spontaneously following carboxylation and partial hydrogenation of the acetylenic bond. Nitromethane anion added smoothly to **98** to form the nitrolactone (**99**). Nef conditions converted the nitrolactone to the tetrahydrofuranyl acetal (**100**), which could be relactonized to the dithiolanyl lactone (**101**) by treatment with ethanedithiol. Ester anion addition to **101** with the mixed ketal ester anion (**102**) followed by hydrolysis and acetonide formation yielded **103**. The thioacetal could then be removed, and the resulting aldehyde was reduced to a primary alcohol, which was subsequently selenated under mild conditions to give **104**. Removal of the actonide was

96 97 98

99 100 101

102 103

104 105

106 → 87

advantagous before conducting the oxidative elimination of the selenide to form the exocyclic double bond. Again, the final steps included amination of the ester group.

Two approaches to pederinal (**88**) have also been reported. In the first, steric control at the ring carbon atoms was achieved by starting with the bicyclic ketone **107**, which could be prepared from furan and 1,3-dibromo-3-methyl-2-butanone [157]. The endo alcohol **108** could be formed selectively. Ozonolysis of the

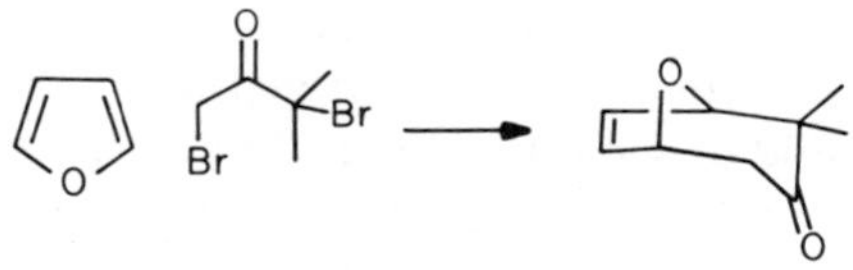

107

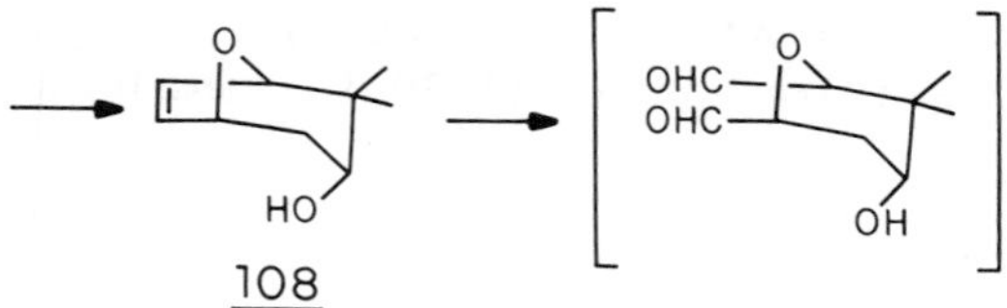

108

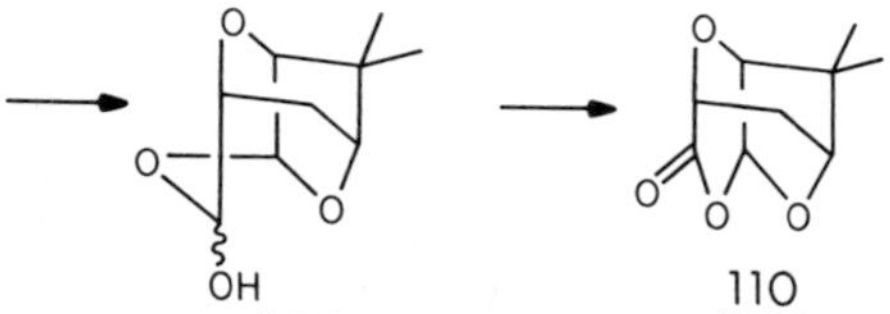

109

110

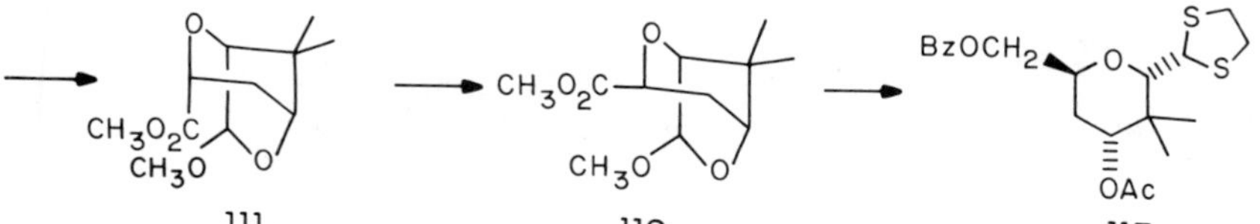

111

112

113

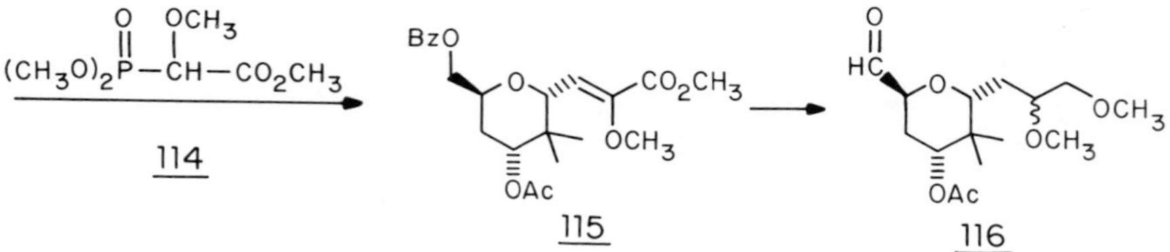

114

115

116

double bond produced an intermediate dialdehyde that cyclized spontaneously to the hemiacetal **109**, which could be oxidized to the acetal lactone **110**. The dithiolane acetate **113**, having all the required ring stereochemistry, could be formed by opening the lactone ring with acidic methanol and epimerization of the resulting ester. This was followed by opening the bicyclic acetal **112** with ethanedithiol, acetylation, reduction, and protection of the resulting primary alcohol. The aldehyde derived from **113** was then condensed with the phosphonate ester **114** to form **115**. A series of simple functional-group transformations converted **115** to pederinal acetate **116**, the planned intermediate for the final formation of pederin. The isomeric mixture of methoxy ethers that was obtained from this sequence could be separated by thin layer chromatography.

The second synthesis of pederinal (**88**) is based on the symmetry of its carbon skeleton about the *gem*-dimethyl group [158]. In this case, the symmetrical dienediol **120** was prepared from 3-hydroxy-2,2-dimethylpropanal (**117**). After protection of the aldehyde, the primary alcohol was oxidized to form the half-masked dialdehyde **118**, which was then treated with allyl Grignard. Hydrolysis of the acetal and a second Grignard reaction formed **120**. The meso form of **120**

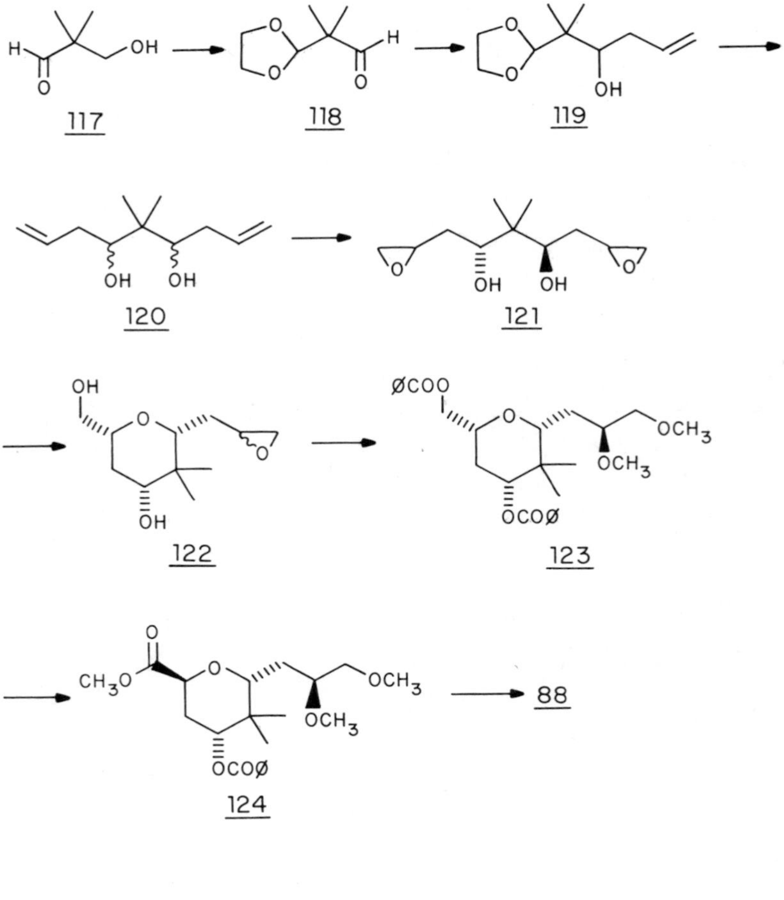

117 118 119

120 121

122 123

124 → 88

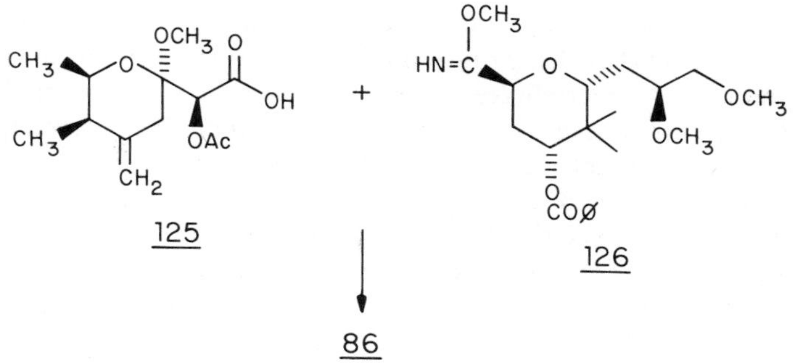

was separated from the *dl* form chromatographically, and the *dl* form was carried through the synthesis. Oxidation of **120** gave a mixture of epoxides, which was cyclized to a mixture of epoxypyrans **122**, epimeric at the secondary carbon–oxygen bond of the epoxide ring. The free hydroxyl groups were protected as benzoate esters, at which point the epimers could be separated, and the epoxide group was opened with methoxide and the resulting alcohol methylated to give **123**. The primary benzoate could be selectively removed, and the resulting alcohol converted to the methyl ester, which could be epimerized to form **124**. Reduction of the latter with diisobutyl aluminum hydride formed pederinal directly.

The final coupling of the two halves of the pederin molecule has been reported quite recently [159]. The acid (**125**) and imino ether (**126**) were coupled, and the pederin obtained after removal of the acetate and benzoate groups was shown to inhibit mitosis in HeLa cells at 1–10 ng/ml.

12.4. Biosynthesis

Pederin is probably formed in *Paederus fuscipes* via the polyketide pathway, since feeding experiments show the incorporation of [1-^{14}C]acetate and (2-^{14}C]acetate at the cis-vicinal methyl groups of **86** [160]. These radiolabels were also incorporated extensively in the 3,3-dimethyl tetrahydropyran half of the molecule. The incorporation of [2-^{14}C]propionate also supports a polyketide biosynthesis.

13. CONCLUSIONS

Arthropods synthesize a wide variety of alkaloids that have been utilized to function both as pheromones and allomones. This is especially true of species in the order Hymenoptera, the members of which have been shown to produce a diversity of pyrazines in their mandibular glands, and pyrrolidines, pyrrolines, piperidines, piperideines, and indolizidines in their poison glands. Although relatively few species of ants and wasps have been analyzed for the presence of nitrogen heterocycles, the results to date augur well for future investigations. The importance of these compounds as both communicative agents and defensive compounds suggests that the evolution of alkaloids was a major development that permitted these invertebrates to exploit a variety of potentially hostile habitats. It will not prove surprising if further analytical investigations of arthropod natural products demonstrate that these animals produce additional novel alkaloids that have been adapted to subserve critical functions for their producers in both interspecific and intraspecific contexts. For both the chemist and the biologist, we believe that when it comes to arthropod alkaloids, the best is yet to come.

REFERENCES

1. M. S. Blum and J. M. Brand, *Am. Zool.* **12,** 553 (1972).
2. B. Tursch, J. C. Braekman, and D. Daloze, *Experientia* **32,** 401 (1976).
3. M. Rothschild, *Symp. R. Entomol. Soc. London* **6,** 59 (1972).
4. W. C. Wildman, in *Chemistry of the Alkaloids*, S. W. Pelletier, Ed., Van Nostrand Reinhold, New York, 1970, p. 119.
5. M. Pavan, *Publ. Ist. Ent. Agr. Univ. Pavia*, 1 (1975).
6. T. H. Jones, M. S. Blum, and H. M. Fales, *Tetrahedron*, **38,** 1949 (1982).
7. K. Hölldobler, *Biol. Zentral.* **48,** 129 (1928).
8. D. J. Pedder, H. M. Fales, T. Jaouni, M. S. Blum, J. MacConnell, and R. M. Crewe, *Tetrahedron* **32,** 2275 (1976).
9. M. S. Blum, T. H. Jones, B. Hölldobler, H. M. Fales, and T. Jaouni, *Naturwissenschaften* **67,** 144 (1980).
10. T. H. Jones, M. S. Blum, and H. M. Fales, *Tetrahedron Lett.*, 1031 (1979).
11. F. J. Ritter and C. J. Persoons, *Neth. J. Zool.* **25,** 261 (1975).
12. J. Smolanoff, A. F. Kluge, J. Meinwald, A. McPhail, R. W. Miller, K. Hicks, and T. Eisner, *Science* **188,** 734 (1975).
13. J. H. Tumlinson, R. M. Silverstein, J. C. Moser, R. G. Brownlee, and J. M. Ruth, *Nature* **234,** 348 (1971).
14. R. G. Riley, R. M. Silverstein, B. Carroll, and R. Carroll, *J. Insect Physiol.* **20,** 651 (1974).
15. J. H. Cross, R. C. Byler, U. Ravid, R. M. Silverstein, S. W. Robinson, P. M. Baker, J. S. de Oliveira, A. R. Jutsum, and J. M. Cherrett, *J. Chem. Ecol.* **5,** 187 (1979).
16. B. Hölldobler, *Oecologia* **11,** 371 (1973).
17. C. B. Urbani and P. B. Kannowski, *Environ. Entomol.* **3,** 755 (1974).
18. F. W. Howard and A. D. Oliver, *J. Georgia Entomol. Soc.* **14,** 259 (1979).
19. T. H. Jones, M. S. Blum, R. W. Howard, C. A. McDaniel, H. M. Fales, M. B. DuBois, and J. Torres, *J. Chem. Ecol.*, **8,** 285 (1982).
20. S. W. Robinson, J. C. Moser, M. S. Blum, and E. Amante, *Insectes Soc.* **21,** 87 (1974).
21. T. H. Jones, J. B. Franko, M. S. Blum, and H. M. Fales, *Tetrahedron Lett.*, **21,** 789 (1980).
22. R. R. Fraser and S. Passannanti, *Synthesis*, 540 (1976).
23. F. J. Ritter and F. Stein, U.S. Patent 4,075,320, Feb. 21 (1978).
24. T. L. MacDonald, *J. Org. Chem.* **45,** 193 (1980).
25. J. H. Tumlinson, J. C. Moser, R. M. Silverstein, R. G. Brownlee, and J. M. Ruth, *J. Insect Physiol.* **18,** 809 (1972).
26. H. Rapoport and J. Bordner, *J. Org. Chem.* **29,** 2727 (1964).

27. P. E. Sonnet, *J. Med. Chem.* **15,** 97 (1972).
28. P. E. Sonnet and J. C. Moser, *Environ. Entomol.* **2,** 851 (1973).
29. J. G. MacConnell, M. S. Blum, and H. M. Fales, *Science* **168,** 840 (1970).
30. J. M. Brand, M. S. Blum, H. M. Fales, and J. G. MacConnell, *Toxicon* **10,** 259 (1972).
31. J. G. MacConnell, M. S. Blum, W. F. Buren, R. N. Williams, and H. M. Fales, *Toxicon* **14,** 79 (1976).
32. J. M. Brand, M. S. Blum, and H. H. Ross, *Insect Biochem.* **3,** 45 (1973).
33. J. M. Brand, M. S. Blum, and M. R. Barlin, *Toxicon* **11,** 325 (1973).
34. H. Schildknecht, D. Berger, D. Krauss, J. Connert, J. Gehlhaus, and H. Essenbreis, *J. Chem. Ecol.* **2,** 1 (1976).
35. J. W. Wheeler, O. Olubajo, C. B. Storm, and R. M. Duffield, *Science* **211,** 1051 (1981).
36. J. W. Wheeler, T. Olagbemiro, A. Nash, and M. S. Blum, *J. Chem. Ecol.* **3,** 241 (1977).
37. T. E. Bellas, W. V. Brown, and B. P. Moore, *J. Insect Physiol.* **20,** 277 (1974).
38. M. R. Caro, V. J. Derbes, and R. Jung, *Arch. Derm.* **75,** 475 (1957).
39. M. S. Blum, J. R. Walker, P. S. Callahan, and A. F. Novak, *Science* **128,** 306 (1958).
40. D. C. Buffkin and F. E. Russell, *Toxicon* **10,** 526 (1972).
41. D. P. Jouvanez, M. S. Blum, and J. G. MacConnell, *Antimicrob. Agents Chemother.* **2,** 291 (1972).
42. G. A. Adrouny, V. J. Derbes, and R. C. Jung, *Science* **130,** 449 (1959).
43. R. B. Koch, D. Desaiah, D. Foster, and K. Ahmed, *Biochem. Pharmacol.* **26,** 983 (1977).
44. E. Y. Cheng, L. K. Cutkomp, and R. B. Koch, *Biochem. Pharmacol.* **26,** 1179 (1977).
45. J. Z. Yeh, T. Narahasi, and R. R. Almon, *J. Pharm. Exp. Therap.* **194,** 373 (1975).
46. G. W. Read, N. K. Lind, and C. S. Oda, *Toxicon* **16,** 361 (1978).
47. J. G. MacConnell, M. S. Blum, and H. M. Fales, *Tetrahedron* **26,** 1129 (1971); M. S. Blum, H. M. Fales, G. Leadbetter, and B. A. Bierl (unpublished results) (1978); J. G. MacConnell, R. N. Williams, J. M. Brand, and M. S. Blum, *Ann. Entomol. Soc. Am.* **67,** 134 (1974).
48. B. P. Mundy, K. B. Lipkowitz, M. Lee, and B. R. Lansen, *J. Org. Chem.* **39,** 1963 (1974).
49. R. K. Hill, and T. Yuri, *Tetrahedron* **33,** 1569 (1977).
50. T. Moriyama, D. Doan-Huynh, C. Monneret, and Q. Khuong-Huu, *Tetrahedron Lett.*, 825 (1977).
51. H. Schildknecht, D. Krauss, J. Connert, H. Essenbreis, and N. Orfanides, *Angew. Chem. Int. Eng. Ed.* **14,** 427 (1975).
52. T. Sakan, A. Fujino, F. Murai, Y. Butsugan, and A. Suzu, *Bull. Chem. Soc. Jpn.* **32,** 315 (1959).
53. G. W. K. Cavill and D. L. Ford, *Austr. J. Chem.* **13,** 296 (1960).
54. G. W. K. Cavill and A. Zeitlin, *Austr. J. Chem.* **20,** 349 (1967).
55. H. Auda, G. R. Waller, and E. J. Eisenbraun, *J. Biol. Chem.* **242,** 4157 (1967).
56. J. Meinwald, G. M. Happ, J. Labows, and T. Eisner, *Science* **151,** 79 (1966).
57. E. Spath and L. Mamoli, *Chem. Ber.* **69,** 1082 (1936).
58. D. Schneider, in *Sensory Physiology and Behavior*, R. Galun, P. Hillman, I. Parnas, and R. Werman, Eds., Plenum, New York, 1975, pp. 173–193.
59. J. A. Edgar and C. C. J. Culvenor, *Nature* **248,** 614 (1974).
60. L. P. Brower, J. V. Z. Brower, and F. P. Cranston, *Zoologica* **50,** 1 (1965).
61. J. A. Edgar, C. J. Culvenor, and T. E. Pliske, *J. Chem. Ecol.* **2,** 263 (1976).
62. J. Meinwald, Y. C. Meinwald, J. W. Wheeler, T. Eisner, and L. P. Brower, *Science* **151,** 583 (1966); J. Meinwald and Y. C. Meinwald, *J. Am. Chem. Soc.* **88,** 1305 (1966).
63. J. Meinwald, Y. C. Meinwald, and P. H. Mazzocchi, *Science* **164,** 1174 (1969); J. Meinwald,

W. R. Thompson, T. Eisner, and D. F. Owen, *Tetrahedron Lett.*, 3485 (1971); J. A. Edgar, C. C. J. Culvenor, and L. W. Smith, *Experientia* **27,** 761 (1971).

64. J. A. Edgar, C. C. J. Culvenor, and G. S. Robinson, *J. Austr. Entomol. Soc.* **12,** 144 (1973).
65. J. Meinwald, C. J. Bonak, D. Schneider, M. Boppré, W. F. Wood, and T. Eisner, *Experientia* **30,** 721 (1974).
66. C. C. J. Culvenor and J. A. Edgar, *Experientia* **28,** 627 (1972).
67. T. Pliske, *Environ. Entomol.* **4,** 455 (1975).
68. J. Meinwald, J. Smolanoff, A. T. McPhail, R. W. Miller, T. Eisner, and K. Hicks, *Tetrahedron Lett.*, 2367 (1975).
69. T. H. Jones, M. S. Blum, H. M. Fales, and C. R. Thompson, *J. Org. Chem.* **45,** 4778 (1980).
70. J. Myers and L. P. Brower, *J. Insect Physiol.* **15,** 2117 (1969).
71. T. E. Pliske and T. Eisner, *Science* **164,** 1170 (1969).
72. D. Schneider and U. Seibt, *Science* **164,** 1173 (1969).
73. U. Seibt, D. Schneider, and T. Eisner, *Z. Tierpsychol.* **31,** 513 (1972).
74. M. Boppré, R. L. Petty, D. Schneider, and J. Meinwald, *J. Comp. Physiol.* **126,** 97 (1978).
75. J. A. Edgar, P. A. Cockrum, and J. L. Frahn, *Experientia* **32,** 1535 (1976).
76. R. L. Petty, M. Boppré, D. Schneider, and J. Meinwald, *Experientia* **33,** 1324 (1977).
77. H. Stetter, W. Basse, H. Kuhlmann, A. Landscheidt, and W. Schlenker, *Chem. Ber.* **110,** 1007 (1977).
78. F. J. Ritter, I. E. M. Rotgans, E. Talman, P. E. J. Verwiel, and F. Stein, *Experientia* **29,** 530 (1973).
79. F. J. Ritter, I. E. M. Bruggeman-Rotgans, E. Verkuil, and C. J. Persoons, in *Proceedings of the Symposium on Pheromones and Defensive Secretions in Social Insects*, Ch. Noirot, P. E. Howse, and G. Le Masne, Eds., University of Dijon Press, 1975, pp. 99–103.
80. J. E. Oliver and P. E. Sonnet, *J. Org. Chem.* **39,** 2662 (1974).
81. P. E. Sonnet and J. E. Oliver, *J. Heterocycl. Chem.* **12,** 289 (1978).
82. P. E. Sonnet, D. A. Netzel, and R. Mendoza, *J. Heterocycl. Chem.* **16,** 1041 (1979).
83. J. M. Pasteels, C. Deroe, B. Tursch, J. C. Braekman, D. Daloze, and C. Hootele, *J. Insect Physiol.* **19,** 1771 (1973).
84. B. P. Moore and W. V. Brown, *Insect Biochem.* **8,** 393 (1978).
85. P. A. Hedin, R. C. Guildner, R. D. Henson, and A. D. Thompson, *J. Insect Physiol.* **20,** 2135 (1974).
86. L. Cuénot, *Arch. Zool. Exp. Gen.* **4,** 655 (1896).
87. D. A. Kendall, *Entomologist*, 233 (1971).
88. B. Tursch, D. Daloze, M. Dupont, J. M. Pasteels, and M. C. Tricot, *Experientia* **27,** 1380 (1971).
89. B. Tursch, D. Daloze, M. Dupont, C. Hootele, M. Kaisin, J. M. Pasteels, and D. Zimmerman, *Chimia* **25,** 307 (1971).
90. R. Karlsson and D. Losman, *J. Chem. Soc. Chem. Commun.*, 626 (1972).
91. B. Tursch, D. Daloze, J. C. Braekman, C. Hootele, D. Losman, A. Cravador, and R. Karlsson, *Tetrahedron Lett.*, 409 (1974).
92. J. T. Purdham, *Can. J. Chem.* **54,** 1807 (1976).
93. B. Tursch, D. Daloze, J. C. Braekman, C. Hootele, and J. M. Pasteels, *Tetrahedron* **31,** 1541 (1975).
94. F. Bohlmann, *Chem. Ber.* **91,** 2157 (1958).
95. B. Tursch, D. Daloze, and C. Hootele, *Chimia* **26,** 74 (1972).
96. R. H. Mueller and M. E. Thompson, *Tetrahedron Lett.*, 1097 (1980).

97. W. A. Ayer, R. Dawe, R. A. Eisner, and K. Furvichi, *Can. J. Chem.* **54,** 473 (1976).
98. W. A. Ayer and K. Furvichi, *Can. J. Chem.* **54,** 1494 (1976).
99. H. C. Brown and W. C. Dickason, *J. Am. Chem. Soc.* **91,** 1226 (1969).
100. R. H. Mueller, *Tetrahedron. Lett.*, 2925 (1976).
101. R. H. Mueller and M. E. Thompson, *Tetrahedron Lett.*, **21,** 1093 (1980).
102. H. Bredereck, F. Effenberger, and G. Simchen, *Chem. Ber.* **98,** 1078 (1965).
103. B. Tursch, J. C. Braekman, D. Daloze, C. Hootele, D. Losman, R. Karlsson, and J. M. Pasteels, *Tetrahedron Lett.*, 201 (1973).
104. B. Tursch, C. Chome, J. C. Braekman, and D. Daloze, *Bull. Soc. Chim. Belg.* **82,** 699 (1973).
105. M. Rothschild, *Trans. R. Entomol. Soc. London* **113,** 101 (1961).
106. E. Gossinger and B. Witkop, *Monatsh. Chem.* **111,** 803 (1980).
107. R. K. Hill and L. A. Renbaum, *Tetrahedron*, **38,** 1959 (1982).
108. J. W. Wheeler and M. S. Blum, *Science* **182,** 501 (1973).
109. C. Longhurst, R. Baker, P. E. Howse, and W. Speed, *J. Insect Physiol.* **24,** 833 (1978).
110. R. M. Duffield, M. S. Blum, and J. W. Wheeler, *Comp. Biochem. Physiol.* **54B,** 439 (1976).
111. J. J. Brophy, G. W. K. Cavill, and W. D. Plant, *Insect Biochem.* **11,** 307 (1981).
112. D. F. Howard, M. S. Blum, T. H. Jones, and M. D. Tomalski, *Insectes Sociaux* (in press) (1982).
113. J. W. Wheeler, J. Avery, O. Olubajo, M. Shamim, C. B. Storm, and R. M. Duffield, *Tetrahedron*, **38,** 1939 (1982).
114. G. W. K. Cavill and E. Houghton, *J. Insect Physiol.* **20,** 2049 (1974).
115. W. V. Brown and P. B. Moore, *Insect Biochem.* **9,** 451 (1979).
116. R. M. Duffield, A Comparative Study of the Mandibular Gland Chemistry of Formicine and Ponerine Ant Species, Ph.D. thesis, University of Georgia (1976).
117. A. Hefetz and S. W. Batra, *Comp. Biochem. Physiol.* **65B,** 455 (1980).
118. A.-K. Borg-Karlson and J. Tengö, *J. Chem. Ecol.* **6,** 827 (1980).
119. E. Brough, *Z. Tierpsychol.* **46,** 279 (1978).
120. G. W. K. Cavill and E. Houghton, *Austr. J. Chem.* **27,** 879 (1974).
121. J. J. Brophy and G. W. K. Cavill, *Heterocycles* **14,** 477 (1980).
122. J. Gelas and R. Rambaud, *C.R. Acad. Sci. Paris C* **266,** 625 (1968).
123. Y. C. Meinwald, J. Meinwald, and T. Eisner, *Science* **154,** 390 (1966).
124. H. Schildknecht and W. F. Wenneis, *Z. Naturforsch.* **21b,** 552 (1966).
125. D. Chakravati, R. N. Chakravati, L. A. Cohen, B. Dasgupta, S. Datta, and H. K. Miller, *Tetrahedron* **16,** 224 (1961).
126. H. Schildknecht and W. F. Wenneis, *Tetrahedron Lett.*, 1815 (1967).
127. J. H. Law, E. O. Wilson, J. A. McCloskey, *Science* **149,** 544 (1965).
128. M. S. Blum, in J. F. Watkins II, F. R. Gehlbach, and J. C. Kroll, *Ecology* **50,** 1098 (1969).
129. R. M. Duffield, M. S. Blum, J. B. Wallace, H. A. Lloyd, and F. E. Regnier, *J. Chem. Ecol.* **3,** 649 (1977).
130. T. Sakan, S. Isoe, and S. B. Heyon, in *Control of Insect Behavior by Natural Products*, D. L. Wood, R. M. Silverstein, and M. Nakajima, Eds., Academic, New York, 1970, pp. 237–247.
131. M. S. Blum, J. B. Wallace, and H. M. Fales, *Insect Biochem.* **3,** 353 (1973).
132. K. Dettner, *Biochem. Syst. Ecol.* **7,** 129 (1979).
133. K. Dettner and G. Schwinger, *Biochem. Syst. Ecol.* **8,** 89 (1980).
134. H. Schildknecht and K. Koob, *Angew. Chem.* **82,** 181 (1970).
135. C. A. Brown, J. Watkins II, and D. W. Eldridge, *J. Kansas Entomol. Soc.* **52,** 119 (1979).

136. H. Schildknecht and K. Koob, *Angew. Chem. Int. Ed.* **10,** 124 (1971).
137. E. Stenhagen, S. S. Abrahamsson, and F. W. McLafferty, *Registry of Mass Spectral Data*, Vol. 1, Wiley, New York, 1974.
138. H. Schildknecht, H. Birringer, and D. Krauss, *Z. Naturforsch.* **24B,** 38 (1969).
139. H. Schildknecht, H. Birringer, and U. Maschwitz, *Angew. Chem.* **79,** 579 (1967).
140. B. P. Moore and W. V. Brown, *Insect Biochem.* **11,** 493 (1981).
141. J. Meinwald, Y. C. Meinwald, A. M. Chalmers, and T. Elsner, *Science* **160,** 890 (1968).
142. H. Schildknecht, *Angew. Chem. Int. Ed.* **9,** 1 (1970).
143. H. Schildknecht, *Endeavour* **30,** 136 (1971).
144. H. Schildknecht, G. Krebs, and H. Birringer, *Chemiker-Zeit.* **95,** 332 (1971).
145. C. Cardani, D. Ghiringhelli, R. Mondelli, M. Pavan, and A. Quilico, *Ann. Soc. Entomol. France* **1,** 813 (1965).
146. C. Cardani, D. Ghiringhelli, R. Mondelli, and A. Quilico, *Tetrahedron Lett.*, 2537 (1965).
147. M. Pavan, *Publ. Ist. Ent. Agr. Univ. Pavia*, 1 (1975).
148. M. Pavan and G. Bo, *Physiol. Comp. Oecol.* **3,** 307 (1953).
149. M. Pavan, *Atti Accad. Naz. Ital. Ent.* **10,** 119 (1963).
150. T. Carollo-Cusimano, *Acta Embryol. Exp.* **1979,** 335 (1979).
151. A. Brega, A. Falaschi, L. de Carli, and M. Pavan, *J. Cell Biol.* **36,** 485 (1968).
152. J. Jiminéz, L. Carrasco, and D. Vásquez, *Biochemistry* **16,** 4727 (1977).
153. T. Matsumoto, M. Yanagiya, S. Maeno, and S. Yasuda, *Tetrahedron Lett.*, 6297 (1968).
154. A. Furusaki, T. Watanabe, T. Matsumoto, and M. Yanagiya, *Tetrahedron Lett.*, 6301 (1968).
155. K. Tsuzuki, T. Watanabe, M. Yanagiya, and T. Matsumoto, *Tetrahedron Lett.*, 4745 (1976).
156. M. A. Adams, A. J. Duggan, J. Smolanoff, and J. Meinwald, *J. Am. Chem. Soc.* **101,** 5364 (1979).
157. J. Meinwald, *Pure Appl. Chem.* **49,** 1275 (1977).
158. K. Tsuzuki, Y. Nakajima, T. Watanabe, M. Yanagiya, and T. Matsumoto, *Tetrahedron Lett.*, 989 (1978).
159. M. Yanagiya, K. Tsuzuki, T. Watanabe, Y. Nakajima, F. Matsuda, K. Hasegawa, and K. Matsumoto, *Koen. Yoshishu Tennen Yuki Kagobutsu Toronkai 22nd* (1979) 635. *Chem. Abstr.* **93,** 7943p (1980).
160. C. Cardani, C. Fuganti, D. Ghiringhelli, P. Grasselli, M. Pavan, and M. D. Valcurone, *Tetrahedron Lett.*, 2815 (1973).

Chapter Three

Biosynthesis and Metabolism of the Tobacco Alkaloids

Edward Leete
Natural Products Laboratory,
School of Chemistry,
University of Minnesota,
Minneapolis, Minnesota 55455

CONTENTS

I am grateful to the National Institutes of Health, U.S. Public Health Service, who have supported our work on the biosynthesis and metabolism of natural products for the last 25 years (Grant GM-13246).

1. INTRODUCTION

Tobacco has been the subject of an enormous number of scientific investigations, and the literature is voluminous. A recent bibliography of monographs relating to tobacco lists 2467 books and pamphlets on this subject [1]. It is safe to say that tobacco has been more thoroughly examined than any other plant product. The main species used commercially for the production of tobacco is *Nicotiana tabacum*; however, much of the biosynthetic work on the tobacco alkaloids has been carried out with other *Nicotiana* species (*glutinosa*, *rustica*, *glauca*). In recent years, using modern analytical techniques, especially GC-mass spectrometry, the tobacco and flavor industries have identified a large number of organic compounds in tobacco [2–12]. Many of these compounds are probably not present in the green leaf and are the product of either enzymatic or chemical reactions that occur during the commercial curing and aging of tobacco leaves. An attempt is being made to identify those alkaloids that are found in the living plant, although in many cases this cannot be done with any certainty. This problem is not unique to tobacco. In many species the alkaloid content and their compositions vary with the methods used for their isolation from the plant. It is beyond the scope of this chapter to discuss the vast number of basic compounds that have been isolated from tobacco smoke and its condensates. Several reviews on this topic are available [2–4, 12–16], and clearly many of these pyrolytic products are derived from the pyridine alkaloids [17]. In this chapter, most of our discussions will be restricted to the biosynthesis and metabolism of pyridine alkaloids of tobacco, most of which are 3-pyridyl derivatives and are illustrated in Fig. 1.

2. SURVEY OF THE ALKALOIDS FOUND IN TOBACCO

Since the last historical survey of the tobacco alkaloids [18], many new alkaloids have been isolated. The main alkaloid found in almost all species of *Nicotiana* [19] is nicotine (**1**), which is levorotatory as the free base. In the older literature, it is referred to as L-nicotine or (−)-nicotine. The configuration at the C(2′) chiral center is (*S*). Nicotine is currently indexed in *Chemical Abstracts* under pyridine-3-(1-methyl-2-pyrrolidinyl). All reports indicate that the nicotine isolated from *Nicotiana* species is optically pure; however, strong bases (e.g., sodium hydride in boiling xylene) cause complete racemization [20]. Nicotine has been found in many other plant species and botanical families remote from the Solanaceae (of which *Nicotiana* is a member). However, the amount present is usually much

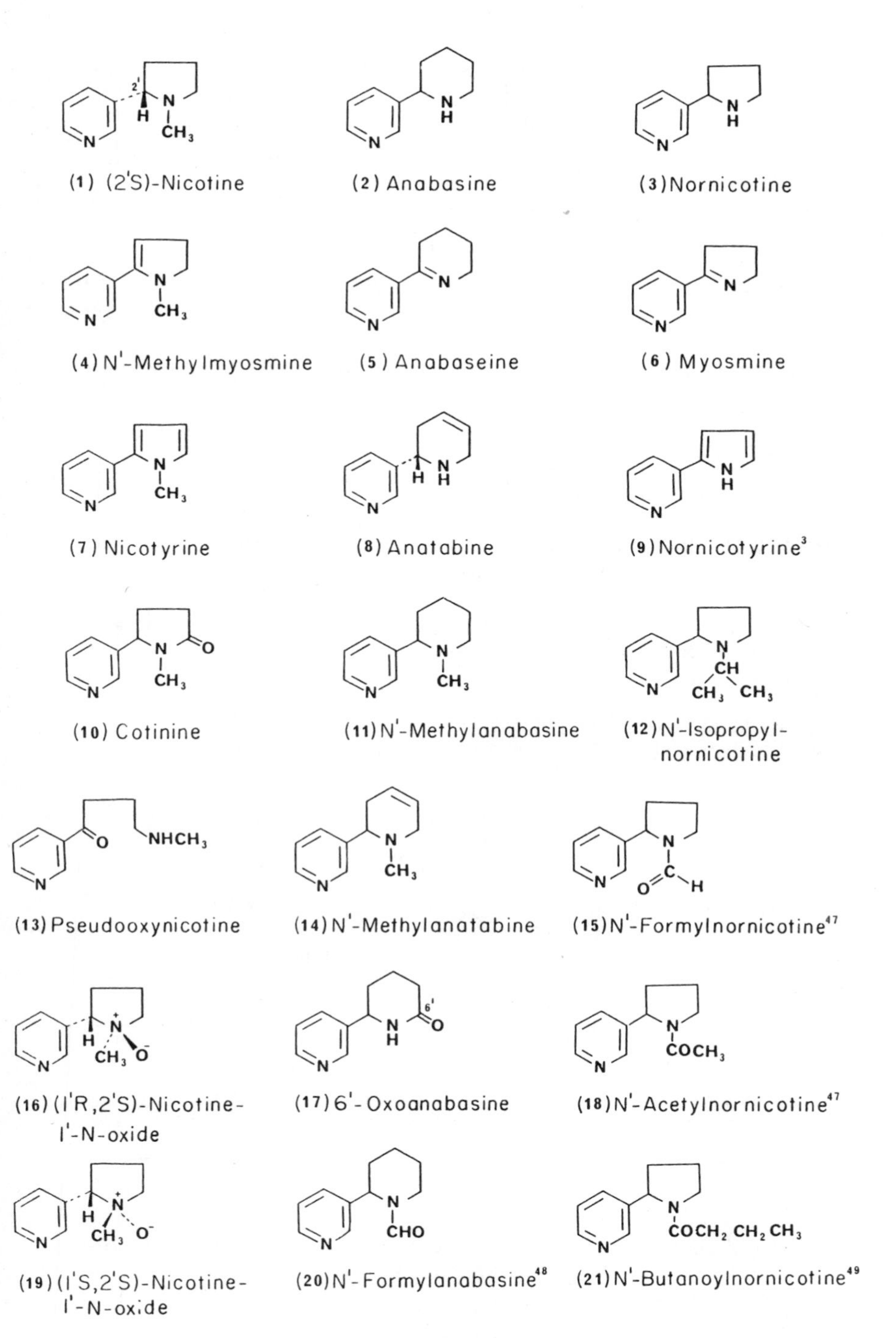
(1) (2'S)-Nicotine
(2) Anabasine
(3) Nornicotine
(4) N'-Methylmyosmine
(5) Anabaseine
(6) Myosmine
(7) Nicotyrine
(8) Anatabine
(9) Nornicotyrine[3]
(10) Cotinine
(11) N'-Methylanabasine
(12) N'-Isopropyl-nornicotine
(13) Pseudooxynicotine
(14) N'-Methylanatabine
(15) N'-Formylnornicotine[47]
(16) (1'R,2'S)-Nicotine-1'-N-oxide
(17) 6'-Oxoanabasine
(18) N'-Acetylnornicotine[47]
(19) (1'S,2'S)-Nicotine-1'-N-oxide
(20) N'-Formylanabasine[48]
(21) N'-Butanoylnornicotine[49]

Figure 1. Alkaloids of Tobacco.

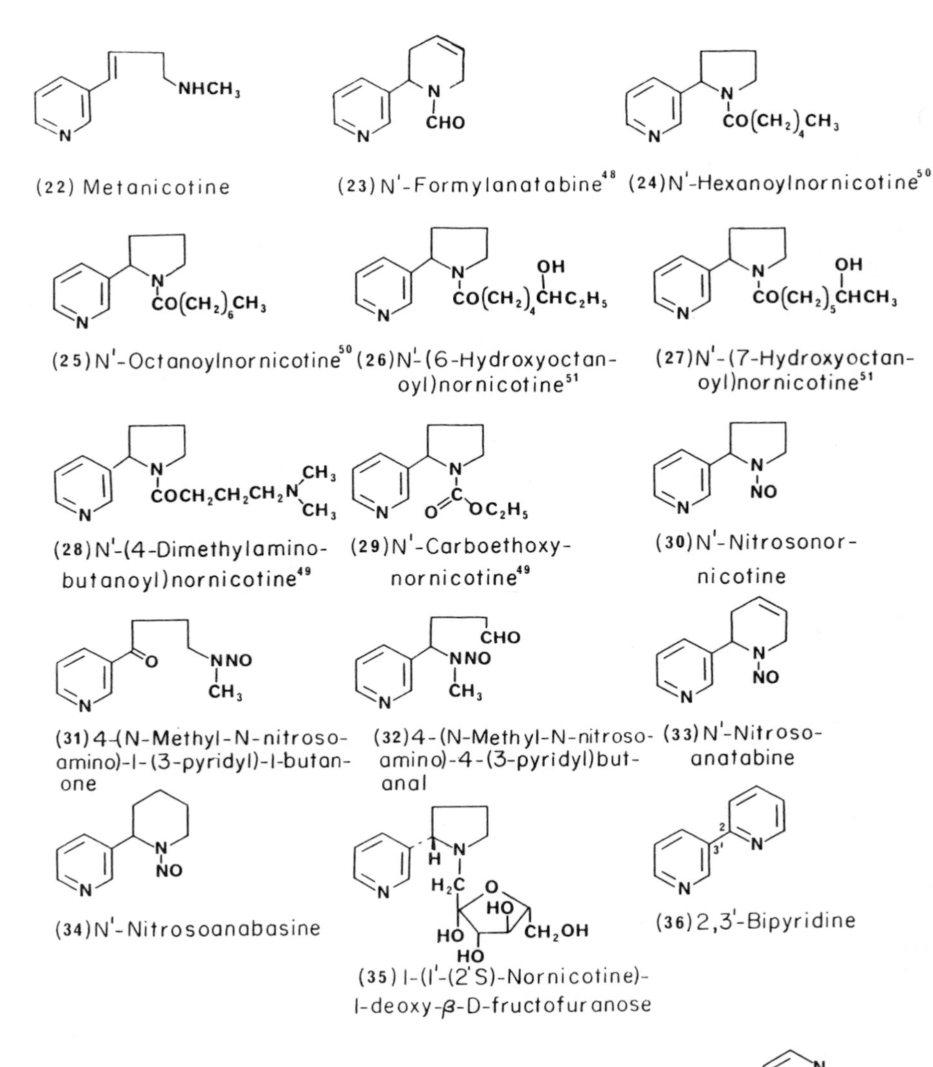

CH$_3$

(37) 3,3'-Bipyridine

(38) 5-Methyl-2,3'-bipyridine

(39) Nicotelline

Figure 1. Continued

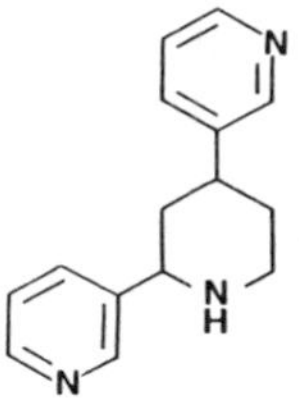

(40) Anatalline

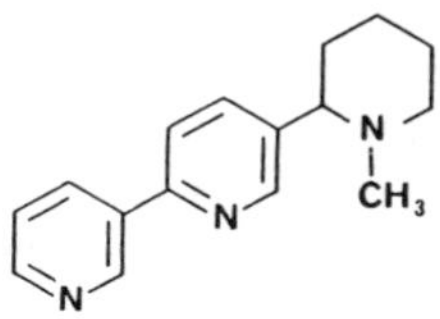

(41) Anabasamine

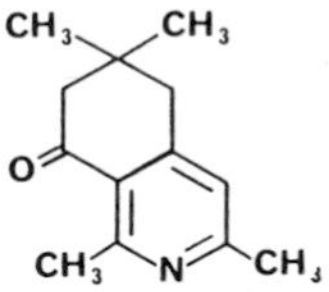

(42) 1,3,6,6-Tetramethyl-5,6,7,8-tetrahydroisoquinolin-8-one

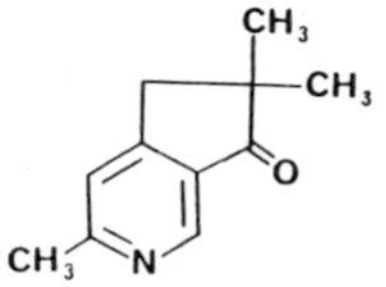

(43) 3,6,6-Trimethyl-5,6-dihydro-7H-2-pyrindin-7-one

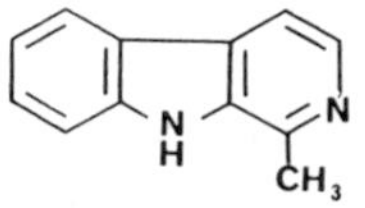

(44) Harman

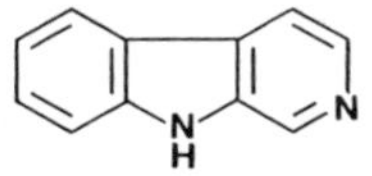

(45) Norharman

Figure 1. Continued

lower than in *N. tabacum*, and none of these plants have been exploited for smoking, except *Cannabis sativa* (marijuana). Also, the very old reference to nicotine in *Cannabis* [21] was not confirmed by later workers [22]. The main alkaloid of *N. glauca* (tree tobacco) is anabasine (**2**), so called since it was first isolated from *Anabasis aphylla* [23], where it is found as the levorotatory isomer, having the same (*S*) configuration as nicotine [24]. However, the anabasine obtained from *N. glauca* is racemic or almost so [25, 26]. Anabasine is also found in many other plant species, and its 1′,2′-dehydroderivative, anabaseine (**5**) has been found in animals (see Section 3). Anabaseine is a very unstable alkaloid, and no definitive reports of its isolation from tobacco have appeared. However, it was obtained from *N. tabacum* plants that has been fed racemic anabasine [27]. Nornicotine (**3**) usually accompanies nicotine as a minor alkaloid, and its main origin is by the demethylation of nicotine, which occurs both in the living plant and during the curing of tobacco leaves (see Section 8). Its dehydrogenation product, myosmine (**6**), was first isolated from tobacco smoke [28, 29], but it has since been found in *N. tabacum*, *glutinosa*, and *glauca* [27, 30, 31]. Anatabine (**8**) (indexed under 2,3′-bipyridine-(1,2,3,6-tetrahydro-) in *Chemical Abstracts*) is fairly widespread in *Nicotiana* species, and in fresh plants of *N. tabacum*, it is more abundant then anabasine or nornicotine. The proportions of the alkaloids found using separation by thin layer chromatography were nicotine 93%, anatabine 3.9%, nornicotine 2.4%, and anabasine 0.5% [32]. Similar proportions were found in cigarette tobacco; the alkaloid fraction assayed by gas chromatography consisted of 94.5% nicotine, 2.7% anatabine, 0.96% nornicotine, and 0.35% anabasine [33].

Other alkaloids that are related to nicotine and almost certainly derived from it in the plant or the cured tobacco leaf are *N*′-methylmyosmine (**4**) [34, 35] and nicotyrine (**7**) [36]. The former alkaloid is an enamine and tautomerism to the Δ^1-pyrrolinium salt followed by hydration of this iminium salt and ring opening affords 4-methylamino-1-(3-pyridyl)-1-butanone (**13**) also known as pseudo-oxynicotine [37]. Cotinine (**10**), a major metabolite of nicotine in humans and other animals, has also been found in cured tobacco [38]. It has also been isolated from the fresh *N. glutinosa* plant [39]. Nicotine-1′-*N*-oxide (oxynicotine) in both the cis (**16**) and trans (**19**) forms has been found in *N. tabacum*, *N. affinis*, and *N. sylvestris* [40]. 4-*N*-Methylamino-1-(3-pyridyl)-1-butene (**22**), known trivially as metanicotine, is a minor component of tobacco [41] and is also formed by the action of hydrobromic acid on nicotine at 280° C [42]. Späth and Kesztler isolated small amounts of *N*′-methylanabasine (**11**) and N′-methylanatabine (**14**) from *N. tabacum* [36]. The presence of the former alkaloid has been confirmed by recent work [10], and it has also been found in tobacco-smoke condensate [43]. 6′-Oxoanabasine (**17**) presumably formed from anabasine by a route analogous to the formation of cotinine from nicotine has been found in *N. glutinosa* [39]. *N*′-Isopropylnornicotine (**12**) has been found in Burley tobacco [44]. This is the first report of any *N*-alkylpyridine alkaloid in tobacco other than *N*′-methyl derivatives. Indeed, the *N*-isopropyl group is very rare in nature, aristomakine and peduncularine are indole alkaloids whose *N*-isopropyl groups are apparently formed from the fragmentation of a terpene precursor [45]. We have shown that *N*′-isopropylnornicotine is formed from nicotine in aged leaves of *N. tabacum* [46] (see Section 9). A large number of *N*′-acyl derivatives of the secondary amines found in *Nicotiana* have been isolated from cured tobacco. These are depicted in Fig. 1 with the appropriate references indicated. Compound **29** is a carbamate, and the related compounds *N*′-carbomethoxy-nornicotine and *N*′-carbomethoxyanabasine have been found in tobacco smoke [52]. The carcinogenic *N*-nitroso compounds (**30–34**) have been found in several unburned tobacco products, especially chewing tobacco [53–58].

I personally claim that the simple pyridines, pyrrolidines, pyrroles, piperidines, pyrazines, and other basic nitrogen heterocycles [4, 9, 11] that have been isolated from tobacco are alkaloids. However, the list is overwhelming, and little would be served by repeating it here. They will only be referred to when their presence in tobacco is relevant to the origin and metabolism of the more "classical" alkaloids of tobacco. A novel alkaloid is **35**, a carbohydrate derivative, apparently the result of an Amadori reaction on the condensation product of nornicotine and D-glucose [59]. The only bipyridine that is well authenticated in unburned tobacco is the 2,3′-isomer (**36**), first known as isonicoteine [60]. The 3,3′-bipyridine (**37**) has been tentatively identified in *N. glutinosa* [39]. The 5-methyl-2,3′-bipyridine has been isolated from *N. tabacum* and its structure proved by synthesis [61]. 2,2′-Bipyridine [43] and 4,4′-bipyridine [62] have been found in cigarette smoke. Nicotelline (**39**) was isolated long ago from tobacco [63]; however, its structure was not determined till 1956 [64]. Anatalline (**40**) was isolated from the roots of *N. tabacum* [65] and was reported to be optically inactive. Anabasamine (**41**) has also been isolated from *N. tabacum* [61], having first been found in *Anabasis aphylla* [66]. Two bicyclic pyridine compounds isolated from Burley tobacco that are probably of terpenoid origin are the isoquinoline **42** and the pyrindone **43** [67]. Small amounts of the β-carboline alkaloids, harman (**44**) and norharman (**45**), have been detected in tobacco [68, 69].

3. THE OCCURRENCE OF NICOTINE AND RELATED PYRIDINE ALKALOIDS IN SPECIES OTHER THAN TOBACCO (*NICOTIANA*)

The wide distribution (24 genera, 12 families) in nature of nicotine is remarkable, especially as it has no known biological role in the plants that produce it. In Table 1 the species are arranged alphabetically, and the family in which these species occurs is also recorded. It must be admitted that some of the reports of the presence of these alkaloids in the indicated species are quite vague, and only occasionally [70] has the optical rotation of nicotine and other chiral alkaloids been determined. Conflicts exist in the literature. For example, Frank [71] was unable to confirm the report of Marion [72] that nicotine is present in *Sedum acre*. However, later workers (see Table 1) agreed with Marion and detected nicotine. Clearly many species, although given the same name, should be subdivided into different varieties on the basis of the secondary natural products that they contain. The amount of nicotine found in some of the species is extremely small. For example, *Datura stramonium* (Jimson weed) was reported [133] to contain 0.0005% nicotine (0.5 mg in 100 g of dried plant). A similar low percentage was found in the intact tomato plant (*Lycopersicum esculentum*). This was apparently confirmed by Solt [134], although the criteria used for identification of the nicotine (spots on paper chromatograms having R_f values similar to those of nicotine) leave something to be desired. A report that mutants of *Datura* species, obtained by exposure to X-rays, yielded nicotine is of considerable interest [135]. However, the lack of any subsequent publication on this subject lead me to question the validity of these observations, which were published as a brief communication with no experimental details. Some earlier citations [18, 19] seem to be in error. Examination of the original literature revealed that nicotine is not a normal component of *Cyphomandra betacea* [80].

4. SUMMARY OF TRACER EXPERIMENTS WITH *NICOTIANA*

All tracer work that has been carried out with *Nicotiana* species relating to the biosynthesis or metabolism of the alkaloids is summarized in Table 2. The method of administration of the labeled compounds is often significant, and this information is added when available in the original publications. The following abbreviations relating to the method of feeding are used:

H *Hydroponics*: The tracer is added to the nutrient solution in which roots of the intact plant are growing.

Table 1. Species (Other Than *Nicotiana*) in Which Nicotine and Related Alkaloids Have Been Found

Species	Family	Reference
	Nicotine (**1**)	
Acacia sp.	Leguminosae	73
Acacia concinna	Leguminosae	74
Anthocercis frondosa	Solanaceae	75
Anthocercis tasmanica	Solanaceae	76
Asclepias syriaca	Asclepiadaceae	77
Atropa belladonna	Solanaceae	41
Cannabis sativa	Moraceae	21, 22
Carica papaya	Caricaceae	78
Cestrum diurnum	Solanaceae	79
Cestrum nocturnum	Solanaceae	79
Datura fastuosa	Solanaceae	41
Datura inermis	Solanaceae	41
Datura metal	Solanaceae	41
Datura meteloides	Solanaceae	41
Datura stramonium	Solanaceae	41, 134
Duboisia hopwoodii	Solanaceae	81, 82
Duboisia myoporoides $[\alpha]_D^{22}$ − 163° (neat)[70]	Solanaceae	70, 83–87
Eclipta alba(*erecta*)	Compositae	88
Equisetum arvense	Equisetaceae	89, 90
Equisetum hyemale	Equisetaceae	90
Equisetum palustre	Equisetaceae	90, 91
Equisetum ramossimum	Equisetaceae	91
Equisetum telmateia	Equisetaceae	91
Erythroxylon coca	Erythroxylaceae	92
Erythroxylon truxillense	Erythroxylaceae	92
Haloxylon persicum	Chenopodiaceae	93
Herpestris monnieria	Scrophulariaceae	94
Lycopersicum esculentum	Solanaceae	95
Lycopodium annotinum	Lycopodiaceae	96
Lycopodium cernum	Lycopodiaceae	97
Lycopodium clavatum	Lycopodiaceae	98, 99
Lycopodium complanatum	Lycopodiaceae	100
Lycopodium flabelliforme	Lycopodiaceae	98, 100
Lycopodium lucidulum	Lycopodiaceae	101
Lycopodium obscurum	Lycopodiaceae	102
Lycopodium sabinaefolium	Lycopodiaceae	103
Lycopodium tristachyum	Lycopodiaceae	104
Mucuna pruriens	Leguminosae	105
Petunia violacea	Solanaceae	136
Sedum acre	Crassulaceae	71, 72, 106, 107
Sedum album	Crassulaceae	107
Sedum carpaticum	Crassulaceae	107

Table 1. Continued

Species	Family	Reference
Sedum oppositifolium	Crassulaceae	107
Sedum pallidum	Crassulaceae	107
Sedum populifolium	Crassulaceae	107
Sedum telephium	Crassulaceae	107
Sempervivum arachnoideum	Crassulaceae	108
Solanum melongena	Solanaceae	134
Solanum tuberosum	Solanaceae	104
Withania somnifera	Solanaceae	109
Zinnia elegans	Compositae	110
	Anabasine (**2**)	
Anabasis aphylla	Chenopodiaceae	23, 111, 112
$[\alpha]_D$ −52° to −80° [111]		
Aniba coto	Lauraceae	113
Duboisia myoporoides	Solanaceae	70, 86
(almost racemic)		
$[\alpha]_D^{21}$ −0.44° (neat)		
Echinochloa sp.	Gramineae	114
Haloxylon persicum	Chenopodiaceae	93
(contained 4.3% on dry weight)		
Haloxylon salicornicum	Chenopodiaceae	115, 116
$[\alpha]_D^{22}$ −58.1°		
Leontice alberti	Berberidaceae	117
Leontice darasica	Berberidaceae	118
Lupinus formosus	Leguminosae	119
(also contains *N*′-methylanabasine)		
Malacocarpus crithmofolius	Zygophyllaceae	120
$[\alpha]_D$ +10°		
Marsdenia rostrata	Asclepiadaceae	121
$[\alpha]_D$ 0°		
Priesteya elliptica	Leguminosae	122
Priesteya tomentosa	Leguminosae	122, 123
Solanum carolinense	Solanaceae	124
Sophora pachycarpa	Leguminosae	125
Verbascum songaricum	Scrophulariaceae	126
Zinnia elegans	Compositae	110
	Nornicotine (**3**)	
Cestrum diurnum	Solanaceae	79
Cestrum nocturnum	Solanaceae	79
Duboisia hopwoodii	Solanaceae	82, 127
Duboisia myoporoides	Solanaceae	85, 87, 127
Sulpiglossis sinuata	Solanaceae	128
Zinnia elegans	Compositae	110

Table 1. Continued

Species	Family	Reference
	Myosmine (***6***)	
Carica papaya	Caricaceae	78
Cestrum nocturnum	Solanaceae	79
	Cotinine (***10***)	
Carica papaya	Carcicaceae	78
Cestrum nocturnum	Solanaceae	79
	Anatabine (***8***)	
Duboisia myoporoides	Solanaceae	86
	Anabaseine (***5***)	
Amphiporus angulatus	Hoplonemertinea	129, 130
Amphiporus lactifloreus	Hoplonemertinea	130
Aphaenogaster fulva	ant (insect)	131
Aphaenogaster tennesseensis	ant (insect)	131
Paranemertes peregrina	Hoplonemertinea	130, 132
Tetrastemma worki	Hoplonemertinea	130

W *Wick Feeding*: A wick, usually cotton, is inserted into the stem of the plant with a sewing needle and the cotton ends placed in a small beaker or vial containing the tracer (usually in water).

ER *Excised Roots*: The excised roots of the plant are grown like a microorganism in a sterile medium, and the tracer is added to this solution.

ES *Excised Stems*: The cut end of an excised stem or leaf of the plant is placed in a solution of the tracer.

CC *Cell Culture*: The tracer is added to a cell-suspension culture or a callus tissue culture of *Nicotiana*.

EN *Enzyme System*: The tracer is incubated with a cell-free system obtained from *Nicotiana*.

A *Atmosphere*: The plant is grown in an atmosphere containing the tracer (usually [^{14}C]carbon dioxide or [^{13}C]carbon dioxide).

SI *Stem Injection*: Tracer is injected into the stems by means of a hypodermic syringe or by fine glass capillaries inserted into the stems [137].

LA *Leaf Administration*: The leaves are sprayed with a solution of the tracer, or it is applied with a paintbrush.

U *Unknown*: Method of feeding is not indicated in the original publication.

These various methods of feeding tracers to plants and their limitations have been discussed by Brown [138]. Most of the early work with *Nicotiana* was carried out by the hydroponics method. It is easy to cultivate tobacco plants in such a soil-free environment, and good incorporations of labeled compounds into nicotine were achieved since the roots are the main site for the synthesis of nicotine, which is then translocated to the leaves. When an excised tobacco leaf is placed in a nutrient solution, roots soon develop in the solution, and this system has sometimes been used for administering compounds to tobacco. Experiments of this type are included in the hydroponics class. The discovery that nicotine synthesis occurs mainly in the roots [139] led Dawson and co-workers to investigate the biosynthesis of nicotine in isolated roots growing in a sterile medium. Very high incorporations of labeled compounds were obtained by this technique. More recently, we and others have used the wick-feeding method described by Comar [140] and first used by us in the study of the biosynthesis of the indole alkaloids of *Rauwolfia serpentina* [141]. This method has the advantage that the plants growing in soil do not have to be disturbed, and the tracer solution is taken into the plant quite rapidly (usually in a few hours). There is thus much less chance of the labeled compounds being modified by microorganisms before it enters the various circulating systems in the plant. The labeled compounds administered are arranged alphabetically under each *Nicotiana* species. Salts of carboxylic acids such as sodium [1-^{14}C]acetate are listed under "[1-^{14}C]Acetate, sodium." The position of the isotope is indicated in accord with recent rules [142]. Thus many compounds are described differently from the names found in the original literature, especially the older citations.

The incorporation of labeled precursor is reported in two ways. The absolute incorporation is defined as the total radioactivity (or excess isotope for nonradioactive atoms) found in the ultimate alkaloid or metabolite divided by the total activity (or excess isotope) in the compound administered to the plant. This value depends on the efficiency of the extraction and purification whereby the alkaloid is isolated. The specific incorporation, which is indicated in Table 2 in square brackets [ ], is defined as the specific activity of the isolated alkaloid (e.g., dpm/mmol, or percentage of excess of a stable isotope) divided by the specific activity of the administered precursor. Some workers have used the term *dilution*, which is the inverse of specific activity, thus a dilution of 100 is equivalent to a 1% specific incorporation. Where the data in the original publication are not sufficient to calculate either of these incorporations (both are recorded if possible), a positive incorporation is recorded with a + sign.

5. BIOSYNTHESIS OF NICOTINIC ACID AND THE MECHANISM WHEREBY IT IS INCORPORATED INTO THE TOBACCO ALKALOIDS

As evidenced by Table 2, a large number of feeding experiments have been carried out with putative precursors of nicotine, anabasine, anatabine, and nornicotine in order to elucidate their biosynthesis. Nicotinic acid (**54**) is the

Table 2. Tracer Experiments Relating to the Biosynthesis and Metabolism of the Tobacco Alkaloids in *Nicotiana* Species

Compound Administered	Method of Feeding	Alkaloids or Related Compounds Labeled	Incorporation (%)	Reference
		Nicotiana glauca		
[2-^{14}C]Acetate, sodium	ES	Anabasine (a lower incorporation when fed with nicotinic acid)	0.032	143
[2-^{14}C]Acetate, sodium	W	Anabasine	2.0	144
[2-^{14}C]Acetate, sodium	W	Anabasine	0.062	145
DL-[2-^{14}C]Alanine	W	Anabasine	0.61	145
(*RS*)-[2′-^{13}C, ^{14}C]Anatabine	W	Anabasine	0	146
		Anatabine	[27]	
DL-[4-^{14}C]Aspartic acid	W	Anabasine	0.078	145
L-[U-^{14}C, ^{15}N]Aspartic acid	H	Anabasine	^{14}C: [1.08] ^{15}N: [0.73]	147
[^{14}C]Bicarbonate, sodium	ES	Anabasine	+	148
[1,5-^{14}C]Cadaverine	H	Anabasine	0.33	149
[1,5-^{14}C]Cadaverine	ES	Anabasine	0.49	149
[^{14}C]Carbon dioxide	A	Anabasine	+	150, 151
5-Fluoro-[5,6-$^{13}C_2$, ^{14}C]Nicotinic acid	W	5-Fluoroanabasine	16.2 [100]	26
D-[U-^{14}C]Glucose	H	Anabasine	+	150
[2-^{14}C]Glycerol	W	Anabasine	0.36 [0.008]	153
L-[U-^{14}C]Lysine	ER	Anabasine	28.6	154
DL-[2-^{14}C]Lysine	H	Anabasine	0.049 [0.086]	155
DL-[2-^{14}C]Lysine	ES	Anabasine	0.039	149
DL-[2-^{14}C]Lysine	W	Anabasine	4.18	145

DL-[2-^{14}C]Lysine	W	Anabasine	6.4 [0.21]	161
		N'-Methylanabasine (added as a carrier)	0.001	
L-[2-^{14}C]Lysine	H	Anabasine	0.075 [2.17]	156
		Pipecolic acid	0.044	
D-[2-^{14}C]Lysine	H	Anabasine	0.0004 [0.072]	156
		Pipecolic acid	0.37	
DL-[4,5-$^{13}C_2$,6-^{14}C]Lysine	W	Anabasine	5.7 [0.24]	157
		Nicotine	0.0003	
		Nornicotine	0.001	
DL-[2-^{14}C,α-^{15}N]Lysine	ER	Anabasine	^{14}C: [38.6] ^{15}N: [2.3]	158
		Nicotine	^{14}C: [0.29] ^{15}N: [0.11]	
DL-[2-^{14}C,ϵ-^{15}N]Lysine	ER	Anabasine	^{14}C: [31.1] ^{15}N: [24.8]	158
		Nicotine	^{14}C: [0.07] ^{15}N: [2.6]	
DL-[6-^{14}C],L-[4,5-$^3H_2$]Lysine	W	Anabasine (The specific incorporation of ^{3}H was 1.89 × that of the ^{14}C).	+	159
		Pipecolic acid (neglible incorporation of ^{3}H)	+	
D-[6-^{14}C],DL-[4,5-$^3H_2$]Lysine	W	Anabasine (neglible incorporation of ^{14}C)	+	159
		Pipecolic acid (48% retention of ^{3}H)	+	

Table 2. Continued

Compound Administered	Method of Feeding	Alkaloids or Related Compounds Labeled	Incorporation (%)	Reference
L-[methyl-^{14}C]Methionine + unlabeled *N′*-methylanabasine	W	*N′*-Methylanabasine	0	160
N-Methyl-[2-^{14}C]-Δ^1-piperideinium chloride	H	Anabasine	1.21 [1.4]	161
		(2′*S*)-*N′*-Methylanabasine	0.71 [104]	
[2′-^{14}C]Myosmine	W	Myosmine	[35]	31
		Nicotine	0	
		Nornicotine	0	
		Anabasine	0	
		Nicotinic acid	0.14	
(2′*S*)-[U-^{14}C]Nicotine (biosynthetic from $^{14}CO_2$)	SI, H	Anabasine	0	162
		Nornicotine	+	
[methyl-^{14}C, ^{15}N]Nicotine	N	Anabasine	^{14}C: [4.4] ^{15}N: [2.2]	163
		Anatabine	^{14}C: [4.9] ^{15}N: [6.4]	
		Nicotine	^{14}C: [21.2] ^{15}N: [37.2]	
		Nornicotine	^{14}C: [17.5] ^{15}N: [65.2]	
(2′*S*)-[^{14}C]Nicotine (biosynthetic, from [U-^{14}C]glycine, thus ^{14}C mostly on *N*-methyl group)	ES	Nicotine	4.0	164
		Nornicotine	0.02	
		Anabasine (no degradations)	1.5	
		Anatabine	0.1	

(RS)-[2′-^{14}C], (2′S)-[2′-^{3}H]Nicotine (3-day feed)	W	Nicotine	^{14}C: 8.0 [19.0]	31
		Nornicotine (> 99% retention of ^{3}H in 2′S isomer. No ^{3}H in 2′R-nornicotine)	^{14}C: 49.5 [26.8]	
		Myosmine (negligible incorporation of ^{3}H)	^{14}C: 2.05	
		Anabasine	^{14}C: 0.01 [0.002]	
		Cotinine	^{14}C: 0.01	
		3-Acetylpyridine	0	
		Nicotinic acid	^{14}C: 0.02	
(RS)[2′-^{14}C]Nicotine (5-day feed)	ES	Nicotine	0.41 [0.52]	165
		Nornicotine	5.7 [1.65]	
		Anabasine	0	
(2′S)[2′-^{14}C]Nicotine	CC	Nicotine	40–45	
		Nornicotine	50	
(2′S)-[2′-^{14}C,2′-^{3}H]Nicotine 1′-N-oxide (cis/trans isomers)	W	Nicotine (104% retention of ^{3}H)	0.3	167
		Nornicotine (99% retention of ^{3}H)	1.4	
		Anabasine	<0.01	
		Nicotine-1′-N-oxide	25	
[2-^{14}C]Nicotinic acid	W	Anabasine	14.3 [1.49]	32
		Anatabine	0.21 [2.76]	
		Nicotine	0.14 [0.54]	
		Nornicotine	0.094 [0.32]	
		2,3′-Bipyridine	0.028	
		Nicotelline	0.0004	

Table 2. Continued

Compound Administered	Method of Feeding	Alkaloids or Related Compounds Labeled	Incorporation (%)	Reference
[5,6-$^{13}C_2$,^{14}C]Nicotinic acid	W	Anabasine	[1.64]	168
		Anatabine	[4.5]	
		Nicotine	[0.3]	
		Nornicotine	[0.92]	
[5,6-$^{13}C_2$,^{14}C,6-^{3}H]Nicotinic acid	W	Anabasine (95% loss of ^{3}H)	[1.98]	169
		Anatabine (53% retention of ^{3}H, essentially all at C-6′ in the *pro-S* position)	[3.8]	
[^{3}H]Nicotinic acid	ER	Anabasine	12.0	154
		Nicotine	4.0	
[^{15}N]Nitrate, potassium	H	Anabasine	+	170
[6-^{14}C]-Δ^1-Piperideine	H	Anabasine	1.2 [1.65]	171
		Nicotine	0	
		Nornicotine	0	
[1,4-^{14}C]Putrescine (fed to *N. glauca* grafted on tomato roots)	H	Anabasine	0	172
		Nicotine	+	
		Nornicotine	+	
[2,3-^{14}C]Succinic acid	W	Anabasine	1.69	145
		Nicotiana glutinosa		
(*RS*)-[2′-^{13}C,^{14}C]Anatabine	W	Anabasine	0	146
		Anatabine	[33]	

		Nicotine	0	
		Nornicotine	0	
[^{14}C]Bicarbonate, sodium	ES	Nornicotine	+	148
[^{14}C]Carbon dioxide	A	Nicotine	+	151, 173–181
		Nornicotine	+	
		Anabasine	+	
		Anatabine	+	
[2-methyl-^{14}C]-1,2-Dimethyl-Δ^1-pyrrolinium chloride	H	2′-Methylnicotine (chirality at C-2′ not determined)	0.04	182
[3-methyl-^{14}C]-1,3-Dimethyl-Δ^1-pyrrolinium chloride	H	(2′*S*,3′*S*)-3′-Methylnicotine	13.8 [100]	182
[carboxyl-^{14}C]Nicotinic acid	W	Anatabine	0	183
		Anabasine	0	
		Nicotine	0	
		Nornicotine	0	
[6-^{14}C]Nicotinic acid (5-day feed)	W	Nicotine	3.6 [0.027]	32, 183
		Nornicotine	0.51 [0.007]	
		Anabasine	0.09 [0.024]	
		Anatabine	0.57 [0.038]	
[6-^{14}C]Nicotinic acid (20 h feed)	W	Nicotine	0.93 [0.25]	32
		Nornicotine	0.039 [0.023]	
		Anabasine	0.046 [0.43]	
		Anatabine	0.13 [0.38]	
[6-^{14}C]Nicotinic acid (3-day feed)	W	Nicotine	1.10 [0.16]	32
		Nornicotine	0.05 [0.016]	
		Anatabine	0.42 [0.44]	
		2,3′-Bipyridine	0.07	
		Nicotelline	0.0006	
[2-^{14}C]Nicotinic acid	W	Nicotine	0.0046 [0.063]	32

Table 2. Continued

Compound Administered	Method of Feeding	Alkaloids or Related Compounds Labeled	Incorporation (%)	Reference
(fed 2 days to plant growing out of doors)		Nornicotine	0.0003 [0.006]	
		Anabasine	0.00007 [0.048]	
		Anatabine (These surprisingly low incorporations could indicate that the environment has a major effect on alkaloid production)	0.0002 [0.070]	
$[2\text{-}^{14}C]$Nicotinic acid (fed 2 days, then the plants air dried for 40 days)	W	Nicotine	0.0004 [0.11]	32
		Nornicotine	0.0008 [0.022]	
		Anabasine	0.00005 [0.061]	
		Anatabine	0.0002 [0.083]	
$(2'S)\text{-}[2'\text{-}^{14}C, {}^{3}H]$Nicotine-1′-*N*-oxide (mixture of cis/trans isomers)	H	Nicotine (97% retention of ^{3}H)	0.35	167
		Nornicotine (64% retention of ^{3}H)	0.12	
		Anabasine	<0.01	
$(2'S)\text{-}[U\text{-}^{14}C]$Nicotine-1′-*N*-oxide	ES	Nicotine	+	184
		Nornicotine	+	
$[^{15}N]$Nitrate, potassium	H	Nornicotine	+	170
DL-$[2\text{-}^{14}C]$Ornithine	W	Nicotine	0.58	185
		Nornicotine	0.09	
		Anabasine	0	
		Anatabine	0	

DL-[2-^{14}C]Ornithine (6-h feed)	H	Nicotine	1.3	178
DL-[5-^{14}C]Ornithine	H	Nicotine	+	174
DL-[2,3-^{13}C, 5-^{14}C]Ornithine	W	Nicotine	1.04 [0.14]	186
		Nornicotine	0.53 [0.051]	
[3,3-dimethyl-$^{14}C_2$]-1,3,3-Trimethyl-Δ^1-pyrrolinium chloride	H	3′,3′-Dimethylnicotine	0.77	182
		Nicotiana rustica		
[1-^{14}C]Acetate, sodium	H	Nicotine	+	187, 188, 191
[2-^{14}C]Acetate, sodium	H	Nicotine	+	175, 179, 181, 187–191
		Glutamic acid	+	179
[1-^{14}C]-4-Aminobutanoic acid (5-h feed)	ER	Nicotine	0	192
		N-Methyl-Δ^1-pyrrolinium salt	0	
[N^{15}]Anabasine	H	Anabasine	[56.0]	163
		Anatabine	[6.5]	
		Nicotine	[4.0]	
		Nornicotine	[6.6]	
DL-[3-^{14}C]Aspartic acid	H	Nicotine	[0.28]	193, 194
[methyl-^{14}C]Betaine	H	Nicotine	+	195
[^{14}C]Bicarbonate, sodium	H	Nicotine	+	196
[^{14}C]Carbon dioxide	A	Nicotine	0.20	197
[^{14}C]Carbon dioxide (6-h exposure)	A	Nicotine	+	151, 179, 181, 198
[^{14}C]Carbon dioxide + [^{3}H]water + [^{15}N]nitrate	A + H	Anabasine	+	199
		Anatabine	+	
		Nicotine	+	
		Nornicotine	+	

Table 2. Continued

Compound Administered	Method of Feeding	Alkaloids or Related Compounds Labeled	Incorporation (%)	Reference
[methyl-^{14}C]-2-Chloroethyltrimethyl-ammonium chloride	U	Nicotine	+	200
[methyl-^{14}C]Choline	H	Nicotine	+	195, 201
[^{14}C]Formaldehyde	H	Nicotine	+	203
[^{14}C]Formate, sodium	H	Nicotine	+	202
DL-[2-^{14}C]Glutamic acid	H	Nicotine	+	204
L-[U-^{14}C]Glutamic acid	ER	Nicotine	0.69	205
L-[U-^{14}C]Glutamic acid + helminthosporic acid	ER	Nicotine	0.18	205
L-[U-^{14}C]Glutamic acid (5-h feed)	ER	Nicotine	3.2	192
		Arginine	0	
		N-Methyl-Δ^1-pyrrolinium salt	0	
D-[3-^{14}C]Glyceraldehyde	H	Nicotine	0.46	206
[1,3-^{14}C]Glycerol	H	Nicotine	[0.70]	188
[2-^{14}C]Glycerol	H	Nicotine	[0.46]	193
[2-^{14}C]Glycine	H	Nicotine	+	207
[2-^{14}C]Glycolate, calcium	H	Nicotine	+	208
[2-^{14}C]Leucine	H	Nicotine	+	209
[3-^{14}C]Malate, sodium	H	Nicotine	+	194
DL-[methyl-^{14}C]Methionine	H	Nicotine	+	202
DL-[methyl-^{14}C]Methionine (fed 2 weeks)	ER	Nicotine	3.5	210
DL-[methyl-^{14}C,$^{2}H_3$]Methionine	H	Nicotine	+	211
DL-[methyl-^{14}C]Methionine (5-h feed)	ER	Nicotine	5.5	192
		Arginine	0	
		N-Methyl-Δ^1-pyrrolinium salt	0.13	

[*N*-methyl-^{14}C]-α-*N*-Methylornithine	H	Nicotine	0.2	212
(c.f. [247])	ER	Nicotine	0.5	212
[*N*-methyl-^{14}C]-δ-*N*-Methylornithine (c.f. [247])	H	Nicotine	0.05	212
[*N*-methyl-^{14}C]-α-*N*-Methyllysine	H	Nicotine	0.008	212
DL-[methyl-^{14}C]Methionine	H	Nicotine (9-day feed)	5.5	152
DL-[methyl-^{14}C]Methionine	ES	Nicotine	0.016	152
[methyl-^{14}C, 1-^{15}N]-*N*-Methylputrescine	H	Nicotine	^{14}C: [0.62] ^{15}N: [0.64]	213
[methyl-^{14}C, 1-^{15}N]-*N*-Methylputrescine	W	Nicotine	^{14}C: [0.13] ^{15}N: [0.28]	213
[1-^{14}C, methylamino-^{15}N]-*N*-Methylputrescine	H	Nicotine	^{14}C: [0.26] ^{15}N: [0.28]	214
[1-^{3}H]-*N*-Methylputrescine	H	Nicotine	+	214
[2,5-^{14}C]-*N*-Methyl-Δ^1-pyrrolinium salt (obtained biosynthetically from [2-^{14}C]ornithine)	ER	Nicotine	7–33	192
[2,3,7-^{14}C]Nicotinamide adenine dinucleotide	W	Nicotine	5.8	215
(2′*S*)-[^{14}C]Nicotine (obtained biosynthetically from [2-^{14}C]acetate or [3-^{14}C]propionate)	H	Nicotine Nicotinic acid	[70–33] [5.6–1.1]	216
(2′*S*)-[methyl-^{14}C]Nicotine	H	Nicotine (the loss of activity was greatest with intact plants, but also occurred in rootless plants)		152
(2′*S*)-[G-^{14}C]Nicotine (biosynthetic, from [2-^{14}C]propionate, activity in both rings).	H	Nicotine (metabolized in both intact plants and rootless ones)		152

Table 2. Continued

Compound Administered	Method of Feeding	Alkaloids or Related Compounds Labeled	Incorporation (%)	Reference
[^{15}N]Nicotine	H	Nicotine	[10.4]	163
		Nornicotine	[2.7]	
		Anabasine	[2.1]	
		Anatabine	[2.2]	
[^{3}H]Nicotine (obtained biosynthetically from [G-^{3}H]nicotinic acid)	ER	Nicotine (20-day incubation)	79	205
[^{3}H]Nicotine + helminthosporic acid	ER	Nicotine (20-day incubation)	83	205
[G-^{3}H]Nicotinic acid	ER	Nicotine	7.4	205
[G-^{3}H]Nicotinic acid + helminthosporic acid	ER	Nicotine	6.8	205
[2,3,7-^{14}C]Nicotinic acid	W	Nicotine	6.9	215
[U-^{3}H]Nicotinic acid (5-h feed)	ER	Nicotine	9.5	192
		Arginine	0	
		N-Methyl-Δ^{1}-pyrrolinium salt	0	
[carboxyl-^{14}C]Nicotinic acid	ER	Nicotine	0	192
[^{15}N]Nicotinic acid	H	Nicotine	+	217
[^{15}N]Nitrate, potassium	H	Nicotine	+	170
[2′,5′-^{14}C]Nornicotine (biosynthetic, from the metabolism of [2′,5′-^{14}C]nicotine in *N. tabacum*) (fed 2 weeks)	ER	Nicotine (this conversion is not considered significant)	0.19	210
[^{15}N]Nornicotine (fed 2 weeks)	ER	Nicotine	0.24	210
		Nornicotine	9.2	

[^{15}N]Nornicotine	H	Nornicotine	[40.5]	163
		Nicotine	[4.9]	
		Anabasine	[71.9]	
		Anatabine	[3.8]	
DL-[2-^{14}C]Ornithine	H	Nicotine	+	218, 355
DL-[2-^{14}C]Ornithine	ER	Nicotine	8.5	205
DL-[2-^{14}C]Ornithine + helminthosporic acid	ER	Nicotine	2.0	205
DL-[2-^{14}C]Ornithine (fed 2 weeks)	ER	Nicotine	5.1	210
DL-[2-^{14}C]Ornithine	ER	Nicotine	1.2	219
		Nornicotine	0.1	
		Anabasine	0	
		Anatabine	0	
DL-[2-^{14}C]Ornithine (5-h feed)	ER	4-Aminobutanoic acid	+	192
		Arginine	0.36	
		Citrulline	+	
		Glutamic acid	+	
		Glutamine	+	
		Nicotine	14.5	
		N-Methyl-Δ^1-pyrrolinium salt	0.90	
		Proline	+	
[1-^{14}C]Propionate, sodium	H	Nicotine	[0.01–0.012]	188
[2-^{14}C]Propionate, sodium	H	Nicotine	[0.43–0.56]	189
[3-^{14}C]Propionate, sodium	H	Nicotine	[0.13]	193
[1-^{14}C]Pseudooxynicotine (^{14}C on the C=O)	H	Nicotine	[0.07]	220
[1,4-^{14}C]Putrescine (5-h feed)	ER	Nicotine	24.3	192
		Arginine	0	
		N-Methyl-Δ^1-pyrrolinium salt	1.1	

Table 2. Continued

Compound Administered	Method of Feeding	Alkaloids or Related Compounds Labeled	Incorporation (%)	Reference
DL-[5-^{14}C]-Δ^1-Pyrroline-5-carboxylic acid (6-h feed)	H	Nicotine	0.04	221
[1,2-^{14}C]Pyruvate, sodium	H	Nicotine	+	187
[3-^{14}C]Pyruvate, sodium	H	Nicotine	+	187
[2,3,7,8-^{14}C]Quinolinic acid	W	Nicotine	5.4	215
DL-[2-^{14}C]Serine	H	Nicotine	+	203
[2,3-^{14}C]Succinate, sodium	H	Nicotine	+	189
DL-[7a-^{14}C]Tryptophan	H	Nicotine	0	222
[^{3}H]Water	H	Nicotine	+	223
		Nornicotine	+	
		Anabasine	+	
		Anatabine	+	
		Nicotiana tabacum		
[1-^{14}C]Acetate, sodium	ER	Nicotine	0.29	224
[1-^{14}C]Acetate, sodium	H	Nicotine	+	225
[2-^{14}C]Acetate, sodium	ER	Nicotine	0.76	224
[2-^{14}C]Acetate, sodium	H	Nicotine	+	225
S-Adenosyl-L-[methyl-^{14}C]methionine	EN	*N*-Methylputrescine	+	226, 227
[G-^{3}H]Agmatine	ER	"Topped" nicotine	3.8 [3.0]	228
	ER	"Nontopped" nicotine	1.7 [2.8]	
DL-[1-^{14}C]Alanine	ER	Nicotine	0.0004	224
DL-[2-^{14}C]Alanine	ER	Nicotine	1.19	224
DL-[3-^{14}C]Alanine	ER	Nicotine	5.00	224

DL-[1,2,3-^{14}C]Alanine	ER	Nicotine	1.8	224
[1-^{14}C]-β-Alanine	ER	Nicotine	0.0002	224
[2-^{14}C]-β-Alanine	ER	Nicotine	4.0–6.33	224
[3-^{14}C]-β-Alanine	ER	Nicotine	1.40	224
[^{14}C]-4-Aminobutanoic acid	H	*N*-Nicotinoyl-2-pyrrolidone	0	229
L-[methyl-^{14}C]-3-Amino-4-methylthiol butanoic acid	H	Nicotine	[0.27]	230
D-[methyl-^{14}C]-3-Amino-4-methylthiol butanoic acid	H	Nicotine	[0.01]	230
(*RS*)-[2′-^{13}C, ^{14}C]Anatabine	H	2,3′-Bipyridine (isolated from air dried leaves)	1.4 [84]	146
[carboxyl-^{14}C]Anthranilic acid	H	Nicotine	0	231
L-[U-^{14}C]Arginine	ER	"Topped" nicotine	2.8 [6.2]	228
	ER	"Nontopped" nicotine	1.1 [5.1]	
L-[U-^{14}C]Arginine	H	Nicotine	+	232
L-[U-^{14}C]Arginine	CC	Putrescine	+	228
		Caffeoylputrescine	+	
		Feruloylputrescine	+	
L-[U-^{14}C]Arginine	EN	Decarboxylase activity measured		233
L-[U-^{14}C]Aspartic acid	ER	"Topped" nicotine	1 [0.19]	228
	ER	"Nontopped" nicotine	0.6 [0.25]	
DL-[2,3-^{14}C]Aspartic acid	ER	Nicotine	0.40	224
DL[3-^{14}C]Aspartic acid	ER	Nicotine	1.10	224
DL-[4-^{14}C]Aspartic acid	ER	Nicotine	0.0025	224
DL-[4-^{14}C]Aspartic acid	H	Nicotinic acid-*N*-glucoside	[0.43]	234
DL-[U-^{14}C]Aspartic acid	ER	Nicotine	0.57	224

Table 2. Continued

Compound Administered	Method of Feeding	Alkaloids or Related Compounds Labeled	Incorporation (%)	Reference
L-[U-^{14}C, ^{15}N]Aspartic acid	H	Nicotine	^{14}C: [1.12] ^{15}N: [0.76]	147
DL-[^{15}N]Aspartic acid	ER	Nicotine	4	224
DL-[^{14}C]Bicarbonate, sodium	H	Nicotine	0.0009	235
DL-[^{14}C]Carbonate, sodium	ER	Nicotine	<0.1	224
[1,4-^{14}C]-*N*-Carbamylputrescine	ER	"Topped" nicotine	11.2 [19.2]	228
	ER	"Nontopped" nicotine	7.3 [36.8]	
[^{14}C]Carbon dioxide (to excised stems)	A	Nicotine	+	236
[^{14}C, ^{13}C]Carbon dioxide	A	Nicotine	4 h [0.1]	237
		Nicotine	10 days [10.2]	238
[1-^{14}C]Citric acid	ER	Nicotine	0.06	224
[5,6-^{14}C]-5-Fluoronicotinic acid	H	5-Fluoronicotine	0.15	239
[2,3-^{14}C]Fumaric acid	ER	Nicotine	0.44	224
D-[1-^{14}C]Glucose	ER	Nicotine	0.05	224
D-[6-^{14}C]Glucose	ER	Nicotine	0.12	224
L-[U-^{14}C]Glutamic acid	H	Nicotine	+	232
DL-[2-^{14}C]Glutamic acid	H	Nicotine	0.0078	240
[1,3-^{14}C]Glycerol	ER	Nicotine	0.94	224
[1,3-^{14}C]Glycerol	H	Nicotine	+	241
[2-^{14}C]Glycerol	ER	Nicotine	2.13	224
[2-^{14}C]Glycerol	H	Nicotine	+	241
[1-^{14}C]Glycine	ER	Nicotine	<0.1	224
[1-^{14}C]Glycine	H	Nicotine	+	232
[2-^{14}C]Glycine	ER	Nicotine	4.33	224
[2-^{14}C]Glycine	H	Nicotine	+	232

[1,2-^{14}C]Glycine	H	Nicotine	+	242
DL-[2-^{14}C]-5-Hydroxylysine	U	Anatabine	0	65
		Anatalline	0	
[3-^{14}C]-3-Hydroxy-3-methylglutaric acid	ER	Nicotine	<0.1	224
[6-^{14}C]-6-Hydroxynicotinic acid	ER	Nicotine	0	243
[^{15}N]-6-Hydroxynicotinic acid	ER	Nicotine	negligible	244
DL-[1-^{14}C]Leucine	H	Nicotine	+	232
DL-[2-^{14}C]Lysine	H	*N'*-Methylanabasine (added as a carrier)	0.0002	161
		Nicotine	0.003 [0.006]	
DL-[2-^{14}C]Lysine	U	Anatabine	0	65
		Anatalline	0	
DL-[2-^{14}C]Lysine	H	Nicotine	0	155
L-[U-^{14}C]Lysine	ER	Nicotine	0.0069	245
L-[ϵ-^{15}N]Lysine	ER	Nicotine	0.72	245
[1-^{14}C]Malonic acid	ER	Nicotine	0.08	224
[2-^{14}C]Malonic acid	ER	Nicotine	0.20	224
DL-[U-^{14}C]Methionine	H	Nicotine	+	232
L-[methyl-^{14}C]methionine	H	Nicotine	[2.1]	230
D-[methyl-^{14}C]methionine	H	Nicotine	[0.26]	230
[1-^{14}C]-4-Methylaminobutanoic acid	W	Nicotine	0	42
[^{14}C]Methylamine	H	Nicotine	+	333
[4-^{14}C]-4-Methylnicotinic acid	W	4-Methylnicotine (added as a carrier)	0	246
DL-[2, *N*-methyl-^{14}C]-α-*N*-Methylornithine	H	Nicotine	0.10	247
DL-[2, *N*-methyl-^{14}C]-δ-*N*-Methylornithine	H	Nicotine	1.25	247
[2-^{14}C]-*N*-Methyl-Δ^1-piperideinium chloride	H	Anabasine	1.28 [38]	161
		(2'*S*)-*N'*-Methyl anabasine	2.25 [102]	
		Nicotine	0.037 [0.03]	

Table 2. Continued

Compound Administered	Method of Feeding	Alkaloids or Related Compounds Labeled	Incorporation (%)	Reference
[1-^{13}C, methylamino-^{15}N]-*N*-Methylputrescine	W	Nicotine	4.0 [0.104]	248
		Nornicotine	0.07 [0.054]	
		Anabasine	0	
		Anatabine	0	
[methyl-^{14}C]-*N*-Methylputrescine	EN	*N*-Methyl-Δ^1-pyrrolinium salt	+	249
[6-^{3}H]-1-Methyl-2-pyridone-5-carboxamide	ER	Nicotine	0.1	244
[2-^{14}C]-*N*-Methyl-Δ^1-pyrrolinium chloride	H	Nicotine	18.2 [3.3]	250
DL-[2-^{14}C]Mevalonic acid	ER	Nicotine	<0.1	224
L-[2-^{14}C]Nicotianine (parallel experiments)	H	Nicotine	0.10	377
[2-^{14}C]Nicotinic acid (parallel experiments)	H	Nicotine	4.0	377
[2-^{14}C]Nicotinic acid + unlabeled nicotianine (parallel experiments)	H	Nicotine	3.4	377
[carboxyl-^{14}C]Nicotinate, ethyl	ER	Nicotine	negligible	255
[^{3}H]Nicotinamide	ER	Nicotine	13	244
(*RS*)-[2-^{14}C]Nicotine	ES	*N′*-Nitrosonornicotine	0.009	57
(2′*S*)-[^{14}C]Nicotine (obtained biosynthetically from [2-^{14}C]acetate) (3-day feed)	H	Nicotine (Activity detected in most α-amino acids)	7.0	251
(2′*S*)-[^{14}C]Nicotine (obtained biosynthetically from [^{14}C]aspartic acid) (3-day feed)	H	Nicotine (activity found in α-amino acids)	6.6	251

(2′*S*)-[^{14}C]Nicotine (obtained biosynthetically from [^{14}C]glutamic acid)	H	Nicotine (low activity in the α-amino acids)	7.0	251
(2′*S*)-[2′-^{14}C]Nicotine	CC	Nicotine	20–80	166
		Nornicotine (up to 80% in this alkaloid)	+	
(*RS*)-[methyl-^{14}C]Nicotine	CC	Extensive incorporation into polymeric material		166
(*RS*)-[methyl-^{14}C]Nicotine	ES	Carbon dioxide	+	252
		Scopolin (activity on OMe)	+	
(2′*S*)-[methyl-^{14}C]Nicotine	ES	Nicotine	[91.5–92.5]	253
(2′*R*)-[methyl-^{14}C]Nicotine (the above two labeled nicotines fed to leaves obtained from *N. tabacum* grafted on tomato)	ES	Nicotine (the reported incorporation is for a 3-day feeding)	[5.6]	253
(2′*S*)-[methyl-^{14}C]Nicotine	H	Nicotine	7 days: 1.1	254
		Nicotine	7 weeks: 1.3	
		Choline	[0.1]	
(2′*S*)-[2,5′-^{14}C]Nicotine (obtained biosynthetically from [2-^{14}C]ornithine	H	Nicotine	7 days: 5.5	254
		Nicotine	7 weeks: 4.4	
(*RS*)-[2′-^{14}C]Nicotine (alkaloids from fresh plants)	LA	Nicotine	4.62	46
		Nornicotine	7.00	
		N′-Isopropylnornicotine	0.003	
(2′*S*)-[2,5-^{14}C]Nicotine (fed 30 days)	ER	Nicotine	61	272
(2′*S*)-[pyridine-^{14}C]Nicotine (fed 30 days)	ER	Nicotine	90	272

Table 2. Continued

Compound Administered	Method of Feeding	Alkaloids or Related Compounds Labeled	Incorporation (%)	Reference
(*RS*)-[2′-^{14}C]Nicotine (alkaloids from excised leaves allowed to dry 4 weeks)	LA	Nicotine	3.95	46
		Nornicotine	3.16	
		N′-Isopropylnornicotine	0.018	
(*RS*)-[2′-^{14}C]Nicotine (alkaloids from excised leaves painted with 20% acetone)	LA	Nicotine	3.40	46
		Nornicotine	7.52	
		N′-Isopropylnornicotine	0.035	
[^{3}H]Nicotinic acid	ES	Nicotine	1.2	134
[^{14}C]Nicotinic acid	ER	Nicotine	9	244, 245, 256
[2-^{3}H]Nicotinic acid } parallel experiments	ER	Nicotine	9.3–11.3	244, 257, 258, 272
[2-^{3}H]Nicotinic acid + gibberellic acid } parallel experiments	ER	Nicotine	7.6–9.3	258
[4-^{2}H]Nicotinic acid	ER	Nicotine	13.0	244, 257
[5-^{3}H]Nicotinic acid	ER	Nicotine	14.2	244, 257
[6-^{3}H]Nicotinic acid	ER	Nicotine	1.1	244, 257
[2,3,7-^{14}C]Nicotinic acid	H	Nicotine	10.0	241
[2,3,7-^{14}C]Nicotinic acid	H	Nicotine	2.8	259
[carboxyl-^{14}C]Nicotinic acid	ER	Nicotine	negligible	255
[2-^{3}H]Nicotinic acid	H	Nicotine	[0.75]	260
[6-^{3}H]Nicotinic acid (fed 6 days to 12-week-old plant)	H	Nicotine	[0.074]	260
[2-^{3}H]Nicotinic acid	H	Nicotine	[1.64]	260
[6-^{3}H]Nicotinic acid (fed 6 days to 6-week-old plant)	H	Nicotine	[0.047]	260
[6-^{14}C, 2-^{3}H]Nicotinic acid (fed 5 days)	H	Nicotinic acid (94% retention of ^{3}H)	+	42

[6-^{14}C, 6-^{3}H]Nicotinic acid (fed 5 days)	H	Nicotinic acid (89% retention of ^{3}H)	+		42
	H	Nicotine (98% loss of ^{3}H)	+		42
[2-^{14}C]Nicotinic acid (fed 7 days)	W	Nicotine	2.13 [0.51]		32
		Nornicotine	0.06 [0.50]		
		Anabasine	0.017 [0.71]		
		Anatabine	0118 [1.05]		
		2,3′-Bipyridine	0.015		
		Nicotelline	0.0021		
(2′*S*)-[^{15}N]Nicotine	SI	Nicotine (considerable degradation occurs)	+		376
[2-^{14}C]Nicotinic acid (fed 7 days, then air dried for 32 days)	W	Nicotine	0.96 [0.81]		32
		Nornicotine	0.40 [0.61]		
		Anabasine	0.018 [0.91]		
		Anatabine	0.25 [1.83]		
		2,3′-Bipyridine	0.05		
		Nicotelline	0.0018		
[5,6-$^{13}C_2$, ^{14}C]Nicotinic acid	W	Nicotine	[0.07]		168
		Nornicotine	[0.06]		
		Anabasine	[0.27]		
		Anatabine	[0.41]		
			Roots	Leaves	
[6-^{14}C]Nicotinic acid	ER or ES	Nicotinic acid-*N*-glucoside	23.4	46.0	243
		6-Hydroxynicotinic acid	8.9	0	
		Nicotinic acid	7.5	11.1	

Table 2. Continued

Compound Administered	Method of Feeding	Alkaloids or Related Compounds Labeled	Incorporation (%)	Reference
[6-^{14}C]Nicotinic acid	ER	Nicotine (parallel experiments)	9.31	243
[6-^{14}C]Nicotinic acid + inactive nicotinic acid-*N*-glucoside	ER	Nicotine (parallel experiments)	8.12	243
[6-^{14}C]Nicotinic acid-*N*-glucoside	ER	Nicotine (parallel experiments)	9.0	243
[6-^{14}C]Nicotinic acid-*N*-glucoside + inactive nicotinic acid	ER	Nicotine (parallel experiments)	0.17	243
[^{15}N]Nitrate, potassium	H	Nicotine	+	170
		Nornicotine	+	
[^{15}N]Nitrate (parallel experiments)	H	Nicotine	[2.7]	261
[^{15}N]Nitrate + 2,4-dichlorophenoxyacetic acid (parallel experiments)	H	Nicotine	[0.29]	261
(*RS*)-[2′-^{14}C]Nornicotine	ES	*N*′-Nitrosonornicotine	0.007	57
(2′*S*)-[U-^{3}H]Nornicotine	ES	Nornicotine (38% recovered after 7 days)		27
(2′*S*)-[U-^{3}H]Nornicotine (in the above two experiments, the [^{3}H]nornicotine was fed to excised leaves of *N. tabacum* that were alkaloid-free, having been grown on a tomato rootstock)	ES	Nornicotine (90% recovered after 7 days)		27
[^{15}N]Nornicotine + L-[methyl-^{14}C]-Methionine (fed at same time)	H	Nicotine (except in one case, the enrichment with ^{15}N was very slow ~ 0.02% excess,	+	262

		compared with the specific incorporation of the ^{14}C)		
L-[1-^{14}C]Ornithine	EN	Ornithine decarboxylase activity measured		233
DL-[2-^{14}C]Ornithine	H	Nicotine	0.48–1.1	263, 264
DL-[2-^{14}C]Ornithine	H	N-Methyl-Δ^1-pyrrolinium salt	1.1	192
		Arginine	1.1	
DL-[2-^{14}C]Ornithine	ES	N-Methyl-Δ^1-pyrrolinium salt	0	192
		Arginine	0.21	
DL-[2-^{14}C]Ornithine	ER	"Topped" nicotine	6.9 [38.1]	228
	ER	"Nontopped" nicotine	2.2 [35.4]	
DL-[2-^{14}C]Ornithine	ER	Nicotine } parallel	15.8–25	258, 265, 272
DL-[2-^{14}C]Ornithine + gibberellic acid	ER	Nicotine } experiments	11.5–11.9	258, 265
DL-[2-^{14}C]Ornithine	CC	Caffeoylputrescine	+	266
		p-Coumaroylputrescine	+	
		Feruloylputrescine	+	
DL-[2-^{14}C,α-^{15}N]Ornithine	ER	Nicotine	^{14}C: [37] ^{15}N: [1.5]	267
DL-[2-^{14}C,δ-^{15}N]Ornithine	ER	Nicotine	^{14}C: [61] ^{15}N: [31]	267
L-[5-^{14}C]Ornithine	CC	Putrescine	+	233
		Caffeoylputrescine	+	
		Feruloylputrescine	+	
		Arginine	+	
L-[U-^{14}C]Phenylalanine	CC	Caffeoylputrescine	+	233
		Feruloylputrescine	+	
L-[U-^{14}C]Proline	H	Nicotine	0.032 [0.05]	240

Table 2. Continued

Compound Administered	Method of Feeding	Alkaloids or Related Compounds Labeled	Incorporation (%)	Reference
DL-[2-^{14}C]Proline	W	Nicotine	0.07	42
[1,4-^{14}C]Putrescine	H	Nicotine	0.12 [0.18]	240
[1,4-^{14}C]Putrescine	CC	Caffeoylputrescine	+	266
		Feruloylputrescine	+	
[1,4-^{14}C]Putrescine	ER	"Topped" nicotine	22.6 [45.8]	228
		"Nontopped" nicotine	8.8 [38.3]	
[2-^{14}C]Pyruvate, sodium	ER	Nicotine	0.48	224
[5-^{14}C]-Δ^1-Pyrroline-5-carboxylic acid	H	Nicotine	+	268
[2,3,7,8-^{14}C]Quinolinic acid	ER	Nicotine	0.78–0.83	224
[2,3,7,8-^{14}C]Quinolinic acid	H	Nicotine	5.6	241
[^{3}H]Quinolinic acid	H	Nicotine	4.6	241
[1,4-^{14}C]Succinic acid	ER	Nicotine	0.04	224
[2,3-^{14}C]Succinic acid	ER	Nicotine	0.42	224
DL-[1′-^{14}C]Tryptophan	W	4,6-Dihydroxyquinoline-2-carboxylic acid (6-Hydroxykynurenic acid)	+	269
DL-[3′-^{14}C]Tryptophan	H	Nicotine	negligible	270
DL-[7a-^{14}C]Tryptophan	H	Nicotine	negligible	271

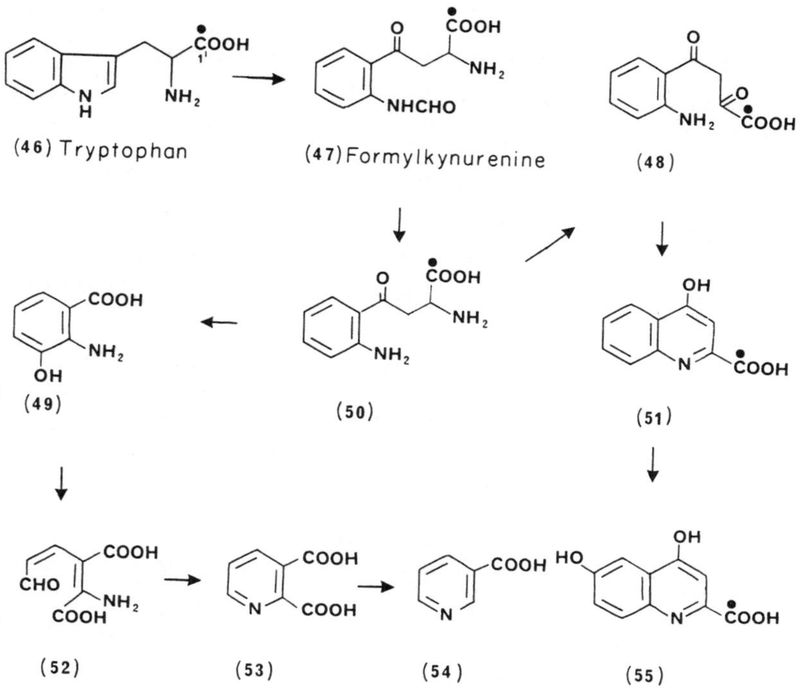

(46) Tryptophan (47) Formylkynurenine (48) (49) (50) (51) (52) (53) (54) (55)

Figure 2. Formation of nicotinic acid and related compounds from tryptophan.

source of the pyridine ring of all these alkaloids, and several investigators have discovered that it is not formed in *Nicotiana* species by the biosynthetic route that has been well established in microorganisms (e.g., *Neurospora*) and animals [i.e., from tryptophan (**46**) via formylkynurenine (**47**), kynurenine (**50**), 3-hydroxyanthranilic acid (**49**), the open-chain aldehyde-acid (**52**), and quinolinic acid (**53**), as illustrated in Fig. 2] [273]. Dawson and co-workers [256] and Tso and Jeffrey [217] independently established in 1956 that nicotinic acid serves as a precursor of the pyridine ring of nicotine. Later it was shown that the pyridine ring of anabasine arises from the same source [154]. Thus the origin of nicotinic acid in *Nicotiana* could be investigated by determining the pattern of labeling in the pyridine rings of these alkaloids after feeding a wide variety of small molecules labeled at specific positions. Complimentary experiments were carried out at the same time in the castor bean (*Ricinus communis*), which produces the alkaloid ricinine, also derived from nicotinic acid [274–276, 312]. The results that emerged led to the discovery of a new biosynthetic pathway for the formation of nicotinic acid in these species; this pathway also seems to operate in some bacteria such as *Mycobacterium tuberculosis* [277–279]. This tentative hypothesis is illustrated in Fig. 3. The primary precursors are considered to be 3-phospho-D-glyceraldehyde (**56**) and aspartic acid, which supplies the nitrogen of the ultimate nicotinic acid. The steps beyond these precursors are hypothetical

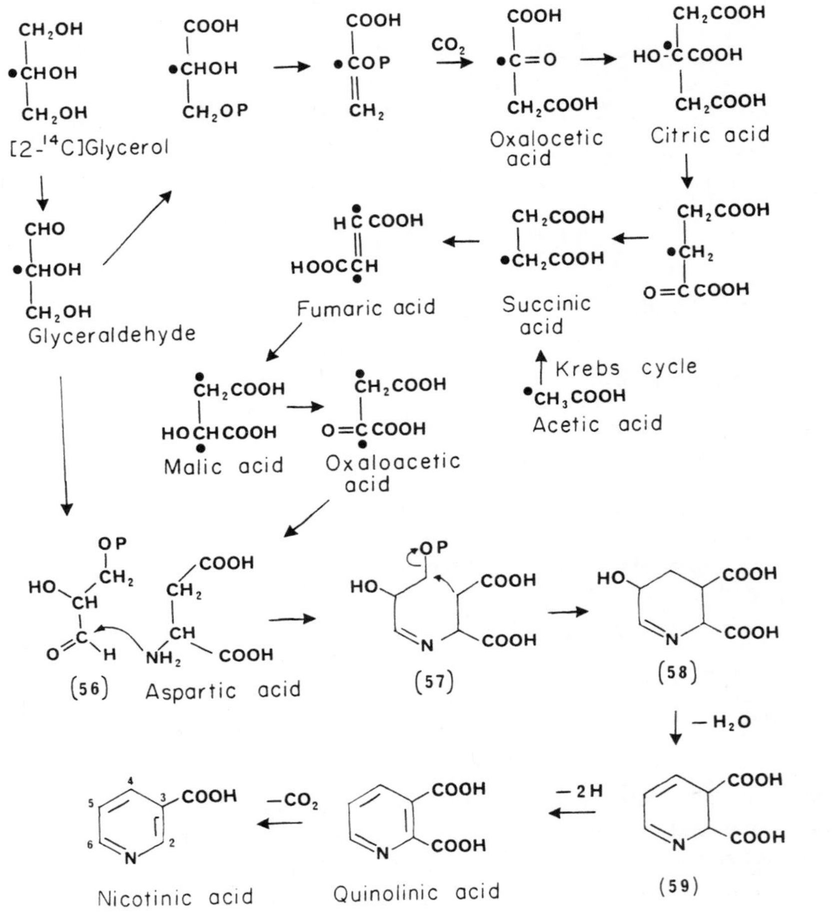

Figure 3. Biosynthesis of nicotinic acid in *Nicotiana*.

but involve plausible biochemical reactions that have precedence. The Schiff base (**57**) cyclizes to yield the tetrahydropyridine derivative (**58**). 2,3-Dihydroquinolinic (**59**) is formed by dehydration. The pyridine ring of quinolinic acid then results by a dehydrogenation. Nicotinic acid is then formed by decarboxylation of quinolinic acid, a reaction that has been shown to occur in plants [324, 336]. The numerous feedings of small molecules are consistent with this scheme and their incorporation via glyceraldehyde and aspartic acid [307]. Thus [2-^{14}C]acetate enters the Krebs cycle, resulting in the labeling of C(2) and C(3) of aspartic acid, which ultimately become C(2) and C(3) of the pyridine ring of nicotine and anabasine. Nicotine and anabasine derived from [2-^{14}C]glycerol are labeled as expected at C(5), but the substantial labeling at C(2) and C(3) requires explanation. A plausible metabolic sequence is illustrated in Fig. 3, a crucial step being the photosynthetic carboxylation of phosphoenolpyruvate to oxaloacetate, which then enters the Krebs cycle. The distribution of activity in the pyridine ring of nicotine obtained from *N. rustica* that had been fed [3-^{14}C]glyceraldehyde [206] was C(2), 13%; C(3), 14%; C(4), 55%; C(5), 4.9%; C(6), 13%. The high level of activity at C(4) indicates that the three-carbon unit that supplies C(4), C(5), and C(6) of nicotinic acid is incorporated unsymmetrically as indicated in Fig. 3. [4-^{14}C]Aspartic acid when fed to *N. tabacum* yielded nicotinic acid-*N*-glucoside with essentially all its activity on the carboxyl group [234]. Nicotine isolated from *N. glutinosa* after exposure to $^{14}CO_2$ for 3 h had the following distribution of activity: C(2), 10.0%; C(3), 9.2%; C(4), 25.0%; C(5), 24.2%; C(6), 26.7% [181]. This pattern of labeling is again consistent with the scheme illustrated in Fig. 3, equal labeling at C(4), C(5), and C(6) being expected from rapid incorporation of CO_2 into 3-phosphoglyceraldehyde via the Calvin photosynthetic cycle. Hess and Tolbert [308] found essentially uniform labeling in 3-phosphoglyceric acid after exposure of *N. tabacum* leaves to $^{14}CO_2$ for 1 min. The different level of activity at C(2) and C(3), which are derived from aspartic acid, could be because the plants contained a larger pool of nonlabeled aspartate at the time of feeding or because the $^{14}CO_2$ was incorporated into the internal carbons of aspartic acid at a slower rate than the formation of glyceraldehyde.

As pointed out in a recent review [280], there is much still to be learned about the origin of nicotinic acid in higher plants, and tobacco is a useful biochemical medium for studying this process in more detail. Even though tryptophan is not a precursor of nicotinic acid in tobacco, it is converted to 6-hydroxykynurenic acid (**55**) in this species [269], presumably via kynurenine, the α-keto acid (**48**), and kynurenic acid (**51**). Thus [1′-^{14}C]tryptophan affords labeled **55** with the bulk of the activity on its carboxyl group, as illustrated in Fig. 2.

When nicotinic acid is incorporated into nicotine, its carboxyl group is lost [255], presumably as carbon dioxide. A particulate fraction that contained mitochondria was isolated from *N. rustica* roots which catalyzed the decarboxylation of nicotinic acid [290], the assay being accomplished by measuring the release of $^{14}CO_2$ from [carboxy-^{14}C]nicotinic acid. Oxygen was required for enzymatic activity. No cofactors for this decarboxylation could be established. Crude extracts of the leaves and stems of tobacco yielded fractions that had only 1–5% the activity of the root extract. Quinolinic acid was not decarboxylated in this system. Several investigators have established [168, 241, 259] that the point of attachment of the second heterocyclic ring (pyrrolidine, piperidine, or piperideine) is always at C(3), where the carboxyl group was attached (with no reaction at C(5), which could hypothetically yield the same products.

The immediate precursor of the *N*-methylpyrrolidine ring that reacts with some derivative of nicotinic acid is the *N*-methyl-Δ^1-pyrrolinium salt (**60**), which is the cyclized form of 4-methylaminobutanal (Fig. 4). This compound labeled with ^{14}C was isolated from *N. rustica* plants that had been fed [2-^{14}C]ornithine [192], and nicotine isolated from *N. tabacum* that had been fed [2-^{14}C]-*N*-methyl-Δ^1-pyrrolinium chloride was labeled solely at C(2′) of its pyrrolidine ring [250]. The piperidine ring of anabasine is derived from Δ^1-piperideine (**62**) [171], which is a metabolite of lysine. Surprisingly the tetrahydropyridine ring of anatabine is

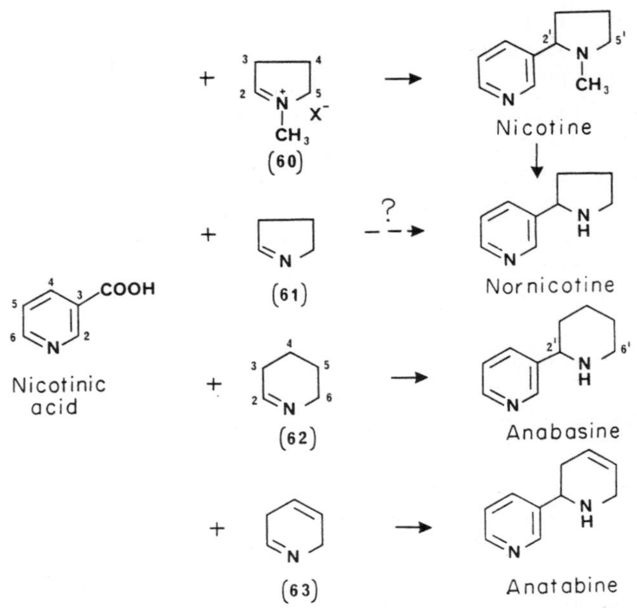

Figure 4. Biosynthetic origin of the second heterocyclic ring of the tobacco alkaloids.

not formed from lysine or its derivatives, and we discovered in 1975 [32, 183] that both rings of this alkaloid are derived from nicotinic acid. The immediate precursor of the tetrahydropyridine ring is considered to be 2,5-dihydropyridine (**63**). The main biosynthetic route to nornicotine is by the demethylation of nicotine, although some of the ambiguities regarding the biosynthesis of this alkaloid could be explained by postulating a direct formation from nicotinic acid and Δ^1-pyrroline (**61**).

The mechanism whereby nicotinic acid is activated at C(3) for reaction with these immediate precursors of the second heterocyclic ring has been the subject of much conjecture [136, 139, 265, 281, 282]. An intermediate is required that is nucleophilic at C(3) for reaction with the electron-deficient C(2) positions of the imines or iminium salts (**60–63**). Wenkert [281] suggested that nicotinic acid forms an *N*-glycoside, for example the *N*-ribosyl derivative (**64**). This pyridinium derivative could then undergo reaction at C(6) with one of the hydroxyl groups of the carbohydrate moiety to yield a 1,6-dihydropyridine derivative (**65**), which is now an enamine, being nucleophilic at C(3) for reaction with **60** to afford **66**. Subsequent steps involve decarboxylation and removal of the carbohydrate residue from **69** (Fig. 5). Nicotinic acid-*N*-glucoside has been found in tobacco [234, 243], and it is formed from nicotinic acid. However, in controlled experiments [243] it does not seem to be a more efficient precursor of nicotine than nicotinic acid. The pyridine moiety of nicotinamide adenine dinucleotide (NAD) was also incorporated in nicotine in *N. rustica*, but again the efficiency of its incorporation was about the same or less than that of nicotinic acid [215].

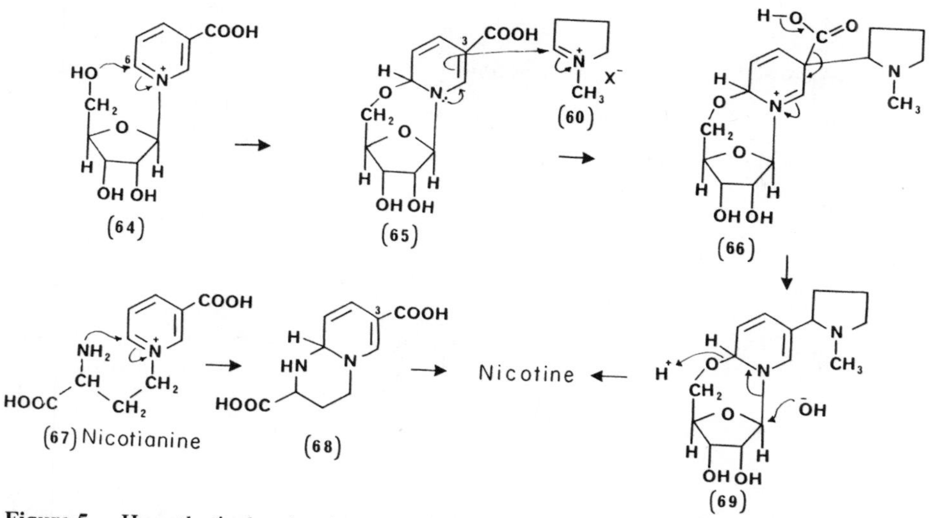

Figure 5. Hypothetical activation of C-3 of nicotinic acid by formation of a quaternary salt.

Another nicotinic-acid-derived quaternary salt that has been found in tobacco [283] is nicotianine (**67**). Activation at C(3) could be achieved in a similar way by formation of the intermediate **68**. However, nicotianine was found to be a much poorer precursor of nicotine than nicotinic acid, and the addition of unlabeled nicotianine had little effect on the incorporation of nicotinic acid into nicotine [377]. This biosynthetic scheme does not account for the removal of hydrogen from C(6) of nicotinic acid, which occurs during the formation of nicotine [244, 257, 260], anabasine [169], and anatabine [169]. This loss of hydrogen, detected by labeling with tritium at C(6), is not the formation of 6-hydroxynicotinic acid, since this compound failed to serve as a precursor of nicotine [243, 244]. We believe that the removal of hydrogen is an integral part of the biosynthetic scheme that ultimately yields nicotine and the other alkaloids, since we discovered [42] that [6-^{14}C, 6-^{3}H]nicotinic acid is incorporated into nicotine with a 98% loss of tritium relative to ^{14}C. However, the nicotinic acid recovered from the plant at the end of the feeding experiment had retained 89% of its tritium.

Our current hypothesis [284] for the biosynthesis of nicotine, which rationalizes the loss of hydrogen from C(6) and also accommodates the formation of anatabine, is illustrated in Fig. 6. The fate of the hydrogen at C(6) of nicotinic acid during this sequence is indicated by means of a tritium label (T). We propose that the nicotinic acid is activated by reduction to 3,6-dihydronicotinic acid (**70**). The reaction is stereospecific at C(6), the incoming hydrogen (source unknown) entering in the *pro-R* position. We suggest, on the basis of stereoelectronic arguments, that the hydrogen introduced at C(3) is cis to the one introduced at C(6). Compound **70** is so far hypothetical, but it is an attractive intermediate

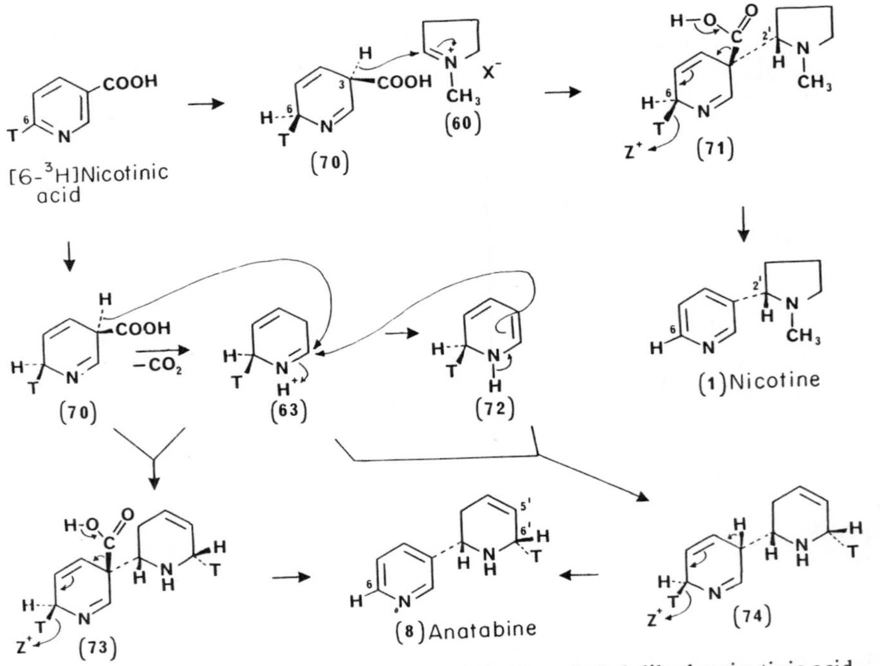

Figure 6. Biosynthesis of nicotine and anatabine via 3,6-dihydronicotinic acid.

for several reasons. The hydrogen at C(3) would be acidic by virtue of the adjacent carboxyl group and the imine function. This position is analogous to the α position in a β-keto acid. Thus a carbanion would be readily generated for reaction with the *N*-methyl-Δ^1-pyrrolinium salt (**60**) and the imines depicted in Fig. 4. Futhermore, **70**, being a β-imino acid, would be expected to undergo facile decarboxylation to yield 2,5-dihydropyridine (**63**) required for the formation of the Δ^3-piperideine ring of anatabine. We propose that the reaction of the carbanion of **70** with **60** is stereospecific (enzyme controlled), leading to the *S* configuration at 2′, the carbon that ultimately becomes C(2′) of nicotine. The product (**71**) then undergoes a concerted decarboxylation and loss of the *pro-S* hydrogen at C(6) to a hydride acceptor (e.g., NAD^+). This reaction is a *syn* elimination, which is characteristic for nonbiochemical 1,4 eliminations in model systems [285]. Reaction of **70** with 2,5-dihydropyridine (**63**) affords **73**, which yields anatabine (**8**) by a similar oxidative decarboxylation. The stereochemistry at C(6′) was determined by means of a degradation by which N′, C(6′), and C(5′) yielded chiral glycine, the stereochemistry of which was elucidated enzymatically [169]. By this means the initial reduction of the nicotinic acid was deduced to be stereospecific, as previously discussed. A slight modification of this biosynthetic scheme for anatabine is also illustrated in Fig. 6. We propose that anatabine is formed by a stereospecific dehydrogenation of **74**, which could arise by a condensation between 2,5-dihydropyridine (**63**) and the enamine, 1,2-dihydropyridine (**72**), formed from **63** by a tautomeric shift of hydrogen. Strong support for this type of activation of the pyridine ring of nicotinic acid was provided by biomimetic syntheses of anatabine and nicotine from dihydropyridines, which are illustrated in Fig. 7. Treatment of baikiain (**75**) with sodium hypochlorite yielded the *N*-chloro compound **76**, which underwent decarboxylation and loss of chloride ion to yield 2,5-dihydropyridine, some of which oxidized to pyridine, but some dimerized, presumably by the second mechanism discussed previously, to anatabine (26% yield at pH 10) [286]. This biomimetic synthesis of anatabine was shown to be completely regiospecific by labeling the baikiain with ^{13}C and deuterium at C(2). The ultimate anatabine was labeled with ^{13}C and deuterium only at the C(2) and C(2′) positions [287]. In particular, there was no scrambling of isotopes between the C(2) and C(6) positions of the pyridine ring of anatabine. This result is consistent with the *in vivo* experiments carried out in *Nicotiana* species. Thus [2-^{14}C]nicotinic acid yielded anatabine labeled equally at its C(2) and C(2′) positions [32]. Analogous results were obtained after feeding [6-^{14}C]nicotinic acid and [5,6-$^{13}C_2$, ^{14}C]nicotinic acid [168, 183]. Dihydropyridines are generated by reaction of glutadialdehyde (**77**) with ammonia. When this condensation was carried out in the presence of air (O_2) and the *N*-methyl-Δ^1-pyrrolinium salt (**60**), nicotine was formed [288]. When 3-methylglutadialdehyde (**78**) was used in this reaction, 4-methylnicotine (**79**) was produced [246].

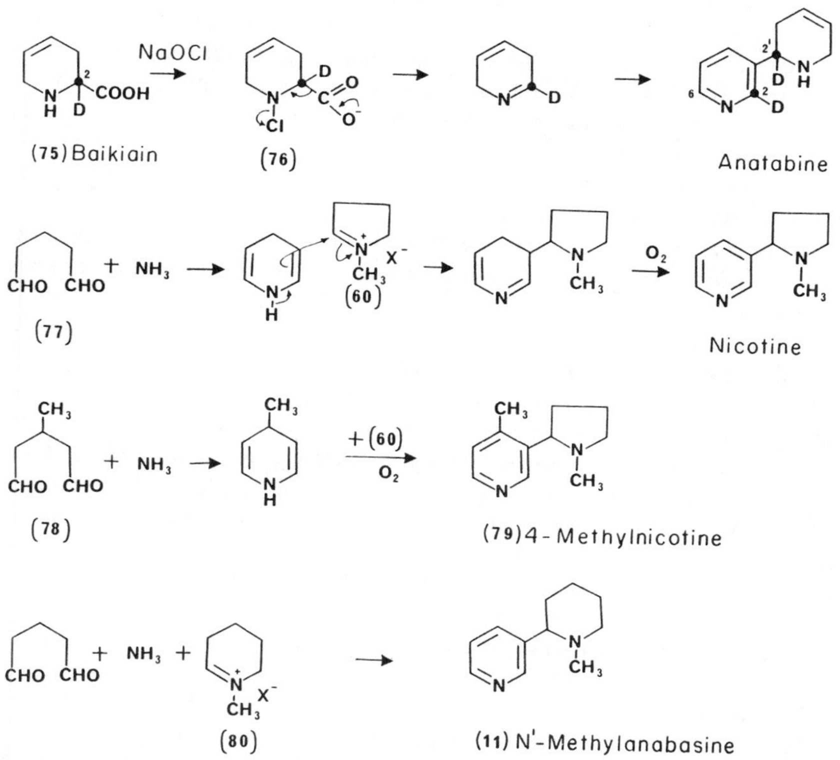

NaOCl
COOH
(75) Baikiain
(76)
Anatabine
CHO CHO + NH₃
(77)
(60)
O₂
Nicotine
(78)
+ (60)
(79) 4-Methylnicotine
(80)
(11) N′-Methylanabasine

Figure 7. Biomimetic syntheses of anatabine, nicotine, and analogs.

Glutadialdehyde, ammonia, and the higher homolog of **60**, i.e., *N*-methyl-Δ^1-piperideinium chloride (**80**), yielded *N'*-methylanabasine (**11**) [42].

Other dihydronicotinic acids were considered as intermediates in the biosynthesis of the tobacco alkaloids but were rejected for several reasons. Reduction at the 1,4 positions is, of course, common in the pyridine nucleotides. However, such intermediates would not account for the stereospecific loss of hydrogen from the C(6) position of nicotinic acid. Also, the presence of NADH and NADPH in all species is probably a good reason for rejecting them as precursors of nicotine, since this alkaloid is certainly not ubiquitous in all species. 1,6-Dihydronicotinic acid is a fairly viable candidate for consideration. However, 2,5-dihydropyridine, required for the biosynthesis of anatabine, cannot be derived from this compound by a reasonable mechanism. We have established that 1,2-dihydropyridine does not equilibrate to 2,5-dihydropyridine in aqueous solution [287]. Molecular-orbital calculations indicate that the former dihydropyridine is thermodynamically more stable [289].

6. PRECURSORS OF THE *N*-METHYLPYRROLIDINE RING OF NICOTINE

The various compounds that are involved in the biosynthesis of this ring are illustrated in Fig. 8. The immediate precursor of the *N*-methyl-Δ^1-pyrrolinium salt (**60**)—or its open-chain form 4-methylaminobutanal (**88**)—is *N*-methylputrescine (**87**). This compound is incorporated directly into nicotine without loss of the *N*-methyl group or equilibration of carbons 1 and 4 [213, 214, 248]. The enzyme responsible for the oxidation was isolated from *N. tabacum* roots [249] and is called *N*-methylputrescine oxidase. The pH optimum for the enzyme is 8.0; its activity is reduced to half at pH 7.2 and pH 8.7. Complete inhibition of the enzyme occurred on treatment with diethyldithiocarbamate, a reagent that is usually considered to be an inhibitor of copper-containing enzymes. The enzyme was found only in the roots of *N. tabacum*, the level of activity being increased by prior decapitation ("topping") of the tobacco plants [227]. The enzyme level reached a maximum 24 h after decapitation. Putrescine and cadaverine are oxidized to Δ^1-pyrroline and Δ^1-piperideine respectively by this enzyme. Other primary amines, histamine, spermine, tyramine, and hexylamine are unaffected.

N-Methylputrescine is formed by the *N*-methylation of putrescine (**86**), the methyl group being derived from the usual source, *S*-adenosyl-L-methionine. The enzyme catalyzing this reaction (putrescine-*N*-methyltransferase) was also isolated from *N. tabacum* roots [226, 227]. Putrescine-*N*-methyltransferase was found in *Nicotiana* cell-suspension cultures that produced nicotine (up to 3.4% on a dry-cell basis), but was absent in cell lines that did not produce nicotine [368]. The first tracer work (1952) on the biosynthesis of nicotine was carried out by Byerrum and co-workers, who showed that the *N*-methyl group of nicotine is derived from the usual sources of one-carbon units: the methyl groups of betaine [195] and choline [195, 201], formaldehyde [203], formate [202], C(2) of glycine [207, 224, 230, 232, 242], C(2) of glycolic acid, [208], the methyl group of

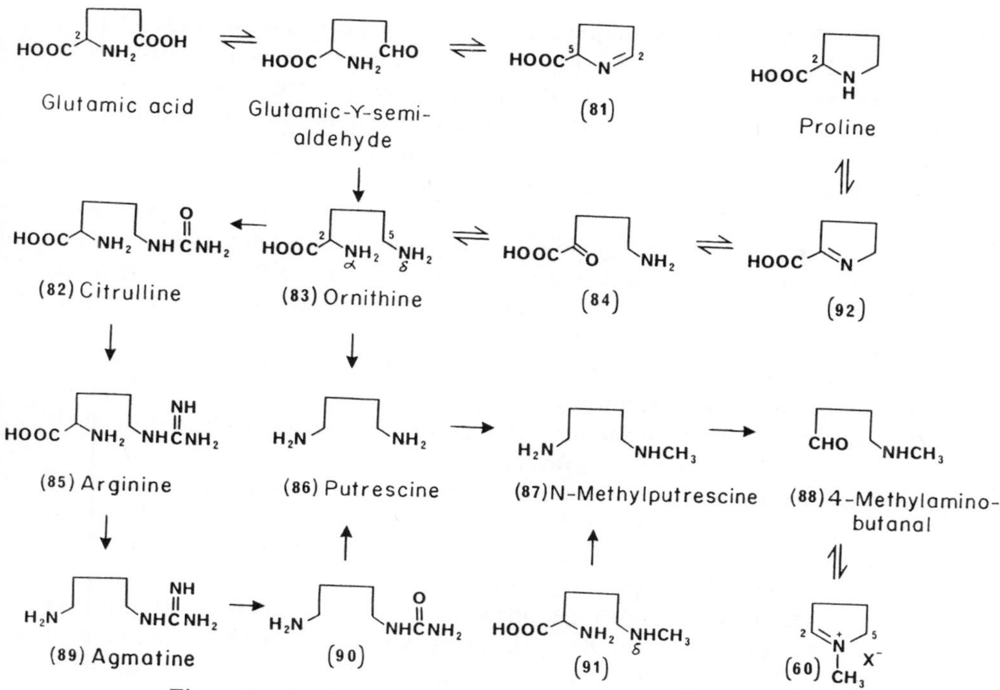

Figure 8. Precursors of the *N*-methyl-Δ^1-pyrrolinium salt (**60**).

methionine [202, 210, 192], and C(2) of serine [203]. Some unnatural compounds served as sources of the *N*-methyl group, e.g., 2-chloroethyltrimethylammonium chloride [200], 3-methyl-4-methylthiolbutanoic acid (β-methionine) [230]. *N*-Methylputrescine can also be formed in *N. tabacum* by the decarboxylation of δ-*N*-methylornithine (**91**). However, this amino acid is not considered to be on the normal pathway between ornithine and *N*-methylputrescine, since the formation of such an intermediate would prevent equilibration of carbons 1 and 4 of *N*-methylputrescine. Thus the feeding of [2-^{14}C]-δ-*N*-methylornithine to tobacco yielded nicotine that had all its activity located at C(2′) with none at C(5′) [247]. In some other species that produce **60** from ornithine (*Atropa belladonna*, *Datura stramonium*), δ-*N*-methylornithine is a true intermediate in the sequence, resulting in unsymmetrical labeling of the pyrrolinium salt and ultimate alkaloids (hygrine, hyoscyamine) that are derived from [2,^{14}C]ornithine [291, 292].

In higher plants it has been shown that putrescine is formed by two routes: one is by the direct decarboxylation of ornithine. The enzyme catalyzing this reaction, ornithine decarboxylase (EC 4.1.17), has been extensively studied in mammalian systems since the product, putrescine, is a precursor of the important polyamines: spermine and spermidine [293, 294]. It has been isolated from *N. tabacum* roots [227], the amount of enzyme increasing in decapitated plants. The second route is from arginine (**85**) via agmatine (**89**) and *N*-carbamylputrescine

(**90**) [295, 296]. The relative importance of these two pathways in tobacco is not clear. Yoshida and Mitake [228] found that arginine and the intermediates **89** and **90** served as precursors of nicotine when applied to excised roots. However, ornithine was also an efficient precursor when fed under the same conditions. Similar results were obtained with tobacco leaf disks [297]. In cell cultures of *N. tabacum*, both routes to putrescine were found to occur [233]. In these particular cell lines, nicotine was not formed, but the putrescine was conjugated with hydroxylated cinnamic acids, yielding caffeoyl and feruloyl putrescine. Arginine decarboxylase and ornithine decarboxylase are two quite distinct enzymes. The arginine decarboxylase was not inhibited by a 100-fold excess of ornithine and visa versa, proving that the enzymes were not competing for the same substrate. Both decarboxylases were inhibited by the products of decarboxylation, ornithine decarboxylase by putrescine, arginine decarboxylase by agmatine. The most effective inhibitor of the ornithine decarboxylase was α-difluoromethylornithine, an enzyme-activated irreversible inhibitor of ornithine decarboxylase. However, in contrast with animal systems, this inhibitor did not reduce the putrescine pool, indicating that the cells compensated by increased synthesis of putrescine from arginine. Arginine is of course produced in nature from ornithine via citrulline (**82**); however, in the experiments with tobacco cell cultures [233], there was little interconversion of these two amino acids during a 5-day incubation. Potassium-deficient plants contain an increased amount of putrescine [372, 373]. Also the enzymes responsible for its formation from arginine increase [374, 375]. However, there have been no reports of increased levels of nicotine in potassium-deficient tobacco plants [6].

The feeding of [2-^{14}C]ornithine of [5-^{14}C]ornithine to *Nicotiana* species yields nicotine that is equally labeled at the C(2′) and C(5′) positions [174, 178, 185, 218, 219, 263–265, 355]. This is the expected result if the biosynthesis proceeds via free putrescine. The incorporation of ornithine via such a symmetrical intermediate was confirmed by feeding [2,3-$^{13}C_2$]ornithine to *N. glutinosa*, the contiguous ^{13}C atoms on either side of the pyrrolidine ring of nicotine giving rise to satellites in the ^{13}C-NMR spectrum of the alkaloid [186]. When [2-^{14}C, δ-^{15}N]ornithine was administered to excised roots of *N. tabacum*, the specific incorporation of the ^{15}N was half that of the ^{14}C [267], a result consistent with the intermediate formation of putrescine, which later loses half its nitrogen in the formation of the pyrrolinium salt **60**. However, nicotine derived from [2-^{14}C, α-^{15}N]ornithine retained only 4% (instead of the expected 50%) of the ^{15}N relative to ^{14}C. This result was rationalized by postulating that ornithine undergoes a more ready transamination of the α-amino group—to yield α-keto-δ-aminovaleric acid (**84**)—than at the δ-amino group. Spenser and co-workers [298] have shown that ornithine is converted to proline via **84** in *N. tabacum*, indicative of facile transamination at the α position. Nicotine derived from [2-^{14}C]glutamic acid [204, 240], [2-^{14}C]proline [42], and [5-^{14}C]-Δ^1-pyrroline-5-carboxylic acid (**81**) [221, 268] was also symmetrically labeled in its pyrrolidine ring. In view of the low incorporation of these compounds relative to ornithine, it is proposed that they are converted to putrescine via ornithine by accepted biochemical pathways, illustrated in Fig. 8. Early reports on the incorporation of α-*N*-methylornithine into nicotine [213] were discounted [247], and it was shown that this amino acid is probably demethylated to ornithine prior to its conversion to nicotine. 4-Aminobutanoic acid failed to serve as a precursor of **60** [192], and the administration of [1-^{14}C]-4-methylaminobutanoic acid to *N. tabacum* failed to yield labeled nicotine [42].

For several years Rapoport and co-workers [173, 174, 176, 177, 180] have been reporting that the short-term feeding of $^{14}CO_2$ to *N. glutinosa* plants affords nicotine that is labeled unsymmetrically in the pyrrolidine ring. Such a result is not in accord with the tracer work involving ornithine and related compounds that has just been described. Rapoport has suggested that his results favor an alternate pathway from CO_2 to the *N*-methyl-Δ^1-pyrrolinium salt **60**, and the route via ornithine and putrescine is an "aberrant" one. This work of Rapoport has been reviewed extensively [299, 300]. The author suggested that the unusual pattern of labeling in the pyrrolidine ring that Rapoport obtained was due to a systematic error in the degradative scheme that was used to determine the distribution of activity [301]. However, it was subsequently claimed that the degradation used was valid [302]. Byerrum and co-workers [179] carried out feeding experiments with $^{14}CO_2$ using both *N. glutinosa* and *N. rustica* plants and obtained essentially uniform labeling of the pyrrolidine ring of nicotine. When [2-^{14}C]acetate was fed to *N. rustica* for only 2 h, the pyrrolidine ring was also symmetrically labeled. The distribution of activity [C(2′), 2.6%; C(3′), 16.4%; C(4′), 15.6%, C(5′), 1.8%] is consistent with the incorporation of acetate via Krebs cycle intermediates (α-ketoglutaric acid) and glutamic acid. There was hope that these conflicting results could be explained by carrying out a feeding experiment with $^{13}CO_2$ and then determinating of the distribution of isotope by ^{13}C-NMR spectroscopy, a nondegradative method. However, the recent work of Hutchinson and co-workers [237, 238] is equivocal. A chemical degradation carried out on nicotine derived from [^{13}C, ^{14}C]carbon dioxide was found to be equally labeled with ^{14}C at C(2′) and C(5′) [185]. However, analysis of the ^{13}C-NMR spectrum of this sample of nicotine and others apparently indicated that there were more contiguous ^{13}C atoms at C(2′) and C(3′) than at C(4′) and C(5′). I find this hard to believe, and I suspect that more intense satellites in the ^{13}C-NMR spectrum (due to coupling of contiguous ^{13}C atoms) on the side attached to pyridine is due to some other unknown factor.

7. PRECURSORS OF THE PIPERIDINE RING OF ANABASINE

The piperidine ring of anabasine is derived from lysine via intermediates that do not allow equilibration of carbons 2 and 6 of the amino acid. Thus [2-^{14}C]lysine when fed to *N. glauca* plants yielded anabasine (**2**), which had all its activity at C(2′) [145, 149, 155]. Anabasine derived from [4,5-$^{13}C_2$, 6-^{14}C]lysine was labeled with ^{13}C at the C(4′) and C(5′) positions (detected by ^{13}C-NMR), and the amount of ^{14}C at C(2′) was negligible (0.2%) [157]. L-Lysine (**93**) is a much more efficient precursor of the alkaloid that the D isomer [156, 159]. Anabasine derived from [2-^{14}C, ϵ-^{15}N]lysine retained most (80%) of the ^{15}N in the piperidine ring relative to ^{14}C. On the other hand, only 6% of the ^{15}N was retained in the anabasine formed from [2-^{14}C, α-^{15}N]lysine [158]. Thus the conversion of lysine to Δ^1-piperideine (**62**) (shown to be a precursor of the piperidine ring of anabasine [171]), unlike the conversion of ornithine to the *N*-methyl-Δ^1-pyrrolinium salt **60**, cannot proceed via a symmetrical intermediate such as cadaverine. [1,5-^{14}C]Cadaverine was incorporated into the piperidine ring of anabasine [149]. However, this is presumably converted to Δ^1-piperideine by a nonspecific amine oxidase such as *N*-methylputrescine oxidase (see Section 6). Initially, it was proposed that lysine is incorporated via ϵ-amino-α-keto caproic acid (**94**) and Δ^1-piperideine-2-carboxylic acid (**95**), established precursors of pipecolic acid (**96**) (Fig. 9) [303]. However, Gupta and Spenser [304] showed that [6-^{14}C, 2-^3H]lysine was incorporated into anabasine and sedamine [103], an alkaloid of *Sedum acre*, with almost complete retention of tritium relative to the ^{14}C. The α-keto acid **94** thus cannot be an intermediate between lysine and Δ^1-piperideine. These authors proposed [304] that unsymmetrical incorporation of lysine was achieved by methylation on the ϵ-amino group to yield ϵ-*N*-methyllysine (**97**), which underwent decarboxylation to yield *N*-methylcadaverine (**98**). A stereospecific oxidation of the primary amino group could then yield 5-methylaminopentanal (**99**) with retention of tritium on the aldehyde carbon. The

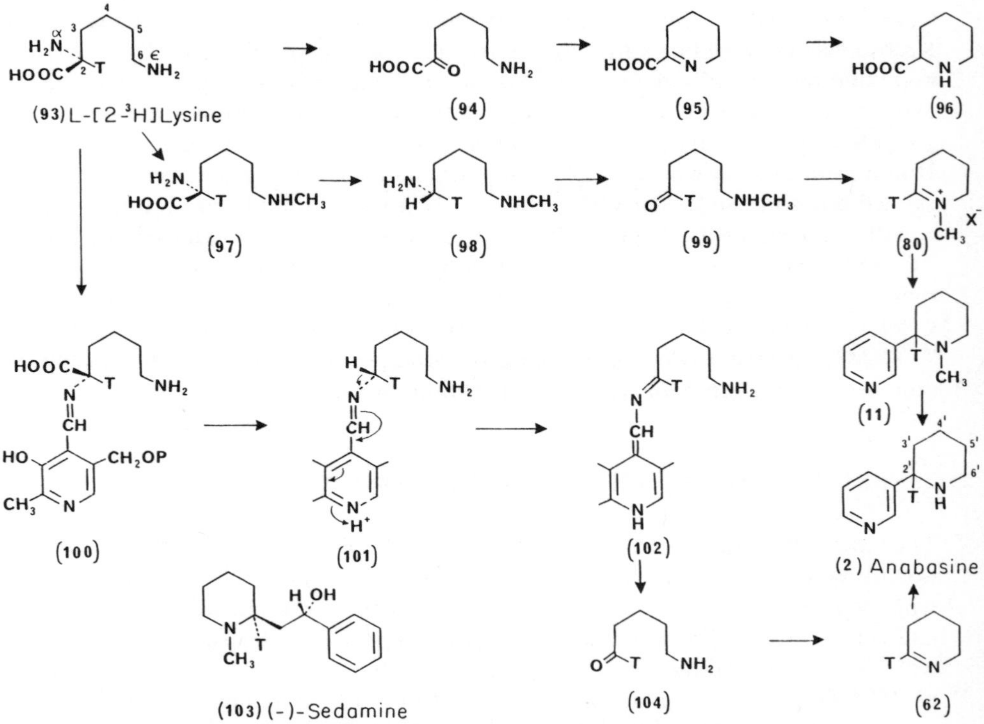

Figure 9. Biosynthetic steps from lysine to anabasine.

cyclized form of **99**, namely, the *N*-methyl-Δ^1-piperideinium salt (**80**), would then condense with nicotinic acid to yield *N'*-methylanabasine (**11**). Anabasine would then be formed by *N* demethylation, analogous to the formation of nornicotine from nicotine (see Section 8). Dawson has, in fact, demonstrated that *N'*-methylanabasine is converted to anabasine in *N. glutinosa* [305]. This hypothesis was tested by feeding [2-^{14}C]-*N*-methyl-Δ^1-piperideinium chloride to *N. tabacum* and *N. glauca* [161]. The alkaloids isolated from *N. tabacum* were nicotine (essentially inactive), anabasine (1.3% absolute incorporation), and *N'*-methylanabasine (2.25% incorporation). However, the *N'*-methylanabasine isolated from both *N. tabacum* and *N. glauca* had the same specific activity as the administered **80**. The conclusion was drawn that the incorporation of **80** to afford *N'*-methylanabasine represented an aberrant synthesis from an unnatural precursor. Although small amounts of *N'*-methylanabasine have been reported in *N. tabacum* [36], none could be detected (by GLC) in the crude alkaloids of *N. tabacum* and *N. glauca* plants that had not been fed **80**. Also, no activity was detected in *N'*methylanabasine that was added as a carrier to *N. tabacum* and *N. glauca* plants that were fed [2-^{14}C]lysine. Leistner and Spenser [160] fed (methyl-^{14}C]methionine along with unlabeled *N'*-methylanabasine to *N. glauca* plants in order to determine whether it was an intermediate. However, the recovered *N'*-methylanabasine was inactive. These authors proposed [160] that lysine forms a Schiff base (**100**) with pyridoxal phosphate. A stereospecific decarboxylation [306] yields bound cadaverine (**101**) that is attached to the pyridoxal by the amino group that was originally the α-amino group of lysine. A stereospecific dehydrogenation involving loss of the *pro-S* at C(1) yields **102**. 5-Aminopentanal (**104**) retaining all the tritium originally at C(2) of lysine is then formed on hydrolysis. We proposed a direct conversion of **101** to **103** by a concerted oxidative decarboxylation [161]. However, Spenser claims that there are no biochemical examples of the direct conversion of an α-amino acid to an aldehyde. Experiments of Griffith and Griffith [145] are consistent with the formation of lysine in *N. glauca* by the diaminopimelic acid pathway, involving the initial condensation of pyruvate (derivable from alanine) and aspartic acid [378], rather than the route via α-aminoadipic acid (involving the initial condensation of acetyl coenzyme A with α-ketoglutaric acid), which is mainly confined to bacteria.

Someone once suggested that anabasine could be formed from nicotine by an enlargement of the pyrrolidine ring and incorporation of the *N*-methyl group into the piperidine ring, and some evidence appeared that apparently substantiated this rather bizarre idea [163, 164]. However, several workers [31, 162, 165] showed that the anabasine isolated from *N. glauca* that had been fed radioactive nicotine was essentially inactive after careful purification.

8. BIOSYNTHESIS OF NORNICOTINE

The amount of nornicotine in tobacco varies with the species and the variety and is also a function of the method of processing of the tobacco leaf. Fresh leaves of

N. glutinosa contain much more nornicotine relative to nicotine than typical samples of *N. tabacum*, and the amount increases in both species after the leaves are harvested and subjected to curing. Generally accepted is the idea that nornicotine is formed by the demethylation of nicotine, a reaction that occurs in both the living plant and during curing and drying of tobacco leaf [6, 31, 32, 170, 198, 309]. Dawson showed that nicotine translocated from the roots of *N. tabacum* was demethylated in an aerial graft of *N. glauca* [310, 311]. Experiments were later carried out in excised leaves of *N. glutinosa*, which were rendered alkaloid-free by grafting on tomato roots [305]. The dealkylation was observed to be nonspecific for the natural (2′*S*)-nicotine. Nornicotine was also produced from (*RS*)-nicotine, (2′*R*)-nicotine, and (*RS*)-*N*′-ethylnornicotine. Anabasine also was formed from (*RS*)-*N*′-methylanabasine (*RS*)-*N*′-ethylanabasine. Kisaki and Tamaki [313, 314] found that pure (2′*S*)-nicotine is partially racemized when it is converted to nornicotine in *N. tabacum* leaves that had been made alkaloid-free by grafting on tomato roots. A surprising observation [315] is that nornicotine isolated from the roots of *N. tabacum* is largely the 2′*R* epimer. When (*RS*)-nicotine was fed to the alkaloid-free *N. tabacum* leaves, the reisolated nicotine was found to become progressively more levorotatory. On the other hand, (*RS*)-nornicotine did not undergo any change in rotation when fed to the same system. However, when (*RS*)-nornicotine was fed to intact *N. tabacum* plants, it became progressively dextrorotatory (excess of the 2′*R* isomer) [27] and the suggestion was made that this was due to the more rapid metabolism of the 2′*S* isomer to myosmine. When (2′*R*)-nicotine was fed to alkaloid-free *N. tabacum* leaves, some demethylation to nornicotine took place, and again there was partial racemization. The recovered (2′*R*)-nicotine was optically pure [253]. Also (2′*R*)-[*N*-methyl-^{14}C]nicotine was observed to be demethylated faster than the natural 2′*S* isomer [253]. In an effort to throw some light on these puzzling observations, (2′*S*)-nicotine labeled with tritium at the chiral center (C-2′) was mixed with an equimolecular amount of (2′*RS*)-[2′-^{14}C]nicotine and administered to intact *N. glauca* plants for 3 days [31]. The ^{3}H:^{14}C ratio in the (2′*S*)-[2′-^{14}C, 2′-^{3}H]nicotine fed was 12.6. The nornicotine isolated from the plant was almost racemic (48% of the 2′*S* isomer) and had a ^{3}H:^{14}C ratio of 11.7 (93% retention of ^{3}H). The (2′*R*)-nornicotine was devoid of tritium. Thus if any (2′*R*)-nornicotine is formed from (2′*S*)-nicotine, the hydrogen at the chiral center C(2′) is lost during this transformation. This experiment thus did not solve the problem of why nornicotine becomes partially racemized during the demethylation of (2′*S*)-nicotine, but it did eliminate a hypothetical scheme proposed to explain this phenomenon [31]. Several investigators have suggested that the demethylation of nicotine proceeds via its 1′-*N*-oxide [316–319]. A Cope-type elimination would afford the unsaturated compound **105** (Fig. 10). Elimination of water from this hydroxylamine yields the Schiff base **107**, which on hydrolysis affords formaldehyde and normetanicotine (**106**). Cyclization of this nonchiral compound could then yield (*R*)- or (*S*)-nornicotine. This reaction could be enzyme controlled to give an excess of the *S* isomer. This hypothesis is unacceptable since the hydrogen originally at C2′ in (2′*S*)-nicotine is still present in the (2′*R*) nornicotine. Other workers have

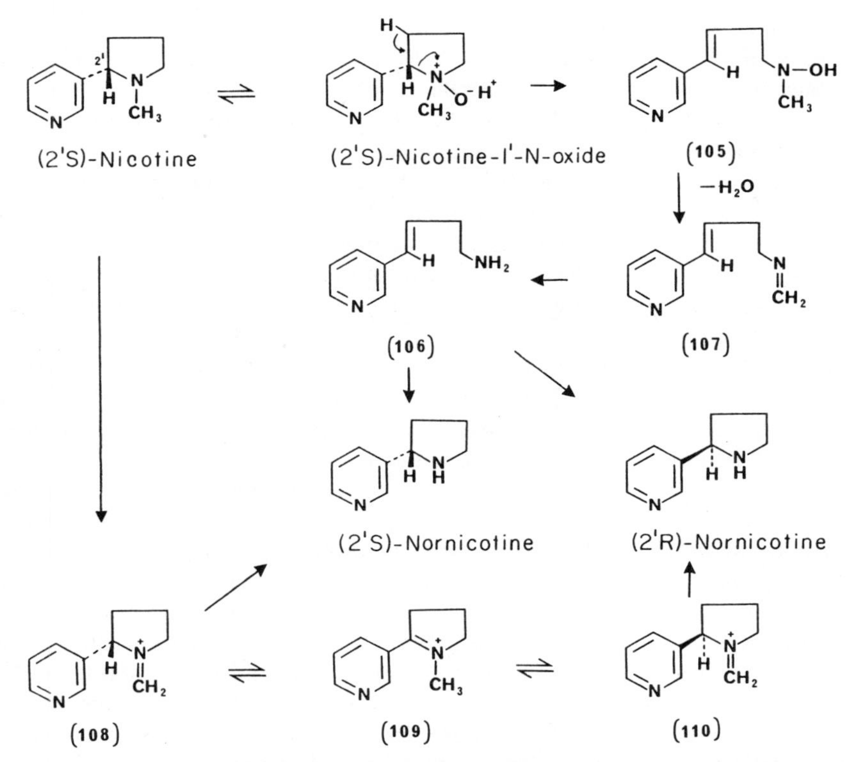

Figure 10. Hypothetical pathways for the demethylation of nicotine to nornicotine.

demonstrated that nicotine-1′-*N*-oxide does not serve as a direct precursor of nornicotine in *N. glutinosa* leaves [184] or intact *N. glutinosa* or *N. glauca* plants [167]. The small conversion observed is attributed to the reduction of the *N*′-oxide to nicotine prior to the formation of nornicotine. The demethylation of nicotine was investigated in *N. alata* (a species of *Nicotiana* that contains negligible alkaloids). Nicotine was infiltrated into the leaves and was demethylated to yield nornicotine, but no nicotine-1′-*N*-oxide was detected in the leaves [322].

Castagnoli and co-workers have investigated the demethylation of nicotine on a rabbit liver extract [320], and they obtained evidence for the intermediate formation of the iminium salt **108**, which on hydrolysis yields nornicotine and formaldehyde. If this iminium salt is the intermediate between nicotine and nornicotine in tobacco, a plausible mechanism for partial racemization of the ultimate nornicotine is available. A tautomeric shift of the iminium salt **108** to **109** eliminates the hydrogen at C(2′) and chirality. A tautomeric shift could yield **110**, which on hydrolysis would afford (*2′R*)-nornicotine. Evidence that this tautomerism is possible *in vitro* can be deduced from the studies of Hecht et al. [56] on the nitrosation of nicotine. They discovered that *N*′-nitrosonornicotine formed by the action of nitrous acid on (*2′S*)-nicotine had undergone considerable racemization and proposed that the racemization takes place by tautomerism of the iminium ion **108**, which is considered to be an intermediate in this reaction.

Little work as been carried out on the fate of the *N*-methyl group of nicotine when it is converted to nornicotine. When [methyl-^{14}C]nicotine was fed to intact *N. tabacum* plants growing hydroponically, there was a fairly rapid loss of activity; after 7 days only 1.1% of the administered activity was recovered in the reisolated nicotine [254]. Some activity was found in the choline isolated from the plant, indicating that the methyl group of nicotine, whatever its oxidation state (formic acid, formaldehyde, etc.), was entering the one-carbon pool. When the same labeled nicotine was applied to excised leaves of *N. tabacum*, activity was detected in the expired carbon dioxide and in the *O*-methyl group of scopolin, a coumarin [252]. Demethylation of nicotine also occurred in a cell culture of *N. tabacum* [166], and activity from [methyl-^{14}C]nicotine was found in "polymeric" material isolated from the culture.

As outlined in Section 5, nornicotine can be formed directly from putrescine via Δ^1-pyrroline. Saunders and Bush [323] have attempted to relate the relative amounts of nicotine and nornicotine found in different genotypes of *N. tabacum* with the levels of the enzymes putrescine-*N*-methyltransferase and *N*-methylputrescine oxidase (see Section 6) in these different varieties. They postulated that varieties have a high nornicotine content would have lower amounts of the putrescine *N*-methyltransferase. Then nornicotine would be formed from Δ^1-pyrroline, which can be formed from the more abundant putrescine by the nonspecific *N*-methylputrescine oxidase. However, no significant difference in enzyme levels was observed in cultivars that differed greatly in their nornicotine/nicotine content.

Also unanswered is the question of the formation of nicotine by the *N* methylation of nornicotine. Because nicotine is found optically pure in *Nicotiana* species, whereas nornicotine is partially racemized, this proposal seems unlikely. Also determination of the levels of radioactivity found in nicotine and nornicotine after short-term exposure to $^{14}CO_2$ does not favor any significant nornicotine → nicotine conversion [176]. Some publications [27, 163, 210, 262, 371] have indicated that the reaction may occur to a limited extent although no definitive experiments have yet been carried out.

The report that nornicotine formed in a root culture of *N. rustica* from [2-^{14}C]ornithine was labeled unsymmetrically in its pyrrolidine ring [219] was not substantiated in later work involving intact *N. glutinosa* plants [185], equal activity being found at C(2′) and C(5′) by an unambiguous chemical degradation.

9. ORIGIN OF OTHER MINOR ALKALOIDS OF TOBACCO

The title of this section was chosen to emphasize the fact that many of these alkaloids are not produced by biosynthetic reactions in the living plant but are formed after harvesting. Indeed, myosmine, cotinine, nicotyrine, and nicotine-1′-

N-oxide were formed by simply bubbling air through a solution of nicotine in water at 30° C for 4 weeks [325]. *N′*-Methylmyosmine (**4**) was also obtained by the chemical oxidation of nicotine with oxygen [366].

Myosmine (**6**) is formed in *N. tabacum* by the dehydrogenation of nornicotine [27]. The myosmine isolated from *N. glauca* plants that had been fed [2′-^{14}C, 2′-^{3}H]nicotine [31] was labeled with ^{14}C solely at C(2′) and was devoid of ^{3}H as expected. Its formation is considered to proceed via nornicotine. This dehydrogenation is apparently not reversible, since nornicotine isolated from *N. glauca* that was fed [2′-^{14}C]myosmine was not labeled. The reisolated myosmine had a reduced specific activity indicative of the normal presence this alkaloid in the growing plant.

2,3′-Bipyridine is also produced by a dehydrogenation reaction from anatabine [32, 146]; however, this reaction does not occur in the growing plant, and this bipyridine is presumably a compound formed during the drying and curing of tobacco leaves [334]. Attempts to show that nicotelline is formed from nicotinic acid in either the living plant or dried leaves were unsuccessful [32].

Cotinine (**10**) and pseudooxynicotine (**13**) were considered to arise from the isomeric iminium salts **111** and **109** produced by the dehydrogenation of nicotine (Fig. 11). These reactions have been shown to take place in a rabbit liver preparation [326], microbiological systems [327], and also by the chemical (mercuric acetate) oxidation of nicotine [328]. Hydration of the iminium salt **111** yields **112**, and further oxidation affords (*2′S*)-cotinine, the major metabolite of (*2′S*)-nicotine in humans and other mammals [329, 330]. We found very little conversion of nicotine to cotinine in the living *N. glauca* plant [31], and it seems to be mainly a product formed in fermented leaves [335]. Both isomers of nicotine-1′-*N*-oxide (oxynicotine) have been isolated from various *Nicotiana* species [40] prior to curing. However, they are also produced to a significant extent during the fermentation of tobacco [337]. It is also formed by the UV irradiation of an aqueous solution of nicotine in the presence of air [338].

We propose that *N′*-isopropylnornicotine (**12**) is formed from nicotine via nornicotine. Reaction of the nornicotine with acetoacetic acid yields the iminium salt **113**, which on decarboxylation affords **114** (also possibly formed from nornicotine and acetone, which has been detected in tobacco leaves). Reduction of **114** then yields *N′*-isopropylnornicotine. In support of this hypothesis labeled *N′*-isopropylnornicotine was obtained after painting the leaves of *N. tabacum* plants with [2′-^{14}C]nicotine [46]. Also, an increased radiochemical yield of **12** was obtained if the leaves were painted with aqueous acetone. These reactions only occurred in the harvested leaf, not in the healthy intact plant.

The *N*-nitroso compounds isolated from tobacco are considered to arise by the action of nitrous acid (either present in the leaf or produced by the microbiological reduction of nitrate) on nornicotine, anatabine, and nicotine. Since nicotine is usually much more abundant in tobacco, this compound is probably the major source of *N′*-nitrosonornicotine [331]. This *N′*-nitroso compound and the related ones, **30** and **34** (Fig. 1), have been demonstrated to be formed by the reaction of nicotine and nitrous acid [321]. [2′-^{14}C]Nicotine and

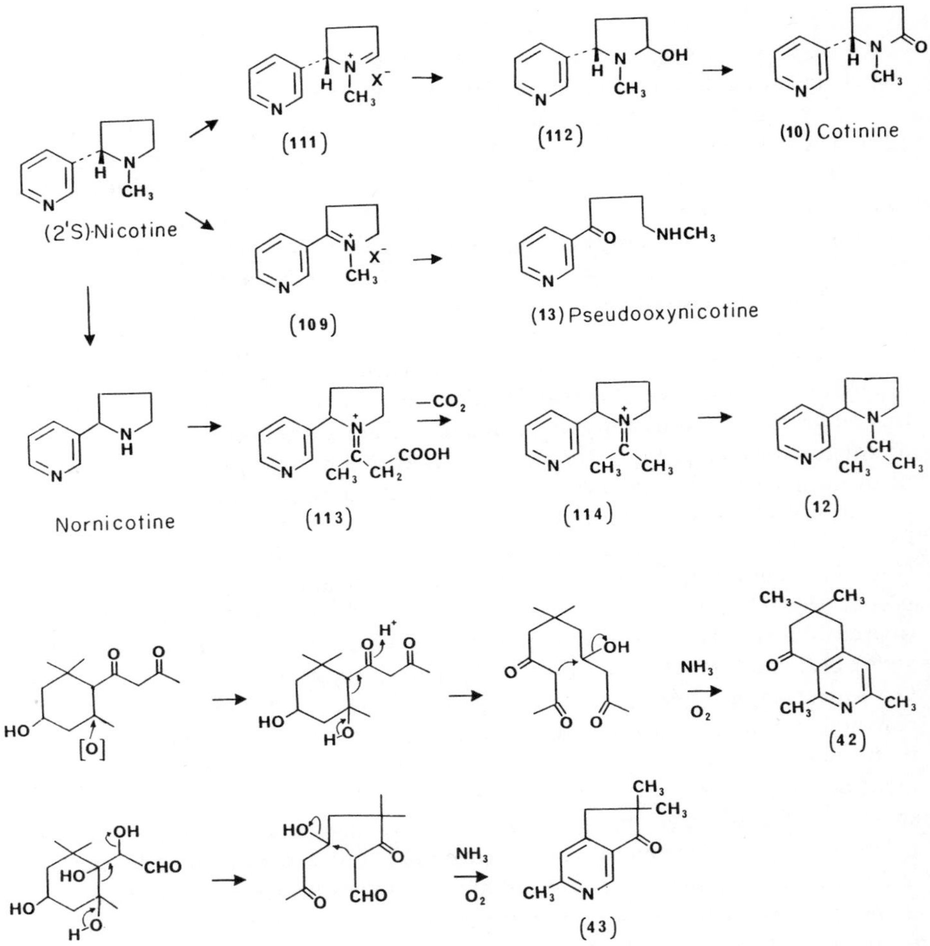

Figure 11. Origin of some minor alkaloids of tobacco.

[2′-^{14}C]nornicotine were applied to *N. tabacum* leaves, which were then air-cured. Each of these compounds was converted to *N′*-nitrosonornicotine with almost the same efficiency [57].

No definitive work in *Nicotiana* species has been reported on any of the other 3-pyridyl alkaloids illustrated in Fig. 1. Kisaki [65] reported that [2-^{14}C]lysine and [2-^{14}C]-5-hydroxylysine failed to serve as precursors of anatalline (**40**).

The proposal [67] that the pyridines **42** and **43** are formed by the reaction of ammonia with degradation products of carotenoids is quite plausible. Mechanistically reasonable routes to these compounds are illustrated in Fig. 11.

The presumed origin of harman (**44**) and norharman (**45**) is the amino acid tryptophan, which reacts with acetaldehyde or formaldehyde *in vitro* to yield tetrahydro-β-carboline-3-carboxylic acids. Oxidation and decarboxylation would then afford **44** and **45**. No work has been described in living *Nicotiana* species, but the addition of radioactive tryptophan to processed tobacco led to the formation of radioactive harman and norharman in the resultant tobacco smoke [332].

10. ABERRANT BIOSYNTHETIC AND METABOLIC REACTIONS IN *NICOTIANA* SPECIES

Aberrant biosynthesis is the term that is applied to abnormal synthetic reactions that occur in biological systems. These aberrant reactions can be divided into two classes. We refer to reactions in which a natural compound is formed from an unnatural compound as Type I. The second class (Type II) involves the conversion of an unnatural precursor to an unnatural compound, presumably utilizing the same enzymes as those involved in the synthesis of a related natural product. Several examples of both types of aberrant reaction have been established in *Nicotiana* species, although it should be realized that reactions of Type I are more difficult to substantiate. Some would claim that many of the tracer experiments designed to elucidate the biosynthesis of alkaloids in general involve aberrant reactions, which the plant is "compelled" to carry out to remove a compound that is injurious or interferes with the normal metabolic reactions.

An example of Type I that has already been discussed (Section 6) is the utilization of δ-*N*-methylornithine to yield nicotine in *N. tabacum*. Another experiment that is considered to be of Type I is the formation of nicotine from pseudooxynicotine (**13**) in *N. rustica* [220]. The pattern of labeling in the nicotine was consistent with a direct conversion, although the specific incorporation (0.07%) was low. Only one significant reaction, namely the reduction of the iminium salt **109** (Fig. 10), is required to form nicotine. However, it was not determined whether this reaction is stereospecific, leading to optically active nicotine. Some of the compounds that yielded *N*-methyl-labeled nicotine are certainly unnatural (chloroethyltrimethylammonium chloride [220], β-methionine [230], and possibly formaldehyde [203]). The incorporation of cadaverine into the piperidine ring of anabasine in *N. glauca* [149] is also considered aberrant, since the intermediacy of this compound is not consistent with the pattern of labeling obtained after feeding [2-^{14}C]lysine to this species. The de-ethylation of *N*′-ethylnornicotine and *N*′-ethylanabasine [305] are also examples of Type I reactions.

Some examples of Type II are illustrated in Fig. 12. Type II reactions are more readily detected, since the product can usually be separated from the related natural alkaloids by some kind of chromatography. Also, in experiments involving isotopically labeled precursors, the aberrant alkaloid should have the same specific activity as the administered compound. In cases where the chirality of the aberrant alkaloid could be determined, it was found to be the same as the natural alkaloid. An interesting example is the conversion of racemic 1,3-dimethyl-Δ^1-pyrrolinium chloride to (*2′S, 3′S*)-3′-methylnicotine, in which the

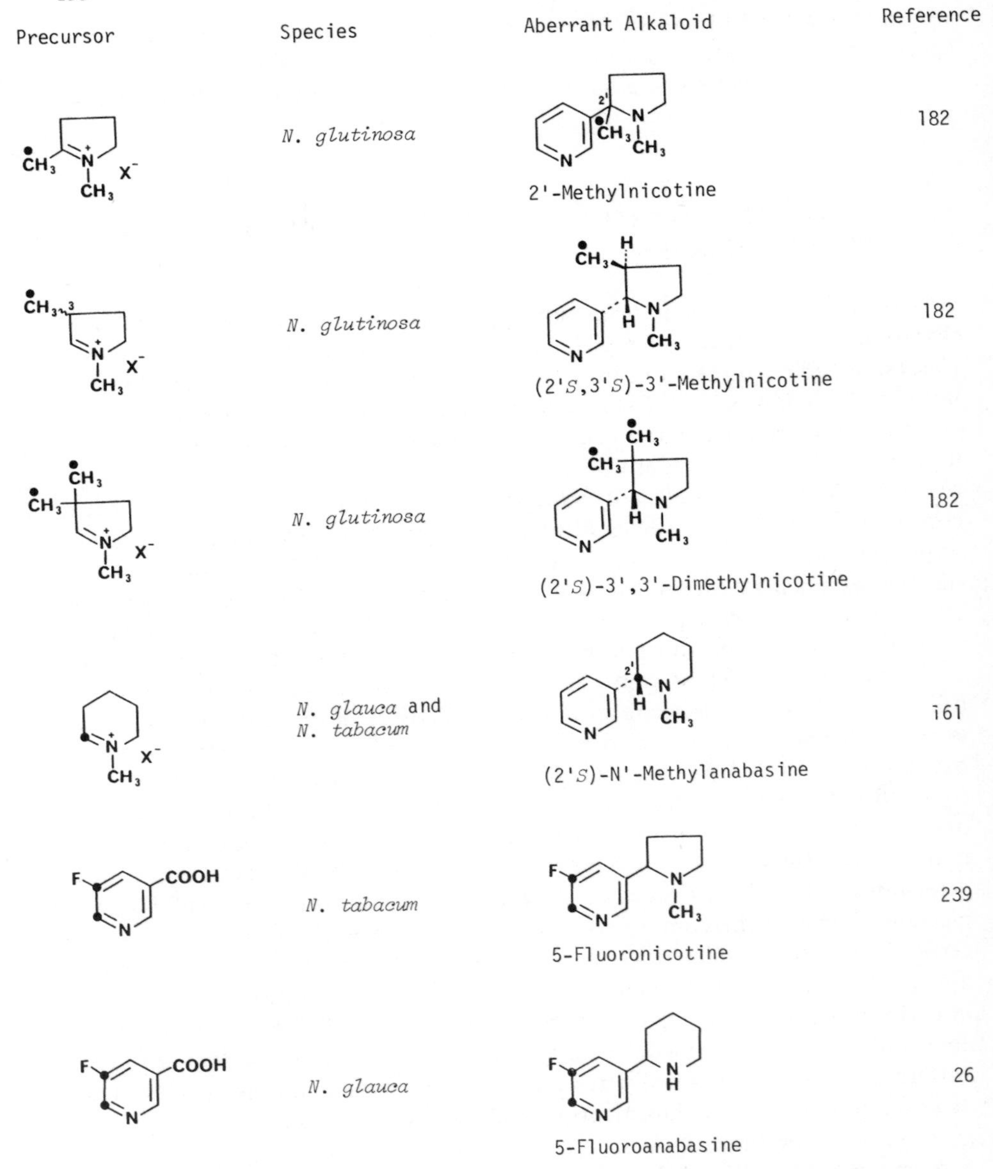

Precursor
Species
Aberrant Alkaloid
Reference
N. glutinosa
2'-Methylnicotine
182
N. glutinosa
(2'S,3'S)-3'-Methylnicotine
182
N. glutinosa
(2'S)-3',3'-Dimethylnicotine
182
N. glauca and
N. tabacum
(2'S)-N'-Methylanabasine
161
N. tabacum
5-Fluoronicotine
239
N. glauca
5-Fluoroanabasine
26

Figure 12. Some examples of aberrant biosyntheses in *Nicotiana* species.

pyridine ring and the 3′-methyl group are in the more stable trans arrangement. Apparently only one of the epimers of the administered precursor is utilized for the synthesis of the aberrant alkaloid. The formation of racemic 5-fluoroanabasine from 5-fluoronicotinic acid in *N. glauca* was unexpected, but then it was realized that the anabasine in this species is also racemic. Although the enzymes responsible for nicotine formation are apparently able to accommodate the relatively large methyl groups on the *N*-methyl-Δ^1-pyrrolinium salt, we have found that nicotinic acid substituted with methyl groups [at C(4), C(5)] apparently cannot be tolerated by the enzymes to yield nicotine with methyl groups attached to the pyridine ring [246, 339].

11. BIOSYNTHESIS OF THE TOBACCO AND RELATED ALKALOIDS IN SPECIES OTHER THAN *NICOTIANA*

Although nicotine has been found in many diverse species (see Section 3), no definitive work on its origin in these species has been reported. We have fed [2-^{14}C]ornithine to *Lycopodium clavatum* [340], *Asclepias syriaca*, and *Equisetum arvense* [341], but no radioactivity was detected in nicotine, which was isolated by dilution with nonlabeled alkaloid, since the amount of nicotine found in these species is very small. We also searched in vain for radioactive nicotine in *Erythroxylon coca* and *Withania somnifera* plants that had been fed [5-^{14}C]ornithine [42, 342].

Several alkaloids that contain the 1-methyl-2-pyrrolidyl moiety found in nicotine are known, and some of these are illustrated in Fig. 13. From a mechanistic point of view, it seems probable that all these are formed by reaction of the *N*-methyl-Δ^1-pyrrolinium salt with another molecule containing a nucleophilic center, analogous to the formation of nicotine. Biosynthetic work that has been carried out on hygrine (**115**) [343], cuscohygrine (**116**) [342], and brevicolline (**117**) [344] is consistent with this hypothesis, their *N*-methylpyrrolidine rings being derived from the same precursors (ornithine, putrescine, arginine) that are incorporated into the pyrrolidine ring of nicotine. Biomimetic

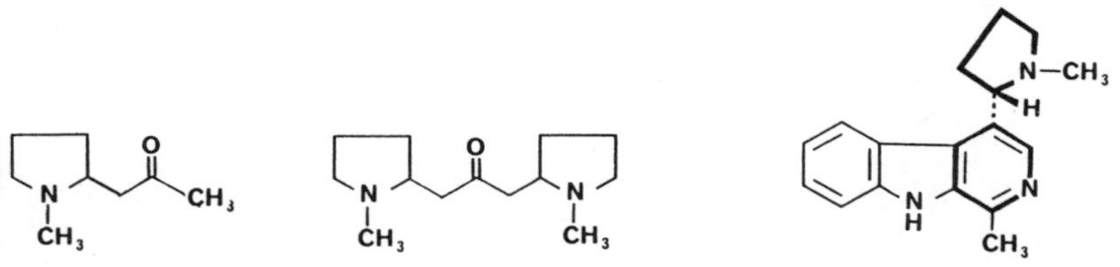

(**115**) Hygrine (**116**) Cuscohygrine (**117**) Brevicolline

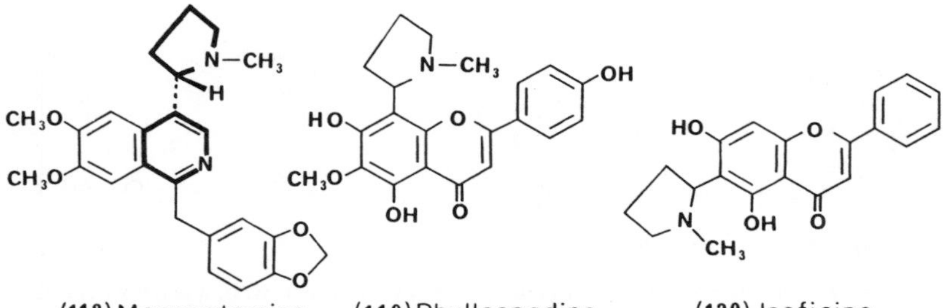

(**118**) Macrostomine (**119**) Phyllospadine (**120**) Isoficine

Figure 13. Some nontobacco alkaloids that contain an *N*-methylpyrrolidine ring.

reactions leading to the formation of brevicolline [345], and analogs of macrostomine (**118**) [346] lend support to this general hypothesis. Indeed, these two alkaloids may be regarded as analogs of nicotine (the nicotine moiety being indicated with thick bonds), and they have the same absolute configuration as (*2′S*) nicotine. The chirality of the rare flavanoid alkaloids phyllospadine (**119**) [347] and isoficine (120) [348] were not recorded.

Considerable work has been carried out on the biosynthesis of anabasine in *Anabasis aphylla*, and the results obtained indicate that it is produced by the same pathway as that occurring in *N. glauca* and other *Nicotiana* species. Thus [2-^{14}C]lysine, [U-^{14}C]lysine, and [1,5-^{14}C]cadaverine served as specific precursors of the piperidine ring of the alkaloid [349–352, 367]. As in *Nicotiana*, the lysine is incorporated unsymmetrically, [2-^{14}C]lysine affording [2′-^{14}C]anabasine [349]. Nicotinic acid has apparently not been tested as a precursor of anabasine in *A. aphylla*, but established precursors of nicotinic acid in *Nicotiana*, [U-^{14}C]aspartic acid [353] and [2-^{14}C]acetate [354], labeled the pyridine ring of anabasine preferentially.

Anabasamine **41**, also found in *A. aphylla*, was obtained labeled after the administration of [^{14}C]lysine and cadaverine, although no degradations were carried out to determine whether the activity was confined to the piperidine ring of this alkaloid [349–351, 356]. [methyl-^{14}C]Methionine was incorporated, presumably into the *N*-methyl group [357].

Anabasine isolated from *Haloxylon salicornicum* plants that had been fed

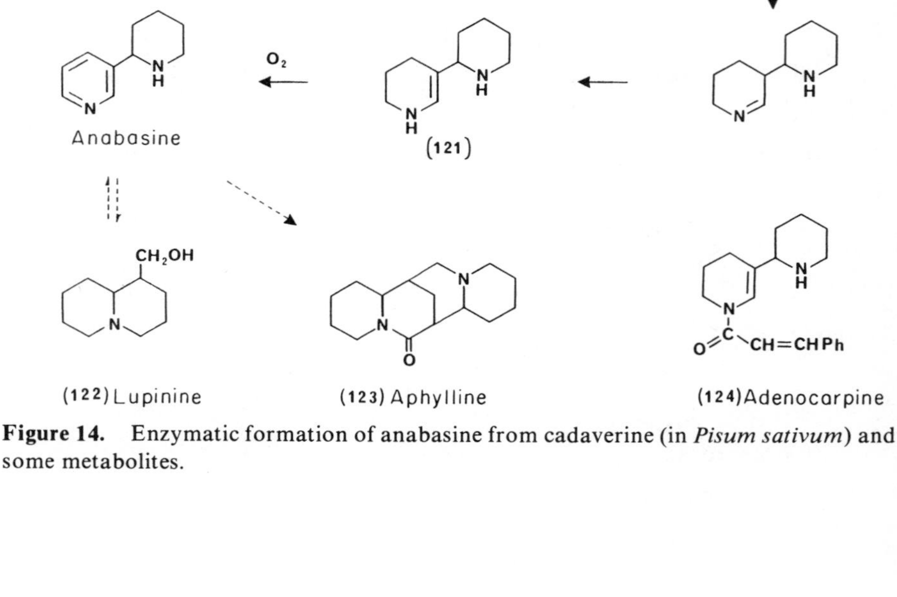

Figure 14. Enzymatic formation of anabasine from cadaverine (in *Pisum sativum*) and some metabolites.

[6-^{14}C]lysine had very low activity [116], and no conclusions could be drawn about its biosynthesis in this species.

Cadaverine is oxidized in the presence of enzymes isolated from pea seedlings (*Pisum sativum*) and lupins to anabasine [369]. In this reaction both heterocyclic rings are derived from cadaverine [358], and the reaction involves the dimerization of Δ^1-piperideine, as illustrated in Fig. 14. The initial product of condensation, 1,4,5,6-tetrahydroanabasine (**121**), has previously been obtained from Δ^1-piperideine [359], and its cinnamoyl derivative is the alkaloid adenocarpine (**124**), which is formed in *Ammodendron conollyi* by this biosynthetic pathway [360]. Formation of anabasine by this route is thus certainly plausible, but does not seem to be a significant route in *Nicotiana* species or *Anabasis aphylla*. I should also point out that anabasine is not found in *Pisum sativum* plants, and the administration of [1,5-^{14}C]cadaverine to intact plants did not yield labeled anabasine [370].

The reports that anabasine is metabolized in *A. aphylla* to lupin alkaloids such as lupinine (**122**) and aphylline (**123**) [361, 362] or that lupinine is converted to anabasine [363] are difficult to interpret or evaluate. One unlikely possibility would be the reduction of anabasine to **121**, which could serve as a source of Δ^1-piperideine, known to be a precursor of the lupin alkaloids [364, 365].

12. SUMMARY: GENERAL REMARKS ON ALKALOID FORMATION IN TOBACCO

A fairly clear picture has now emerged regarding the biosynthesis of nicotine and related alkaloids. The progress that has been achieved in this area in the last 30 years was made possible by the availability of radioactive and stable isotopes of carbon, hydrogen, and nitrogen. The techniques that have been developed for administering labeled compounds to tobacco have wide applicability and have been used for studying the origin of other natural products, many of which are biologically more important and significant than nicotine. However, as pointed out in Section 5, work with tobacco led to the discovery of a new biosynthetic route to the ubiquitous nicotinic acid that may operate in all higher plants. Aberrant biosynthesis of nicotine and anabasine analogs has been accomplished in *Nicotiana* species. The exploitation of this type of aberrant biosynthesis in plants for the production of useful medicinal agents will probably be more practical with cell-suspension cultures derived from plants.

The synthesis of nicotine in such systems has not been discussed to any appreciable extent. Early work with callus tissues of *Nicotiana* was not too exciting, much lower levels of nicotine being found in such systems [379–381] than in the intact plants. However, recently callus tissues have been obtained that contain >2% nicotine (on a dry-weight basis) [382]. Also cell-suspension cultures of *N. tabacum* and *N. rustica* have been developed [383–385] that yield up to 150 mg/liter of nicotine. Not unexpectedly, the nicotine content of a callus tissue is directly related to the nicotine content of the cultivar of *N. tabacum* from which the cullus was produced [386]. Nicotine production in both cell cultures and intact plants is influenced by growth promoters (2,4-dichlorophenoxyacetic acid, gibberellic acid, indole-3-acetic acid) in both a positive and negative direction [205, 258, 261, 387, 388]. These compounds apparently can affect the production of enzymes responsible for nicotine formation. Thus the administration of indole-3-acetic acid to the roots of decapitated plants increased the levels of ornithine decarboxylase, putrescine-*N*-methyltransferase, and *N*-methylputrescine oxidase [227].

Even after all these investigations, no satisfying answer can be given to the question: Why is nicotine produced in tobacco? Needless to say, this same question applies to all the other alkaloids (now over 5000) that are found in other species.

This chapter is dedicated to Ray Fields Dawson, who celebrated his 71st birthday on February 11, 1982. He was a pioneer in studies of the biosynthesis of nicotine and I have long admired him as a scholar and a gentleman. Remarks that he made in 1960 [272] are still germane:

> The rate stability of nicotine biosynthesis in the excised root culture, even in the presence of excess precursors, is noteworthy. Very little is known of the special chemistry of growth in plants insofar as intermediary metabolism is concerned. It would appear that the rate limiting step must lie somewhere between nicotinic acid and ornithine on one hand and nicotine on the other. If this were not so, additions of nicotinic acid and/or ornithine should increase nicotine yields. It becomes a matter of first importance, therefore, to define the individual steps that intervene between precursors and product. If this effort should succeed, it may turn out that the biochemistry of alkaloid formation has provided us with a backdoor into the intermediary biochemistry of growth. Since alkaloids have customarily been regarded as wastes of metabolism, it is likely that an entirely new concept of their biological significance may be in the making.

REFERENCES

1. *Tobacco Literature: A Bibliography*, compiled by Carmen M. Marin, North Carolina. Agricultural Research Service, 1979.
2. R. A. W. Johnstone and J. R. Plimmer, *Chem. Rev.* **59,** 885 (1959).
3. R. L. Stedman, *Chem. Rev.* **68,** 153 (1968).
4. I. Schmeltz and D. Hoffmann, *Chem. Rev.* **77,** 295 (1977).
5. C. R. Enzell, I. Wahlberg, and A. J. Aasen, *Fortschr. Chem. Org. Naturst.* **34,** 1 (1977).
6. T. C. Tso, *Physiology and Biochemistry of Tobacco Plants*, Dowden, Hutchinson and Ross, Stroudsburg, Pa., 1972.
7. G. Neurath and M. Dünger, *Beitr. Tabakforsch.* **5,** 1 (1969).
8. G. Neurath, M Dünger, and I. Kustermann, *Beitr. Tabakforsch.* **6,** 12 (1971).
9. E. Demole and D. Berthet, *Helv. Chim. Acta* **55,** 1866 (1972).
10. E. Demole and D. Berthet, *Helv. Chim. Acta* **55,** 1898 (1972).
11. R. A. Lloyd, C. W. Miller, D. L. Roberts, J. A. Giles, J. P. Dickerson, N. H. Nelson, C. E. Rix, and P. H. Ayers, *Tob. Sci.* **20,** 43 (1976).

12. B. J. Gudzinowicz, *Analysis of Drugs and Metabolites by Gas Chromatography-Mass Spectrometry*, Vol. 7, Dekker, New York, 1980, pp. 1–269.
13. G. B. Neurath, *Beitr. Tabakforsch.* **5,** 115 (1969).
14. H. Kuhn, in *Tobacco Alkaloids and Related Compounds*, U. S. Von Euler, Ed., MacMillan, New York, 1965, p. 37.
15. I. Schmeltz, Ed., *The Chemistry of Tobacco and Tobacco Smoke*, Plenum, New York, 1972.
16. E. L. Wender and D. Hoffmann, *Tobacco and Tobacco Smoke, Studies in Experimental Carcinogenesis*, Academic, New York, 1967.
17. I. Schmeltz, A. Wenger, D. Hoffmann, and T. C. Tso, *Agric. Food Chem.* **27,** 602 (1979).
18. M. Pailer in *Tobacco Alkaloids and Related Compounds*, U. S. Von Euler, Ed., MacMillan, New York, 1965, p. 15.
19. (a) J. J. Willaman and B. G. Schubert, *Alkaloid-Bearing Plants and Their Contained Alkaloids*, Technical Bulletin No. 1234, Agricultural Research Service, U. S. Department of Agriculture, 1961. (b) J. J. Willaman and H.-L. Li, *Alkaloid-Bearing Plants and Their Contained Alkaloids*, 1957–1968, *Lloydia*, **33** (Suppl.) (1970).
20. H. McKennis and E. R. Bowman, Abstract No. 62, 32nd Tobacco Chemists' Research Conference, October 30–November 1, 1978, Montreal, Canada; also E. Leete and T. Zebovitz, unpublished observations.
21. W. Preobraschensky, *Chem. Ber.* **9,** 1024 (1876).
22. L. Siebold and T. Bradbury, *Pharm. J. Trans. III* **12,** 326 (1881).
23. A. Oreykhov and G. Menshikhov, *Chem. Ber.* **64,** 266 (1931).
24. R. Lukes, A. A. Arojan, J. Kovar, and K. Blaha, *Coll. Czech. Chem. Commun.* **27,** 751 (1962).
25. C. R. Smith, *J. Am. Chem. Soc.* **57,** 959 (1935).
26. E. Leete, *J. Org. Chem.* **44,** 165 (1979).
27. T. Kisaki and E. Tamaki, *Phytochemistry* **5,** 293 (1966).
28. A. Wenusch and R. Scholler, *Fachliche Mitt. Osterr. Tabakregie* **2,** 15 (1933).
29. E. Späth, A. Wenusch, and E. Zajic, *Chem. Ber.* **69,** 393 (1936).
30. O. Fejer-Kossey, *Phytochemistry* **11,** 415 (1972).
31. E. Leete and M. R. Chedekel, *Phytochemistry* **13,** 1853 (1974).
32. E. Leete and S. A. Slattery, *J. Am. Chem. Soc.* **98,** 6326 (1976).
33. H. Elmenhorst, *Bull. Inf. Coresta,* Special Issue, Abstr. 513, p. 120 (1978).
34. E. Späth, J. P. Wibaut, and F. Kesztler, *Chem. Ber.* **71,** 100 (1938).
35. W. G. Frankenburg and A. M. Gottscho, *J. Am. Chem. Soc.* **77,** 5728 (1955).
36. E. Spàth and F. Kesztler, *Chem. Ber.* **70,** 2450 (1937).
37. S. Maeda, H. Matsushita, Y. Mikami, and T. Kisaki, *Agric. Biol. Chem.* **42,** 2177 (1978); **44,** 1643 (1980).
38. W. G. Frankenburg and A. A. Vaitekunas, *J. Am. Chem. Soc.* **79,** 149 (1957).
39. W. L. Alworth, The Biosynthesis of the Nicotine Alkaloids in *Nicotiana glutinosa*, Ph.D. dissertation, University of California, Berkeley, 1965. *Diss. Abstr.* **29,** 4582B (1969).
40. J. D. Philipson and S. S. Handa, *Phytochemistry* **14,** 2683 (1975).
41. R. Wahl, *Tabak-Forsch. Sonderheft,* 36 (1953).
42. E. Leete, unpublished work.
43. E. V. Brown and I. Ahmad, *Phytochemistry* **11,** 3485 (1972).
44. M. Miyano, H. Matsushita, N. Yasumatsu and K. Nishida, *Agric. Biol. Chem.* **43,** 2205 (1979).
45. (a) H.-P. Ros, R. Kyburz, N. W. Preston, R. T. Gallagher, I. R. C. Bick, and M. Hesse, *Helv. Chim. Acta* **62,** 481 (1979). (b) I. R. C. Bick and M. A. Hai, *Tetrahedron Lett.*, 3275 (1981).
46. E. Leete, *Phytochemistry* **20,** 1037 (1981).

47. A. H. Warfield, W. D. Galloway, and A. G. Kallianos, *Phytochemistry* **11,** 3371 (1972).
48. M. Miyano, H. Matsushita, N. Yasumatsu, and K. Nishida, *Agric. Biol. Chem.* **43,** 1607 (1979).
49. H. Matsushita, Y. Tsujino, D. Yoshida, A. Saito, T. Kisaki, K. Kato, and M. Noguchi, *Agric. Biol. Chem.* **43,** 193 (1979).
50. A. J. N. Bolt, *Phytochemistry* **11,** 2341 (1972).
51. M. Miyano, N. Yasumatsu, H. Matsushita, and K. Nishida, *Agric. Biol. Chem.* **45,** 1029 (1981).
52. S. S. Hecht, R. M. Ornaf, and D. Hoffmann, *J. Natl. Cancer Inst.* **54,** 1237 (1974).
53. D. Hoffmann, S. S. Hecht, R. M. Ornaf, and E. L. Wynder, *Science* **186,** 265 (1974).
54. K. D. Burnnemann and D. Hoffmann, 35th Tobacco Chemists Research Conference 1981, Winston-Salem, NC.
55. V. P. Bharadwaj, S. Takayama, T. Yamada, and A. Tanimura, *Gann* **66,** 585 (1975).
56. S. S. Hecht, C. B. Chen, M. Dong, R. M. Ornaf, D. Hoffmann, and T. C. Tso, *Beitr. Tabakforsch.* **9,** 1 (1977).
57. S. S. Hecht, C. B. Chen, N. Hirota, R. M. Ornaf, T. C. Tso, and D. Hoffmann, *J. Natl. Cancer Inst.* **60,** 819 (1978).
58. D. Hoffmann, J. D. Adams, K. D. Brunnemann, and S. S. Hecht, *Cancer Res.* **39,** 2505 (1979).
59. A. Koiwai, Y. Mikami, H. Matsushita, and T. Kisaki, *Agric. Biol. Chem.* **43,** 1421 (1979).
60. E. Späth and E. Zajic, *Chem. Ber.* **69,** 2448 (1936).
61. A. H. Warfield, W. D. Galloway, and A. G. Kallianos, *Phytochemistry* **11,** 3371 (1972).
62. Y. Saint-Jalm and P. Moree-Testa, *J. Chromat.* **198,** 188 (1980).
63. A. Pictet and A. Rotschy, *Chem. Ber.* **34,** 696 (1901).
64. F. Kuffner and N. Faderl, *Monatsh. Chem.* **87,** 71 (1956).
65. T. Kisaki, S. Mizusaki, and E. Tamaki, *Phytochemistry* **7,** 323 (1968).
66. S. Z. Mukhamedzhanov, A. Aslanov, A. S. Sadykov, V. B. Leont'ev, and V. K. Kiryukhin, *Khim. Prir. Soedin.* **4,** 158 (1968).
67. E. Demole and C. Demole, *Helv. Chim. Acta* **58,** 523 (1975).
68. E. H. Poindexter and R. D. Carpenter, *Phytochemistry* **1,** 215 (1962).
69. E. H. Poindexter and R. D. Carpenter, *Chem. Ind. (London)*, 176 (1962).
70. P. I. Mortimer and S. Wilkinson, *J. Chem. Soc.*, 3967 (1957).
71. B. Frank, *Chem. Ber.* **91,** 2803 (1958).
72. L. Marion, *Can. J. Res.* **B23,** 165 (1945).
73. L. H. Fikenscher, *Pharm. Weekblad* **95,** 233 (1960).
74. G. L. Gupta and S. S. Nigam, *Planta Med.* **19,** 55 (1971).
75. W. C. Evans and K. P. A. Ramsey, *J. Pharm. Pharmacol.* **31** (Suppl.) 9P (1979).
76. I. R. C. Bick, J. B. Bremner, J. W. Gillard, and K. N. Winzenberg, *Austr. J. Chem.* **27,** 2515 (1974).
77. L. Marion, *Can. J. Res.* **B17,** 21 (1939).
78. T. M. Smalberger, J. H. G. Rall, and H. L. DeWaal, *Tydskr. Natuurwetensk.* **8,** 156 (1968).
79. A. F. Halim, R. P. Collins, and M. S. Berigari, *Planta Med.* **20,** 44 (1971).
80. D. Neumann and K. H. Tschoepe, *Flora (Jena)* **156,** 521 (1966).
81. A. C. H. Rothera, *Biochem. J.* **5,** 193 (1910).
82. W. Bottomley, R. A. Nottle, and D. E. White, *Austr. J. Sci.* **8,** 18 (1945).
83. K. L. Hills, W. Bottomley, and P. I. Mortimer, *Nature* **171,** 435 (1953).
84. P. I. Mortimer, *Austr. J. Sci.* **20,** 87 (1957).

85. E. Dufva, G. Loison, and B. Holmstedt, *Toxicon* **14,** 55 (1976).
86. Y. Kitamura, S. Hasegawa, H. Miura, and M. Sugii, *Shoyakugaku Zasshi* **34,** 117 (1980).
87. L. Cosson and J. C. Vaillant, *Phytochemistry* **15,** 818 (1976).
88. S. N. Pal and M. Narasinham, *J. Ind. Chem. Soc.* **20,** 181 (1943).
89. C. H. Eugster, R. Griot, and P. Karrer, *Helv. Chim. Acta* **36,** 1387 (1953).
90. P. Karrer, C. H. Eugster, and D. K. Patol. *Helv. Chim. Acta* **32,** 2397 (1949).
91. T. Bayton and E. Gurkan, *Istanbul Univ. Eczacilik Fak. Mecm.* **8,** 63 (1972); *Chem. Abstr.* **79,** 15817 (1973).
92. L. N. Fikenscher, *Pharm. Weekblad* **93,** 932 (1958).
93. A. A. M. Habib, M. M. A. Hassan, and F. J. Muhtadi, *J. Pharm. Pharmacol.* **26,** 837 (1974).
94. P. K. Das, C. L. Malhotra, and N. S. Dhalla, *Ind. J. Physiol. Pharmacol.* **5,** 136 (1961).
95. K. Blaim, *Flora (Jena)* **152,** 171 (1962).
96. O. Achmatowicz and W. Rodewald, *Rocz. Chem.* **32,** 485 (1958).
97. L. Marion and R. H. F. Manske, *Can. J. Res.* **B26,** 1 (1948).
98. L. Marion and R. H. F. Manske, *Can. J. Res.* **B22,** 137 (1944).
99. W. J. Rodewald and G. Grynkiewicz, *Rocz. Chem.* **51,** 1271 (1977).
100. R. H. F. Manske and L. Marion, *Can. J. Res.* **B20,** 87 (1942).
101. R. H. F. Manske and L. Marion, *Can. J. Res.* **B24,** 57 (1946).
102. R. H. F. Manske and L. Marion, *Can. J. Res.* **B22,** 53 (1944).
103. L. Marion and R. H. F. Manske, *Can. J. Res.* **B24,** 63 (1946).
104. L. Marion and R. H. F. Manske, *Can. J. Res.* **B22,** 1 (1944).
105. D. N. Majumdar and G. B. Paul, *Ind. Pharmacist* **10,** 79 (1954); *Chem. Abstr.* **49,** 9881 (1955).
106. A. Z. Fulubov and Z. I. Bozhova, *Nauch. Tr. Plovdivski Univ., Mat., Fiz., Khim., Biol.* **10,** 105 (1972); *Chem. Abstr.* **78,** 13763 (1973).
107. S. Gill, W. Raszeja, and G. Szynkiewicz, *Farm. Pol.* **35,** 151 (1979).
108. R. R. Paris and P. Frigot, *C. R. Hebd. Seances Acad. Sci. C. R.* **248,** 1849 (1959).
109. D. N. Majumdar, *Current Sci. (India)* **21,** 46 (1952).
110. H.-B. Schröter, *Arch. Pharm.* **288,** 141 (1955).
111. N. Z. Nazrullaeva, Kh. A. Aslanov, A. A. Abuvakhabov, and S. Z. Mukhamedzhanov, *Khim. Prir. Soedin.* **12,** 552 (1976).
112. L. I. Brutko, P. S. Massagetov, and L. M. Utkin, *Rast, Res.* **4,** 334 (1968); *Chem. Abstr.* **70,** 17564 (1969).
113. W. B. Mors and O. R. Gottlieb, *An. Assoc. Bras. Quim.* **18,** 185 (1959).
114. M. I. Goryaev, K. A. Shurov, U.S.S.R. Patent 207,233, *Chem. Abstr.* **69,** 21905 (1968).
115. K. H. Michel, F. Sandberg, F. Haglid, and T. Norin, *Acta Pharm. Suecia* **4,** 97 (1967).
116. D. G. O'Donovan and P. B. Creedon, *Tetrahedron Lett.*, 1341 (1971).
117. D. Kamalitdinov, S. Iskandarov, S. Yu Yunusov, *Khim Prir. Soedin.* **5,** 409 (1969).
118. A. Zunnunzhanov, S. Iskandarov, and S. Yu Yunusov, *Khim. Prir. Soedin.* **10,** 373 (1974).
119. W. L. Fitch, P. M. Dolinger, and C. Djerassi, *J. Org. Chem.* **39,** 2974 (1974).
120. V. Kh. Zharekeev, M. V. Telezhenetskaya, and S. Yu Yunusov, *Khim. Prir. Soedin.* **7,** 538 (1971).
121. R. E. Summons, J. Ellis, and E. Gellert, *Phytochemistry* **11,** 3335 (1972).
122. E. Schlunegger and E. Steinegger, *Pharm. Acta Helv.* **46,** 147 (1970).
123. E. Steinegger, E. Schlunegger, F. Schnyder, and R. Frehner, *Pharm. Weekblad* **106,** 245 (1971).
124. W. C. Evans and A. Somanabandhu, *Phytochemistry* **16,** 1859 (1977).

125. A. L. Markman and A. I. Glushenkova, *Uzeksk. Khim. Zh.* **7,** 81 (1963); *Chem. Abstr.* **58,** 14326 (1963).
126. R. Ziyaev, A. Abdusamatov, and S. Yu. Yunasov, *Khim. Prir. Soedin.* **7,** 853 (1971).
127. E. Späth, C. S. Hicks, and E. Zajic, *Chem. Ber.* **68,** 1388 (1935).
128. H.-B. Schröter, *Naturwissenschaften* **45,** 338 (1958).
129. W. R. Kem, K. N. Scott, and J. H. Duncan, *Experientia* **32,** 684 (1976).
130. W. R. Kem, *Toxicon* **9,** 23 (1971).
131. J. W. Wheeler, O. Olubajo, C. B. Storm, and R. M. Duffield, *Science* **211,** 1051 (1981).
132. W. R. Kem, B. C. Abbott, and R. M. Coats, *Toxicon* **9,** 15 (1971).
133. R. Wahl, *Tabak Forsch.* issue No. 8, p. 3 (1952); issue No. 10, p. 3 (1953).
134. M. L. Solt, *Plant Physiol.* **32,** 484 (1957).
135. K. Mothes, A. Romeike, and H.-B. Schröter, *Naturwissenschaften* **42,** 214 (1955).
136. W. R. Brode, *Science in Progress*, Vol. 12, R. F. Dawson, Ed. Yale University Press, New Haven, 1962, p. 117.
137. G. R. Waller, K. S. Yang, R. K. Gholson, L. A. Harwiger, and S. Chaykin, *J. Biol. Chem.* **241,** 4411 (1966).
138. S. A. Brown, *Biosynthesis*, Vol. 1, p. 1, 1972, Specialist periodical report of the Chemical Society.
139. R. F. Dawson, *Adv. Enzymol.* **8,** 203 (1948).
140. C. L. Comar, *Radioisotopes in Biology and Agriculture*, McGraw-Hill, New York, 1955, p. 151.
141. E. Leete, *J. Am. Chem. Soc.* **82,** 6338 (1960).
142. *Eur. J. Biochem.* **86,** 9 (1978).
143. E. Leete, *Chem. Ind.* (*London*), 1477 (1958).
144. A. R. Friedman and E. Leete, *J. Am. Chem. Soc.* **85,** 2141 (1963).
145. T. Griffith and G. D. Griffith, *Phytochemistry* **5,** 1175 (1966).
146. E. Leete, K. C. Ranbom, and R. M. Riddle, *Phytochemistry* **18,** 75 (1979).
147. M. Ya. Lovkova, G. S. Iljin, and N. I. Klimentyeva, *Fiziol. Rast.* **17,** 409 (1970).
148. G. S. Iljin, *Dokl. Akad. Nauk SSSR* **105,** 777 (1955).
149. E. Leete, *J. Am. Chem. Soc.* **80,** 4393 (1958).
150. S. Aronoff, *Plant Physiol.* **31,** 355 (1956).
151. T. C. Tso, R. N. Jeffrey, and T. P. Sorokin, *Arch. Biochem. Biophys.* **92,** 241 (1961).
152. G. D. Griffith and T. Griffith, *Plant Physiol.* **39,** 970 (1964).
153. E. Leete and A. R. Friedman, *J. Am. Chem. Soc.* **86,** 1224 (1964).
154. M. L. Solt, R. F. Dawson, and D. R. Christman, *Plant Physiol.* **35,** 887 (1960).
155. E. Leete, *J. Am. Chem. Soc.* **78,** 3520 (1956).
156. T. J. Gilbertson, *Phytochemistry* **11,** 1737 (1972).
157. E. Leete, *J. Nat. Prod.* **45,** 197 (1982).
158. E. Leete, E. G. Gros, and T. J. Gilbertson, *J. Am. Chem. Soc.* **86,** 3907 (1964).
159. E. Leistner, R. N. Gupta, and I. D. Spenser, *J. Am. Chem. Soc.* **95,** 4040 (1973).
160. E. Leistner and I. D. Spenser, *J. Am. Chem. Soc.* **95,** 4715 (1973).
161. E. Leete and M. R. Chedekel, *Phytochemistry* **11,** 2751 (1972).
162. H.-B. Schröter, *Z. Naturforsch.* **12B,** 334 (1957).
163. T. C. Tso and R. N. Jeffrey, *Arch. Biochem. Biophys.* **80,** 46 (1959).
164. M. Ya. Lovkova, G. S. Iljin, and N. S. Minozhedinova, *Prikl. Biokhim. Mikrobiol.* **9,** 595 (1973).
165. E. Leete, *Tetrahedron Lett.*, 4433 (1968).

166. W. Barz, M. Kettner, and W. Husemann, *Planta Med.* **34,** 73 (1978).

167. E. Leete, in *Recent Advances in the Chemical Composition of Tobacco and Tobacco Smoke, Proceedings of the American Chemical Society Symposium*, The 173rd Am. Chem. Soc. Meeting, Agric. and Food Chem. Div., 1977, New Orleans, La, p. 363.

168. E. Leete, *Bioorganic Chem.* **6,** 273 (1977).

169. E. Leete, *J. Chem. Soc. Chem. Commun.*, 610 (1978).

170. T. C. Tso and R. N. Jeffrey, *Plant Physiol.* **32,** 86 (1957).

171. E. Leete, *J. Am. Chem. Soc.* **91,** 1697 (1969).

172. K. Mothes and H.-B. Schröter, *Arch. Pharm.* **294,** 99 (1962).

173. W. L. Alworth, R. C. DeSelms, and H. Rapoport, *J. Am. Chem. Soc.* **86,** 1608 (1964).

174. W. L. Alworth, A. A. Liebman, and H. Rapoport, *J. Am. Chem. Soc.* **86,** 3375 (1964).

175. G. S. Iljin, *Fiziol. Rast.* **10,** 79 (1963).

176. W. L. Alworth and H. Rapoport, *Arch. Biochem. Biophys.* **112,** 45 (1965).

177. A. A. Liebman, F. Morsingh, and H. Rapoport, *J. Am. Chem. Soc.* **87,** 4399 (1965).

178. A. A. Liebman, B. P. Mundy, and H. Rapoport, *J. Am. Chem. Soc.* **89,** 664 (1967).

179. H. R. Zielke, R. U. Byerrum, R. M. O'Neal, L. C. Burns, and R. E. Koeppe, *J. Biol. Chem.* **243,** 4757 (1968).

180. M. L. Rueppel, B. P. Mundy, and H. Rapoport, *Phytochemistry* **13,** 141 (1974).

181. H. R. Zielke, C. M. Reinke, and R. U. Byerrum, *J. Biol. Chem.* **244,** 95 (1969).

182. M. L. Rueppel and H. Rapoport, *J. Am. Chem. Soc.* **93,** 7021 (1971).

183. E. Leete, *J. Chem. Soc. Chem. Commun.*, 9 (1975).

184. W. L. Alworth, L. Liberman, and J. A. Ruckstahl, *Phytochemistry* **8,** 1427 (1969).

185. E. Leete, *J. Org. Chem.* **41,** 3438 (1976).

186. E. Leete and M.-L. Yu, *Phytochemistry* **19,** 1093 (1980).

187. T. Griffith and R. U. Byerrum, *Science* **129,** 1485 (1959).

188. T. Griffith, K. P. Hellman, and R. U. Byerrum, *J. Biol. Chem.* **235,** 800 (1960).

189. T. Griffith and R. U. Byerrum, *Biochem. Biophys. Res. Comm.* **10,** 293 (1963).

190. P.-H. L. Wu and R. U. Byerrum, *Biochemistry* **4,** 1628 (1965).

191. P.-H. L. Wu, T. Griffith, and R. U. Byerrum, *J. Biol. Chem.* **237,** 887 (1962).

192. S. Mizusaki, T. Kisaki, and E. Tamaki, *Plant Physiol.* **43,** 93 (1968); *Arch. Biochem. Biophys.* **117,** 677 (1966).

193. T. Griffith, K. P. Hellman, and R. U. Byerrum, *Biochemistry* **2,** 336 (1962).

194. T. M. Jackanicz and R. U. Byerrum, *J. Biol. Chem.* **241,** 1296 (1966).

195. R. U. Byerrum, C. S. Sato, and C. D. Ball, *Plant Physiol.* **31,** 374 (1956).

196. R. U. Byerrum and H. W. Culp, Abstracts 126th Meeting of the American Chemical Society, New York, September 12–17, 1954, p. 16C.

197. A. Ganz, F. E. Kelsey, and E. M. K. Geiling, *Bot. Gaz.* **113,** 195 (1951).

198. T. C. Tso and R. N. Jeffrey, *Plant Physiol.* **31,** 433 (1956).

199. T. C. Tso, *Phytochemistry* **5,** 287 (1966).

200. U. Stephan and H. R. Schütte, *Biochem. Physiol. Pflanzen* **161,** 499 (1970).

201. R. U. Byerrum and R. E. Wing, *J. Biol. Chem.* **205,** 637 (1953).

202. S. A. Brown and R. U. Byerrum, *J. Am. Chem. Soc.* **74,** 1523 (1952).

203. R. U. Byerrum, R. L. Ringer, R. L. Hamill, and C. D. Ball, *J. Biol. Chem.* **216,** 371 (1955).

204. B. L. Lamberts and R. U. Byerrum, *J. Biol. Chem.* **233,** 939 (1958).

205. N. Yasumatsu, A. Sakurai, and S. Tamura, *Agric. Biol. Chem.* **31,** 1061 (1967).

206. J. Fleeker and R. U. Byerrum, *J. Biol. Chem.* **242,** 3042 (1967).

207. R. U. Byerrum, R. L. Hamill, and C. D. Ball, *J. Biol. Chem.* **210,** 645 (1954).

208. R. U. Byerrum, L. J. Dewey, R. L. Hamill, and C. D. Ball, *J. Biol. Chem.* **219,** 345 (1956).
209. L. J. Dewey, referred to by R. A. Steinberg and T. C. Tso, *Ann. Rev. Plant Physiol.* **9,** 163 (1958).
210. S. Mizusaki, T. Kisaki, and E. Tamaki, *Agric. Biol. Chem.* **29,** 719 (1965).
211. L. J. Dewey, R. U. Byerrum, and C. D. Ball, *J. Am. Chem. Soc.* **76,** 3997 (1954).
212. H.-B. Schröter and D. Neumann, *Tetrahedron Lett.*, 1279 (1966).
213. H. R. Schütte, W. Maier, and K. Mothes, *Acta Biochem. Pol.* **13,** 401 (1966).
214. H. R. Schütte, W. Maier, and U. Stephan, *Z. Naturforsch.* **23B,** 1426 (1968).
215. G. M. Frost, K. S. Yang, and G. R. Waller, *J. Biol. Chem.* **242,** 887 (1967).
216. G. D. Griffith, T. Griffith, and R. U. Byerrum, *J. Biol. Chem.* **235,** 3576 (1960).
217. T. C. Tso, and R. N. Jeffrey, Tobacco Chemists Research Meeting, Washington, D.C. (1956).
218. L. J. Dewey, R. U. Byerrum, and C. D. Ball, *Biochim. Biophys. Acta* **18,** 141 (1955).
219. S. Mizusaki, T. Kisaki, and E. Tamaki, *Agric. Biol. Chem.* **29,** 714 (1965).
220. H. R. Schütte and U. Stephan, *Z. Naturforsch.* **22B,** 1335 (1967).
221. V. Krampl, H. R. Zielke, and R. U. Byerrum, *Phytochemistry* **8,** 843 (1969).
222. L. M. Henderson, J. F. Someroski, D. R. Rao, P.-H. L. Wu, T. Griffith, and R. U. Byerrum, *J. Biol. Chem.* **234,** 93 (1959).
223. T. C. Tso and R. N. Jeffrey, *Arch. Biochem. Biophys.* **97,** 4 (1962).
224. D. R. Christman and R. F. Dawson, *Biochemistry* **2,** 182 (1963).
225. G. S. Iljin, *Dokl. Akad. Nauk SSSR* **119,** 544 (1958).
226. S. Mizusaki, Y. Tanabe, M. Noguchi, and E. Tamaki, *Plant Cell Physiol.* **12,** 633 (1971).
227. S. Mizusaki, Y. Tanabe, M. Noguchi, and E. Tamaki, *Plant Cell Physiol.* **14,** 103 (1973).
228. D. Yoshida and T. Mitake, *Plant Cell Physiol.* **7,** 301 (1966).
229. B. P. Mundy, K. B. Lipowitz, M. Lee, and B. R. Larsen, *J. Org. Chem.* **39,** 1963 (1974).
230. B. Ladesic, Z. Devide, N. Pravdic, and D. Keglevic, *Arch. Biochem. Biophys.* **97,** 556 (1962).
231. J. Grimshaw and L. Marion, *Nature* **181,** 112 (1958).
232. M. Ya. Lovkova and G. S. Iljin, *Biokhimya* **26,** 82 (1958).
233. J. Berlin, *Phytochemistry* **20,** 53 (1981).
234. T. A. Scott and A. L. Devonshire, *Biochem. J.* **124,** 949 (1971).
235. G. S. Iljin, *Biokhimya* **21,** 108 (1956).
236. A. M. Kuzin and V. I. Merenova, *Dokl. Akad. Nauk* **85,** 393 (1952).
237. C. R. Hutchinson, M.-T. S. Hsia, and R. A. Carver, *J. Am. Chem. Soc.* **98,** 6006 (1976).
238. M. Nakane and C. R. Hutchinson, *J. Org. Chem.* **43,** 2122 (1978).
239. E. Leete, G. B. Bodem, and M. F. Manuel, *Phytochemistry* **10,** 2687 (1971).
240. E. Leete, *J. Am. Chem. Soc.* **80,** 2162 (1958).
241. K. S. Yang, R. K. Gholson, and G. R. Waller, *J. Am. Chem. Soc.* **87,** 4184 (1965).
242. G. S. Iljin, Proceedings 4th International Tobacco Scientific Conference Athens, Greece 1966, p. 172.
243. S. Mizusaki, Y. Tanabe, T. Kisaki, and E. Tamaki, *Phytochemistry* **9,** 549 (1970).
244. R. F. Dawson, D. R. Christman, A. D'Adamo, and A. P. Wolf, *J. Am. Chem. Soc.* **82,** 2628 (1960).
245. A. A. Bothner-By, R. F. Dawson, and D. R. Christman, *Experientia* **12,** 151 (1956).
246. E. Leete and S. A. Leete, *J. Org. Chem.* **43,** 2122 (1978).
247. T. J. Gilbertson and E. Leete, *J. Am. Chem. Soc.* **89,** 7085 (1967).
248. E. Leete and J. A. McDonell, *J. Am. Chem. Soc.* **103,** 658 (1981).
249. S. Mizusaki, Y. Tanabe, M. Noguchi, and E. Tamaki, *Phytochemistry* **11,** 2757 (1972).

250. E. Leete, *J. Am. Chem. Soc.* **89,** 7081 (1967).

251. G. S. Iljin and M. Ya. Lovkova, *Abh. Dtsch. Akad. Wiss. Berlin, Kl. Chem. Geol. Biol.*, No. 4, 207 (1969).

252. L. J. Dewey and W. Stepka, *Arch. Biochem. Biophys.* **100,** 91 (1963); *Plant Physiol.* **36,** 592 (1961).

253. T. Kisaki and E. Tamaki, *Agric. Biol. Chem.* **28,** 492 (1964).

254. E. Leete and V. M. Bell, *J. Am. Chem. Soc.* **81,** 4358 (1959).

255. R. F. Dawson, D. R. Christman, and R. C. Anderson, *J. Am. Chem. Soc.* **75,** 5114 (1953).

256. R. F. Dawson, D. R. Christman, R. C. Anderson, M. L. Solt, A. F. D'Adamo, and U. Weiss, *J. Am. Chem. Soc.* **78,** 2645 (1956).

257. R. F. Dawson, D. R. Christman, S. F. D'Adamo, M. L. Solt, and A. P. Wolf, *Chem. Ind. (London)*, 100 (1958).

258. M. L. Solt, R. F. Dawson, and D. R. Christman, *Tob. Sci.* **5,** 95 (1960).

259. T. A. Scott and J. P. Glynn, *Phytochemistry* **6,** 505 (1967).

260. E. Leete and Y.-Y. Liu, *Phytochemistry* **12,** 593 (1973).

261. N. Yasumatsu, *Agric. Biol. Chem.* **31,** 1441 (1967).

262. B. Ladesic and T. C. Tso, *Phytochemistry* **3,** 541 (1964).

263. E. Leete, *Chem. Ind. (London)*, 537 (1955).

264. E. Leete and K. J. Siegfried, *J. Am. Chem. Soc.* **79,** 4529 (1957).

265. R. F. Dawson, *Am. Scientist* **48,** 321 (1960), and private communication.

266. S. Mizusaki, Y. Tanabe, M. Noguchi, and E. Tamaki, *Phytochemistry* **10,** 1347 (1971).

267. E. Leete, E. G. Gros, and T. J. Gilbertson, *Tetrahedron Lett.*, 587 (1964).

268. V. Krampl and C. A. Hoppert, *Fed. Proc.* **20,** 375 (1961).

269. M. Slaytor, L. Copeland, and P. K. Macnicol, *Phytochemistry* **7,** 1779 (1968).

270. K. Bowden, *Nature* **172,** 768 (1953).

271. E. Leete, *Chem. Ind. (London)*, 1270 (1957).

272. R. F. Dawson, D. R. Christman, A. D'Adamo, M. L. Solt, and A. P. Wolf, *Arch. Biochem. Biophys.* **91,** 144 (1960).

273. R. F. Dawson, D. R. Christman, and R. U. Byerrum, *Meth. Enzymol.* **18B,** 90 (1971).

274. E. Leete and F. H. B. Leitz, *Chem. Ind. (London)*, 1572 (1957).

275. G. R. Waller and L. M. Henderson, *Biochem. Biophys. Res. Commun.* **5,** 5 (1961); *J. Biol. Chem.* **236,** 1186 (1961).

276. G. R. Waller, K. S. Yang, R. K. Gholson, L. A. Harwiger, and S. Chaykin, *J. Biol. Chem.* **241,** 4411 (1966).

277. D. Gross, H. R. Schütte, G. Hubner, and K. Mothes, *Tetrahedron Lett.*, 451 (1963).

278. E. Mothes, D. Gross, H. R. Schütte, and K. Mothes, *Naturwissenschaften* **48,** 623 (1961).

279. F. Lingens, P. Vollprecht, and V. Gildemeister, *Biochem. Z.* **344,** 462 (1966).

280. J. Arditti and J. B. Tarr, *Am. J. Bot.* **66,** 1105 (1979).

281. E. Wenkert, *Accts. Chem. Res.* **1,** 78 (1968).

282. R. F. Dawson and T. S. Osdene, *Rec. Adv. Phytochem.* **5,** 317 (1972).

283. M. Noguchi, H. Sakuma, and E. Tamaki, *Phytochemistry* **7,** 1861 (1968).

284. E. Leete in *Biosynthesis*, Vol. 6, specialist periodical report, J. D. Bu'Lock, Ed., The Royal Society of Chemists, 1980, p. 181.

285. R. K. Hill and M. G. Bock, *J. Am. Chem. Soc.* **100,** 637 (1978).

286. E. Leete, *J. Chem. Soc. Chem. Commun.*, 1055 (1978).

287. E. Leete and M. E. Mueller, *J. Am. Chem. Soc.* (in press).

288. E. Leete, *J. Chem. Soc. Chem. Commun.*, 1091 (1972).

289. N. Bodor and R. Pearlman, *J. Am. Chem. Soc.* **100,** 4946 (1978).
290. J. L. R. Chandler and R. K. Gholson, *Phytochemistry* **11,** 239 (1972).
291. E. Leete, *Planta Med.* **36,** 97 (1979).
292. S. H. Hedges and R. B. Herbert, *Phytochemistry* **20,** 2064 (1981).
293. T. A. Smith, *Phytochemistry* **18,** 1447 (1979).
294. D. B. Maudsley, *Biochem. Pharmacol.* **28,** 153 (1979).
295. M. R. Suresh, S. Ramakrishna, and P. R. Adiga, *Phytochemistry* **17,** 57 (1978).
296. M. J. Montague, T. A. Armstrong, and E. Jaworski, *Plant Physiol.* **63,** 341 (1979).
297. D. Yoshida, *Plant Cell Physiol.* **10,** 393, 923 (1969).
298. L. J. J. Mestichelli, R. N. Gupta, and I. D. Spenser, *J. Biol. Chem.* **254,** 640 (1979).
299. E. Leete in *Biosynthesis*, Vol. 4, specialist periodical report, J. D. Bu'Lock, Ed., The Chemical Society, 1976, p. 98.
300. E. Leete in *Biosynthesis*, Vol. 5, specialist periodical report, J. D. Bu'Lock, Ed., The Chemical Society, 1977, p. 137.
301. E. Leete, *J. Chem. Soc. Chem. Commun.*, 1524 (1971).
302. A. A. Liebman, B. P. Mundy, M. L. Rueppel, and H. Rapoport, *J. Chem. Soc. Chem. Commun.*, 1022 (1972).
303. A. Meister, in *Biochemistry of the Amino Acids*, Vol. 2, 2nd ed., Academic, New York, 1965, p. 939.
304. R. N. Gupta and I. D. Spenser, *Phytochemistry* **9,** 2329 (1970). A private communication from Spenser described the retention of tritium when [6-^{14}C, 2-^{3}H]lysine was incorportaed into anabasine.
305. R. F. Dawson, *J. Am. Chem. Soc.* **73,** 4218 (1951).
306. E. Leistner and I. D. Spenser, *J. Chem. Soc. Chem. Commun.*, 378 (1975).
307. I. D. Spenser in *Comprehensive Biochemistry*, Vol. 20, M. Florkin and E. H. Stotz, Eds., Elsevier, 1968, pp. 275–281.
308. J. L. Hess and N. E. Tolbert, *J. Biol. Chem.* **241,** 5705 (1966).
309. E. Wada, *Arch. Biochem. Biophys.* **62,** 471 (1956).
310. R. F. Dawson, *J. Am. Chem. Soc.* **67,** 503 (1945).
311. R. F. Dawson, *Am. J. Bot.* **32,** 416 (1945).
312. J. M. Essery, P. F. Juby, L. Marion, and J. Trumbull, *Can. J. Chem.* **41,** 1142 (1963).
313. T. Kisaki and E. Tamaki (née Wada), *Arch. Biochem. Biophys.* **92,** 351 (1961).
314. T. Kisaki and E. Tamaki, *Arch. Biochem. Biophys.* **94,** 252 (1961).
315. T. Kisaki and E. Tamaki, *Naturwissenschaften* **47,** 541 (1960).
316. E. Wenkert, *Experientia* **10,** 346 (1954).
317. J. C. Craig, N. Y. Mary, N. L. Goodman, and L. Wolf, *J. Am. Chem. Soc.* **86,** 3866 (1964).
318. L. Egri, *Fachliche Mitt. Osterr. Tabakregie*, 19 (1957).
319. G. S. Iljin and K. B. Serebrovskaya, *Dokl. Akad. Nauk SSSR* **118,** 139 (1958).
320. T.-L. Nguyen, L. D. Gruenke, and N. Castagnoli, *J. Med. Chem.* **19,** 1168 (1976).
321. S. S. Hecht, C. B. Chen, R. M. Ornaf, E. Jacobs, J. D. Adams, and D. Hoffmann, *J. Org. Chem.* **43,** 72 (1978).
322. K. Blain and R. Ciszewska, *Acta Agrobot.* **26,** 303 (1973); *Chem. Abstr.* **81,** 60961 (1974).
323. J. W. Saunders and L. P. Bush, *Plant Physiol.* **64,** 236 (1979).
324. L. A. Hadwiger, S. E. Badiei, G. R. Waller, and R. K. Gholson, *Biochem. Biophys. Res. Comm.* **13,** 466 (1963).
325. E. Wada, T. Kisaki, and K. Saito, *Arch. Biochem., Biophys.* **79,** 124 (1959).
326. P. J. Murphy, *J. Biol. Chem.* **248,** 2796 (1973).

327. M. Bruhmuller, H. Mohler, and K. Decker, *Eur. J. Biochem.* **29,** 143 (1972).

328. E. B. Sanders, J. F. DeBardeleben, and T. J. Osdene, *J. Org. Chem.* **40,** 2848 (1975).

329. H. McKennis in *Tobacco Alkaloids and Related Compounds*, U.S. Von Euler, Ed., MacMillan, New York, 1965, p. 53.

330. J. W. Gorrod and P. Jenner, The Metabolism of Tobacco Alkaloids, in *Assays in Toxicology*, Vol. 6, Academic, New York, 1975, p. 35.

331. S. S. Mirvish, J. Sams, and S. S. Hecht, *J. Natl. Cancer Inst.* **59,** 1211 (1977).

332. E. H. Poindexter, A. Bavley, and H. Wakeham, in *Proceedings 3rd World Tobacco Scientific Congress* (Salisbury, Rhodesia), 1963, p. 550.

333. G. S. Iljin, in *Proceedings 3rd World Tobacco Scientific Congress* (Salisbury, Rhodesia), 1963.

334. W. G. Frankenburg, A. M. Gottscho, I. W. Maynard, and T. C. Tso, *J. Am. Chem. Soc.* **74,** 4309 (1952).

335. W. G. Frankenburg and A. A. Vaitekunas, *J. Am. Chem. Soc.* **79,** 149 (1957).

336. J. Arditti, *Am. J. Bot.* **54,** 291 (1967).

337. W. G. Frankenburg and A. M. Gottscho, *J. Am. Chem. Soc.* **77,** 5728 (1955).

338. C. H. Rayburn, W. R. Harlan, and H. R. Hammer, *J. Am. Chem. Soc.* **63,** 115 (1941).

339. C.-H. Yiu, Investigation of Aberrant Biosynthesis in Tobacco, M.Sc. dissertation, University of Minnesota, 1976.

340. E. Leete and M. C. L. Louden, in *Abstracts American Chemical Society Meeting, September 1963*, American Chemical Society, New York, 1963, p. 1C.

341. E. Leete and C. F. Fyrand, unpublished work, 1967–1968.

342. E. Leete, *J. Am. Chem. Soc.* **104,** 1403 (1982).

343. D. G. O'Donovan and M. F. Keogh, *J. Chem. Soc. C*, 223 (1969).

344. M. Ya. Lovkova, N. I. Klimentjeva, and G. V. Lazurjevskii, *Prikl. Biokhim. Mikrobiol.* **15,** 775 (1979).

345. E. Leete, *J. Chem. Soc. Chem. Commun.*, 821 (1979).

346. E. Leete, *Tetrahedron Lett.*, 4527 (1979).

347. M. Takagi, S. Funahashi, K. Ohta, and T. Nakabayashi, *Agric. Biol. Chem.* **44,** 3019 (1980).

348. S. R. Johns, J. H. Russel, and M. L. Heffernan, *Tetrahedron Lett.*, 1987 (1965).

349. M. Ya. Lovkova, E. Nurimov, and G. S. Iljin, *Biokhimiya* **39,** 388 (1974).

350. E. Nurimov, M. Ya. Lovkova, and B. A. Abdusalamov, *Prikl. Biokhim. Mikrobiol.* **10,** 785 (1974).

351. N. I. Klimentjeva and L. K. Klyschev, *Trudy Inst. Bot. Akad. Nauk Kaz. SSR* **29,** 164 (1971); *Chem. Abstr.* **76,** 70243 (1972).

352. L. K. Klyschev, R. K. Moiseev, N. I. Klimentjeva, and B. S. Ibraeva, *Izvest. Akad. Nauk. SSSR, Ser. Biol.*, 1 (1977).

353. M. Ya. Lovkova, E. Nurimov, and G. S. Iljin, *Biokhimiya* **39,** 523 (1974).

354. E. Nurimov, M. Ya. Lovkova, and G. S. Iljin, *Prikl. Biokhim. Mikrobiol.* **11,** 149 (1975).

355. B. L. Lamberts, L. J. Dewey, and R. U. Byerrum, *Biochim. Biophys. Acta.* **33,** 22 (1959).

356. M. Ya. Lovkova and E. Nurimov, *Prikl. Biokhim. Mikrobiol.* **13,** 129 (1977).

357. E. Nurimov, M. Ya. Lovkova, and B. L. Abusalamov, *Prikl. Biokhim. Mikrobiol.* **13,** 628 (1977).

358. K. Mothes, H. R. Schütte, H. Simon, and F. Weygand, *Z. Naturforsch.* **14B,** 49 (1959).

359. C. Schöpf, F. Braun, and A. Komzak, *Chem. Ber.* **89,** 1821 (1956).

360. H. R. Schütte, K. L. Kelling, D. Knofel, and K. Mothes, *Phytochemistry* **3,** 249 (1964).

361. M. Ya. Lovkova, B. S. Ibraeva, and L. K. Klyschev, *Prikl. Biokhim. Mikrobiol.* **13,** 919 (1977).

362. M. Ya. Lovkova, E. Nurimov, and B. A. Abdusalamov, *Isvest. Akad. Nauk. SSSR, Ser. Biol.*, 509 (1977).

363. L. K. Klyschev, R. K. Moiseev, N. I. Klimentjeva, and B. S. Ibraeva, *Isvest. Akad. Nauk SSSR, Ser. Biol.* 67 (1977).

364. M. Golebiewski and I. D. Spenser, *J. Am. Chem. Soc.* **98,** 6726 (1976).

365. M. Wink, T. Hartmann, and H.-M. Schiebel, *Z. Naturforsch.* **34C,** 704 (1979).

366. E. Werle and K. Koekbe, *Ann.* **562,** 60 (1949).

367. N. I. Klimentjeva, L. K. Klyschev, and M. Ya. Lovkova, *Isvest. Akad. Nauk. SSSR, Ser. Biol.* 627 (1970).

368. S. Ohta and M. Yatazawa, *Biochem. Physiol. Pflanz.* **175,** 382 (1980).

369. K. Hasse and P. Berg, *Naturwissenschaften* **44,** 584 (1957); *Biochem. Z.* **331,** 349 (1959).

370. K. Mothes and H. R. Schütte, *Angew. Chem., Int. Ed.* **2,** 341 (1963).

371. T. C. Tso, *Bot. Bull. Acad. Sinica* **4,** 75 (1963).

372. T. A. Smith and F. J. Richards, *Biochem. J.* **84,** 292 (1962).

373. T. Takahashi and D. Yoshida, *J. Soil Sci. Manure* (*Japan*) **31,** 39 (1960).

374. T. A. Smith, *Phytochemistry* **2,** 241 (1963).

375. T. A. Smith and J. L. Garraway, *Phytochemistry* **3,** 23 (1964).

376. D. Yoshida, *Plant Cell Physiol.* **3,** 391 (1962).

377. C. R. Hutchinson, Biosynthetic Studies in *Convalaria majalis, Nicotiana tabacum,* and *Tulipa generiana,* Ph.D. dissertation, University of Minnesota, Minneapolis, 1970.

378. H. J. Vogel, *Proc. Natl. Acad. Sci. USA* **45,** 1717 (1959).

379. M. Tabata, H. Yamamoto, H. Hiraoka, Y. Marumoto, and M. Konoshima, *Phytochemistry* **10,** 723 (1971).

380. T. Furuya, H. Kojima, and K. Syono, *Chem. Pharm. Bull.* **14,** 1189 (1966), **15,** 901 (1967).

381. T. Furuya, H. Kojima, and K. Syono, *Phytochemistry* **10,** 1529 (1971).

382. S. Ohta, *Agr. Biol. Chem.* **42,** 873, 1245, 1733 (1978).

383. T. Ogino, N. Hiraoka, and M. Tabata, *Phytochemistry* **17,** 1907 (1978).

384. M. Tabata and N. Hiraoka, *Physiol. Plant.* **38,** 19 (1976).

385. H. Smith and D. W. Pearson, Eur. Patent. Appl. 7244, *Chem. Abstr.* **92,** 194784 (1980).

386. A. M. Kinnersley and D. K. Dougall, *Planta* **149,** 205 (1980).

387. I. Shiio and S. Ohta, *Agr. Biol. Chem.* **37,** 1857 (1973).

388. M. Takahashi and Y. Yamada, *Agric. Biol. Chem.* **37,** 1755 (1973).

Chapter Four

The Toxicology and Pharmacology of Diterpenoid Alkaloids

M. H. Benn
Chemistry Department
University of Calgary
Alberta T2N 1N4 Canada

John M. Jacyno
Department of Pharmaceutical Chemistry
Medical College of Virginia, Virginia Commonwealth University, Richmond, Virginia 23298

CONTENTS

1. INTRODUCTION

At the outset, the reader should note some restrictions in the scope of this chapter.

A *diterpenoid alkaloid* is generally understood to be a naturally occurring nitrogenous base, or derivative, whose nitrogen-functionalized skeleton can be identified as formed from some C_{20}-terpenoid precursor. However, in this review, we have largely restricted ourselves to those diterpenoid alkaloids that are found in plants of the genera *Aconitum* and *Delphinium* and rarely, if at all, elsewhere. (Here, and later, we are following the taxonomically questionable [1] but common practice of not distinguishing *Consolida* from *Delphinium*.) Some closely related bases from *Garrya* receive attention, but those from *Anopteris* [2] and *Spiraea* [3] do not, since we had little information about their pharmacological properties. For the same reason, the structurally rather different alkaloids from *Incacina* [4] and *Daphniphyllum* [5] have also been excluded. (An outline of the pharmacology of the *Daphniphyllum* alkaloids was provided by Yamamura and Hirata [5] in their review of the chemistry of these bases.)

Within the set of diterpenoid alkaloids thus defined, not all have been subjected to toxicological and pharmacological studies. We have summarized what is known for some 64 compounds (natural alkaloids and chemical transformation products thereof).

Notwithstanding these limitations, since the recent reviews of the subject [6–8] are not readily available, we hope that our synopsis will prove useful to those interested in the toxicology and pharmacology of diterpenoid alkaloids.

2. THE TOXICOLOGY AND PHARMACOLOGY OF DITERPENOID ALKALOIDS

2.1. Ethnopharmacology of Aconitum, Delphinium, and Garrya

There is a long and fascinating history of the use by various civilizations of species of *Aconitum* and *Delphinium* as sources of poisons and medicinals. What we provide here is a very bare outline of this story, designed to provide the reader interested in such matters with some leads to the literature.

Aconitum has had an evil reputation since antiquity. In the mythology of ancient Greece, the plants were said to have been born of the salival spatterings of Cerberus, gate hound of Hell [9, 10b]. The generic name *Aconitum* comes to us from the ancient Greeks and may derive from ακοντιον, a "javelin," in recognition of the use of some of these plants to poison their points [11], although other derivations relating to a place of origin or habitat have also been suggested [9b, 10a]. Whatever, it is clear that *Aconitum* preparations have been long and widely used as weapon poisons and seem to have been effective in this role [10e, 11] (at least against mammals up to the size of bears and tigers; the whale-lance poisons of the natives of the Pacific Northwest were probably more wishful than proved in their efficacy [11c]). The use of *Aconitum* to poison arrows and baits for killing wolves has given us the species name *lycoctonum* (from the ancient Greek λυκο-κτονο = "wolf slayer") and the old common name in English (with the equivalent in many other European languages): wolfsbane. (Monkshood, the common name now used in English, derives from the characteristic cowl shape of the upper sepal and came into use in the Middle Ages following the extirpation of wolves in Britain.)

Apart from employing *Aconitum* to do away with such undesirables as wolves, boars, tigers, panthers, and rodents, and in warfare, man has used it for homicide. Ovid and others were of the opinion that Medea used *Aconitum*, as in the poisoned cup prepared for Theseus [9]. Other old reports tell us that decoctions of *Aconitum* were used to dispose of the aged and infirm on Khios [10a, b] and, as an alternative to hemlock, for the similar execution-by-suicide of criminals [9b]. That poisoning by *Aconitum* may have got out of hand in Imperial Rome is suggested by Trajan's proscription that made cultivation of the plants a capital offense [10b].

The history of *Delphinium* is more peaceful. The genus name is taken from the fanciful resemblance of the unopened flower bud to a little dolphin (δελφινιον), and so too with the common name Larkspur [10a–c]. Although various species have been recognized since antiquity as poisonous toward mammals, this has usually been in the context of accidental poisonings, particularly of children and cattle. The latter is a problem that still bedevils the ranchers of the western foothills of the Rocky Mountains [6, 12]. Various cultures have used *Delphinium* as pediculicides. Thus both Pliny [10a] and Dioscorides [10c] noted the effectiveness of crushed seed preparations of the plant we now know as *D. staphisagria* L. to kill body lice. For the same purpose, a similar preparation was apparently a standard issue in the British Army at the time of the battle of Waterloo [10f], and the practice is still recommended [10g].

The medicinal use of *Aconitum* and *Delphinium* also spans many centuries. Typically [10a–d], *Aconitum* preparations have been used in cardiotonics and also as febrifuges, sedatives, and anodynes. Anticancer use has also been reported [13] but remains of unproved value. *Delphinium* extracts have also been employed [10a–e, 14] in analgesic balms (whence the old name "King's consound" for the plant now described as *Consolida regalis* or *D. consolida*) and also as sedatives, emetics, and anthelmintics.

Among the miscellaneous folk uses of these plants, *Aconitum* has been used in aphrodisiacs [10e] and in potions used by witches (it has been suggested [10h] that the toxic effects contributed toward a sensation of flying!).

The New World genus *Garrya* immortalizes Nicholas Garry of the Hudson's

Bay Company, who helped David Douglas with his botanizing in western America [15]. The ethnopharmacology of *Garrya* is not as well documented as that of *Aconitum* or *Delphinium*, but in Mexico *G. laurifolia* has had some use in treating chronic diarrhea [16].

The use of *Aconitum* preparations as mammalian poisons and *Delphinium* as pediculicides has been shown in a number of cases, and may confidently be inferred in many others, to be due to the diterpenoid alkaloids present in the plants. So too with their therapeutic use, especially the use of *Aconitum* in anodynes, and as cardiotonics, cf. the properties of the individual alkaloids as described in the section that follows.

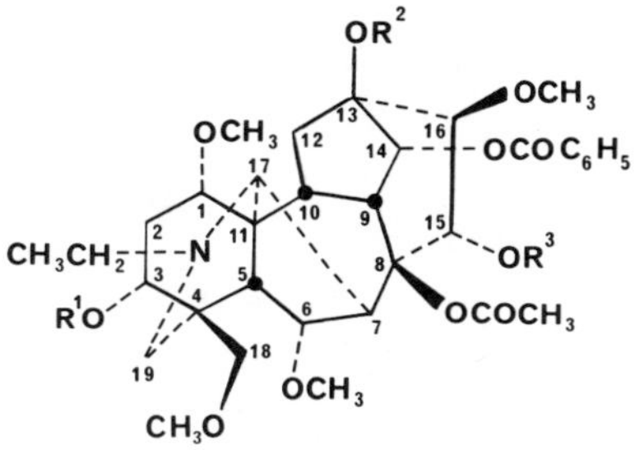

1 $R^1 = R^2 = R^3 = H$

2 $R^1 = R^3 = COCH_3 \quad R^2 = H$

3 $R^1 = R^2 = R^3 = COCH_3$

2.2. Properties of Individual Alkaloids

2.2.1. Aconitine. Of all the diterpenoid alkaloids, most attention has been given to aconitine (**1**), a reflection of its extremely high toxicity (see Table 1).

Much of the early work, which has been reviewed [17], is fraught with questions concerning the identity and chemical homogeneity of the alkaloid used. Thus the important studies of Cash and Dunstan [18, 19] with "aconitine" isolated from English *A. napellus* L. may well have been conducted with material that was in part mesaconitine (**11**). (European *A. napellus* spp. contain varying proportions of aconitine and mesaconitine [20].) However, the pharmacological properties of these two alkaloids are very similar, and the original observations of Cash and Dunstan have largely been substantiated by later workers using authenticated aconitine.

Cash and Dunstan [18] reported that the subcutaneous administration of aconitine hydrobromide to cats, anesthetized with ether, produced bradycardia; larger doses caused tachycardia and pronounced cardiac irregularities, with the development of ventricular extrasystoles and fibrillation, terminating in cardiac arrest. The arrhythmias were partially antagonized by the intravenous adminis-

Table 1. Acute Toxicities of Diterpenoid Alkaloids (LD_{50} as mg/kg)[a]

Drug	Mouse I.V.	Mouse I.P.	Frog	Rabbit	Guinea Pig	Dog
Aconitine (**1**)	0.12–0.17	0.33–0.38	0.075–1.65	0.04–0.05	0.06–0.12	0.07
Diacetylaconitine (**2**)	—	—	39	4.2	0.6–4.2	—
Triacetylaconitine (**3**)	125	—	—	—	—	—
Benzoylaconine (**4**)	23	70	289	27.2	24–30	20
Aconine (**8**)	120	—	1000–1750	—	275–300	200
Pyroaconitine (**9**)	—	—	48–50	3.8–4.0	3.8	—
Mesaconitine (**11**)	0.1	0.21–0.3				
Benzoylmesaconine (**12**)	21	240				
Pyromesaconitine (**13**)	—	60–80				
Mesaconine (**14**)	—	300–330				
Pyromesaconine (**15**)	—	900–1000				
Hypaconitine (**16**)	0.47	1.1				
Benzoylhypaconine (**17**)	23	120				
Pseudaconitine (**19**)	0.36	—	0.1–1.2	0.04–0.05	0.45	—
Pseudaconine (**21**)	—	1700	—	—	—	—
Bikhaconitine (**23**)	—	—	1.25	0.08	—	—
Jesaconitine (**25**)	—	0.35	0.25	0.05	0.018	—
Aconifine (**27**)	0.22	—	—	—		—
Delphinine (**28**)	—	—	0.05–0.1	1.5–3.0		1.5–3.0
Delphinine methochloride (**29**)	1–2	—		1–2		1–2
Delphonine methochloride (**30**)	1–2	—		1–2		
Lappaconitine (**31**)	6.1–11.5	9.1				4.8
Lappaconitine diacetate (**32**)	145	—				

Lappaconine (**34**)	142–144	—	8–16			
Talatizamine (**35**)	115	—	—			
Karakoline (**36**)	51.5	298.3	—			
Methyllycaconitine (**37**)	3.0–3.5	18	3.0–3.5	2.0–2.5		
Delsemine (**38**)	—	37.4	—			
Elatine (**42**)	5.0–5.5	—	1.5–2.0			
Anthranoyllycoctonine (**43**)	—	75 (LD_{80})				
Lycoctonine (**46**)						
hydrochloride	—	350 (LD_{80})				
Delsoline (**51**)	175	550	—	—	—	20–25
Delcosine (**52**)	108.7	—				
Condelphine (**54**)	(73: rat)					
Deltaline (**55**)	>300	—	300–400			
Heteratisine (**56**)	180–192					
Veatchine (**59**)						
hydrochloride	12.5					
Garryfoline (**60**)						
oxalate	16.2					
Garryine (**61**)						
hydrochloride	12.8					
Cuauchichicine (**62**)						
hydrochloride	<10					
Napelline (**63**)	87.5					
Songorine (**64**)	142.5					

[a] The figures, extracted by Jacyno [6] from a variety of sources, are offered as a guide to relative toxicities.

tration of atropine. Hypothermia and a fall in blood pressure were also noted. In guinea pigs, additional symptoms observed included retching, salivation, diarrhea, clonicotonic convulsions, muscular weakness, and marked dyspnea. Respiratory failure appeared to precede heart failure. Experiments with frogs indicated a marked involvement of the central nervous system in the action of aconitine. Sensory nerves were at first stimulated then depressed, but motor nerves were unaffected except by relatively high doses of the drug. Skeletal muscle showed fibrillation after quite large doses also, but the effects were not observed in preparations pretreated with tubocurarine, benzaconine (Section 2.2.4), or aconine (Section 2.2.9) (q.v.).

The results of further investigations of some of these, and other effects are described below.

Cardiovascular Effects. Despite the broad range of physiological effects elicited by aconitine, most attention has been focused on aconitine-induced cardiac arrhythmias. The current widespread use of aconitine for the generation of model cardiac arrhythmias was introduced and developed by Scherf [21], who first used the drug to probe the mechanisms of flutter and fibrillation in the heart. (The meanings of the terms *atrial flutter* and *atrial fibrillation* in the context of aconitine-induced arrhythmias have been discussed by Mendez and Kabela [22].)

The arrhythmias were generally induced by topical application of aconitine to the exposed dog heart. The results of these investigations, together with much related material of a clinical nature, have been analyzed by Scherf and Schott [23].

Other studies relating to the mechanism of aconitine-induced cardiac arrhythmias have been summarized by Tanz et al. [24, 25, 26]. These workers themselves investigated the effects of aconitine on an *in vitro* cat papillary muscle preparation. They were able to establish a dose–response relationship between the time of onset of automaticity and drug concentration and also to show that normally quiescent papillary muscle, when exposed to graded doses of aconitine in the absence of electrical stimulation, failed to exhibit automaticity. Furthermore, after exposure to the drug at a concentration of 3.9×10^{-4} *M*, the muscle was subsequently unresponsive to electrical stimulation. They also found that neither catecholamine depletion (by prior reserpinization) nor beta-adrenergic blockade (with practolol) antagonized the aconitine-induced automaticity. In confirmation of experiments in other physiological systems (including non-cardiac tissue preparations), they observed that reduced sodium concentration in the bathing medium or addition of tetrodotoxin antagonized the automaticity and tachycardia caused by aconitine. Tetrodotoxin, a specific blocker of sodium channels, also reversed aconitine-induced tachycardia in the whole-heart preparation of Langendorff. They concluded that aconitine-induced tachycardia and automaticity were related to an enhancement of sodium permeability in cell membranes. This conclusion is in good agreement with those drawn from many related investigations and is supported especially by the work of Catterall with cultured mouse neuroblastoma cells [27]. Further discussion of this process is deferred to Section 3.1.1.

Smooth Muscle Effects. The effect of aconitine on smooth muscles other than the heart has not been extensively investigated. Early work in Japan showed that the drug elicited contractions of the rabbit ileum [28, 29, 30] and potentiated the spontaneous contractions of isolated rabbit uterus [28, 31]. A recent Soviet review [8], however, stated that aconitine in concentrations of 1.5×10^{-8}–1×10^{-6} *M* did not have any distinct effects on the spontaneous contractions of isolated sections of intestine (the animal was not specified, but much of the Soviet work of this kind has been with rat intestine), on the intestinal contractions induced with acetylcholine, or on a guinea pig uterus preparation.

Recently, a Japanese group [7] has undertaken a reexamination of aconitine pharmacology using scrupulously purified and characterized drug material. These workers found that at a concentration of 1.5×10^{-7} *M*, aconitine did not affect the isolated guinea pig ileum, but at a concentration of 1.5×10^{-6} *M*, the drug caused prolonged contractions that could be completely blocked by atropine. (Why this result differs from that of Tulyaganov et al. [8] is not clear unless it is because of differences in the species used.) Contractions of the ileum induced by acetylcholine, histamine, or barium chloride were unaffected.

The Japanese workers [7] also showed that, at concentrations of 1.5×10^{-5} *M*, aconitine induced contractions of the isolated guinea pig vas deferens, which could be blocked by phentolamine. The contractions induced by noradrenaline or acetylcholine were unaffected by aconitine, but those induced by tyramine were slightly potentiated. In the electrically stimulated hypogastric nerve–vas deferens preparation of the guinea pig, aconitine partially suppressed the response at a concentration of 1.5×10^{-7} *M*. At a concentration of 10^{-5} *M*, the drug initially potentiated the contractions, then caused a prolonged inhibition, which could not be overcome with repeated washings. The response of the isolated rabbit jejunum was markedly potentiated by aconitine at a concentration of 1.5×10^{-6} *M*, whereas the inhibition of contractions induced by electrical stimulation of the mesenteric nerve was almost completely abolished by the same concentration of this drug. From these results, the authors concluded that aconitine had no direct effect on smooth muscle, but exerted a blocking action on sympathetic nerves. The mechanism by which mesaconitine [11] acts on the guinea pig vas deferens was investigated by these workers in greater detail and is discussed in Section 2.2.12.

Skeletal Muscle Effects. There is a paucity of data regarding the action of aconitine on skeletal muscle. Soviet workers [8] have reported that aconitine in doses of 1.0×10^{-2}–2.0×10^{-2} mg/kg did not change the amplitude of the electrically excited contractions in a cat sciatic nerve–gastrocnemius muscle preparation. In contrast, Sato et al. [7] have shown that contractions of the rat diaphragm induced by electrical stimulation of the phrenic nerve are inhibited by aconitine at a concentration of 1.5×10^{-6} *M*. This inhibition is not relieved by physostigmine.

In our hands [6], a graded response of the rat phrenic nerve–diaphragm preparation was observed in response to aconitine. The resulting log dose–response curve was essentially parallel to those of tubocurarine and methyllycaconitine with an ED_{50} of 6.6×10^{-7} *M*, and ED_{100} of 9.0×10^{-7} *M*. Tissues poisoned with concentrations of aconitine of $\sim 10^{-6}$ *M* did not respond to direct electrical stimulation of the muscle.

An electrophysiological study of the effects of aconitine in skeletal muscle of the frog *Rana pipiens* was made by Ellis and Bryant [32]. When the drug was applied to a sartorius muscle preparation in concentrations ranging from 1.5×10^{-8} to 1.5×10^{-4} *M*, bursts of repetitive firing were produced. After this activity, the muscle fibers remained quiescent, in a depolarized state. The firing induced by low concentrations of aconitine (1.5×10^{-8}–1×10^{-7} *M*) could be blocked by tubocurarine (10^{-4} *M*), but higher concentrations of aconitine (1.5×10^{-6} *M*) caused repetitive activity even in the presence of tubocurarine. The authors suggested that in low concentrations, aconitine caused its effects by stimulating repetitive activity in the motor nerve, whereas at higher concentrations, it exerted direct action on muscle cells as well as on nerve tissue. Additional experiments led these workers to conclude that aconitine acted on the sodium activation or inactivation mechanisms in such a way that each action potential produced a sustained increase in the sodium conductance (see Section 3.1.1).

Central Effects. Although the ability of aconitine to affect the central nervous system (CNS), including consequential cardiovascular effects [18, 33], has been recognized for some time, exploration of these effects is relatively recent. Thus Bhargava et al. [34, 35] were first to show that the intracerebroventricular administration of aconitine in low doses ($\sim 2.0 \times 10^{-3}$ mg/kg) produced hypertension and cardiac arrhythmias in dogs. On the basis of extensive surgical and chemical modification of various neural pathways, these workers concluded that aconitine exerted its cardiovascular effects both by exciting central beta-adrenergic neurons, with subsequent release of catecholamines, and by exciting central cholinergic neurons of the muscarinic type.

Similar conclusions were reached by a group in Kenya [36, 37] working with cats, and by others in Egypt [38, 39], with rabbits. The latter group demonstrated that a wide range of physiological responses, including ventricular fibrillation, hypotension, respiratory paralysis, and convulsive siezures, could be elicited by the intracerebroventricular administration of aconitine. Conversely, the intravenous administration of the drug markedly altered brain biopotentials in a dose-dependent manner. Thus injection of 0.02 mg/kg of aconitine produced immediate stimulation, which was characterized by a general decrease in the mean voltage, accompanied by an increase in the total number of discharges and the number of rapid beta waves. A dose of 0.04 mg/kg evoked the same pattern of response, but with convulsive discharges. A dose of 0.06 mg/kg caused complete cessation of electrical activity a few minutes before cardiac arrest. Intracerebroventricular administration of aconitine produced brain (EEG) effects that were similar to, but more pronounced than, those seen after intravenous administration.

Sato et al. [7] have found that aconitine, unlike mesaconitine, when given subcutaneously in doses up to 0.06 mg/kg did not potentiate barbiturate-induced sleeping time in mice, and Soviet workers [8] have shown that intravenous doses of up to 0.01 mg/kg aconitine did not affect conditioned reflexes in dogs. However, there was a marked inhibition of activity on wheel-cage revolution by mice (given the alkaloid at 0.3–1.0 mg/kg p.o.), and the alkaloid also induced (at 0.3–1.0 mg/kg) a significant drop in the rectal temperature of mice, suggesting an inhibitory effect on the CNS.

Other Effects. Aconitine was shown to activate mechanoreceptors in the lung, carotid sinus, and heart of cats, and in the slowly adapting stretch receptors of crayfish by Wellhoner et al. [40, 41, 42]. Fairly low concentrations (1.5×10^{-6}-$-1 \times 10^{-4} M$) of aconitine produced a steady-state activity attributed to a depolarization of the nerve endings through enhanced sodium permeability. Haegerstam [43] found that the drug apparently stimulated sensory units in the tooth of the cat in a similar way.

Other workers [44, 45] have reported the excitation of certain chemoreceptors by aconitine, and that sensory receptors in the skin were paralyzed after a brief increase in pain excitability [46]. In fact, various *Aconitum* extracts and preparations containing aconitine have been reported [10, 47, 48, 49, 50] to have analgesic activity, but the experimental methodology and results have been somewhat inconsistent. However, a recent study [51] of aconitine and some closely related compounds in assays for analgesia in the mouse showed that aconitine at a dose of 0.06 mg/kg s.c. did have analgesic effects toward tail pinching and acetic acid–induced writhing—although not in a hot-plate (contact) test. Mesaconitine was somewhat more active, and hypaconitine less so. Aconitine was also found to be anti-inflammatory in two tests (acetic acid i.p. in the rat; carrageenin on the mouse hind paw), but not so in some chronic cases, or on UV-light-induced erythema (guinea pig) [51b] (see also Sections 2.2.12 and 2.2.17).

Effects of Metabolism. The biochemical correlates of aconitine treatment are virtually unstudied. Barankova and Sorm [52] incubated aconitine with homogenates and slices of rat brain cortex, kidney cortex, and liver under various conditions. These workers found that in brain tissue, the oxidation of glucose, pyruvate, and lactate was stimulated, but that of succinate, α-ketoglutarate, and L-glutamate was not. Metabolism was not stimulated in kidney or liver tissues, but their dry weights were increased due to the retention of an unidentified crystalline substance.

Recently, Bekemeier and co-workers [53] observed that aconitine had no influence on prostaglandin biosynthesis in the incubated homogenate of the rabbit renal medulla.

Gendenshtein et al. [54] have reported on the metabolic changes in cat heart during aconitine-induced arrhythmia; they were found to be accompanied by decreases in the levels of glycogen and nucleic acids, and in the activities of acid phosphatase and cholinesterase.

Summary. Aconitine emerges as a toxin that exerts its action both centrally and peripherally, with predominant effects on the cardiovascular and respiratory systems. There is evidence that a considerable part of the peripheral activity observed *in vivo* is centrally mediated. Aconitine interacts with nerve and muscle tissue, although relatively high concentrations are generally required to affect the latter. It appears to be capable of evoking transmitter release.

Numerous electrophysiological studies [55–67] have shown that aconitine causes prolonged depolarization after stimulation of a nerve, and that this depolarization can be blocked by tetrodotoxin. Thus at the molecular level aconitine appears to be a neurotoxin by virtue of an ability to bind to the nerve-cell membrane in such a way as to prevent the normal closing of sodium channels, a process that may be visualized as an allosteric interaction [27, 68, 69] (see also Section 3.1.1).

2.2.2. Diacetylaconitine. Dunstan and Carr [70] prepared diacetylaconitine by treatment of aconitine with acetyl chloride. The structure of the compound was not proved (and there are no later references in the literature to its synthesis), but it probably corresponds to the 3,15-diacetate (**2**).

Diacetylaconitine was found [18] to be somewhat less toxic (by a factor of 40–50) than aconitine, and although it caused strong bradycardia, arrhythmia was not produced as consistently as by aconitine. The compound was also reported to exhibit marked inhibition of motor-nerve terminals and skeletal muscle.

2.2.3. Triacetylaconitine. Triacetylaconitine, presumably (**3**), was also first obtained by Dunstan and Carr [70] by treating aconitine with acetyl chloride at room temperature. They indicated that the compound exhibited the same physiological actions as aconitine, whereas Tulyaganov et al. [8] have reported that "aconitine triacetate" was relatively inactive.

2.2.4. Benzaconine. Mild hydrolysis of aconitine results in cleavage of the 8-acetate group to give a compound known as benzaconine—or benzoylaconine (**4**). Cash and Dunstan [18] carried out this reaction and subjected the product to a number of pharmacological tests. The most notable difference in activity, compared with their aconitine, was a drastic reduction of the acute toxicity by a factor of about 300.

Cardiovascular Effects. Small doses of benzaconine produced tachycardia, but larger doses caused a slowing of the heart, with some arrhythmia. Benzaconine showed some antagonism toward the effects of aconitine upon the heart, but not to the same extent as atropine. Respiration was slowed, without any prior acceleration, and blood pressure was markedly reduced. Death occurred from respiratory failure, without spasm or convulsions.

Other Effects. Paresis and clonic movements were also observed in guinea pigs, but there was no salivation or hypothermia. The sensory nerves were not greatly affected by ordinary doses, but the presence of central effects was clearly

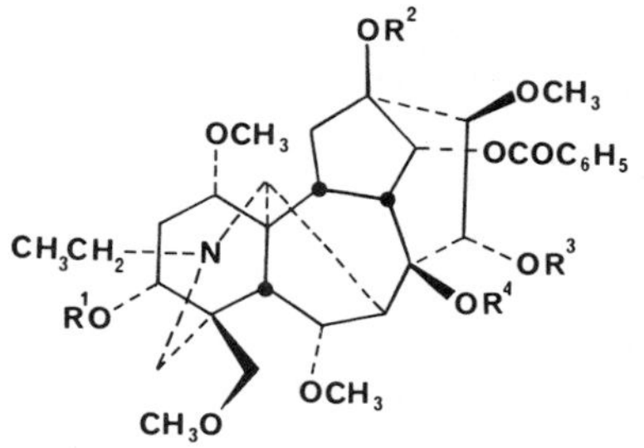

4 $R^1 = R^2 = R^3 = R^4 = H$

5 $R^1 = R^3 = COCH_3$ $R^2 = R^4 = H$

6 $R^1 = R^2 = R^3 = COCH_3$ $R^4 = H$

7 $R^1 = R^2 = R^3 = H$ $R^4 = CH_3$

indicated by a lethargic, seminarcotized state. The alkaloid also extended hexobarbital-induced sleep in mice and showed hyperthermic properties (at 30 mg/kg s.c.). Benzaconine also exhibits slight analgesic [51a] and anti-inflammatory effects [51b] in the tests described in Section 2.2.1.

2.2.5. Diacetylbenazconine. Diacetylbenzaconine (**5**) has only been reported once in the literature, by Dunstan and Carr [70], who obtained the compound by treating a solution of benzaconine (**4**) in chloroform with acetic anhydride. Although the structure has not been confirmed, we suggest that it corresponds to 3,15-diacetylbenzaconine, on the grounds that the secondary alcohols at C(3) and C(15) in benzaconine would be expected to acetylate in preference to the tertiary alcohols at C(8) and C(13). Diacetylbenzaconine was described by Dunstan and Carr as apparently nontoxic.

2.2.6. Triacetylbenzaconine. "Triacetylbenzaconine," like diacetylbenzaconine, has only been reported once, by Dunstan and Carr [70], and its structure has not been rigorously established. This compound was obtained from benzaconine by the action of acetic anhydride, in chloroform solution, at 100° C. We suggest that the alcohol at C(13) would be less sterically hindered toward acetylation than that at C(8), resulting in the structure **6**. Triacetylbenzaconine was described [70] as apparently nonpoisonous in small doses, and as lacking the ability to produce any tingling of the tongue, a property at one time used for assaying aconitine.

2.2.7. Tetraacetylbenzaconine. In common with diacetylbenzaconine and triacetylbenzaconine, "tetraacetylbenzaconine" has only been investigated once and is a somewhat mysterious substance. Dunstan and Carr reported [70] that they obtained the compound by heating benzaconine with acetyl chloride in a sealed tube at 100° C, and they claimed that it was not identical to triacetylaconitime, either in physical or biological properties. It was described as nontoxic.

2.2.8. 8-*O*-Methylbenzaconine. The product formed when aconitine was heated with methanol in a sealed tube at 120–130° C was named [71] "methylbenzaconine," since it appeared to result from the replacement of the acetyl group by methyl. The most reasonable structure for this product is the 8-*O*-methyl ether (**7**), and we [6] have confirmed this by ^{13}C-NMR spectroscopy.

Cash and Dunstan [71] discussed the pharmacology of their "methylbenzaconine" in some detail. In small doses it caused slowing of the heart, and larger doses produced arrhythmias. Atropine did not act as an antagonist. Respiration was depressed without any initial acceleration. However, the predominant effect observed in mammals was muscular paralysis, apparently similar to that produced by curare. Direct application of the drug to muscle resulted in fibrillation more commonly than with any of the other aconitine derivatives examined by the investigators.

The acute toxicity was reported to be intermediate between that of diacetylaconitine and benzaconine, and about one-hundredth that of aconitine (see Table 1).

2.2.9. Aconine. Complete saponification of aconitine results in the formation of the alkanolamine aconine (**8**). Aconine obtained in this way was subjected to pharmacological tests by Cash and Dunstan [18], who found it to be less toxic than their benzaconine by a factor of 5–10, depending on animal species.

Cardiovascular Effects. Blood pressure was not significantly altered by aconine. The drug alone produced a slight bradycardia but was found to be strongly antiarrhythmic, antagonizing the effects of aconitine or diacetylaconitine to such an extent that death could be prevented in animals poisoned with lethal doses of the latter drugs.

Skeletal Muscle Effects. The most characteristic effect of aconine was its pronounced curarelike action on neuromuscular transmission, resulting in death from respiratory paralysis.

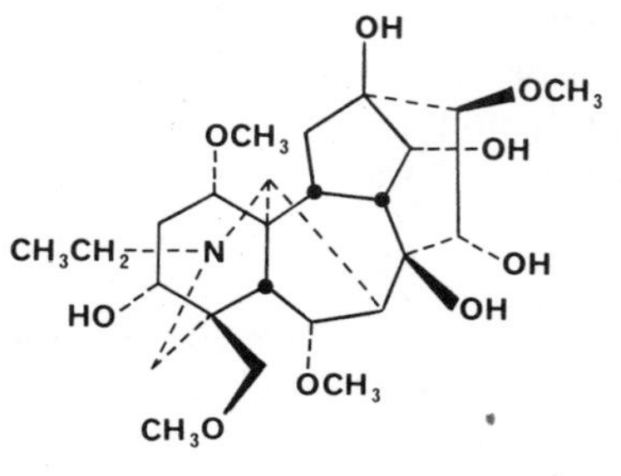

OH OCH$_3$ OCH$_3$ OH CH$_3$CH$_2$ N OH HO OH OCH$_3$ CH$_3$O

8

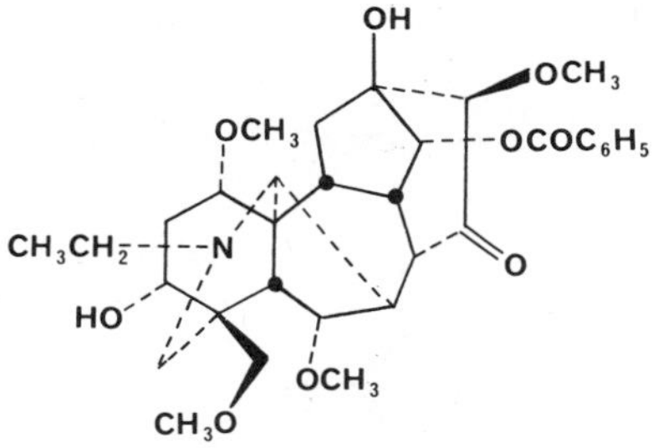

9

Other Effects. Relatively high doses of aconine were said to cause an unusual type of muscular tremor in guinea pigs, but not to significantly alter the body temperature.

2.2.10. Pyroaconitine. The pharmacology of pyroaconitine, a rearranged thermolysis product of aconitine now known to have the structure **9**, was also investigated by Cash and Dunstan [71].

Cardiovascular Effects. The compound caused lowering of blood pressure and respiratory depression (apparently of central origin). Small doses of the drug produced tachycardia, whereas larger doses caused bradycardia, which could be partially alleviated by vagotomy or atropinization.

Neuromuscular Effects. Peripheral motor nerves or skeletal muscle were unaffected by pyroaconitine.

Toxicology. In experiments with small animals, the symptoms of poisoning generally resembled those of aconitine. Guinea pigs exhibited paresis and clonic movements; death was preceded by convulsions. The acute toxicology of pyroaconitine was about the same as that of diacetylaconitine (**2**) (see Table 1).

2.2.11. Triacetylpyroaconitine. Treatment of pyroaconitine hydrochloride in chloroform solution with acetyl chloride yielded a crystalline product that Dunstan and Carr [70] identified as pyroaconitine triacetate. If that is correct, the compound may be **10**.

It did not show the characteristic physiological effects of aconitine and was apparently nontoxic [70].

2.2.12. Mesaconitine. Mesaconitine (**11**) is known to occur in a number of *Aconitum* species, often together with aconitine. (Indeed, mesaconitine may be the predominant alkaloid in certain subspecies of *A. napellus*, the traditional source of European "aconitine" [20].) The two alkaloids have very closely related structures (mesaconitine having an *N*-methyl function rather than the *N*-ethyl of

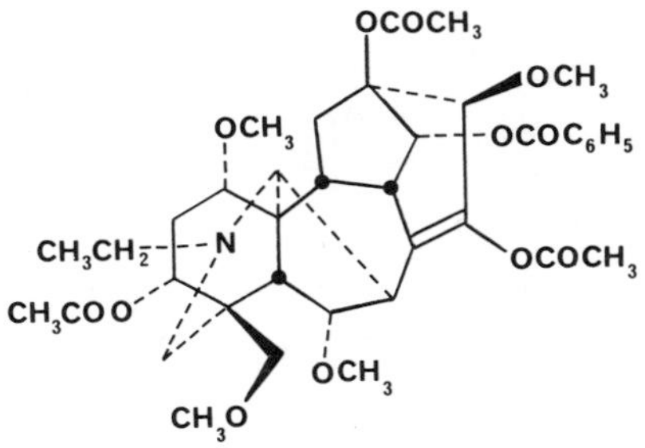

10

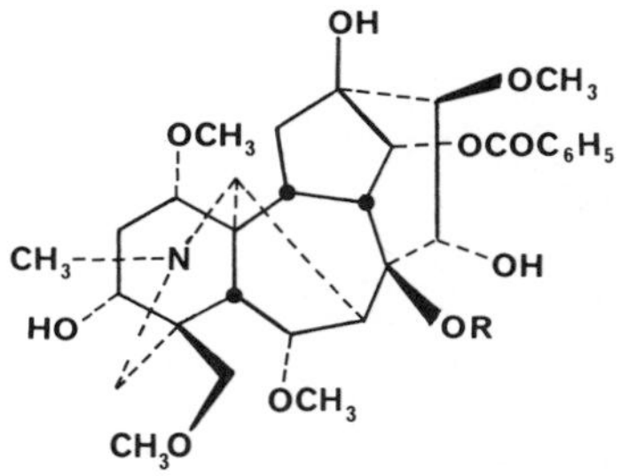

11 $R = COCH_3$

12 $R = H$

aconitine), and are difficult to separate. Pharmacological studies with mesaconitine of proven purity have only recently appeared in the literature.

An early investigation of the pharmacology of mesaconitine (of unknown purity) was made by Goto [72, 73]. Among the symptoms reported after intraperitoneal administration of the drug to mice were the following: respiratory inhibition, paralysis of the extremities, fluctuation in blood pressure, emesis, salivation, dilation of the ear vein, and lachrymation. The effects on isolated frog or toad heart were described as excitation and ventricular constriction.

The recent studies by Hikino's group [7, 74] with very pure mesaconitine showed it to have pharmacological properties almost identical with those of aconitine (Section 2.2.1) in most cases, but with minor differences in potency.

Cardiovascular Effects. The cardiovascular effects of mesaconitine were virtually identical with those of aconitine: after an intravenous dose of 5×10^{-2} mg/kg to rats, mesaconitine caused a temporary fall in blood pressure that was not inhibited by propranolol but was antagonized to some extent by atropine.

The pressor effects of adrenaline, tyramine, and 1,1-dimethyl-4-phenylpiperazinium were unaltered by the drug. On the isolated guinea pig atrium, mesaconitine at a concentration of 4.5×10^{-8} M showed a positive inotropic action; at a concentration of 1.5×10^{-7} M, a negative inotropic effect replaced the initial positive inotropic action, and bradycardia set in.

Smooth Muscle Effects. Unsurprisingly, the effects of mesaconitine on smooth muscle were also found to be very similar to the effects of aconitine. The effects of the drug on a guinea pig vas deferens and on ileum preparations were subjected to detailed scrutiny. Thus Sato et al. [74] found that the mesaconitine-induced contractions of the vas deferens preparation were dose dependent, but that tachyphylaxis developed on repeated application of the drug. Pretreatment of the vas deferens with atropine or hexamethonium had no effect on the response to mesaconitine, whereas pretreatment with bretylium, procaine, or tetrodotoxin blocked the response significantly. When guinea pigs were pretreated with reserpine or 6-hydroxydopamine, the response of the isolated vas deferens to mesaconitine was abolished. Denervation also prevented response to the drug. The response of the preparation to noradrenaline, acetylcholine, histamine, and tyramine were nonspecifically potentiated by mesaconitine. At a concentration of 7.5×10^{-5} M, mesaconitine elicited a release of noradrenaline from the vas deferens that could be reduced by depletion of calcium from the bathing medium and prevented completely by treatment with tetrodotoxin, or prior reserpinization of the animal.These effects were interpreted as indicating that mesaconitine caused presynaptic release of noradrenaline and acted postsynaptically in some way that potentiated the effects of the transmitter.

Sato et al. [74] have also demonstrated that mesaconitine induced dose-dependent contractions of an isolated guinea pig ileum preparation. As with the vas deferens system, repeated applications of the alkaloid resulted in tachyphylaxis, and at low drug concentrations ($\sim 1.5 \times 10^{-8}$–4.5×10^{-7} M), the contractions could be inhibited by prior treatment of the tissue with atropine, strychnine, or morphine. At higher concentrations of mesaconitine, these antagonists were only partially effective. However, tetrodotoxin, cocaine, noradrenaline, or papaverine completely suppressed the ability of mesaconitine to induce contraction of the ileum, at high or low concentrations of the alkaloid. Hexamethonium was without effect, but the contractions could also be totally suppressed by using a calcium-ion-free bathing medium for the ileum. Mesaconitine was observed to increase the release of acetylcholine from the ileum; that effect was also inhibited by prior treatment of the tissue with tetrodotoxin, or cocaine, or use of a calcium-ion-free bath solution.

Mesaconitine was found to slightly potentiate contractions of the ileum induced by electrical stimulation, acetylcholine, or histamine.

These authors concluded [74] that mesaconitine had some action on the postganglionic nerves and at high concentrations, induced prostaglandin release. The latter effect might explain the anti-inflammatory properties of aconitine (see Section 2.2.1).

Skeletal Muscle Effects. The effects of mesaconitine on skeletal muscle were probed using the rat phrenic nerve–diaphragm preparation and were found to resemble very closely those of aconitine.

Other Effects. Mesaconitine slightly prolonged barbiturate-induced sleeping time in mice in doses at which aconitine was completely ineffective, and it was about twice as potent as aconitine in the analgesic tests referred to in Section 2.2.1.

2.2.13. Benzoylmesaconine. The hydrolysis product of mesaconitine (**11**), benzoylmesaconine (**12**), is obtained in the same way as benzaconine from aconitine (see Section 2.2.4). The clinical symptoms produced after intraperitoneal injection of this drug into mice were first reported by Goto [72]. Respiratory inhibition, motor paralysis, and emesis were observed, as with the parent compound. The derivative was said to have an effect similar to that of mesaconitine on the isolated toad heart and to cause strong respiratory inhibition in rabbits.

A pharmacological study of benzoylmesaconine of high purity was recently reported by Sato et al. [74]. The drug was found to have relatively little effect on the heart, either *in vivo* (rat) or *in vitro* (guinea pig atrium), although high concentrations ($\sim 5 \times 10^{-4}$ *M*) in the latter assay produced negative inotropic and chronotropic effects. Benzoylmesaconitine had virtually no action on smooth or skeletal muscle preparations and showed little activity in analgesic tests [51a], except at high doses. At doses of 30–60 mg/kg s.c. in mice, hexabarbital-induced sleeping time was increased slightly. Lethal doses (Table 1) slowly caused respiratory paralysis in mice.

2.2.14. Pyromesaconitine. Pyromesaconitine (**13**) resembles pyroaconitine (**9**) and is prepared in an analogous manner. The only published report concerning the pharmacology of this compound is that of Goto [73], who noted that the drug had some effect on the respiration and blood pressure of experimental animals, but less than did benzoylmesaconine.

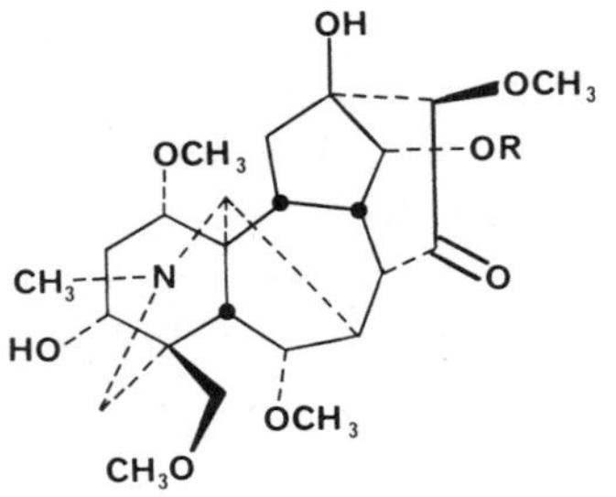

13 R = COC_6H_5

15 R = H

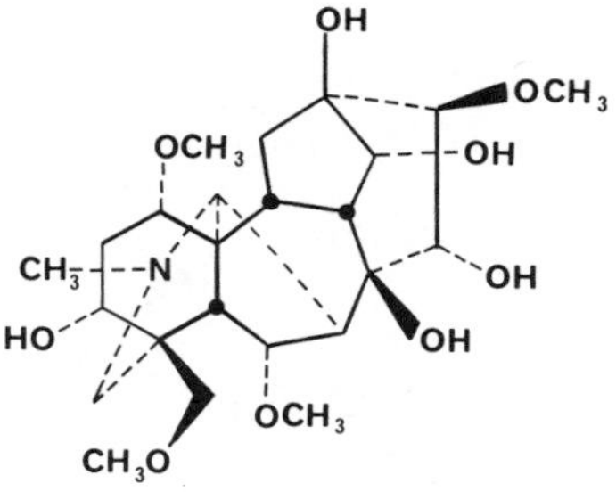

14

2.2.15. Mesaconine. Mesaconine (**14**) the parent alkanolmine of mesaconitine, was also the subject of pharmacological testing by Goto [72]. The symptoms of poisoning in mice included slight respiratory inhibition, motor paralysis, and emesis. The acute toxicity of mesaconine was slightly less than that of benzolymesaconine (see Table 1).

Cardiovascular Effects. Mesaconine had practically no effect on the toad heart or on rabbit blood pressure.

Other Effects. Histological studies revealed that mesaconine caused profound pathological changes in various tissues.

2.2.16. Pyromesaconine. Pyromesaconine is the hydrolysis product of pyromesaconitine and by analogy with related compounds very probably has the structure **15**, although this has not been confirmed. Goto [73] reported that pyromesaconine had no effect on respiration or blood pressure.

2.2.17. Hypaconitine. Hypaconitine (**16**), which may be considered as 3-deoxymesaconitine, has been isolated from several *Aconitum* species [75]. Pharmacological studies of the drug were first reported by Sato et al. [7]. The effects observed were very similar to those seen with aconitine and mesaconitine, except that hypaconitine showed lower potency in some of the assays. Thus hypaconitine was five to eight times less effective in eliciting contractions of the isolated vas deferens, in potentiating barbiturate-induced sleeping time, and in analgesic activity. This lower potency was also reflected in the acute toxicity of the drug (see Table 1). As an anti-inflammatory agent, it was said [51b] to be about as effective as aconitine.

2.2.18. Benzoylhypaconine. The hydrolysis product of hypaconitine, benzoylhypaconine (**17**) was also studied by Sato et al. [7, 51]. This drug was found to have a very similar pharmacological profile to that described for benzoylaconine (see Section 2.2.4) and benzoylmesaconine (see 2.2.13).

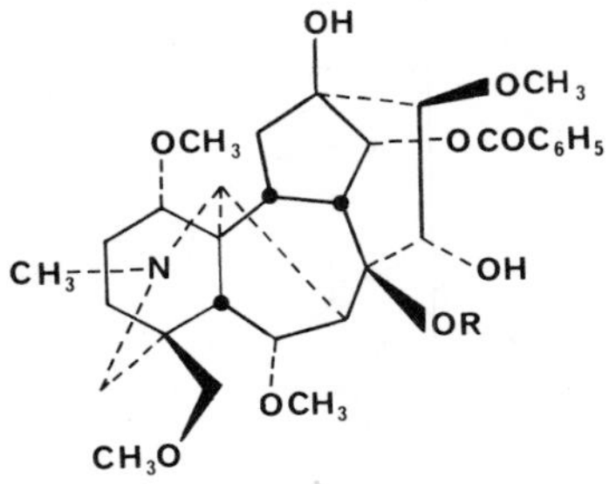

16 R = $COCH_3$

17 R = H

2.2.19. Indaconitine. First isolated from *A. chasmanthum* Stapf and subsequently from other *Aconitum* spp. [75], indaconitine (**18**) (which corresponds to 15-deoxyaconitine) was included by Cash and Dunstan [76] in their pharmacological studies. They concluded that the acute toxicity of the alkaloid toward small mammals was very close to that of aconitine. Qualitatively and quantitatively the profile of pharmacological effects of indaconitine hydrobromide was very similar to that of the effects produced by aconitine hydrobromide. Atropine antagonized the cardiovascular effects.

2.2.20. Pseudaconitine. Dunstan and Cash [77] described pseudaconitine (**19**), an alkaloid isolated from *A. ferox* and other *Aconitum* sp., as the most toxic of the aconitines that they studied. From some comparative studies, they concluded that pseudaconitine was about twice as toxic as aconitine and slightly more so than bikhaconitine in the small mammals and birds that they used for test purposes. Death resulted from respiratory failure. (If artificial respiration was maintained, the toxic differential between pseudaconitine and aconitine was reduced.) As these observations suggest, pseudaconitine is more effective as a respiratory depressant than aconitine, but the cardiovascular potencies of the

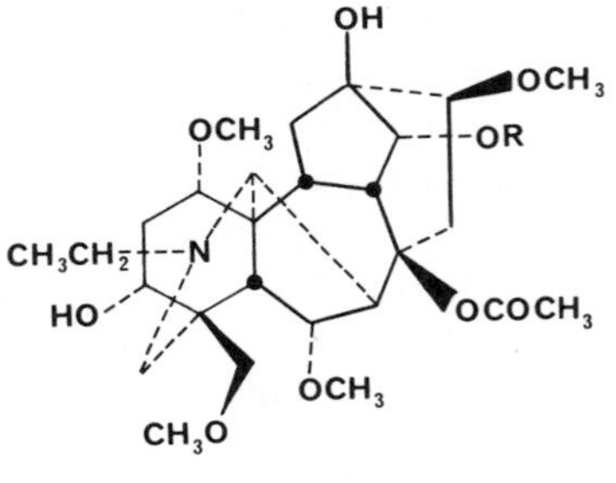

18 R = COC_6H_5

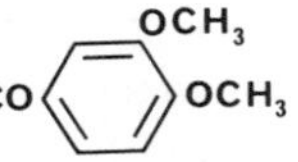

19 R = CO–(3,4-dimethoxyphenyl) (OCH_3, OCH_3)

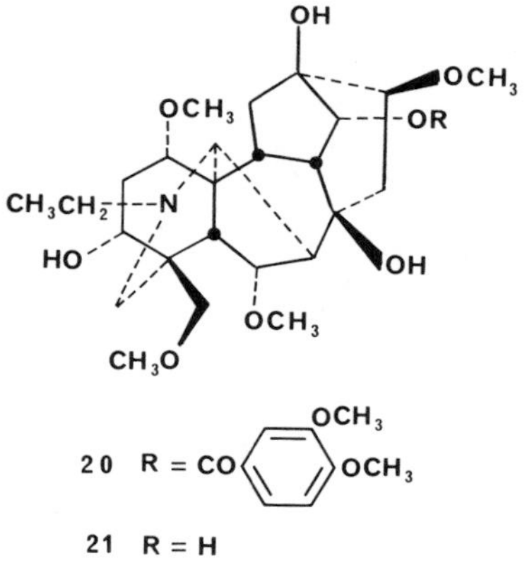

20 R = CO

21 R = H

two alkaloids are very similar. (Pseudaconitine was found to be slightly more powerful than aconitine in its action on cat heart, but less on frog heart.) The neuromuscular effects of pseudaconitine in *Rana* spp. preparations were similar to, but slightly weaker than, those produced by aconitine.

2.2.21. Veratroylpseudaconine. Hydrolytic deacetylation of pseudaconitine yields veratroylpseudaconine (**20**), and this compound has also been isolated from *Aconitum* spp. [75]. Cash and Dunstan [77] noted that the alkaloid, like benzaconine, did not produce the tingling sensation of tongue and lips when the compound was tasted (a response "characteristic" of the aconitines), nor did it have the toxicity of the parent alkaloid.

2.2.22. Pseudaconine. Complete saponification of pseudaconitine (or indaconitine) yields pseudaconine (**21**). Cash and Dunstan [77] again noted the absence of the tingling response when this compound was tasted. The acute toxicity was vastly reduced compared with that of the parent alkaloid and was about the same as that of aconine (see Table 1). The major physiological effects observed were attributed to a curarelike action of the drug, which could (like the other aconines) partially antagonize the cardiovascular effects of aconitine(s).

2.2.23. Pyropseudaconitine (Falaconitine). Produced from pseudaconitine in an analogous manner to the production of pyroaconitine from aconitine, and recently isolated as the natural product falaconitine from *A. falconeri* [75], pyropseudaconitine (**22**) was mentioned by Cash and Dunstan [77] as resembling pyraconitine in its failure to give the tingling-tongue test (see also Table 1).

2.2.24. Bikhaconitine. Bikhaconitine (**23**) (3-deoxypseudaconitine), another diterpenoid alkaloid first isolated from Indian *Aconitum* spp. [75], was also studied by Cash and Dunstan [76]. It was described as the second most toxic toward cats, rats, and rabbits of the aconitines that they examined, being only slightly less potent than pseudaconitine (see also Table 1). Bikhaconitine was

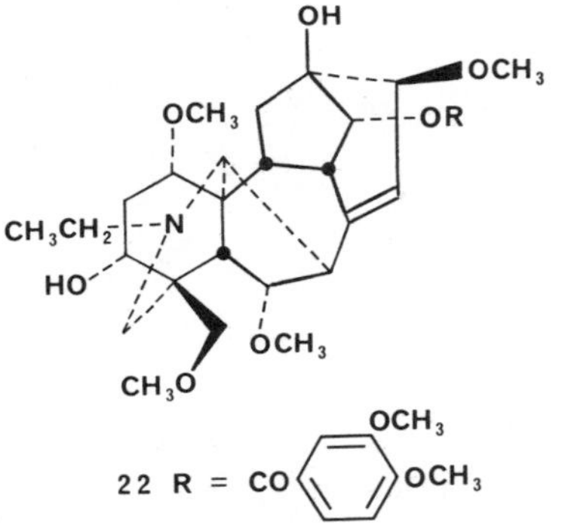

22 R = CO–(dimethoxyphenyl: OCH_3, OCH_3)

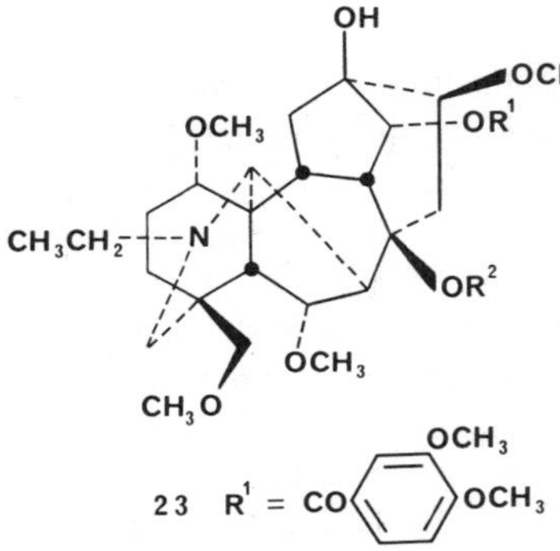

23 R^1 = CO–(dimethoxyphenyl: OCH_3, OCH_3) $R^2 = COCH_3$

24 $R^1 = R^2 = H$

found to be somewhat more effective than indaconitine (see Section 2.2.19) in depressing respiration in rats and frog, but somewhat less active in frog muscle and neuromuscular preparations.

2.2.25. Bikhaconine. Bikhaconine (**24**), the parent alkanolamine of bikhaconitine, was also examined by Cash and Dunstan [76]. Its pharmacological properties resembled those of the other aconines (see Sections 2.2.9, 2.2.15, and 2.2.22).

2.2.26. Jesaconitine. Jesaconitine (**25**) [75], has been reported [78] to possess pharmacological properties similar to those of "japaconitine" (aconitine), i.e., it is highly toxic to mammals (see Table 1) and affects blood pressure, heart rate, and respiration. Small doses were said to increase the contractions of a mammalian intestine preparation.

2.2.27. Aconifine. The Tashkent group isolated an alkaloid from *A. karakolicum* Rapcs., which they have named aconifine [79]. The pharmacology of this base most closely resembled that of aconitine in its spectrum of effects, but it had only about half that alkaloid's toxicity in mice (Table 1) [8, 80].

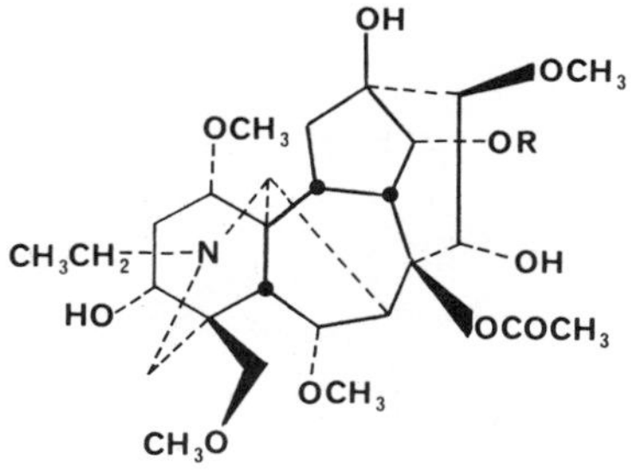

25

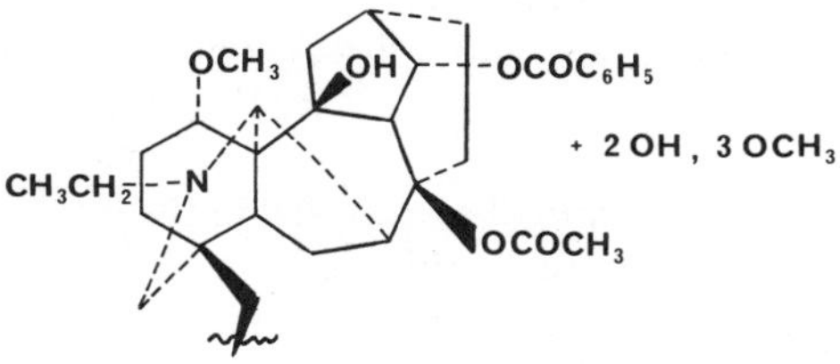

26

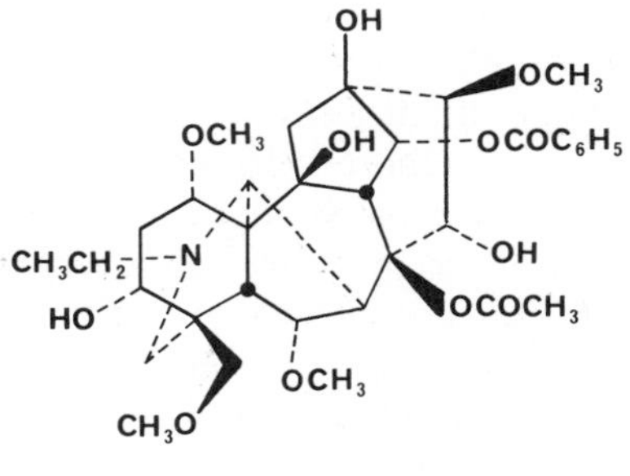

27

At a time when only the part structure (**26**) was known to us, we predicted [6] from this pharmacological resemblance to aconitine that the unassigned functionalities could be distributed as in **27**. Indeed, Yunusov et al. [81] have very recently reported that aconifine has this structure.

2.2.28. Delphinine. As one of the earlier alkaloids to be discovered, delphinine (**28**) was examined extensively during the 19th century, but pharmacological investigations thereafter have been very sparse. Descriptions of the drug's physiological effects may be found in the papers of Marsh et al. [82] and Beath [83] but were excerpted mainly from the earlier publications of Boehm and Serck

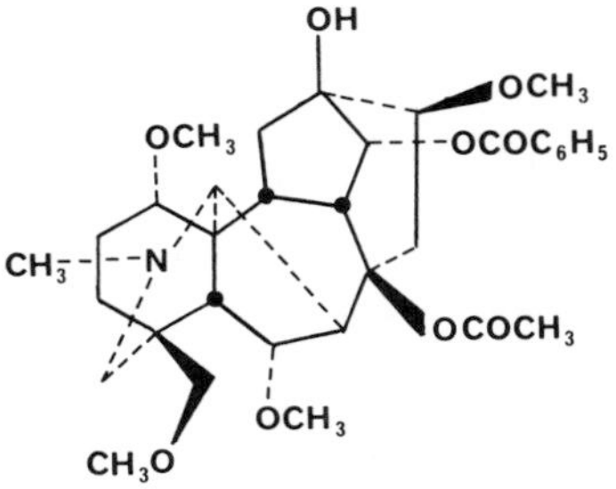

28

[84]. Slighly more modern summaries of the effects of delphinine are given in the books of Waugh and Abbott [85], and Solis-Cohen and Githens [86]. The former provides a useful account of the "historical" studies and discusses the therapeutic utility of the drug.

As with aconitine, caution must be exercised in the interpretation of the early pharmacological observations on delphinine, since they were made with material of uncertain homogeneity. Waugh and Abbott [85], for example, noted that Rabuteau's [87] sample of delphinine was heavily contaminated by "staphisagrine" (some other alkaloid or mixture of alkaloids, not the bisditerpenoid alkaloid now known by that name [88]).

The physiological effects of delphinine appear to resemble closely those of aconitine (Section 2.1.1). In humans, the drug produces the "characteristic" tingling and burning sensation of the mouth and lips. The reported symptoms of poisoning in animals include emesis, salivation, diarrhea, muscular fibrillation, clonicotonic convulsions, loss of motor control, paralysis of the sensory nerves, respiratory depression, hypotension, bradycardia, and finally, cardiac arrest. The most prominent symptom seems to be respiratory depression, which is the primary cause of death. Motor-nerve terminals are not paralyzed until last, and the curarelike properties attributed to delphinine by some authors are most likely due to the use of particularly impure drug samples or extremely high doses. Lohmann [89], for example, tested delphinine on the frog preparation of Bernard at doses of about 1.3 g/kg.

A more recent study [23, 90] of delphinine has shown that topical application of the drug to the dog heart produces cardiac flutter, and that the arrhythmia is temporarily antagonized by diphenylhydantoin [23, 91].

Delphinine has also been found to exert a destructive effect on epithelial calls of the mouse colon [92, 93].

In general, the toxicity of delphinine seems to be slightly lower than that of aconitine (see Table 1).

2.2.29. Delphinine Methochloride. Schneider and Enders [94] prepared delphinine methochloride (**29**) by the alkylation of delphinine with methyl chloride and studied the pharmacology of this quaternary salt.

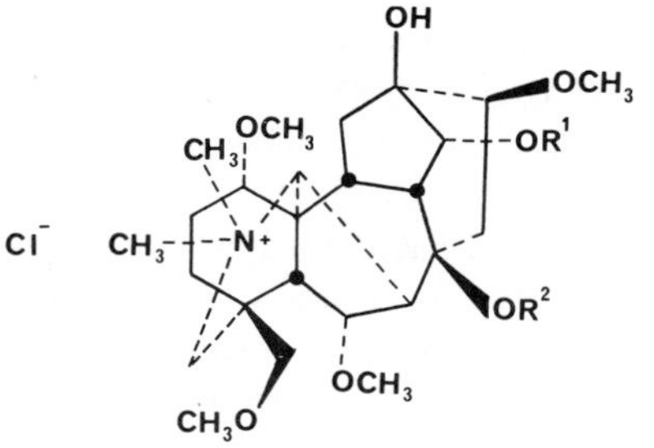

29 $R^1 = COC_6H_5$ $R^2 = COCH_3$

30 $R^1 = R^2 = H$

The compound was found to be toxic in mice (see Table 1), with death resulting from respiratory failure. In cats, dogs, and rabbits, sublethal doses depressed respiration and lowered blood pressure. However, even at $\sim 8 \times 10^{-7}$ *M*, the drug did not effect the activity of an isolated frog heart preparation; and i.p. administration of the salt to rats did not affect their ECGs. There was also very little effect on smooth muscles as judged by the slight fall in tonus and reduced mobility caused by the drug in an isolated guinea pig colon preparation. The methochloride, however, had pronounced curarelike properties in neuromuscular preparations and was judged to be about one half to one quarter as effective as curare in Bernard's frog muscle preparation.

2.2.30. Delphonine Methochloride. Like delphinine methochloride, delphonine methochloride (**30**) was prepared by Schneider and Enders [94]. The pharmacological profile of this drug was found by them to be virtually identical with that of the esterified analog, delphinine methochloride.

2.2.31. Lappaconitine. The alkaloid lappaconitine (**31**) occurs in both *Aconitum* and *Delphinium* species [75]. The structure of this alkaloid differs considerably from that of aconitine, and a toxicological study of this compound by Dybing et al. [95] showed it to be some 40 times less toxic than aconitine on

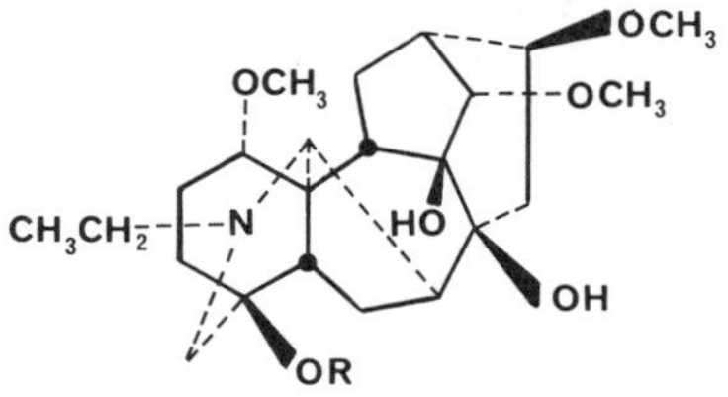

31 R = CO(C_6H_4)-NHCOCH$_3$ (o-acetamidobenzoyl)

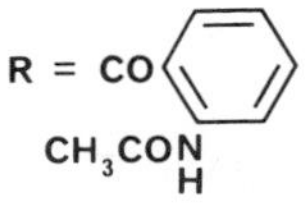

intravenous administration to mice; lethal doses in other animal species have also been determined [96, 97] (see Table 1). Lappaconitine did not induce the vomiting reflex in mice that is normally associated with aconitine [95].

Cardiovascular Effects. The Tashkent group reported [8] that lappaconitine had no effect on respiration or blood pressure in anesthetized dogs at doses of 0.1–0.5 mg/kg. However, the drug did evoke the range of ECG responses that are usually observed with aconitine in anesthetized rats. Recent work by a group in the United States [98] showed that, in anesthetized rabbits, lappaconitine (1 mg/kg i.v.) induced cardiac arrhythmias, preceded by bradycardia, and accompanied by a marked fall in blood pressure. At a concentration of 10^{-4} *M*, lappaconitine produced arrhythmia in the isolated guinea pig atrium, but did not alter beat rate. The drug could be washed out, and tachyphylaxis apparently did not develop. In marked contrast, the aconitine-induced arrhythmia in the same preparation was preceded by tachycardia, and the drug could not be washed out; tachyphylaxis was observed.

Smooth Muscle Effects. At concentrations of $\sim 2 \times 10^{-8}$–2×10^{-6} *M*, lappaconitine had no effect on the spontaneous contractions of intestinal segments, on acetylcholine-induced intestinal contractions, or on guinea pig uterus [8].

Skeletal Muscle Effects. The administration of 0.5–2.0 mg/kg doses of lappaconitine to cats or dogs did not affect the contractions of the gastrocnemius muscle resulting from electrical stimulation of the sciatic nerve [8].

Central Nervous System. Effects on the CNS were noted [8], in as much as the alkaloid caused prolonged disruption of conditioned reflexes in dogs after intravenous doses of 0.1–1.0 mg/kg.

Other Effects. The excretion of lappaconitine and its metabolites has received some attention [95]. Lappaconitine and the deacetylated base anthranoyllappaconine (picrolappaconitine) were detected by paper chromatography in the urine of rats given an i.p. dose of the alkaloid. Lappaconine (Section 2.2.34) was not detected. Fifteen percent of the administered dose of lappaconitine was estimated to be excreted as an approximately 1:1 mixture of anthranoyllappaconine and unchanged lappaconitine.

2.2.32. Diacetyllappaconitine. There is no report of the chemical characterization of diacetyllappaconitine per se, but Tulyaganov et al. [8] noted a medium lethal dose (see Table 1) for "lappaconitine triacetate," which could be either lappaconine triacetate or lappaconitine diacetate (**32**). From the context of the original discussion, the latter possibility seems more likely, since the compound is mentioned in comparison with aconitine and aconitine triacetate to exemplify the loss of toxicity that occurs on peracetylation of the free hydroxyl groups in the molecules.

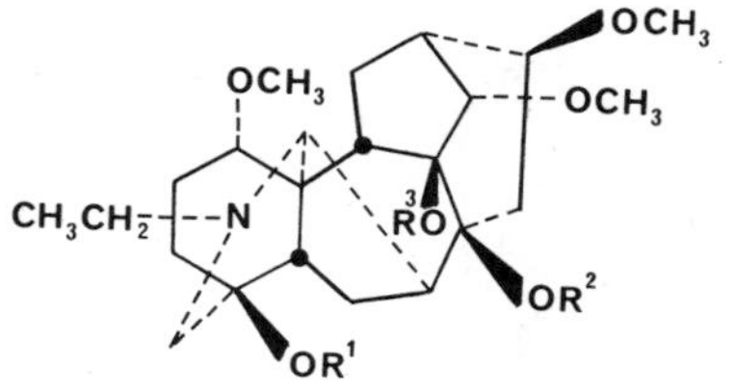

32 R^1 = CO-(2-acetamidophenyl) (CH_3CONH) $R^2 = R^3 = COCH_3$

33 R^1 = CO-(2-aminophenyl) (NH_2) $R^2 = R^3 = H$

34 $R^1 = R^2 = R^3 = H$

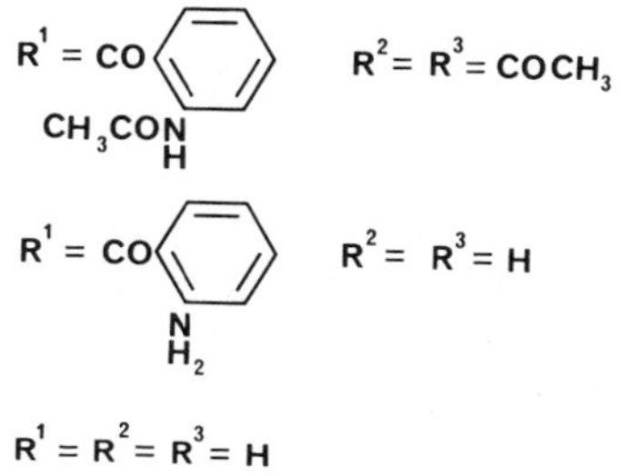

2.2.33. Anthranoyllappaconine (Picrolappaconitine) (Deacetyllappaconitine). Anthranoyllappaconine (**33**), referred to in the older literature as picrolappaconitine, is obtained from lapaconitine by acidic hydrolysis [97, 99] and as a catabolite of lappaconitine (see Section 2.2.31). The drug was found [98] to produce cardiac arrhythmias accompanied by hypotension at a dose of 1 mg/kg i.v. in anesthetized rabbits. However, bradycardia was not evident, as it was in the case with lappaconitine (Section 2.2.31).

2.2.34. Lappaconine. Saponification of lappaconitine gives the alkanolamine lappaconine (**34**). The pharmacology of this compound was briefly examined by Shamma et al. [98], who reported that doses of 0.05–0.5 mg/kg had no effect on the heart rate of anesthetized rabbits nor on the blood pressure. Lappaconine was found to have about a tenth the toxicity of lappaconitine (Table 1).

2.2.35. Talatizamine (Talatisamine). Talatizamine (talatisamine) (**35**) has been found in several species of *Aconitum* [75]. The pharmacology of the alkaloid has been outlined by Tulyaganov et al. [8].

Cardiovascular Effects. In small mammals, talatizamine produced brief hypotension at doses of 5–15 mg/kg and intensified the hypotensive response to adrenaline and noradrenaline. No effect was observed on heart rhythm.

Smooth Muscle Effects. The drug increased the amplitude of spontaneous contractions of isolated intestinal preparations at a concentration of 2.5×10^{-4} *M*. This concentration of talatizamine also induced contractions of the isolated guinea pig uterus.

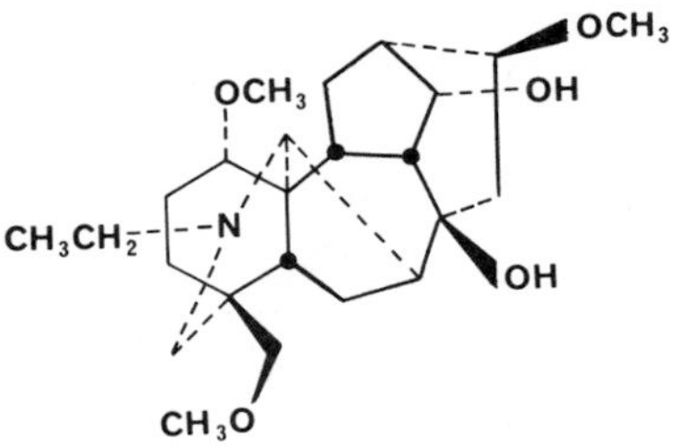

35

Skeletal Muscle Effects. Talatizamine inhibited acetylcholine-induced contractions of the frog rectus abdominis preparation and also inhibited the response of the cat sciatic nerve–gastrocnemius preparation at a dose of 30–40 mg/kg.

Central Effects. Talatizamine was reported to have no effect on conditioned reflexes in test animals.

2.2.36. Karakoline (Karacoline). Karakoline (karacoline) (**36**) was first obtained from *A. karakolicum* Rapcs. [75]. It is the least oxygenated of the C_{19}-diterpenoid alkaloids for which pharmacological information is available. The pharmacology of this compound has been described by Dzhakhangirov et al. [80] and bears close resemblance to that of talatizamine.

Intraperitoneal injection of karakoline into mice produced muscular weakness and death from respiratory depression. The acute toxicity of karakoline was about twice that of talatizamine (see Table 1).

Cardiovascular Effects. Intravenous administration of karakoline to cats or dogs lowered blood pressure for 15–30 min. This effect was attributed to ganglionic blockade.

Smooth Muscle Effects. At a concentration of 2.5×10^{-6} *M*, the drug had no effect on the isolated rat intestine. The same concentration did induce contractions of the isolated guinea pig ileum.

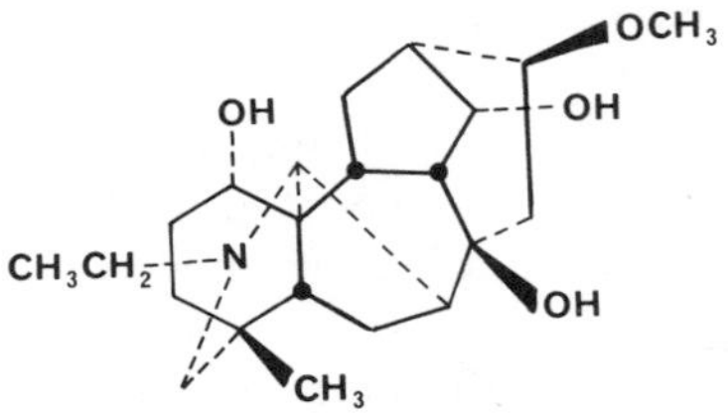

36

Skeletal Muscle Effects. The contractions of the isolated frog rectus abdominis preparation elicited by acetylcholine were reduced by karakoline in concentrations of 2.5×10^{-5}–1×10^{-4} *M*. In the cat sciatic nerve–gastrocnemius muscle preparation, a dose of 25 mg/kg i.v. of karakoline caused some reduction in the amplitude of the contraction, and a dose of 30–40 mg/kg caused a complete block of neuromuscular transmission lasting 30–40 min.

Central Effects. Like talatizamine, karakoline appeared to have no distinct effects on conditioned reflexes in rats.

2.2.37. Methyllycaconitine. Methyllycaconitine (**37**)* occurs in numerous *Delphinium* species [75] but despite its name, has so far not been isolated from any species of *Aconitum*.

In marked contrast with aconitine (Section 2.2.1) and its congeners and lappaconitine (Section 2.2.31), methyllycaconitine shows a much more pronounced action on the skeletal neuromuscular system than on the cardiovascular system. The curarelike properties seem to have been first commented upon by Kuzovkov and Bocharnikova [102], who also noted similar properties associated with the "delphinium" alkaloids elatine (Section 2.2.42), avad(k)haridine (Section 2.2.40), delsemine (Section 2.2.38), and ajacine (*N*-acetylanthranoyllycoctonine). These authors stated that ". . . curare-like properties are inherent in esters of lycoctonine or amino alcohols like it, esterified with *N*-acylated anthranilic acid."

A full paper on the pharmacology of methyllycaconitine (studied as its hydroiodide salt, "mellictine") was subsequently published by Dozortseva [103]. She found that i.v. administration of the drug to mice resulted in temporary

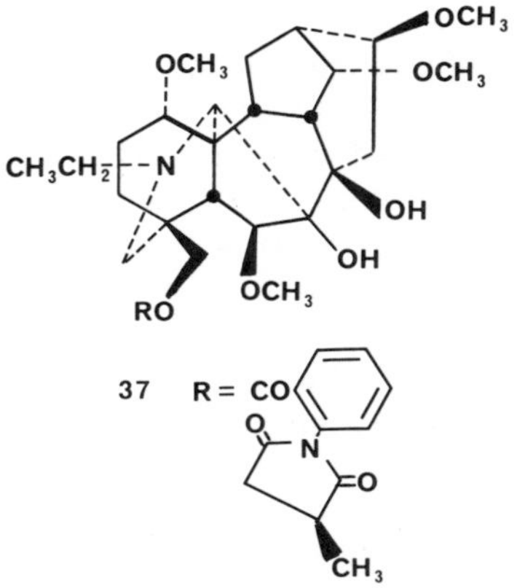

OCH_3 OCH_3 OCH_3 CH_3CH_2–N OH OH OCH_3 RO

37 R = CO–(N-substituted phenyl, 3-methyl-2,5-dioxopyrrolidin-1-yl; CH_3)

*Very recently, Edwards et al. [100] and Pelletier et al. [101] found that the structures previously written [75] for lycoctonine and some alkaloids correlated with that base must be revised so as to change the configuration of the oxygenation at C(1) from β to α. Here and elsewhere in this chapter, this structural correction has been made.

general depression and disturbed motor coordination. A dose of 2 mg/kg produced muscular relaxation, respiratory depression, and clonicotonic convulsions, with return to normal after 15–20 min. Larger doses, 3.0–3.5 mg/kg, caused complete motor paralysis and respiratory arrest, with death from asphyxia. No abnormalities were observed in cardiac rhythm. The drug was also found to be active after stomach-tube or rectal administration.

Cardiovascular Effects. In experiments with anesthetized cats, 1 mg/kg i.v. of mellictine caused a brief fall in blood pressure; higher doses caused more pronounced hypotension. No significant changes were observed in cardiac rhythm or amplitude. Perfusion of the drug at concentrations of $\sim 1 \times 10^{-6}$–2.5×10^{-6} M produced no changes in the blood vessels of the isolated rabbit ear, kidney, or heart (after its arrest by strophanthin).

Smooth Muscle Effects. Mellictine, in unstated concentration, had no effect on rabbit intestine or guinea pig uterus *in vitro*, nor on the intestine or uterus of rabbit and cat *in vivo*. Our experiments [6, 104], too, showed that methyllycaconitine had relatively little effect on the electrically induced contractions of the guinea pig ileum in concentrations up to 5×10^{-4} M.

Skeletal Muscle Effects. Injection of 13–14 mg/kg of mellictine into the abdominal lymph sac of the frog blocked the response of the gastrocnemius muscle to electrical stimulation of the sciatic nerve [103]. Direct muscle excitability was not altered. The same effect was observed *in vitro*: the muscle became completely unresponsive to electrical stimulation of the sciatic nerve after exposure to a drug concentration of 2.5×10^{-5} M. This concentration also blocked the response of the isolated frog rectus abdominis to acetylcholine.

Our own experiments [6, 104] with nerve–muscle preparations from the frog showed that within 2 min, methyllycaconitine at a concentration of 10^{-8} M caused 80–90% inhibition of postsynaptically recorded action potentials in the sartorius muscle after electrical stimulation of the sciatic nerve. A drug concentration of 10^{-7} M caused complete inhibition, which was partially antagonized by physostigmine (10^{-5} M); direct stimulation of the paralyzed muscle elicited a normal response. At a concentration of 10^{-7} M, methyllycaconitine also caused 50% inhibition in the amplitude of miniature end-plate potentials in the extensor digitus longus IV muscle of the frog; 100% inhibition occurred at a drug concentration of 4×10^{-6} M.

In anesthetized cats, a dose of 2 mg/kg i.v. of methyllycaconitine produced threshold inhibition of the sciatic nerve-gastrocnemius muscle preparation; a sharp decrease in muscle contractions, gradually returning to normal over 20–25 min, was obtained with a dose of 3 mg/kg [103]. Administration of 3.5 mg/kg completely inhibited the muscle response, with subsequent death. However, the neuromuscular block could be lifted by i.v. administration of 0.15 mg/kg neostigmine. Almost identical results were obtained with methyllycaconitine on the cat sciatic nerve–anterior tibialis muscle preparation [105].

We found [6, 104] that a 50% decrease in the response of the electrically stimulated rat phrenic nerve–diaphragm preparation was produced by a concentration of 2×10^{-5} *M* of methyllycaconitine; total inhibition was caused by a drug concentration of 3×10^{-5} *M*. The poisoned preparation responded normally to direct stimulation of the muscle, and the inhibition of muscle contractions could be partially antagonized by physostigmine.

Ganglion-Blocking Effects. The ganglion-blocking properties of methyllycaconitine were demonstrated by Dozortseva [103] with the nictitating membrane preparation of the cat. Threshold inhibition was observed with a dose of 1–2 mg/kg i.v., and complete inhibition of response followed the administration of 4 mg/kg. The effect of cytisine on this preparation was completely prevented by prior administration of 5 mg/kg of methyllycaconitine. The same dose also blocked the response of cardiac muscle fibers to electrical stimulation of the vagus nerve. The hypotensive effects of methyllycaconitine were attributed to its ganglion-blocking action.

Central Effects. The effects of methyllycaconitine on the central nervous system have not been clearly defined, although there is no doubt that the drug is centrally active. Mellictine is used clinically in the Soviet Union in the treatment of various neurological disorders [106], and details of its use in the treatment of spastic paralyses have been described by Kabelyanskaya [107]. Alyautdin [108] has shown that methyllycaconitine in doses of 3.5–4.5 mg/kg i.v. increases the amplitude of monosynaptic potentials in the spinal cord of decerebrate cats, as a result of its inhibition of the nicotinic receptors of the Renshaw cells.

Anesthetic Effects. Using the sciatic nerve–sartorius muscle preparation of the frog, Wilkens [109] found that electrical conduction along the sciatic nerve or over directly stimulated sartorius muscle was not affected by methyllycaconitine at a concentration of 10^{-7} *M*.

Other Effects. Dozortseva [110] reported that mellictine at a dose of 5 mg/kg temporarily reduced the tremor in mice caused by the i.v. administration of iodipamide sodium (2.0–2.5 mg/kg).

The administration of 3.0–3.5 mg/kg of methyllycaconitine to atropinized cats was found [103] to inhibit the effect of acetylcholine on the chromaffin tissue of the adrenal glands.

The effects of methyllycaconitine on muscle reflexes and cerebral blood circulation have been reported by Dmitrieva [111].

Summary. Methyllycaconitine thus appears to be a potent, nondepolarizing neuromuscular poison, acting as a competitive inhibitor of acetylcholine at nicotinic receptors, in the same way as tubocurarine. This conclusion is considered further in Section 3.1.2.

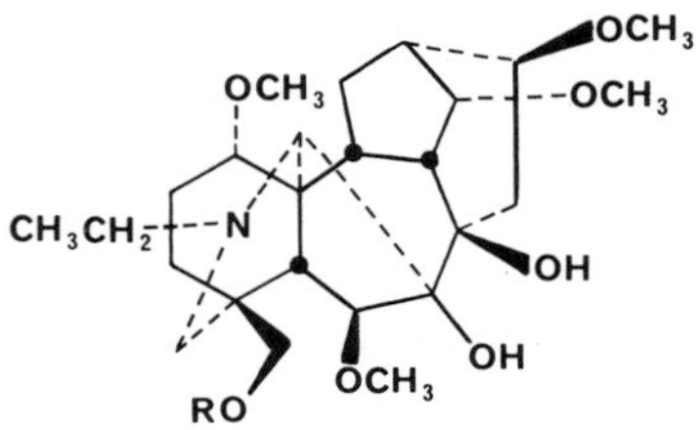

38a R = CO-C6H4-NHCO-C(CH3)(H)-CH2CONH2

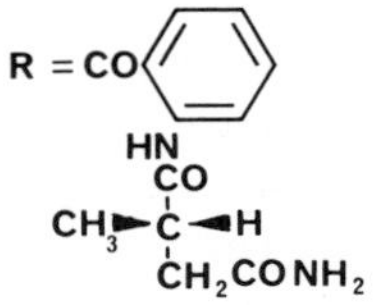

38b R = CO-C6H4-NHCO-CH2-C(H)(CH3)-CONH2

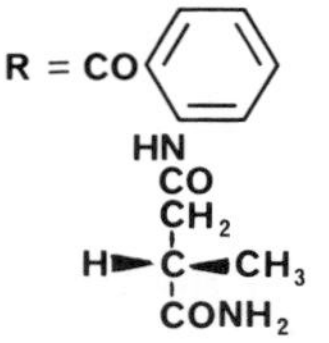

2.2.38. Delsemine. "Delsemine" is a mixture of isomeric amides (**38a** and **38b**) produced by the action of ammonia on melthyllycaconitine [75]. As such it is usually an artifact of the isolation procedure, but there is some evidence that this ammonolysis mixture may occur naturally [112].

Little published information is available on the pharmacology of delsemine. The acute toxicity in mice is somewhat lower than methyllycaconitine (see Table 1). Gubanow [106] reported that "delsemine" is a neuromuscular blocker and a hypotensive agent and that it has been used clinically in the Soviet Union as a substitute for tubocurarine in surgery.

We have found [6] that delsemine inhibits the response of the rat diaphragm to electrical stimulation of the phrenic nerve and reduces the amplitude of the miniature end-plate potentials in the frog extensor digitus longus IV muscle at concentrations very close to those of methyllycaconitine.

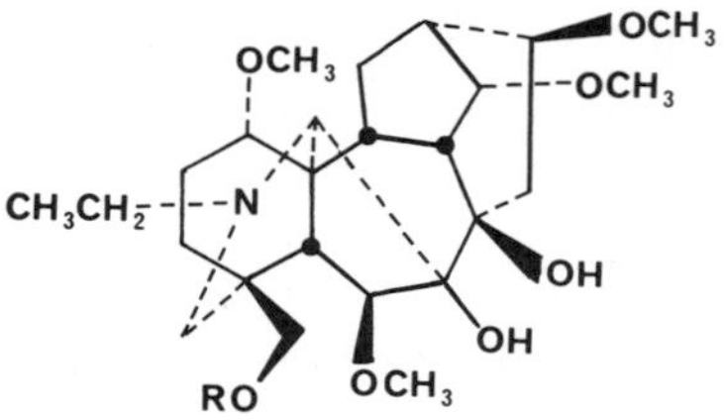

39 R = CO-C6H4-N(succinimide)

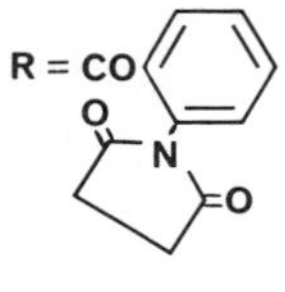

2.2.39. Lycaconitine. Lycaconitine (**39**), originally isolated from *A. lycoctonum*, has been found in both *Aconitum* and *Delphinium* species [75]. Early references to the biological activity of this compound should be viewed with caution, since there has been some confusion concerning the identity of both the alkaloid and the plant from which it was obtained [113].

Recently, lycaconitine isolated from *D. cashmirianum* was found [98] to have little effect on the heart rate of anesthetized rabbits in doses up to 5 mg/kg i.v. However, drug doses above 0.25 mg/kg produced a small rise in blood pressure.

2.2.40. Avadharidine (Avadkharidine, Awadcharidine). Although avadharidine (avadkharidine, awadcharidine) (**40**) is considered [75] to be an isolation artifact, derived by ammonolysis of lycaconitine (in the same way that delsemine is obtained from methyllycaconitine), it has recently been isolated [98] from *D. cashmirianum* under conditions that did not involve the use of ammonia.

The biological properties of avadharidine have been alluded to only very briefly. Thus Kuzovkov and Bocharnikova [102] noted that this alkaloid possessed curarelike properties. Shamma et al. [98] found the drug to have a negligible effect on the heart rate of anesthetized rabbits in doses up to 5 mg/kg i.v. They also reported that avadharidine was inactive on the isolated guinea pig atrium in concentrations up to 10^{-4} *M*.

2.2.41. Septentriodine (Cashmiradelphine). The alkaloid cashmiradelphine (**41**) was isolated recently from *D. cashmirianum* [98] and subsequently recognized to be identical with septentriodine, previously isolated from *A. septentrionale* [114]. In doses up to 5 mg/kg i.v., cashmiradelphine was found [98] to have negligible activity on the heart rate of anesthetized rabbits, although doses greater than 0.25 mg/kg produced a slight decrease in blood pressure.

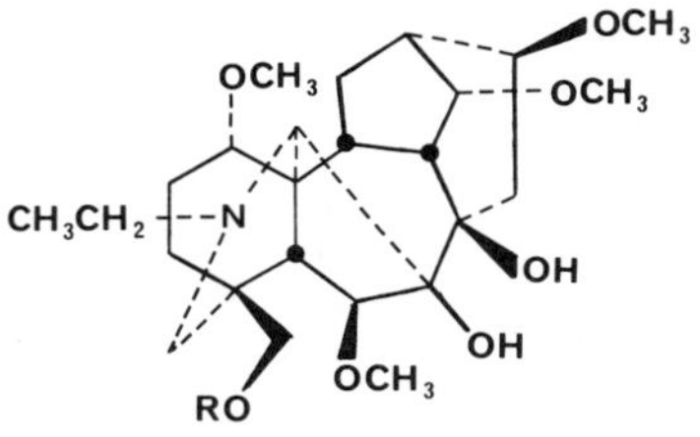

OCH_3, OCH_3, OCH_3, CH_3CH_2–N, OH, OH, OCH_3, RO

40 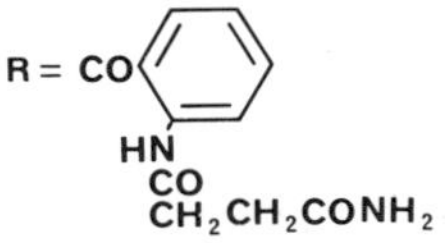 R = CO–C_6H_4–HN–CO–$CH_2CH_2CONH_2$

41 R = CO–C_6H_4–HN–CO–$CH_2CH_2COOCH_3$

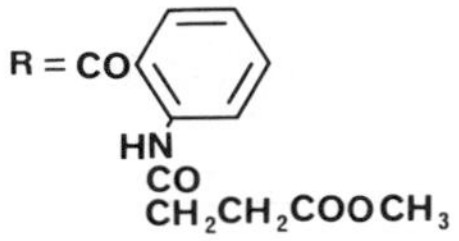

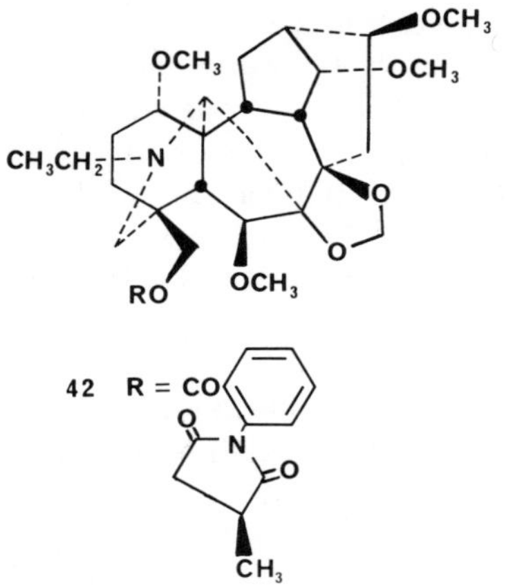

42 R = CO

2.2.42. Elatine. The alkaloid elatine (**42**) has so far been found only in *D. elatum* L. [75]. The pharmacology of this compound was described by Dozortseva [115], and it seems to differ little from that of methyllycaconitine (Section 2.2.37), except that the latter drug exhibits slightly higher potency. Elatine has apparently been used in the Soviet Union for the treatment of neurological disorders [106, 116]. In comparison with tubocurarine, elatine proved to be seven to eight times less toxic, but is claimed [117] to have a five to six times wider therapeutic spectrum.

2.2.43. Anthranoyllycoctonine (inuline). Anthranoyllycoctonine (**43**) has been isolated [75, 98] from species of *Delphinium* and *Inula royleana* (Asteraceae). The pharmacology of this compound has received some attention.

Cook [118], after performing a small number of toxicity tests with mice, noted that the first symptoms of poisoning following i.p. injection of the drug were a

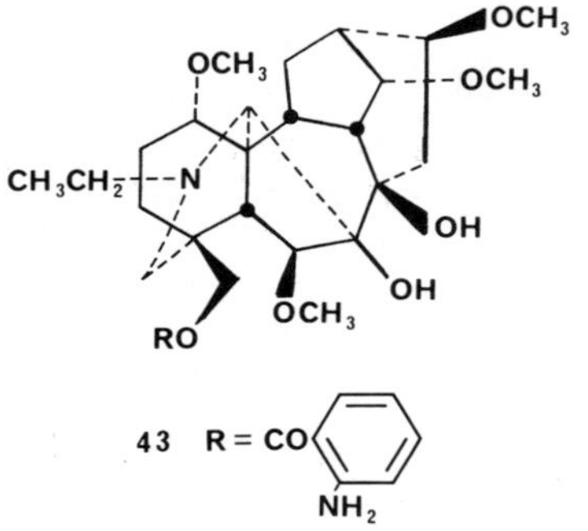

43 R = CO

44 R = COC_6H_5

45 R = $COCH_3$

46 R = H

general attitude of lethargy and rapid respiration. These were followed by partial loss of motor control and respiratory paralysis, culminating in death, preceded by a short period of violent convulsions. The heart was observed to continue beating for a short time after breathing ceased. An "LD_{80}" was recorded (see Table 1).

Shamma et al. [98] reported that anthranoyllycoctonine did not appreciably alter the heart rate of anesthetized rabbits when given intravenously in doses up to 5 mg/kg.

In our own experiments [6], we found that anthranoyllycoctonine decreased the amplitude of miniature end-plate potentials in the frog extensor digitus longus IV muscle, with an approximate ED_{50} of 2×10^{-6} *M*, i.e., the drug had some potency as a neuromuscular blocking agent.

2.2.44. Benzoyllycoctonine (Lycoctonine 18-Benzoate). Lycoctonine 18-benzoate (**44**) is so far unknown in nature. It has been synthesized [6] in a direct fashion from lycoctonine (**46**) and benzoic anhydride. Preliminary pharmacological data [6] on this compound indicate that it decreases the amplitude of miniature end-plate potentials in the frog extensor digitus longus IV preparation in the approximate concentration range of 10^{-5}–10^{-4} *M*. Studies with the rat phrenic nerve–diaphragm preparation showed that the drug at a concentration of 2×10^{-4} *M* reduced the amplitude of the responses by $\sim 50\%$.

2.2.45. Tricornine (Acetyllycoctonine) (Lycoctonine 18-Acetate). Lycoctonine 18-acetate (**45**) was originally prepared as a synthetic derivative of lycoctonine [119] but has more recently been found [120] to occur naturally in *D. tricorne*; it was named tricornine.

In our hands [6], this compound produced rather erratic responses in the rat phrenic nerve–diaphragm preparation: low doses—$(2–3) \times 10^{-4}$ *M*—produced only a slight enhancement of the response, but higher doses—$(3–8) \times 10^{-4}$ *M*—caused an initial enhancement followed by depression. In the frog sciatic nerve-sartorius preparation at a concentration of 10^{-7} *M*, lycoctonine 18-acetate produced a 90% inhibition in postsynaptic action potentials within 2 min. Similarly, the drug decreased the amplitude of miniature end-plate potentials in the frog extensor digitus longus IV muscle preparation, with an approximate ED_{50} of 5×10^{-6} *M*.

2.2.46. Lycoctonine (Delsine, Royline). Lycoctonine (delsine, royline) (**46**) is the "parent" amino alcohol of methyllycaconitine (**37**), lycaconitine (**39**), and elatine (**42**) and may be obtained from them by saponification. It also occurs naturally [75] in species of *Delphinium* and in *Inula royleana*.

Toxicity tests with this compound were first conducted by Beath [121] in rabbits but the identity of Beath's toxin was only clarified by the later studies of Cook [118], who established an approximate lethal dose in mice (see Table 1) and noted that death occurred from respiratory paralysis, preceded by loss of motor control and convulsions. The general biological properties of lycoctonine were studied by Chopra et al. [122], and later by Tulyaganov et al. [8].

Cardiovascular Effects. Chopra et al. [122] reported that lycoctonine caused a sudden fall in blood pressure when given i.v. to cats in doses of 2–5 mg/kg; some tachyphylaxis was observed. The Soviet workers [8], using anesthetized cats, obtained a hypotensive response with drug doses of 5–15 mg/kg i.v. and attributed it to ganglionic blockade. Shamma et al. [98] showed that lycoctonine did not affect the beating rate of the isolated guinea pig atrium in concentrations up to 10^{-4} *M*. The drug also had no specific effect on the heart rate of anesthetized rabbits in doses up to 5 mg/kg i.v. A similar result was reported by Tulyaganov et al. [8], who did not observe any alteration of cardiac rhythm in their animals at all doses up to the lethal.

Smooth Muscle Effects. It was reported by Chopra et al. [122] that the administration of 2 mg/kg i.v. of lycoctonine to anesthetized cats resulted in stimulation of the tone and motility of the intestines. The guinea pig uterus *in vitro* was said to be unaffected by drug concentrations of $(1.5\text{–}2.0) \times 10^{-4}$ *M*. The Soviet workers [8] claimed that the spontaneous contractions of isolated intestinal preparations of the rat and rabbit were not affected by lycoctonine, but concentrations were not given. Contrary to the results of Chopra et al. [122], Tulyaganov et al. [8] reported that lycoctonine elicited contractions of the isolated guinea pig uterus at concentrations of $\sim 1.2 \times 10^{-5}$ *M*. In our own experiments [6], lycoctonine had no effect on the electrically induced contractions of the isolated guinea pig ileum at concentrations up to 10^{-5} *M*.

Skeletal Muscle Effects. Brief respiratory depression was observed [8] when lycoctonine was administered to anesthetized cats in doses of 5–15 mg/kg i.v. However, the drug had no effect on the cat sciatic nerve–gastrocnemius muscle preparation in doses of 10–20 mg/kg. We [6] have observed that lycoctonine had no effect on the contractions of the rat phrenic nerve–diaphragm preparation in concentrations up to 10^{-5} *M*. In the frog sciatic nerve–sartorius preparation, however, lycoctonine produced a 70% inhibition of postsynaptic action potentials in 2 min at a concentration of 10^{-5} *M*. Similarly, the drug caused a decrease in the amplitude of miniature end-plate potentials in the frog extensor digitus longus IV muscle, with an ED_{50} in the range of $10^{5}\text{–}10^{-4}$ *M*.

Central Effects. Lycoctonine had no effect on conditioned reflexes in rats in doses of 10–50 mg/kg i.p. or in dogs in doses of 5–20 mg/kg i.v. [8].

Other Effects. The alkaloid had no local, irritant action on unbroken skin, nor did it affect the conjunctiva or pupil when instilled into the eye [122].

2.2.47. Lycoctonal. Lycoctonal (**47**) may be obtained by the oxidation of lycoctonine [98, 123] but is not known to occur naturally. It was subjected to brief pharmacological testing by Shamma et al. [98], who found the drug to be inactive in altering the beat rate of the isolated guinea pig atrium at concentrations up to 10^{-4} *M*. Similarly, lycoctonal did not affect the heart rate of anesthetized rabbits in doses up to 5 mg/kg.

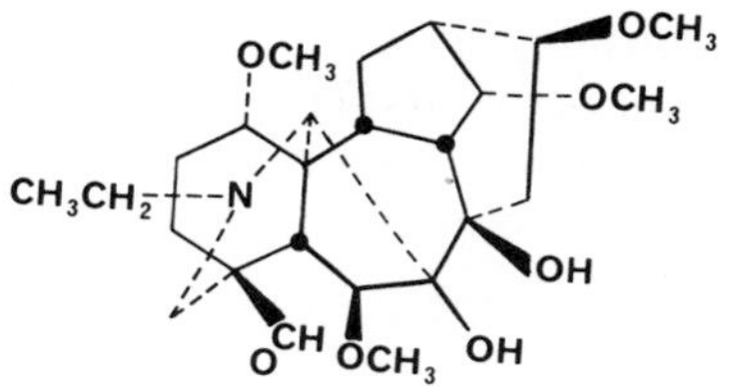

47

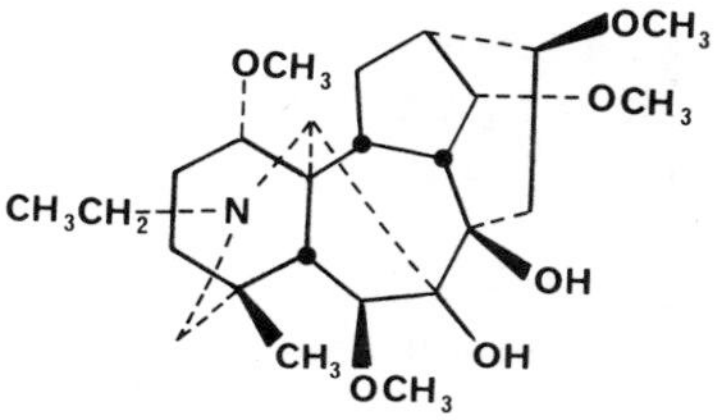

48

2.2.48. Deoxylycoctonine. Also as yet unknown as a naturally occurring compound, deoxylycoctonine (**48**) is prepared [123] by the reduction of lycoctonal. What little work has been done [109] on the compound suggests that it has curariform properties at about the same level of potency as lycoctonine in an *R. pipiens* extensor digitus longus IV muscle system.

2.2.49. Acetylbrowniine (Browniine 14-Acetate). Browniine 14-acetate (**49**) occurs naturally [75] and may also be readily prepared by the acetylation of browniine with acetic anhydride in pyridine. Our studies [6] with this compound showed it to have negligible activity in the rat phrenic nerve–diaphragm assay at concentrations up to 3×10^{-4} *M*. However, the drug exhibited some activity in the guinea pig ileum assay, causing 50% inhibition of the response at a concentration of 2×10^{-4} *M*.

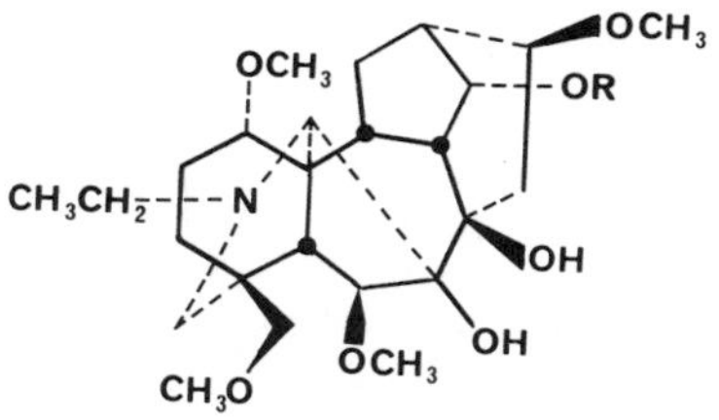

49 $R = COCH_3$

50 $R = H$

2.2.50. Browniine. Browniine (**50**) occurs in a number of species of *Aconitum* and *Delphinium* [75]. We have found [6] the alkaloid to be moderately active in depressing the response of the rat phrenic nerve–diaphragm preparation (ED_{50} 3.5×10^{-4} *M*) and also the response of the guinea pig ileum preparation (ED_{50} 2.6×10^{-4} *M*).

2.2.51. Delsoline (Acomonine). Pelletier's group [124] has recently shown that an alkaloid recently isolated by Yunusov et al. [125] from *A. monticola*, named by them acomonine and assigned a novel structure, is in fact the long-known (from *Delphinium* spp.) delsoline (**51**), an alkaloid with the same oxygenation pattern as lycoctonine and browniine.

The first pharmacological report (on "acomonine") was made by Khamdamov et al. [126], and these findings were further summarized, with some additional data, a year later by the same group [8].

Parenteral administration of delsoline to mice caused weakness in the muscles of the extremities, clonic convulsions, and respiratory depression, with death resulting from asphyxiation. In cats and dogs, respiratory failure was produced by 20–25 mg/kg of acomonine. The acute toxicity of this drug was somewhat lower than that of delcosine (see Table 1).

Cardiovascular Effects. The blood pressure of anesthetized cats and dogs was lowered by 5–15 mg/kg doses of delsoline. Doses of 20–30 mg/kg caused prolongation of the QRST and PR intervals in the ECG recordings of anesthetized rats and rabbits.

Smooth Muscle Effects. Delsoline (dose not stated) had no effect on the spontaneous contractions of the isolated rat or rabbit intestine. However, a drug concentration of $\sim 2.5 \times 10^{-5}$ *M* induced contractions of the isolated guinea pig uterus.

Skeletal Muscle Effects. Acetylcholine-induced contractions of the frog rectus abdominis preparation were inhibited by delsoline at concentrations ranging from $\sim 2.5 \times 10^{-6}$–10^{-4} *M*, with a pronounced response at the higher

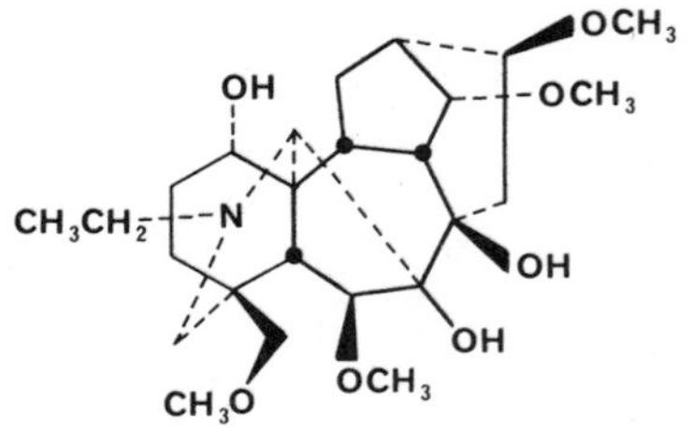

OH OCH$_3$ OCH$_3$ CH$_3$CH$_2$–N OH OH CH$_3$O OCH$_3$

51

concentration. Contractions of the cat sciatic nerve–gastrocnemius muscle preparation were reduced in amplitude by 20 mg/kg (the animal being kept alive by artificial respiration). The neuromuscular block could be dispelled by the i.v. administration of neostigmine.

Central Effects. The drug had no effect on conditioned reflexes in rats or dogs.

Other Effects. Delsoline has also been reported [127] to possess insecticidal properties.

2.2.52. Delcosine (Iliensine, Ilienzine). Delcosine (**52**) corresponds to 14-desmethyldelcosine [75] and often accompanies that alkaloid. In their revision of the structures assigned by the Tashkent group [125] to some *A. monticola* alkaloids, Pelletier et al. [124] demonstrated that iliensine (ilienzine) was delcosine.

In the first pharmacological studies, Schneider [128] noted that delcosine was the most poisonous of the bases from *D. consolida* to cold-blooded animals, but the least toxic to warm-blooded creatures. Subcutaneous injection of the alkaloid into frogs caused strong, curarelike paralysis, as well as effects on the gut, central nervous system, and heart.

Dzhakhangirov and Sadritdinov [129], describing the pharmacology of iliensine, reported that the acute toxicity of the alkaloid was about three times that of lycoctonine (see Table 1); in mice, doses of 75–125 mg/kg caused weakness of the extremities, exophthalmos, and respiratory depression, with death from asphyxiation.

Cardiovascular Effects. A dose of 10 mg/kg of delcosine given to anesthetized cats produced a substantial fall in blood pressure; threshold effects were observed with 0.5–0.1 mg/kg doses. However, drug doses of 2–10 mg/kg had no effect on the ECG of anesthetized rats [129].

Skeletal Muscle Effects. In anesthetized cats, doses of 10–15 mg/kg of delcosine reduced the response of the electrically stimulated sciatic nerve–

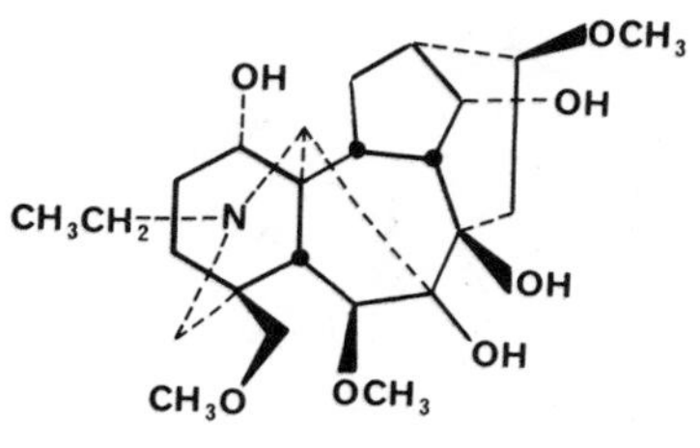

OCH3 OH OH CH3CH2 N OH OH CH3O OCH3

52

gastrocnemius muscle preparation; complete neuromuscular block in this preparation was achieved with 25 mg/kg of the drug, but this could be antagonized with neostigmine [129]. *In vitro*, delcosine concentrations of $\sim 2.5 \times 10^{-5}$–5×10^{-5} *M* inhibited the contractions of the frog rectus abdominis induced by acetylcholine.

Other Effects. A dose of 5 mg/kg of delcosine caused 50% reduction in the response of the cat nictitating membrane preparation. The same dose also blocked ganglionic transmission in the vagus, and the hypotensive effect of the drug was attributed [129] to its ganglion-blocking properties.

Davidson [127] has also reported delcosine to be insecticidal.

2.2.53. Delcorine. The pharmacology of delcorine **(53)**, an alkaloid found in *D. corumbosum* [130], has also been outlined by the Tashkent group [8].

Cardiovascular Effects. Intravenous doses of 5–15 mg/kg of delcorine in cats and dogs resulted in a fall in blood pressure and a temporary respiratory depression.

Smooth Muscle Effects. Delcorine was found to inhibit the spontaneous contractions of isolated rat and rabbit intestine at a concentration of $\sim 2.5 \times 10^{-4}$ *M* and to induce contractions of the isolated guinea pig uterus at a concentration of $\sim 2.5 \times 10^{-5}$ *M*.

Skeletal Muscle Effects. In doses of 10–20 mg/kg i.v. delcorine had no effect on neuromuscular transmission in the sciatic nerve–gastrocnemius preparation of the dog or cat.

Central Effects. Delcorine did not affect conditioned reflexes in rats or dogs.

Other Effects. The drug was a ganglionic blocking agent.

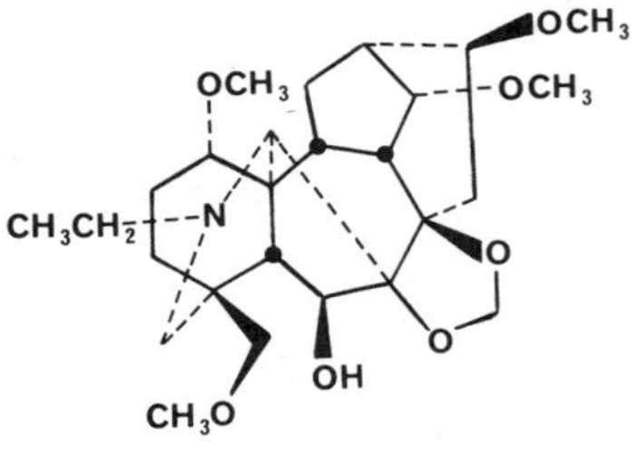

53

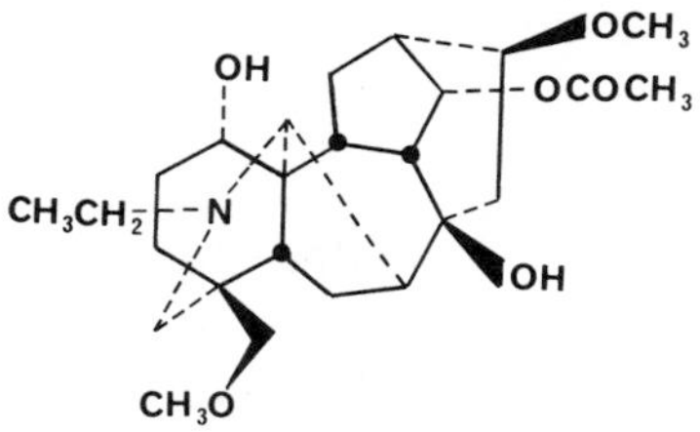

54

2.2.54. Condelphine. The alkaloid condelphine (**54**) occurs in a number of *Delphinium* species [75] and has recently also been isolated from *A. delphinifolium* [131]. The drug is said [106] to have similar pharmacological properties to those of methyllycaconitine, in that it is a curarelike neuromuscular blocker and a hypotensive agent, and to have been used clinically in the Soviet Union in the treatment of various neurological disorders.

A recent study [132] of condelphine showed that in rats it has an age-dependent toxicity (LD_{50} 550–73 mg/kg; neonates to 3–4 months old), the younger animals being the least susceptible. It was implied that this effect might be due to low cholinesterase activity and low levels of catecholamines in the young animals.

2.2.55. Deltaline (Eldeline). Deltaline (eldeline) (**55**) has been isolated from several *Delphinium* species [75]. It is an example of the relatively uncommon 10-hydroxylated diterpenoid alkaloids. The pharmacology of deltaline was first described by Dozortseva [115] and has more recently been briefly described by Tulyaganov et al. [8].

Parenteral administration of eldeline to frogs (200–500 mg/kg by injection into the abdominal lymph sac), mice (250–300 mg/kg i.v.), cats and rabbits (80–100 mg/kg i.v.) produced general depression. In the latter two species, doses of 300–400 mg/kg resulted in death from respiratory failure.

Cardiovascular Effects. Doses up to 15 mg/kg i.v. did not have any distinct effects on circulation or respiration in anesthetized cats. Intravenous administra-

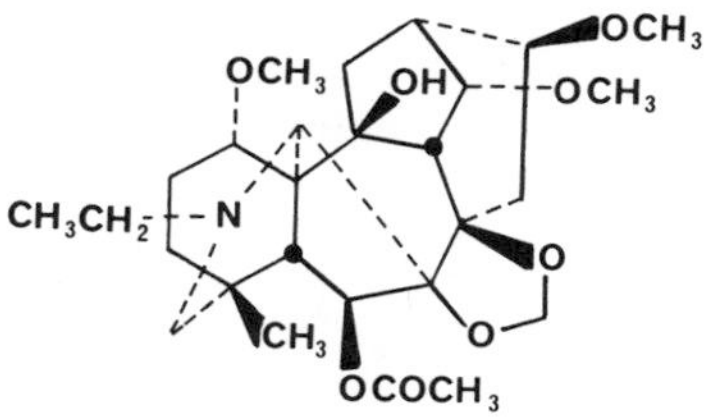

55

tion of 20 mg/kg to cats lowered the blood pressure, but respiration was not affected by this dose.

Smooth Muscle Effects. At a concentration of 4×10^{-4} *M*, eldeline reduced the tone and amplitude of contractions in rabbit intestinal segments. The same concentration also inhibited the response of guinea pig uterus to histamine. In whole animals, relatively high doses (50 mg/kg) were required to abolish the intestinal spasm created by carbachol.

Skeletal Muscle Effects. The cat sciatic nerve–gastrocnemius muscle preparation was unaffected by eldeline in doses of 1–15 mg/kg.

Other Effects. In general, eldeline was found to have no effect on autonomic ganglia nor on cholinergic or adrenergic receptors.

2.2.56. Heteratisine. Heteratisine (**56**), an alkaloid occurring in *A. heterophyllum* and *A. zeravschanicum* Steinb. [75], typifies a further subclass of C_{19} diterpenoid alkaloids, having a lycoctonine-type skeleton modified by the expansion of ring-C to a δ-lactone.

The pharmacology of heteratisine has been described in publications from the Soviet Union [8, 133] and shows features in common with napelline (see Section 2.2.63), despite the structural differences between the "atisinoid" (see Section 2.2.57) and "heteratisinoid" skeletons.

Cardiovascular Effects. The effects of heteratisine on blood pressure and respiration were similar to those of napelline, i.e., brief hypotension and disturbed respiration. Like napelline, heteratisine also antagonised aconitine-induced and calcium chloride–induced cardiac arrhythmias but was only about half as potent as napelline. Heteratisine itself also prolonged PQ and QRST intervals.

Smooth Muscle Effects. The general effects of heteratisine on isolated rabbit intestine and guinea pig uterus were identical with those noted for napelline (see Section 2.2.63).

OCH3 O O CH3CH2 N OH CH3 OH

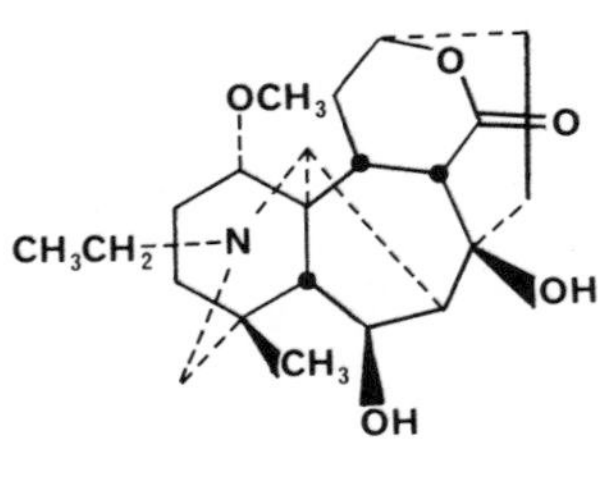

56

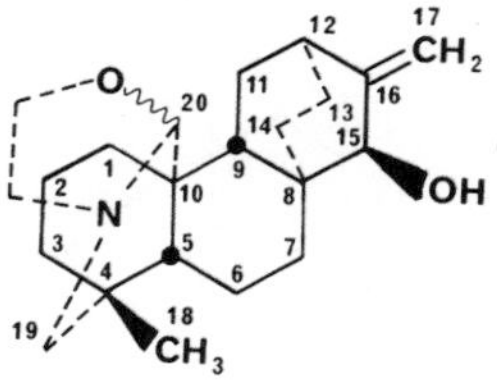

57

2.2.57. Atisine. Atisine (**57**), the first of the C_{20} diterpenoid alkaloids to be discussed in this catalog, is one of the alkaloids of *A. heterophyllum* Wall. [75]. Although preparations of this plant are important in Indian folk medicine [10, 134] and there have been pharmacological investigations of the crude drug [135], there are relatively few published studies of atisine itself.

Cardiovascular Effects. Chopra et al. [134] found in experiments with cats, rabbits, and dogs that the administration of atisine in doses up to 15 mg/kg i.v. caused only a slight, transient fall in blood pressure. This study was extended by Raymond-Hamet [136, 137], who observed the same weak and transient hypotension with doses of 4 mg/kg in dogs, but a marked, prolonged fall in blood pressure with 16 mg/kg of the drug. The lower dose of atisine enhanced the hypertensive effect of adrenaline but reduced the bradycardia. A dose of 20 mg/kg of atisine administered after 0.02 mg of adrenaline resulted in considerable enhancement of hypertension, completely abolished the bradycardia, and enhanced bradypnea.

2.2.58. Denudatine. The alkaloid denudatine (**58**) is found in the roots of *D. denudatum* Wall. [75] and has an atisine-type skeleton. A report on the pharmacology of this compound was published by Singh and Chopra [138].

Cardiovascular Effects. Intravenous doses of 0.5–5.0 mg/kg of denudatine had no effect on blood pressure or respiration in anesthetized dogs.

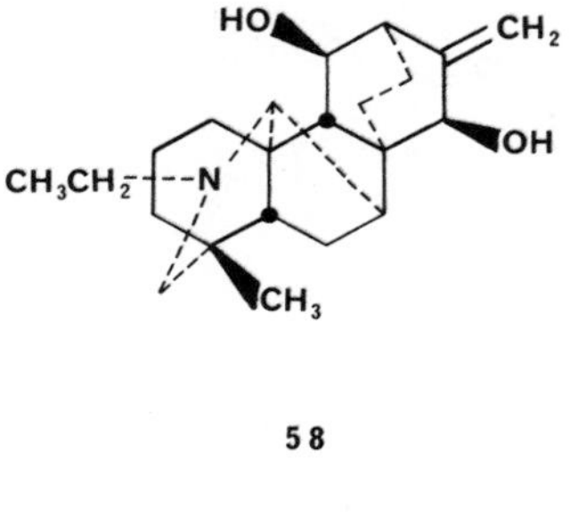

58

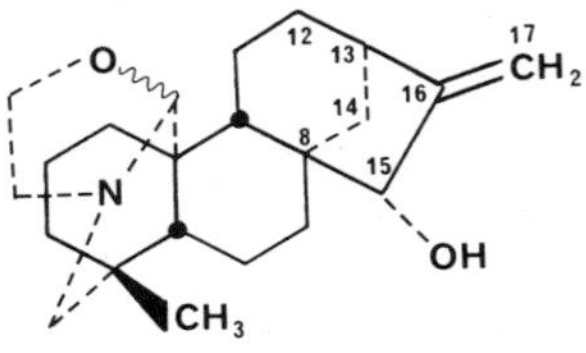

59

Smooth Muscle Effects. In doses of 0.5–5.0 mg/kg, the drug decreased tone and inhibited peristalsis of the dog intestine *in situ*. At a concentration of 2×10^{-5} *M*, denudatine produced a marked inhibition of the contraction of the isolated rabbit duodenal strip. The same concentration did not affect the acetylcholine- and histamine-induced contractions of guinea pig intestinal strips but did stimulate the response of guinea pig uterus strips. Atropinization did not alter the effect of the drug.

Skeletal Muscle Effects. Doses of denudatine up to 200 mg/kg i.v. had no effect on the sciatic nerve–gastrocnemius muscle preparation of the dog.

2.2.59. Veatchine. Veatchine (**59**), isolated from *Garrya veatchii* Kellogg [75], exemplifies another structural type. The pharmacology of this compound has been studied by Powell et al. [139]. Its toxicity was relatively high in comparison with other alkanolamines of either the C_{19} or C_{20} type (See Table 1). In mice, toxic doses caused increased respiration, gasping, and clonic convulsions; lethal doses caused death by respiratory failure.

Cardiovascular Effects. In the anesthetized cat, veatchine hydrochloride (0.5–1.0 mg/kg i.p.) had little effect on blood pressure and caused a small decrease in heart rate.

Smooth Muscle Effects. Veatchine hydrochloride in concentrations of (3–6) $\times 10^{-4}$ *M* had relatively little effect on the contractions of segments of small intestine from rabbit or guinea pig. The same concentration also exerted no appreciable effects on the isolated uterus of rabbit, or guinea pig.

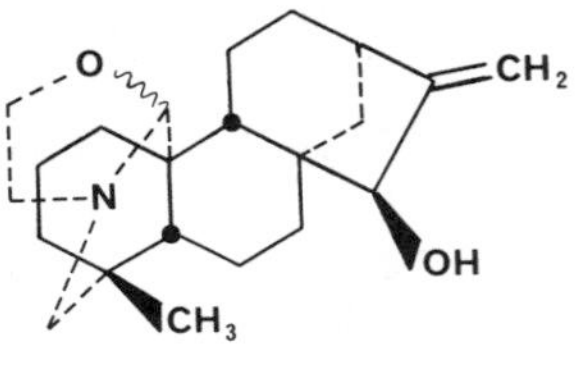

60

2.2.60. Garryfoline. The alkaloid garryfoline **(60)**, isolated from *G. laurifolia* Hartw. [75], is the C(15) epimer of veatchine. The pharmacology of this compound has also been studied by Powell et al. [139], who found it to be slightly less toxic than veatchine (see Table 1). As in the case of the latter drug, toxic symptoms included gasping, convulsions, and respiratory failure.

Smooth Muscle Effects. Garryfoline oxalate at a concentration of 4×10^{-4} *M* partially reduced the spasm of guinea pig ileum caused by histamine or methacholine. At concentrations of $(2\text{–}4) \times 10^{-4}$ *M*, the drug had negligible effects on rabbit or guinea pig uterus.

Other Effects. No significant signs of histopathological damage were observed in the following tissues taken from mice that had been dosed with garryfoline oxalate: heart, lung, liver, pancreas, spleen, gastrointestinal tract, kidney, adrenal, and thymus.

2.2.61. Garryine. The alkaloid garryine **(61)** was first isolated from *G. veatchii* [75]. This compound may also be obtained by the isomerization of veatchine in alcohol solution at room temperature.

Powell et al. [139] reported that the effects of garryine hydrochloride administered i.v. to mice were virtually identical with those obtained with veatchine hydrochloride. The LD_{50} values of the two drugs were also virtually identical (see Table 1).

Cardiovascular Effects. In the anesthetised cat, a dose of 5 mg/kg i.p., of garryine hydrochloride decreased blood pressure, and also decreased the heart rate.

Other Effects. Garryine was found [140] to be inactive as an antimalarial in ducks.

2.2.62. Cuauchichicine. Cuauchichicine **(62)** accompanies garryfoline in *G. laurifolia* [75]. Powell et al. [139] included cuauchichicine in their study of the

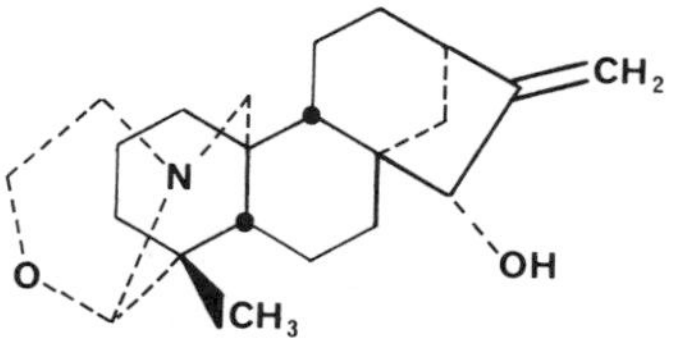

61

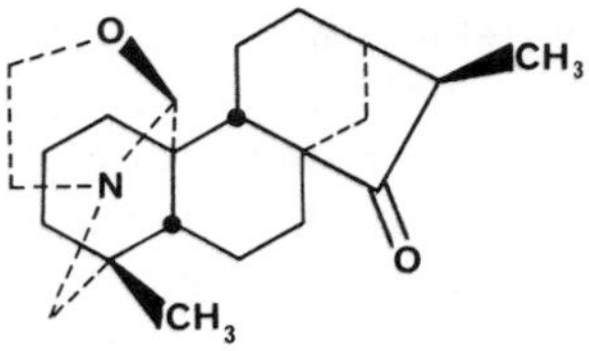

62

pharmacology of *Garrya* alkaloids and reported that in mice this alkaloid was the most toxic of the veatchinelike alkaloids (see Table 1); as usual, death resulted from respiratory failure. Given intravenously, cuauchichicine caused tremors, followed by prostration. Other effects noted were as follows.

Cardiovascular Effects. Doses of 1 mg/kg of cuauchichicine hydrochloride briefly lowered the blood pressure and increased the heart rate.

Smooth Muscle Effects. Doses of $(3–6) \times 10^{-4}$ M did not produce any responses in gut or uterus of the guinea pig or rabbit.

2.2.63. Napelline (Luciculine). The alkaloid napelline (**63**) occurs in several *Aconitum* species [75], including *A. napellus*, from which it derives its name. It has an atisine-type skeleton, with some structural resemblance to denudatine. Studies on the pharmacology of napelline were reported by Dzhakhangirov and Sadritdinov [133].

Cardiovascular Effects. In experiments with cats, napelline (10–20 mg/kg) caused a brief lowering of blood pressure and disturbed respiration for rather longer. In doses of 10–40 mg/kg, the drug was found to prolong the PQ and QRST intervals in the ECG of anesthetized rats. However, smaller doses (5–20 mg/kg) of napelline were sufficient to antagonize the cardiac arrhythmias induced by the i.v. administration of calcium chloride or aconitine, surpassing the effect of novocaine (30 mg/kg). The i.p. injection of 25–100 mg/kg of napelline into mice 30 min before the administration of a normally lethal dose (0.2 mg/kg) of aconitine resulted in a dose-related reduction in mortality. The highest dose prevented the death of almost all the animals.

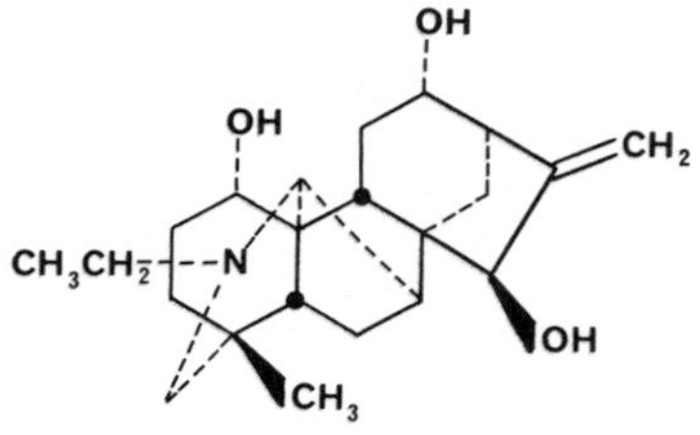

63

Smooth Muscle Effects. The drug (concentration not stated) did not have any effect on the contractions of the isolated rabbit intestine induced by barium chloride or acetylcholine. At a concentration of $\sim 3 \times 10^{-4}$ *M*, it reduced the amplitude of acetylcholine-induced contractions in the isolated frog rectus abdominis preparation by 50%.

Other Effects. A dose of 20 mg/kg of napelline blocked impulse transmission in sympathetic and parasympathetic ganglia. The drug did not affect the hypotensive response to acetylcholine or histamine, but reinforced the pressor effect of adrenaline.

2.2.64. Songorine (Napellonine). The alkaloid songorine (**64**) was first isolated from *A. soongaricum* Stapf [75]. It is the 12-keto analog of napelline, and its pharmacology has been discussed in two Soviet reviews [8, 141]. The physiological effects of this drug appear to differ somewhat from those of the other diterpenoid alkaloids examined so far, for although cardiovascular activity is seen at high doses, the most noticeable effects are on CNS activity. Songorine 1,15-diacetate resembles the parent base in its pharmacological properties [142a].

In mice and rats, 50–100 mg/kg doses of songorine produced a syndrome of decreased motor activity, coupled with a heightened response to external stimuli. Larger doses of the drug led to respiratory difficulties, tremor, increased tone in the skeletal musculature, and clonicotonic convulsions.

Cardiovascular Effects. A dose of 20 mg/kg of songorine caused a fall in blood pressure.

Smooth Muscle Effects. Songorine had no effect on the spontaneous contractions of isolated rabbit intestine in concentrations up to $\sim 3 \times 10^{-4}$ *M*. However, the latter concentration did evince contractions of the isolated guinea pig uterus.

Skeletal Muscle Effects. In doses of 10–20 mg/kg, the drug did not affect the cat sciatic nerve–gastrocnemius muscle preparation.

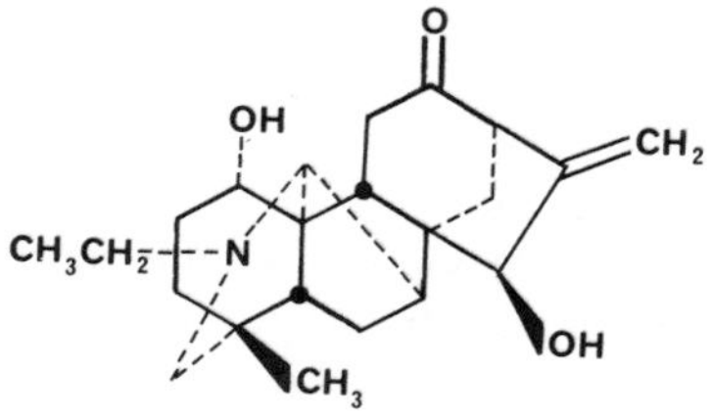

O
OH
CH_2
CH_3CH_2–N
OH
CH_3

64

Central Effects. Subcutaneous administration of 50–150 mg/kg of songorine to cats resulted in a stereotypic rocking motion of the head. In dogs, 10–15 mg/kg i.v. of the drug disturbed conditioned reflexes for several hours. Barbiturate-induced sleeping time was prolonged by doses of 25–100 mg/kg of songorine. The drug produced hypothermia in rabbits and antagonized resperpine-induced hypothermia. The sedative effect of bulbocapnine was antagonized by songorine, but the effects of tryptamine and phenamine were enhanced. The overall effect of the drug on EEG was that of desynchronization. (A detailed discussion of the EEG responses to songorine is given in [8].)

Other Effects. In the cat nictitating membrane preparation, doses of 20–25 mg/kg of songorine were sufficient to cause total inhibition of the response. The hypotension resulting from administration of songorine was attributed to its ganglion-blocking properties.

Although the acute toxicity of songorine was found to be relatively low (see Table 1), repeated administration of the drug to rats was observed to induce dose-dependent histomorphological changes. Thus doses of 50 mg/kg produced destructive effects in cerebral, cardiac, hepatic, and nephrotic tissue. On the other hand, when songorine hydrochloride (10–12 mg/kg day) was administered over a period of 2 weeks orally or i.p. to rats with sarcoma 45, sarcoma M1, Walker carcinosarcoma, or leukosarcoma, no specific antineoplastic activity was seen [142b].

3. STRUCTURE–ACTIVITY RELATIONSHIPS: OVERVIEW

From the data presented in Table 1, one can see that the greatest mammalian toxicity resides in the aconitinelike group of alkaloids. Scrutiny of the pharmacological properties of the individual alkaloids (Section 2.2) reveals a broad range of symptoms including impairment of the cardiovascular system (hypotension, cardiac arrhythmias), respiratory inhibition, muscular paralysis, and disturbances of the central nervous system. These effects appear to be due to the drugs acting as neurotoxins, and two main types can be characterized: those with aconitinelike (aconitiform) activity and those with curarelike (curariform) activity.

3.1. Aconitiform Activity

Aconitine exemplifies a group of diterpenoid alkaloids that includes bikhaconitine, hypaconitine, indaconitine, jesaconitine, mesaconitine, pseudaconitine, aconifine, and delphinine. As demonstrated for aconitine, one may infer that these compounds exhibit neurotoxicity principally as a consequence of their abilities to interact with excitable membranes in such a way as to hold open sodium channels following an action potential, so that prolonged depolarization results.

Catterall [27] has suggested that aconitine (and, by implication, the other diterpenoid alkaloids with similar pharmacological activities) thus very closely resembles three other natural, lipophilic neurotoxins: batrachotoxin (**65**), veratridine (**66**), and grayanotoxin-I (**67**). Working with mouse neuroblastoma cells, he has produced compelling evidence [69, 143] that all four toxins act reversibly at the same receptor site. In this system batrachotoxin was a full agonist (i.e., capable of activating all the sodium channels), whereas aconitine

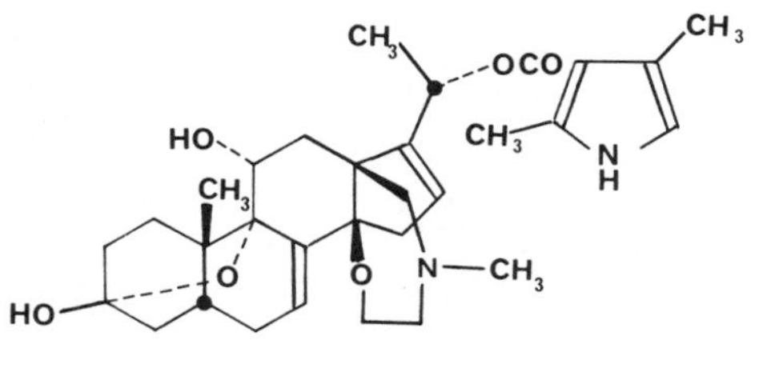

65

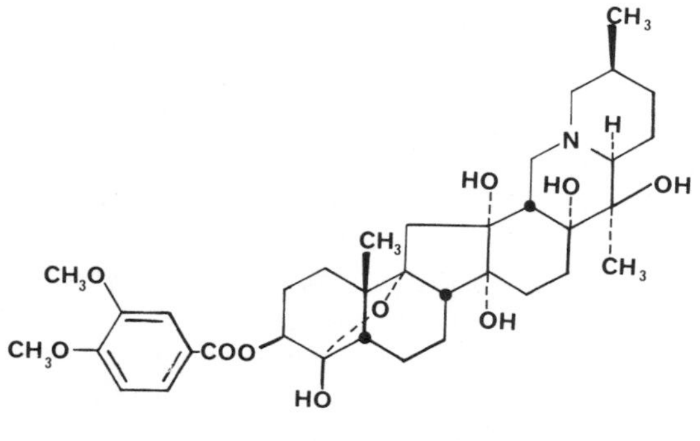

66

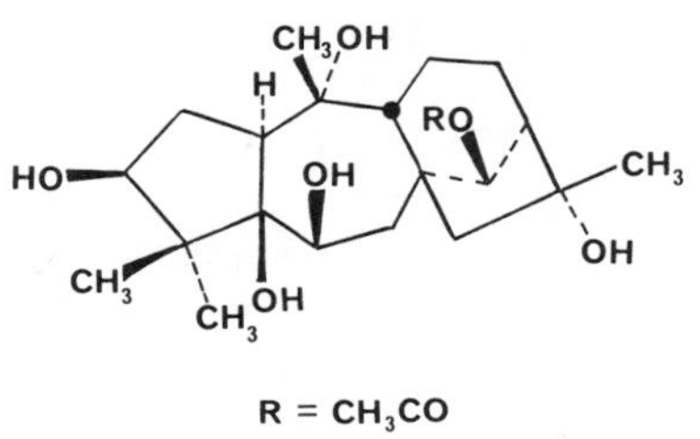

$R = CH_3CO$

67

and the other two neurotoxins were partial agonists. In line with these findings, Schmitt and Schmidt have shown [62] that two populations of sodium channels could be seen in nodes of Ranvier treated with aconitine: one was incapable of inactivation (closure), and its activation potential was shifted by more than 50 mV to a more negative membrane potential; the other had near normal properties, but the voltage dependence for activation and deactivation was shifted by only about 10 mV to a more negative membrane potential.

Catterall's [27, 69, 143] simple and reasonable model for the mode of action of aconitine is that the receptor is coupled to the sodium channel: modifications involving the latter in its switching from "closed" to "open" also affect the former, and vice versa. If the affinity of aconitine for the receptor is greatest when the sodium channel is open, the consequence will be persistent activation. The binding energy of the alkaloid to the receptor will also be manifested in the shift in the activation potential.

Turning to a consideration of structure–activity relationships (SARs), since grayanotoxin-I is nonalkaloidal, it is possible that the key structural features of the aconitinelike diterpenoid alkaloids (as also batrachotoxin, and veratridine) that confer the receptor-binding properties do not include the nitrogen atom. This line of thought, in which the nitrogen functionality of aconitine is either unimportant or a minor component of the haptophore, seems to have been considered by Smythies [144]. While discussing a model for the molecular structure of part of the sodium channel, he likened the shape of aconitine to a brick with a lipophilic center, surrounded by a ring of hydrogen-bonding functions (of which the nitrogen is but one of many). Matsutani et al. [145] have also proposed a nitrogen-free model to account for the depolarizing properties of aconitine. (They also proposed similar, nitrogen-free, models for batrachotoxin and veratridine, as well as the grayanotoxins.) They suggested that it is the relative positioning of the 16β-methoxyl, 15α-hydroxyl, 8β-acetoxy oxygen (or 14α-benzoyloxy oxygen) and possibly the methyl of the acetoxy group that endows aconitine with the ability to interact with the same receptor as grayanotoxin. (But see the comment below regarding the 15α-OH.)

What is certain, and has been recognized since the very earliest SAR studies, is that the aconitiform alkaloids are esters, and that removal of the ester functions (e.g., by saponification to aconines) results in an enormous drop in their acute toxicity and essentially complete loss of the neurotoxicity characteristic of the parent alkaloids. We have adopted a conservative approach to SAR models and simply note that all the aconitiform alkaloids possess the part structure **68**: a hexacyclic, aconitane skeleton [75] carrying an acetate function at C(8), an aryl ester at C(14) (α), one hydroxyl at C(13), four methoxyl groups at C(1) (α), C(6) (α), C(16) (β), and C(18), and methyl or ethyl substitution on the nitrogen. Other functions, such as the hydroxyls at C(3) (α) and C(15) (α) as found in aconitine, jesaconitine, and mesaconitine, may increase the potency of this system, but are not essential (c.f. bikhaconitine and delphinine), i.e., are haptophoric. Thus we can predict that chasmaconitine (**68**: $R = CH_3CH_2$, $Ar = C_6H_5$) [146], chasmanthinine (**68**: $R = CH_3CH_2$, $Ar =$ (E)-$C_6H_5CH{=}CH$) [146], and yunaconi-

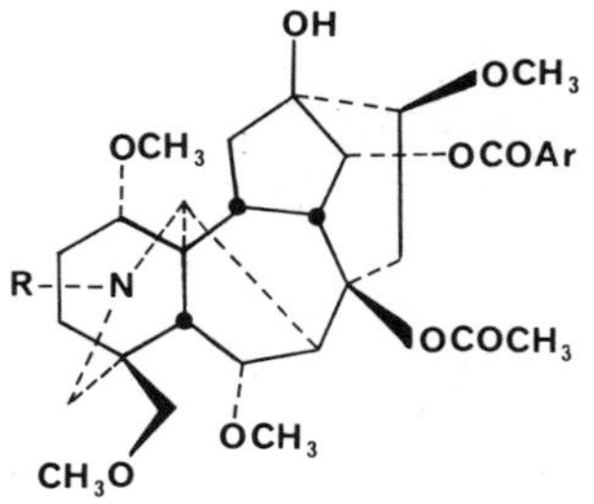

68

tine (3α-hydroxy **68**: $R = CH_3CH_2$, $Ar = p\text{—}CH_3OC_6H_4$) [147] will resemble aconitine very closely in their pharmacological effects. Further experiments are required to refine the structure of the aconitiform pharmacophore beyond **68**, e.g., to examine the pharmacological properties of foresaconitine [148] (which corresponds to 2,13-dideoxyjesaconitine) and isodelphinine (13-deoxy-15α-hydroxydelphinine) [149] to see if the 13α-OH is, or is not, an essential part of **68**.

3.2. Curariform Activity

A second kind of neurotoxicity encountered in the diterpennoid alkaloids is that shown by methyllycaconitine (see Section 2.2.37). This is classical curariform activity, in which the drug is a competitive, postsynaptic inhibitor of the neurotransmitter acetylcholine, acting at nicotinic sites. As such, it is a potent neuromuscular poison in mammals. It also acts in the same way at cholinergic ganglia in sympathetic or parasympathetic nerves.

It is apparent from our catalog (Section 2.2) that the pharmacophore associated with disturbance of cholinergic neurotransmission is the lycoctonine skeleton itself. The pattern of oxygenation and the electronic nature of the oxygen-bearing functionalities appear to determine the physiological manifestations of this disturbance (e.g., whether ganglionic or neuromuscular blockade is predominant). The most potent competitive antagonist of acetylcholine at the skeletal neuromuscular junction among the natural diterpenoid alkaloids is methyllycaconitine (**37**). (In passing it may be noted that most competitive inhibitors of this type are quaternary ammonium salts like (+)-tubocurarine chloride (**69**); the only tertiary amine rivaling **37** in effectiveness is (+)-dihydro-β-erythroidine (**70**)). Methyllycaconitine differs from the weakly active alkanolamine diterpenoid alkaloids—e.g., lycoctonine (**46**)—most obviously in the complex ester grouping at C(18).

Since methyllycaconitine so closely resembled (+)-tubocurarine chloride (**69**) in its mode of action, we compared [6] CPK space-filling molecular models of these two compounds. Inspection showed similarity in overall size and restricted

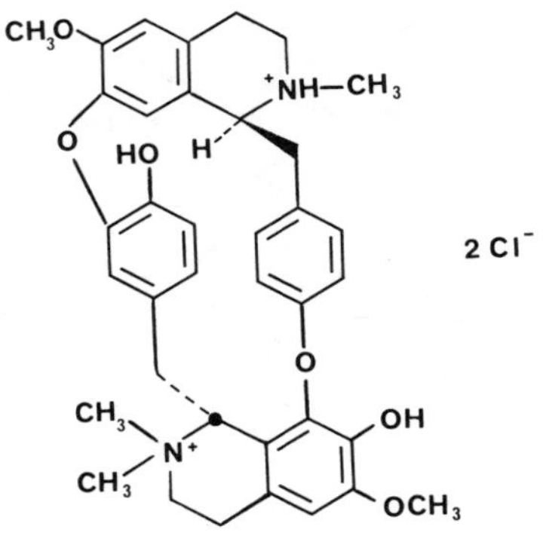

69

70

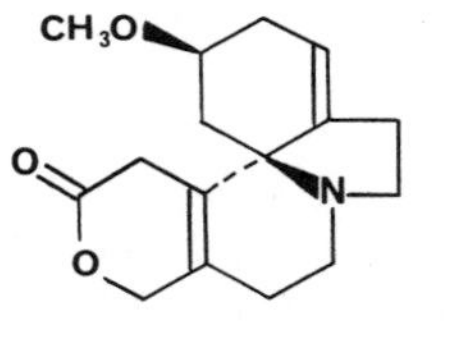

71

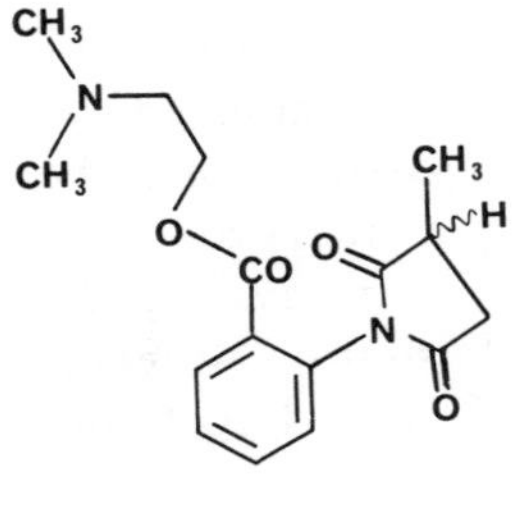

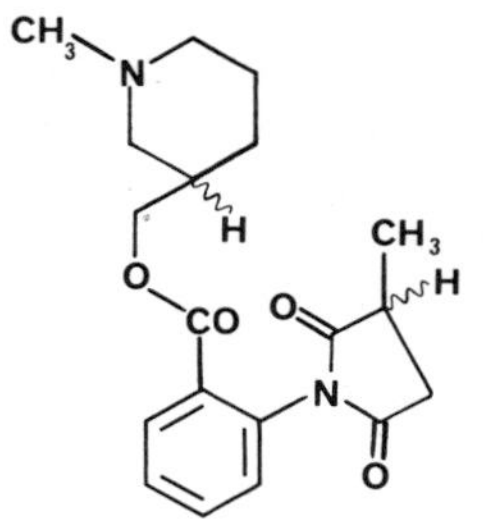

72

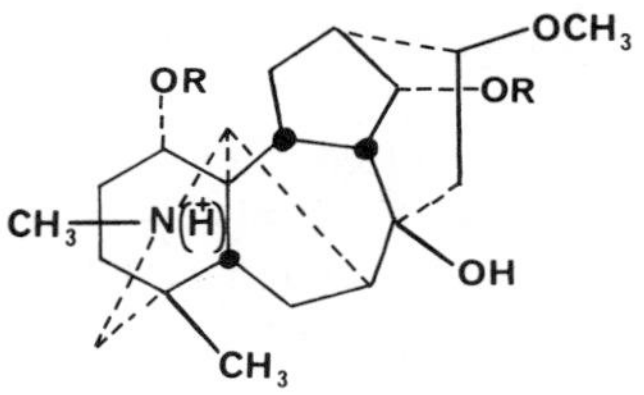

73

conformational mobility. Superposition of nitrogen and oxygen atoms revealed considerable congruity, and it indeed appears that methyllycaconitine would well fit the "curariform template" described by Pauling and Petcher [150].

To probe the importance of the complex ester functionality in determining the curariform activity of methyllycaconitine, we examined [6, 109] a series of derivatives, measuring their effects on the miniature end-plate potentials in an extensor digitus longus IV muscle preparation from *R. pipens.* Delsemine (**38a**, **38b**) was thus found to be only slightly less potent than methyllycaconitine, whereas anthranoyllycaconitine (**43**) was 1/10, and lycoctonine about a 1/200 as potent: 18-*O*-benzoyllycoctonine was only about twice as active as lycoctonine. We also examined the synthetic drugs **71** and **72**: they were much lower in potency than lycoctonine.

Our conclusion [6] is that the ester function in methyllycaconitine is a haptophore, providing secondary binding sites or aiding in orientation, whereas the toxophore involves other groups. An examination of molecular models together with the data provided in the catalog (Section 2.2) resulted in the suggestion [6] that the key acetylcholine receptor "recognition sites" are the nitrogen and the C(8) oxygen atoms. An SAR analysis (as in 3.1.1) leads to 73 as the curariform pharmacophore in the diterpenoid alkaloids.

Here also, further studies are indicated, but some support for the conclusions is provided by the observation [94] that the semisynthetic drugs delphinine methochloride (**29**) and delphonine methochloride (**30**) are highly potent (apparently more active than methyllycaconitine) as curariform neuromuscular blockers.

REFERENCES

1. C. S. Keener, *Castanea* **41,** 12 (1976).
2. N. K. Hart, S. R. Johns, J. A. Lamberton, H. Suares, and R. I. Willing, *Austr. J. Chem.* **29,** 1295 (1976), **29,** 1319 (1976).
3. O. E. Edwards in *Specialist Periodical Report: The Alkaloids*, Vol. 1, J. E. Saxton, Ed., The Chemical Society, London, 1971, p. 348.
4. (a) P. On'okoko, M. Hans, B. Colan, C. Hootele, J. P. Declerco, G. Germain, and M. van Meersche, *Bull. Soc. Chim. Belg.,* **86,** 655 (1977). (b) P. On'okoko and M. Vanhaelen, *Phytochemistry* **19,** 303 (1980).
5. S. Yamamura and Y. Hirata in *The Alkaloids: Chemistry and Physiology*, Vol. 15, R. H. F. Manske, Ed., Academic, New York, 1975, p. 41.
6. J. M. Jacyno, Ph. D. thesis, University of Calgary, 1981.
7. H. Sato, C. Yamada, C. Konno, Y. Ohizumi, K. Endo, and H. Hikino, *Tohoku J. Exp. Med.* **128,** 175 (1979).
8. N. Tulyaganov, F. N. Dzhakhangirov, F. S. Sadritdinov, and I. Khamdamov in *Farmakol. Rastit. Veshchestv*, M. B. Sultanov, Ed., Academy of Sciences, Uzbek SSR, Tashkent, USSR, 1976, p. 76; *Chem. Abstr.* **89,** 140208 (1978).
9. (a) E. Tripp, *Crowell's Handbook of Classical Mythology*, Crowell, New York, 1970, and references therein. (b) Anon, *Enciclopedia Universal Ilustrado*, Vol. 2, p. 327, Espasa-Calpe S.A., Madrid, 1958, and references therein.
10. (a) M. Grieve, *A Modern Herbal*, Vols. 1 and 2, Hafner, New York, 1961. Vol. 1, p. 6, and Vol. 2, pp. 464 and 770. (b) R. Le Strange, *A History of Herbal Plants*, Angus and Robertson, London, 1977. (c) R. T. Gunther, Ed., *The Greek Herbal of Dioscorides*, illustrated by a Byzantine A.D. 512, Englished by J. Goodyer A.D. 1655, Hafner, New York, 1959. (d) M. Woodward, Ed., *Gerard's Herbal*, from the edition of Th. Johnson 1636, Spring Books, London, 1964, pp. 227 and 249. (e) R. N. Chopra, P. L. Badhawar, and S. Ghosh, *Poisonous Plants of India*, Vol. 1, Government of India Press, Delhi, 1949, pp. 97 and 125. (f) F. Minton and V. Minton, *Practical Modern Herbal*, W. Foulsham, London, 1976, p. 102. (g) M. Moore, *Medicinal Plants in the Mountain West*, Museum of New Mexico Press, Sante Fe, 1979, p. 95. (h) W. A. Emboden, *Narcotic Plants*, MacMillan, New York, 1979, p. 18 and 127.
11. (a) N. G. Bisset, *J. Ethnopharmacol.* **4,** 247 (1981) and references therein. (b) N. G. Bisset, *J. Ethnopharmacol.* **1,** 325 (1979) and references therein. (c) N. G. Bisset, *Llyodia* **39,** 87 (1976) and references therein.
12. (a) J. M. Kingsbury, *Poisonous Plants of the United States and Canada*, Prentice-Hall, Englewood Cliffs, NJ, 1964. (b) J. D. Olsen in *Effects of Poisonous Plants on Livestock*, R. F. Keeler, K. R. van Kampen, and L. F. James, Eds., Academic, New York, 1978, p. 535, and references therein.
13. (a) J. L. Hartwell, *Lloydia* **34,** 103 (1971). (b) S. von R. Altschul, *Drugs and Foods from Little Known Plants*, Harvard University Press, Cambridge, MA 1973, p. 69.
14. (a) R. Jaretzsky and H. Jaenecke, *Arch. Pharm.* **278,** 156 (1940), and references therein. (b) M. Gatty-Kostyal and A. Stawowczyk, *Diss. Pharm. Pharmacol.* **8,** 257 (1956), and references therein.

15. B. D. Morley, *Wild Flowers of the World*, Putnam, New York, 1970, p. 155.
16. M. Martinez, *Plantas Medicinales de Mexico*, Ediciones Botas, Mexico City, 1959, p. 96.
17. E. E. Swanson, H. W. Youngken, C. J. Zufall, W. J. Husa, J. C. Munch, and J. B. Wolffe, *Aconitum*, American Pharmaceutical Association Monograph No. 1, Washington, D.C., 1938.
18. J. T. Cash and W. R. Dunstan, *Phil. Trans.* **190B**, 239 (1898).
19. J. T. Cash and W. R. Dunstan, *Proc. R. Soc. London* **68,** 384 (1901).
20. A. Katz and E. Staehelin, *Pharm. Acta Helv.* **54,** 253 (1979).
21. D. Scherf, *Proc. Soc. Exp. Biol.* **64,** 233 (1947).
22. R. Mendez and E. Kabela, *Ann. Rev. Pharmacol.* **10,** 291 (1970).
23. D. Scherf and A. Schott, *Extrasystoles and Allied Arrhythmias*, 2nd ed., Heinemann, London, 1973.
24. R. D. Tanz, J. B. Robbins, K. L. Kemple, and P. A. Allen, *J. Pharm. Exp. Ther.* **185,** 427 (1973).
25. R. D. Tanz, *J. Pharm. Exp. Ther.* **191,** 232 (1974).
26. R. D. Tanz, *Proc. West. Pharmacol. Soc.* **17,** 22 (1974).
27. W. A. Catterall, *Ann. Rev. Pharmacol. Toxicol.* **20,** 15 (1980).
28. H. Ishikawa, *Kyoto Igaku Zasshi* **18,** 1496 (1921); cited in [74].
29. M. Yokota, *Tohoku Igaku Zasshi* **6,** 162 (1922); *Chem. Abstr.* **17,** 3376 (1923); cited in [74].
30. F. Kataki, *Okayama Igakkai Zasshi* **54,** 1735 (1942); cited in [74].
31. T. Inoue, *Igaku Kenkyu* **20,** 48 (1950); cited in [74].
32. K. O. Ellis and S. H. Bryant, *Life Sci.* **13,** 1607 (1973).
33. E. S. Stern in *The Alkaloids: Chemistry and Physiology*, Vol. 4, R. H. F. Manske and H. L. Holmes, Eds., Academic, New York, 1954, p. 275.
34. K. P. Bhargava and R. P. Kohli, J. N. Sinha, and G. Tayal, *Br. J. Pharmacol.* **36,** 240 (1969).
35. K. P. Bhargava and R. K. Srivastava, *Neuropharmacology* **11,** 123 (1972).
36. B. V. Telang and J. N. Nganga, *Ind. J. Physiol. Pharmacol.* **19,** 1 (1975).
37. B. V. Telang, B. B. Gaitonde, M. S. Kekre, and G. V. Sukthankar, *Ind. J. Physiol. Pharmacol.* **19,** 11 (1975).
38. M. Abdelaal, E. M. Ammar, A. M. Afifi, and A. M. Zohdy, *J. Drug Res. (Egypt)* **7,** 147 (1975).
39. E. M. Ammar, M. Abdelaal and A. M. Afifi, *Acta Pharm. Jugosl.* **26,** 223 (1976); *Chem. Abstr.* **86**, 229g (1977).
40. H.-H. Wellhoner and B. Conrad, *Naunyn-Schmiedebergs Arch. Exp. Pathol. Pharmakol.* **252,** 269 (1965).
41. H.-H. Wellhoner and D. Haferkorn, *Naunyn-Schmiedebergs Arch. Exp. Pathol. Pharmakol.* **255,** 407 (1966).
42. H.-H. Wellhoner, *Pfluegers Arch.* **304,** 104 (1968).
43. G. Haegerstam, *Acta Physiol. Scand.* **98,** 1 (1976).
44. S. F. Maksimenko, *Byull. Eksp. Biol. Med.* **65,** 60 (1968); *Chem. Abstr.* **69**, 1632u (1968).
45. L. L. Grechishkin and V. E. Ryzhenkov, *Farmakol. Toksikol.* **26,** 578 (1963); *Ex. Med. Sect. II* **17,** 9290 (1964); *Chem. Abstr.* **60,** 12549c (1964).
46. R. Weigmann, *Naunyn-Schmiedebergs Arch. Exp. Pathol. Pharmakol.* **212,** 116 (1950).
47. O. Eichler, F. Hertle, and I. Staib, *Arzneim.-Forsch.* **7,** 349 (1957).
48. T.-M. Chang, K.-C. Ch'ao, and F.-H. Lu, *Sheng Li Hsueh Pao* **22,** 98 (1958); *Chem. Abstr.* **53,** 13390 (1959).
49. D. G. Patel, S. S. Karbhari, O. D. Gulatti, and S. D. Gokhale, *Arch. Int. Pharmacodyn* **157,** 22 (1965).
50. J. M. Melon and A. Buzas, *Fr. Pat. Appl.* 2,255,887 (1975); *Chem. Abstr.* **84,** 22106a (1975).

51. (a) H. Hikino, T. Ito, C. Yamada, H. Sato, C. Konno and Y. Ohizumi, *J. Pharmacobio-Dyn.* **2,** 78 (1979). (b) H. Hikino, C. Konno, H. Takara, Y. Yamada, C. Yamada, Y. Ohizumi, K. Sugio, and H. Fujimura, *J. Pharmacobio-Dyn.* **3,** 514 (1980).
52. Z. Berankova and F. Sorm, *Chem. Listy* **50,** 637 (1956).
53. H. Bekemeier, A. J. Geissler, and E. Vogel, *Pharmacol. Res. Comm.* **9,** 587 (1977).
54. E. I. Gendenshtein, R. E. Kiseleva, and L. I. Igol'nikova, in *Materialy Povolzhskaya Konferentsiya, Fiziologov s Uchasteim Biokhimikov, Farmakologov i Morfologov*, 6th ed., Vol. 2, G. D. Anikin, Ed., Chuvash State University, Cheboksary, USSR, 1973, p. 18; *Chem. Abstr.* **82,** 80656 (1975).
55. H. T. Graham and H. H. Gasser, *J. Pharm. Exp. Ther.* **43,** 163 (1931).
56. K. Matsuda, T. Hoshi, and S. Kameyama, *Jpn. J. Physiol.* **9,** 419 (1959); *Chem. Abstr.* **59,** 8024d (1963).
57. R. F. Schmidt, *Pfluegers Arch.* **271**, 526 (1960).
58. P. Heistracher and B. Pillat, *Naunyn-Schmiedebergs Arch. Exp. Pathol. Pharmakol.* **244,** 48 (1962).
59. W. H. Herzog, R. M. Feibel, and S. H. Bryant, *J. Gen. Physiol.* **47,** 719 (1964).
60. W. H. Herzog, Ph.D. thesis, University of Cincinnati, 1962; *Diss. Abstr.* **24,** 2514 (1963).
61. K. Peper and W. Trautwein, *Pfluegers Arch.* **296,** 328 (1967).
62. H. Schmitt and O. Schmidt, *Pfluegers Arch.* **349,** 133 (1978).
63. E. M. Peganov, S. V. Revenko, B. I. Khodorov, and L. D. Shishkova, *Mol. Biol.* (*Kiev*) **15,** 42 (1976); *Chem. Abstr.* **86,** 165113 (1977).
64. G. N. Mozhaeva, A. P. Naumov, and Y. A. Negulyaev, *Neirofiziologiya* **8,** 152 (1976); *Chem. Abstr.* **85,** 87982 (1976).
65. Z. I. Kruteskaya, A. V. Lonskii, G. N. Mozhaeva, and A. P. Naumov, *Neirofiziologiya* **9,** 320 (1977); *Chem. Abstr.* **87,** 162161 (1977).
66. G. N. Mozhaeva, A. P. Naumov, Y. A. Negulyaev, and E. D. Nosyreva, *Biochim. Biophys. Acta* **466,** 461 (1977).
67. A. P. Naumov, Y. A. Negulyaev, and E. D. Nosyreva, *Dokl. Akad. Nauk SSSR* **224,** 229 (1979); *Chem. Abstr.* **90,** 101241 (1979).
68. W. A. Catterall and R. Ray, *J. Supramol. Struct.* **5,** 397 (1976).
69. (a) W. A. Catterall, *J. Biol. Chem.* **252,** 8669 (1977). (b) J. C. Lawrence and W. A. Catterall, *J. Biol. Chem.* **256,** 6213 (1981). (c) J. C. Lawrence and W. A. Catterall, *J. Biol. Chem.* **256,** 6223 (1981).
70. W. R. Dunstan and F. H. Carr, *J. Chem. Soc. Trans.,* **67,** 459 (1896).
71. J. T. Cash and W. R. Dunstan, *Proc. R. Soc. London*, **68,** 384 (1901).
72. H. Goto, *Nippon Yakurigaku Zasshi* **52,** 496 (1965); *Chem. Abstr.* **51,** 14986d (1957).
73. H. Goto, *Nippon Yakurigaku Zasshi* **52,** 511 (1956); *Chem. Abstr.* **51,** 14986h (1957).
74. H. Sato, Y. Ohizumi, and H. Hikino, *Eur. J. Pharmacol.* **55,** 83 (1979).
75. S. W. Pelletier and N. V. Mody in *The Alkaloids: Chemistry and Physiology*, Vol. 17, R. H. F. Manske and R. G. A. Rodrigo, Eds., Academic, New York, 1979, p. 1; and references therein.
76. J. T. Cash and W. R. Dunstan, *Proc. R. Soc. London*, **76,** 468 (1905).
77. J. T. Cash and W. R. Dunstan, *Proc R. Soc. London*, **68,** 378 (1901).
78. A. Berni, *Arch. Farm. Specim.* **50,** 110 (1930); *Chem. Abstr.* **24,** 4861 (1930).
79. M. N. Sultankhodzhaev, M. S. Yunusov, and S. Y. Yunusov, *Khim. Prir. Soedin.* **9,** 127 (1973).
80. F. N. Dzhakhangirov, I. Khamdamov, and F. S. Sadritdinov, *Dokl. Akad. Nauk Uzb. SSR*, 32 (1976); *Chem. Abstr.* **85,** 103953 (1976).
81. M. N. Sultankhodzhaev, L. V. Beshitashvili, M. S. Yunusov, M. R. Yagudov, and S. Y. Yunusov, *Khim. Prir. Soedin.*, 665 (1980); *Chem. Abstr.* **94,** 153427 (1981).

82. C. D. Marsh, A. B. Clawson, and H. Marsh, *U.S. Dept. Agric. Bull.* **365,** 1 (1916).
83. O. A. Beath, *Wyoming Agric. Exp. Stn. Bull.* **143,** 49 (1925).
84. R. Boehm and J. Serck, *Archiv. Exp. Path. Pharmakol.* **5,** 311 (1876).
85. W. F. Waugh and W. C. Abbott, *A Textbook of Alkaloidal Therapeutics*, 3rd ed., Abbott, Chicago, 1911.
86. S. Solis-Cohen and T. S. Githens, *Pharmacotherapeutics, Materia Medica and Drug Action*, Appleton, New York, 1928.
87. A. Rabuteau, *C. R. Soc. Biol.* **26,** 286 (1874).
88. S. W. Pelletier and N. V. Mody, *Heterocycles* **5,** 771 (1976).
89. A. Lohmann, *Pfluegers Archiv.* **92,** 473 (1902).
90. D. Scherf, S. Blumenfeld, D. Taner, and M. Yildiz, *Arch. Kreislaufforsch.* **33,** 4 (1960); *Chem. Abstr.* **55,** 789h (1961).
91. D. Scherf, S. Blumenfeld, D. Taner, and M. Yildiz, *Am. Heart J.* **60,** 936 (1960).
92. H. Fischer and P. Huber, *Helv. Phys. Pharmacol. Acta* **2,** C38 (1944).
93. P. Huber, *Naturforsch. Ges. Zürich* **90,** 1 (1945); *Chem. Abstr.* **41,** 213 (1947).
94. W. Schneider and A. Enders, *Arzneim. -Forsch.* **5,** 324 (1955).
95. F. Dybing, O. Dybing, and K. B. Jensen, *Acta Pharmacol. Toxicol.* **7,** 337 (1951).
96. K. U. Aliev and M. B. Sultanov in *Farmakologiya Alkaloidov Serdechnykh Glikozidov*, M. B. Sultanov, Ed., FAN, Tashkent, USSR, 1971, p. 202; *Chem. Abstr.* **78,** 92484 (1973).
97. Vishwapaul and K. L. Handa, *Planta Med.* **12,** 177 (1964).
98. M. Shamma, P. Chinnasmy, G. A. Miana, A. Khan, M. Bashir, M. Salazar, A. Patil, and J. L. Beal, *J. Nat. Prod.* **42,** 615 (1979).
99. H. Schulze and F. Ulfert, *Arch. Pharm.* **260,** 230 (1922); *Chem. Abstr.* **18,** 2899 (1924).
100. M. Cygler, M. Przybylska, and O. E. Edwards, *Acta Cryst.* (in press); O. E. Edwards, *Can. J. Chem.* **59,** 3039 (1981).
101. S. W. Pelletier, N. V. Mody, K. I. Varughese, J. A. Maddry, and H. K. Desai, *J. Am. Chem. Soc.* **103,** 6536 (1981).
102. A. D. Kuzovkov and A. V. Bocharnikova, *J. Gen. Chem. USSR* **28,** 546 (1958).
103. P. M. Dozortseva, *Farmakol. Toksikol.* **22,** 34 (1959).
104. V. N. Aiyar, M. H. Benn, T. Hanna, J. Jacyno, S. H. Roth, and J. L. Wilkens, *Experientia* **35,** 1367 (1979).
105. M. N. Mats, *Rast. Res.* **8,** 249 (1972).
106. I. A. Gubanow, *Planta Med.* **13,** 200 (1965).
107. P. G. Kabelyanskaya, *Farmakol. Toksikol.* **22,** 38 (1959).
108. R. N. Alyautdin, *Farmakol. Toksikol.* **41,** 397 (1978).
109. J. L. Wilkens, unpublished results; private communication.
110. P. M. Dozortseva, *Farmakol. Toksikol.* **28,** 206 (1965).
111. T. P. Dmitrieva, *Tr. Volgograd Gos. Med. Inst.* **21,** 165 (1968); *Chem. Abstr.* **74,** 41009 (1971).
112. S. W. Pelletier, R. S. Sawhney, H. K. Desai, and N. V. Mody, *J. Nat. Prod.* **43,** 395 (1980).
113. T. A. Henry, *The Plant Alkaloids*, 4th ed., Churchill, London, 1949.
114. S. W. Pelletier, R. S. Sawhney, and A. J. Aasen, *Heterocycles* **12,** 377 (1979).
115. P. M. Dozortseva, *Farmakol. Toksikol.* **19,** 42 (1956).
116. M. P. Serkova, *Farmakol. Toksikol.* **19,** 48 (1956).
117. P. M. Dozortseva, *Med. Prom. SSSR* **10,** 31 (1956).
118. W. B. Cook, Ph. D. thesis, University of Wyoming, 1950.
119. H. Suginome and K. Ohno, *J. Fac. Sci. Hokkaido Ser. III* **4,** 36 (1950); cited in [33].
120. S. W. Pelletier and J. Bhattacharrya, *Phytochemistry* **16,** 1464 (1977).

121. O. A. Beath, *Wyoming Agric. Exp. Stn. Bull.* **120,** 53 (1918).
122. I. C. Chopra, J. D. Kohli, and K. L. Handa, *Ind. J. Med. Res.* **33,** 139 (1945).
123. O. E. Edwards and L. Marion, *Can. J. Chem.* **30,** 627 (1952).
124. S. W. Pelletier and N. V. Mody, *Tetrahedron* **22,** 207 (1981).
125. Y. E. Nezhevenko, M. S. Yunusov, and S. Y. Yunusov, *Khim. Prir. Soedin.*, 409 (1974).
126. I. Khamdamov, F. N. Dzhakhangirov, and F. S. Sadritdonov, *Dokl. Akad. Nauk Uzb. SSR* **32,** 37 (1975); *Chem. Abstr.* **84,** 84349 (1976).
127. W. M. Davidson, *J. Econ. Entomol.* **22,** 226 (1929).
128. W. Schneider, *Pharm. Zentralhalle* **90,** 151 (1951).
129. F. N. Dzhakhangirov and F. S. Sadritdinov *Dokl. Akad. Nauk Uzb. SSR*, 33 (1977); *Chem. Abstr.* **87,** 78519 (1977).
130. A. S. Narzullaev, M. S. Yunusov, and S. Y. Yunusov, *Khim. Prir. Soedin.*, 468 (1973).
131. V. N. Aiyar, unpublished results.
132. M. V. Kalitina, *Lek. Sredstva Dal'nego Vostoka* **11,** 173 (1972); *Chem. Abstr.* **82,** 51572 (1975).
133. F. N. Dzhakhangirov and F. S. Sadritdinov, *Dokl. Akad. Nauk Uzb. SSR,* 50 (1977); *Chem. Abstr.* **87,** 194033 (1977).
134. I. C. Chopra, J. D. Kohli, and K. L. Handa, *Ind. J. Med. Res.* **33,** 157 (1945).
135. Raymond-Hamet, *C. R. Soc. Biol.* **215,** 247 (1942).
136. Raymond-Hamet, *C. R. Soc. Biol.* **168,** 1221 (1954).
137. Raymond-Hamet, *C. R. Acad. Sci.* **243,** 324 (1956).
138. N. Singh and K. L. Chopra, *J. Pharm. Pharmacol.* **14,** 288 (1962).
139. C. E. Powell, R. G. Herrmann, and K. K. Chen, *J. Am. Pharm. Assoc. Sci. Ed.* **45,** 733 (1956).
140. F. Y. Wiselogle in *A Survey of Antimalarial Drugs*, Vol. 2, J. W. Edwards, Ed., Ann Arbor, 1945, p. 1573; cited in [138].
141. F. Sadridtinov, *Farmakol. Alkaloidov* **2,** 312 (1965); *Chem. Abstr.* **66,** 93772 (1967).
142. (a) F. N. Dzhakangirov, in *Farmakol. Rastit. Veshchestv*, M. B. Sultanov, Ed., FAN, Tashkent, USSR, 1976, p. 92; *Chem. Abstr.* **89,** 173584j (1978). (b) R. A. Ashrafova, in *Farmakol. Rastit. Veshchestv*, M. B. Sultanov, Ed., FAN, Tashkent, USSR, 1976, p. 156; *Chem. Abstr.* **89,** 158258 (1978).
143. W. A. Catterall, *Proc. Natl. Acad. Sci. USA* **72,** 1782 (1975).
144. J. R. Smythies, *Adv. Cytopharmacol.* **3,** 317 (1979).
145. T. Matsutani, I. Seyama, T. Narahashi, and J. Iwasa, *J. Pharmacol. Exp. Ther.* **217,** 812 (1981).
146. (a) K. B. Birnbaum, K. Wiesner, E. W. K. Jay, and L. Jay, *Tetrahedron Lett.*, 867 (1971). (b) O. Achmatowicz and L. Marion, *Can. J. Chem.* **42,** 154 (1964).
147. S. Y. Chen, *Hua Hsueh Hsueh Pao* **37,** 15 (1979); *Chem. Abstr.* **91,** 20833 (1979).
148. W. S. Chen and E. Breitmaier, *Chem. Ber.* **114,** 394 (1981).
149. S. W. Pelletier, N. V. Mody, and N. Katsui, *Tetrahedron Lett.*, 4027 (1977).
150. P. Pauling and T. J. Petcher, *Chem.-Biol. Interact.* **6,** 351 (1973).

Chapter Five

A Chemotaxonomic Investigation of the Plant Families of **Apocynaceae**, **Loganiaceae**, and **Rubiaceae** by Their Indole Alkaloid Content

M. Volkan Kisakürek
Organisch-Chemisches Institut der Universität Zürich, Winterthurerstrasse 190, CH-8057 Zürich, Switzerland

Anthony J. M. Leeuwenberg
Department of Plant Taxonomy and Plant Geography, Agricultural University, NL-6700 Wageningen, The Netherlands

Manfred Hesse
Organisch-Chemisches Institut der Universität Zürich, Winterthurerstrasse 190, CH-8057 Zürich, Switzerland

CONTENTS

This chapter is part of the Ph. D. thesis of M. V. Kisakürek, Universität Zürich, 1981. Part 177, papers on organic natural products. Dedicated to Professor Dr. K. Mothes on the occasion of his 80th birthday.

All the indole alkaloids [1] with a C_9- or C_{10}-monoterpene moiety isolated from the plants of families Apocynaceae, Rubiaceae, and Loganiaceae are summarized in tables (Appendixes 1–9). We included considerations about their structural features as well as their absolute configurations. To describe absolute configurations we use the symbols α and β depending on the configuration at C-15 (Fig. 5). The huge number of alkaloids (a total of 3450 reports covers the isolation of 1200 different alkaloids) have been classified into eight different skeletal types: corynanthean (C), vincosan (D), vallesiachotaman (V), strychnan

(S), aspidospermatan (A), eburnan (E), plumeran (P), and ibogan (J) (Fig. 2). In this way, we have been able to demonstrate that those skeletal types with a nonrearranged secologanin (**V**, Fig. 1) skeleton constituting the C_{10}-monoterpene moiety have the same absolute configuration as secologanin itself. On the other hand, alkaloids with a rearranged secologanin component (E, F, J) can occur with either absolute configuration.

Each of the eight skeletal types has been subdivided according to increasing "chemical complexity" compared with the basic type of each group as given in Fig. 6. Comparison of the eight skeletal types and their distribution in plant families can be summarized as follows: skeletal types with a rearranged secologanin moiety (E, P, and J types) occur exclusively in the subfamily Plumerioideae of the Apocynaceae. The occurrence of alkaloids of A and S types is restricted to the Loganiaceae and Apocynaceae. On the other hand, alkaloids of C, D, and V types have been detected in all of the three plant families. Further detailed examination of the skeletons of alkaloids of type C shows that only alkaloids of relatively simple structure occur in all three families. The more complex alkaloids tend to be specific to particular families, tribes, genera, and species (see Fig. 21).

As a result of this kind of investigation, a relationship can be established between chemotaxonomic patterns on the one hand and botanical differentiation and the evolution of plant families on the other. The botanical classification and relationships illustrated in Fig. 24, may be regarded as a summary of this study.

1. INTRODUCTION

1.1. General Introduction

The number of indole alkaloids of known structure today amounts to approximately 1400. Indole alkaloids are defined as the natural organic products containing either the indole nucleus or an oxidized, reduced, or substituted equivalent of it [e.g., oxindole, pseudoindoxyl (ψ-indoxyl), dihydroindole, *N*-acylindole].

With respect to their structural features, indole alkaloids can be divided into two main classes. The first comprises the simple indole alkaloids. Their structure is not uniform, having only the indole nucleus or a direct derivative of it as a common feature. Depending on the constitution of the rest of the molecule, they occur in many plant families [e.g., harman (**I**), obtained from the families **Apocynaceae**, **Chenopodiaceae**, **Elaeagnaceae**, **Leguminosae**, **Loganiaceae**, **Passifloraceae**, **Polygonaceae**, **Rubiaceae**, **Symplocaceae**, and **Zygophyllaceae**] or they are restricted to very few families or to only one [e.g., koenigine (**II**) obtained only from the **Rutaceae**; Fig. 1]. Because of the relatively small number of known compounds in this type, a chemotaxonomic examination of this heterogeneous class of plant bases would seem to be irrelevant at present.

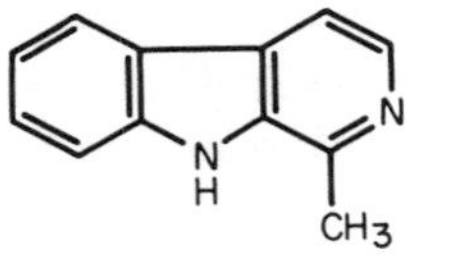

I harman

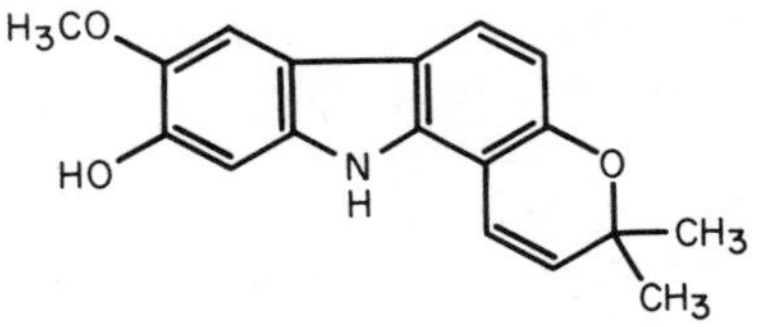

II koenigine

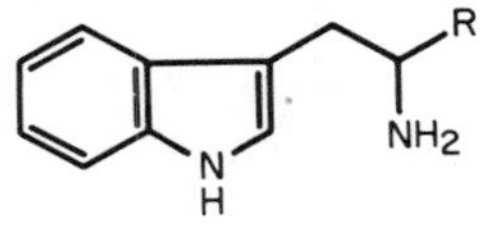

III tryptamine R=H

IV tryptophan R=COOH

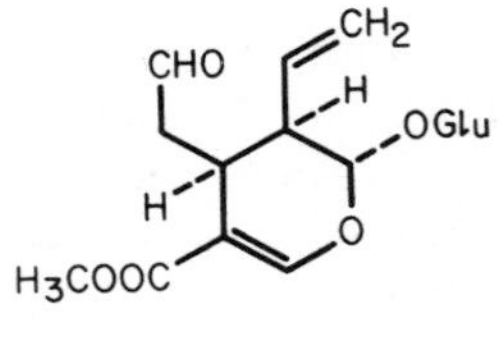

V (−)-secologanin

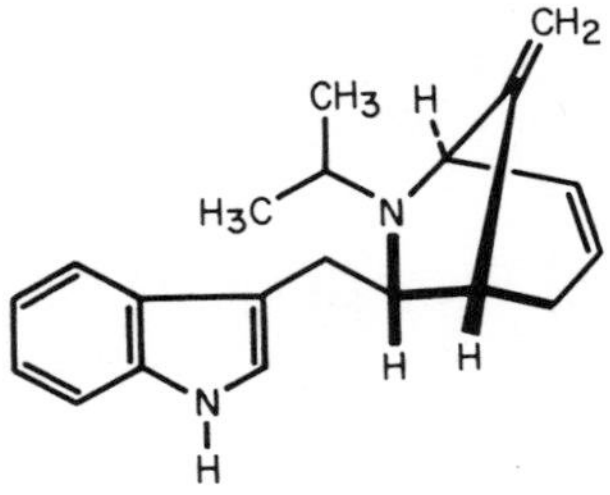

VI peduncularine

Figure 1.

Indole bases of the second class contain two structure elements: tryptamine or tryptophan (**III, IV**, Fig. 1) with an indole nucleus and a C_9- or C_{10}-monoterpene moiety, derived from secologanin (**V**, Fig. 1). Probably because they are constructed from two common components and because they are biogenetically interrelated, indole alkaloids of this second class have a more specific distribution and are therefore more suitable as a vehicle for a comparative chemotaxonomic investigation.* For this study it appeared desirable to investigate the relationships between indole alkaloids and their distribution in nature. Features

*In the following, we discuss only those C_9- or C_{10}-monoterpene indole alkaloids that are derived from secologanin (**V**) or its biogenetic equivalent. Indole alkaloids recently isolated from *Aristotelia* species (**Elaeocarpaceae**)—e.g., peduncularine (**VI**) [2]—are not considered because the C_9- or C_{10}-monoterpene moiety they contain in addition to the tryptamine (**III**) unit is not derived from secologanin (**V**).

of the alkaloids that are important in this respect are their structures (including absolute configurations) and biogenesis.

Several chemotaxonomic investigations of the plant families **Apocynaceae**, **Loganiaceae**, and **Rubiaceae** were published [3] between 1964 and 1973, but since then a tremendous number of new alkaloids have been discovered. Furthermore, the absolute configurations and the biogenesis of the alkaloids were not taken into consideration in these earlier works. Also, not all of the naturally occurring bases isolated at that time were listed because the authors wanted only to illustrate the different kinds of natural products isolated from many plant families. Hitherto, only the plumeran alkaloids [4] have been examined from a chemotaxonomic point of view, as in the present chapter.

For the present study, the chemotaxonomy of the following types are considered: the corynanthean, vincosan, vallesiachotaman, strychnan, aspidospermatan, eburnan, and ibogan.* Their structures are illustrated in Fig. 2.

*We have attempted to consider the literature on alkaloid isolation and structural elucidation up to the end of 1978. If any relevant papers have not been mentioned, the omission was not intentional.

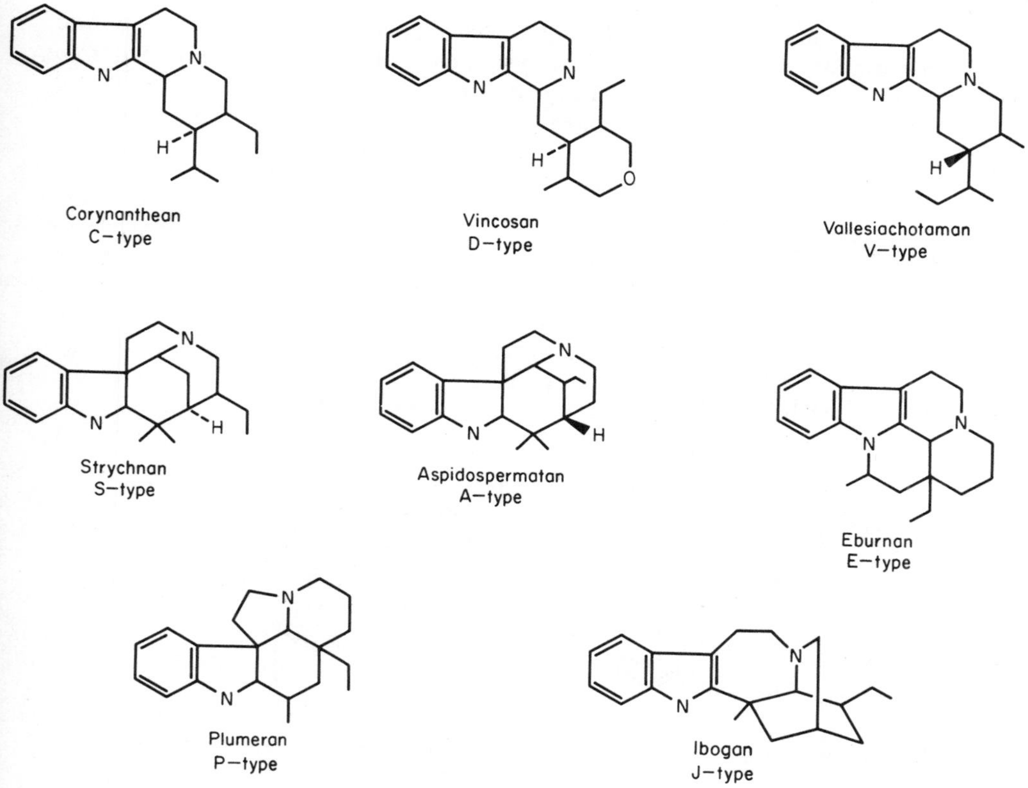

Corynanthean C–type

Vincosan D–type

Vallesiachotaman V–type

Strychnan S–type

Aspidospermatan A–type

Eburnan E–type

Plumeran P–type

Ibogan J–type

Figure 2.

The following abbreviations are used for the skeletal types of the alkaloids isolated: C = corynanthean, D = vincosan, V = vallesiachotaman, S = strychnan, A = aspidospermatan, E = eburnan, P = plumeran, J = ibogan. The numbers after the abbreviations correspond to the number of steps (oxidation, bond formation, cyclization, rotation followed by bond formation or cyclization) necessary to derive a certain skeletal variation starting from the hypothetical compound **1** (Fig. 6). Chemical or enzymatic reactions follow, and as a result, a compound chemically more complex than **1** results. The number of steps can be seen immediately by the structure numbers: e.g., by this definition **J7** is more complex compared with **1** than **P4** (Fig. 6). But **P4** is higher than **1** in the rank of chemical complexity.

As already noted, we consider only indole alkaloids with a C_9- or C_{10}-monoterpene moiety for the purpose of this study. Those alkaloids that contain additional carbon atoms in their skeleton are also included (e.g., strychnine; **VII**, Fig. 3). Biogenetically related bases containing less than nine carbon atoms—e.g., apparicine (**VIII**, Fig. 3) [5]—are not taken into consideration. We do not specially differentiate between alkaloids with or without a carboxyl group in position 5—e.g., 5α-carboxytetrahydroalstonine) (**IX**) [6]—or between an alkaloid and its corresponding glycoside—e.g., 10β-D-glucosyloxyvincoside-lactam (**X**), Fig. 3) [7].

On the basis of published papers (see Appendix 9), the biogenetic and chemotaxonomic relationships for approximately 1200 alkaloids, obtained from 3450 alkaloid isolations, are considered in this study. The C, D, V, S, A, E, P, and J alkaloids examined were obtained from plant species distributed between three

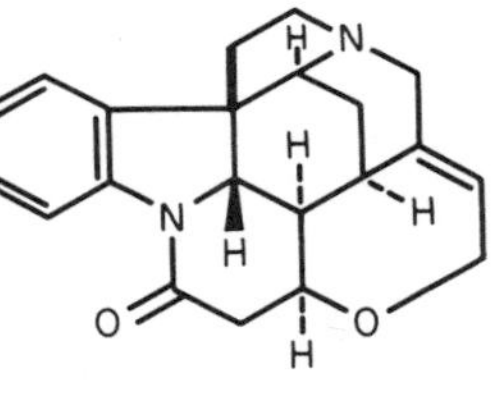

VII, strychmne

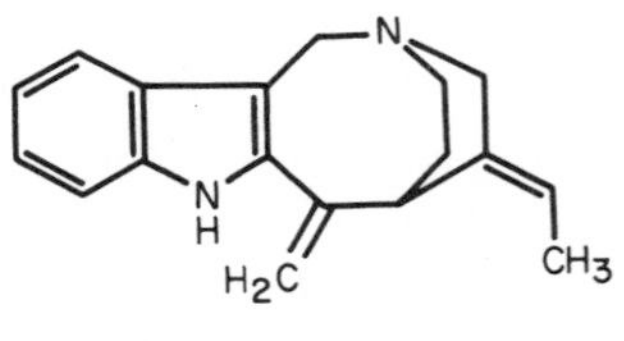

VIII, apparicine

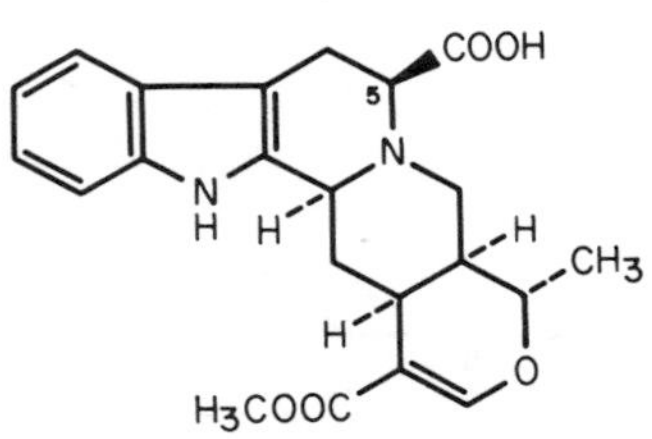

IX, 5α-carboxytetrahydroalstonine

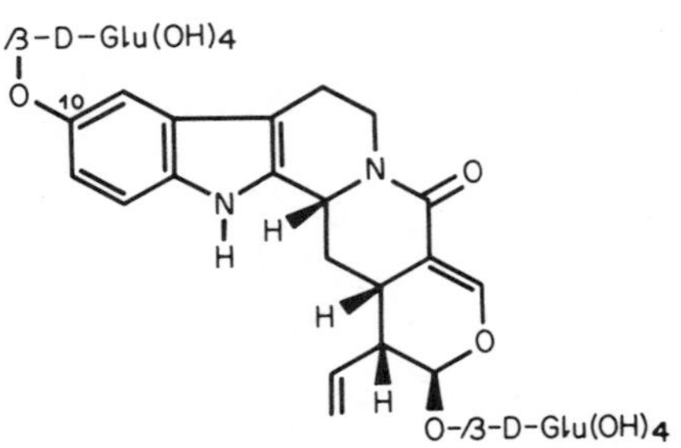

X, 10β-D-glucosyloxyvincosidelactam

Figure 3.

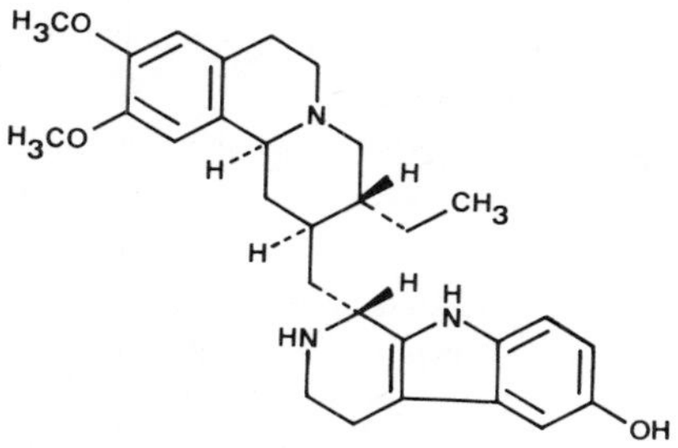

XI, tubulosine

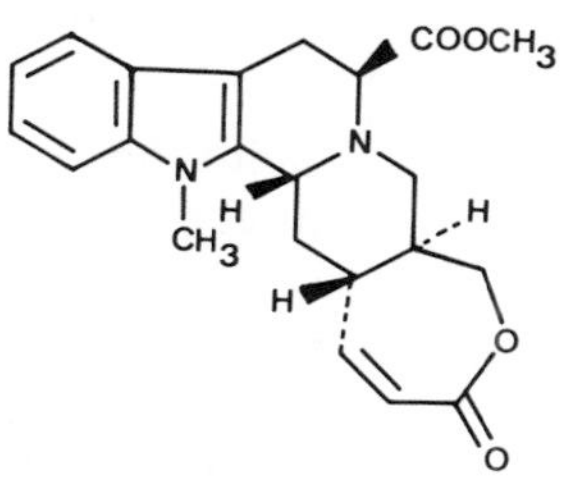

XII, cannagunine B

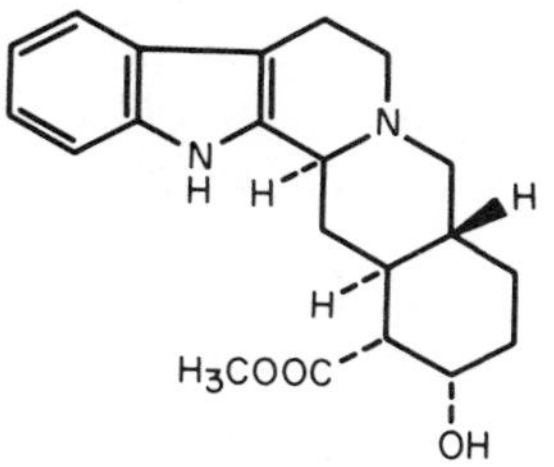

XIII, yohimbine

Figure 4.

plant families: the Apocynaceae, the Loganiaceae, and the Rubiaceae. A few alkaloids of the corynanthean (C) type have also been detected in some species of the *Alangiaceae* (e.g., tubulosine, **XI**, Fig. 4, from *Alangium lamarckii* Thw. [8, 9]), *Ericaceae*, (e.g., cannagunine B, **XII** from *Vaccinum oxycoccus* L., [10]), and *Euphorbiaceae* (e.g., yohimbine **XIII**, Fig. 4, from *Alchornea floribunda* Müll Arg. [11]). The chemotaxonomy of the above-mentioned subgroups of indole alkaloids containing a C_9- or C_{10}-monoterpene moiety isolated from the **Apocynaceae, Loganiaceae,** and **Rubiaceae** will now be discussed from the point of view of their structural, stereochemical, and biogenetical relationships.

1.2. Assumptions and Definitions

The following assumptions and definitions are used in this study.

1. Papers were accepted without any further examination of the published results regarding the botanical identification of the plant species or the physical data given for the determination of the isolated alkaloids.
2. The detection of *a particular alkaloid* from the *same* plant source including synonyms reported by different authors is considered only one isolation.
3. The bisindole alkaloids are taken into consideration either as one or two distinct alkaloids, according to whether they contain one or two units, respectively, of the skeletal type that is being discussed.

4. The natural indole bases occuring with (±)- and (+)- or (−)-$[\alpha]_D$ values have been regarded as two distinct alkaloids.
5. In addition to structural and chemical characteristics, the absolute configuration of the alkaloid is considered as an additional chemotaxonomic criterion. Unfortunately, we could use only about 46% of isolated alkaloids in this respect, since the necessary indications were completely or partly neglected in the original papers. In these cases, the absolute configurations are still unknown. In most cases, the alkaloids examined have several chiral centers that are partly interdependent. In contrast to common usage, we use the symbols α and β to designate absolute configurations. The conventional *R*, *S* nomenclature [12] cannot be used, since changes in substituents at centers with the same basic absolute configuration would result in a reversal of the stereochemical description (*R* or *S*) simply through the logical working of the sequence rules.

   The absolute configurations of the alkaloid groups taken into consideration in this study are defined on the basis of biogenetic correlations as shown in Fig. 5.
6. The names of the plant species were checked, and in some cases corrected and replaced, if given under synonymous names in the papers.
7. The so-called plumeran alkaloids are considered in so far as they should be taken into account for a chemotaxonomic evaluation of the indole alkaloids with a C_9- or C_{10}-monoterpene moiety. (For further details, Ref. [4] is recommended.) We considered the literature up until the end of 1978.
8. Numbers have been assigned to the various alkaloid skeletons in accordance with the biogenetic system [13].
9. The botanical classification of the three aforementioned plant families accepted by Leeuwenberg [14] is given in Table 1. Only those genera are indicated (in alphabetical order) that contain indole alkaloids under discussion.
10. Unless otherwise stated, the structures represent skeletal types and not complete constitutions. Accordingly we have partly omitted *N*-substituents and have, in general, only shown the absolute configuration at C(15) or the equivalent position. Also, the oxidation state of individual carbon atoms is not always shown correctly.

In order to demonstrate the significance of the indole alkaloid content of species of the plant families **Apocynaceae**, **Loganiaceae**, and **Rubiaceae**, the alkaloid structures are first given (Section 2). They are discussed in terms of the basic skeletons given in Fig. 2. Their biogenetic relationships and botanical occurrence (compare Appendix 1 with Appendix 9) are also given. In Section 3, the absolute configurations of the alkaloids are considered. These two sections are the basis for the chemotaxonomic considerations properly given in Section 4.

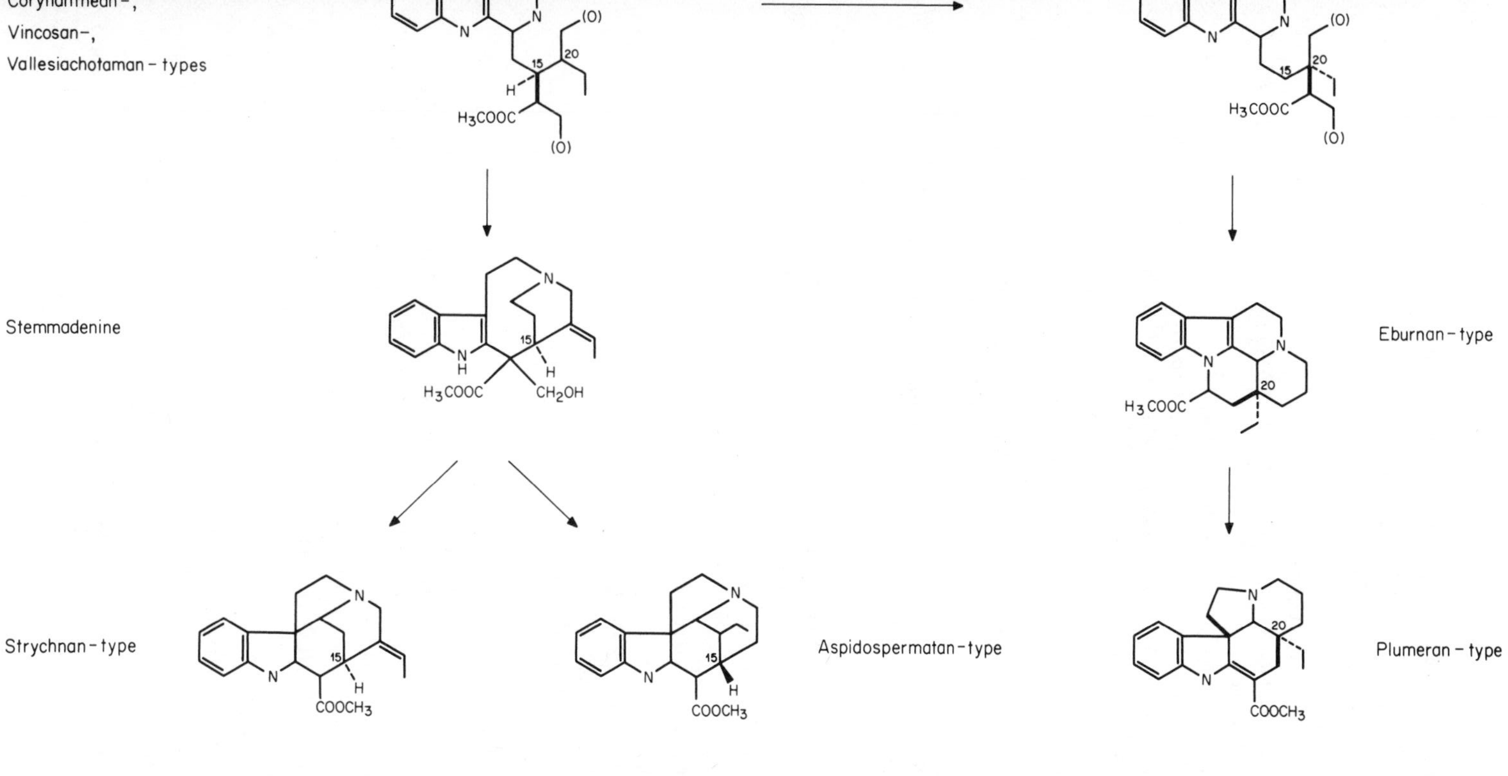

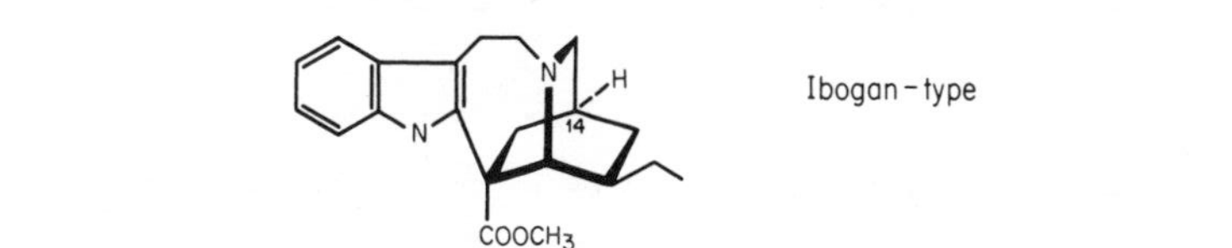

Figure 5. Configurations of the indole alkaloids [Unless otherwise stated, the structures in this figure represent skeletal types and not complete constitutions. Accordingly we have partly omitted *N*-substituents and have, in general, only shown the absolute configuration at C(15) or the equivalent position. Also, the oxidation state of individual carbon atoms is not always shown correctly.] with a C_9- or C_{10}-monoterpene moiety.

Table 1. The Botanical Classification of the Plant Families Apocynaceae, Loganiaceae, and Rubiaceae[a]

Family	**Apocynaceae** (APO)
Subfamily	Plumerioideae
Tribe	CARISSEAE (CAR)
	Carpodinus, Hunteria, Landolphia, Melodinus, Picralima, Pleiocarpa, Polyadoa
Tribe	TABERNAEMONTANEAE (TAB)
	Anacampta, Bonafousia, Callichilia, Capuronetta, Conopharyngia, Crioceras, Ervatamia, Gabunia, Hazunta, Hedranthera, Muntafara, Pagiantha, Pandaca, Peschiera, Phrissocarpus, Rejoua, Schizozygia, Stemmadenia, Stenosolen, Tabernaemontana, Tabernanthe, Voacanga
Tribe	PLUMERIEAE (PLU)
	Alstonia, Ammocallis, Amsonia, Aspidosperma, Catharanthus, Craspidospermum, Diplorhynchus, Geissospermum, Gonioma, Haplophyton, Lochnera, Plumeria, Rhazya, Tonduzia, Vinca
Tribe	RAUVOLFIEAE (RAU)
	Bleekeria, Cabucala, Excavatia, Kopsia, Neiosperma, Ochrosia, Rauvolfia, Vallesia
Family	**Loganiaceae** (LOG)
Tribe	STRYCHNEAE (STR)
	Gardneria, Strychnos
Tribe	GELSEMIEAE (GEL)
	Gelsemium, Mostuea
Family	**Rubiaceae** (RUB)
Subfamily	Rubioideae
Tribe	PSYCHOTRIEAE (PSY)
	Palicourea
Tribe	UROPHYLLEAE (URO)
	Pauridiantha
Subfamily	Cinchonoideae
Tribe	NAUCLEEAE (NAU)
	Adina, Anthocephalus, Cephalanthus, Haldina, Mitragyna, Nauclea, Neonauclea, Ourouparia, Pertusadina, Sarcocephalus, Uncaria
Tribe	CINCHONEAE (CIN)
	Cinchona, Corynanthe, Pausinystalia, Pseudocinchona, Remijia
Tribe	MUSSAENDEAE (MUS)
	Isertia
Subfamily	Guettardoideae
Tribe	GUETTARDEAE (GUE)
	Antirhea, Guettarda
Subfamily	Hillioideae

[a]Only those genera that contain indole alkaloids, together with their synonyms, are given. See also reference [4].

2. SKELETAL TYPES OF INDOLE ALKALOIDS AND THEIR BOTANICAL OCCURRENCE

Indole alkaloids with a C_9- or C_{10}-monoterpene moiety are classified into the following subgroups: corynanthean (C), vincosan (D), vallesiachotaman (V), strychnan (S), aspidospermatan (A), eburnan (E), plumeran (P), and ibogan (J) types. In a simplified manner, the biogenetic relationships of these main skeletal types are shown in Fig. 6.* As an established fact [15] compound **D3a** is obtained from condensation of tryptamine (**III**)—or in some other cases tryptophan (**IV**)—with secologanin (**V**). All of the main skeletal types can be derived from **D3a**. Skeleton **D3a** can be converted into compound **1** by opening of the C(17)—O—C(21) bond via $\mathbf{2b}^3$. From compound **1**, compounds $\mathbf{2b}^1$, $\mathbf{2b}^2$, $\mathbf{2b}^3$, and **2c** can be obtained without rearrangement, or structure **2a** by rearrangement of the secologanin portion of the molecule. Ring formation between C(2) and C(3) leads to compound **2b**. Intermediates $\mathbf{2b}^1$, $\mathbf{2b}^2$, and $\mathbf{2b}^3$ differ from each other only through rotation about the C(14)—C(15) and C(15)—C(16) bonds respectively. Ring closures between C(21) and N(b) in $\mathbf{2b}^1$, and between C(17) and N(b) in $\mathbf{2b}^2$ give rise to the main corynanthean-type skeleton **C3a** and the main vallesiachotaman-type **V3**, respectively. A new additional bond between C(17)—*O*H and C(21) in $\mathbf{2b}^3$ yields the basic skeleton of vincosan group **D3a**. Intermediate **2c** is obtained by ring closure between C(21) and N(b) in **1**. An additional ring closure between C(16) and C(2) in **2c** yields **A3**, the fundamental skeleton of the aspidospermatan group. Starting with **A3**, **S4** is obtained by another ring formation between C(3) and C(7). On the other hand, ring closure between C(21) and C(7) yields **A4**. In addition to the biogenetic pathways as shown in Fig. 5, other proposals are worthy of consideration [16–19].

Intermediate **2a** is derived from **1** by cleavage of the C(15)—C(16) bond followed by the formation of a new bond at C(17)—C(20). Ring closure between C(21) and N(b) leads to **3a**, from which **4a** and the main skeleton of plumeran group **P4** can be derived by additional ring closures [C(2)—C(21) and C(2)—C(16), respectively]. Ring closure [N(a)—C(16)] in **4a** yields **E5a**, the main skeleton of the eburnan group. Cleavage of the C(17)—C(20) bond in **P4** forms **P3**. By further reactions, the main skeletons of ibogan groups **J7a** and **J7b** can be derived from **P3**. Further reactions are necessary, starting from **C3a**, **D3**, **V3**, **S4**, **A4**, **E5a**, **P4**, and **J7a**, to form derivatives of various other skeletal types.

2.1. Alkaloids of the Corynanthean (C) Type

In Fig. 7, skeletal types of the corynanthean group, which presents the most extensive class of indole alkaloids, are given according to their botanical occurences in plant families. Only 26 of the 41 known skeletal variations have

*Compounds representing intermediates that have not been isolated from plants are designated only by numbers (e.g., **2a**, **4a**).

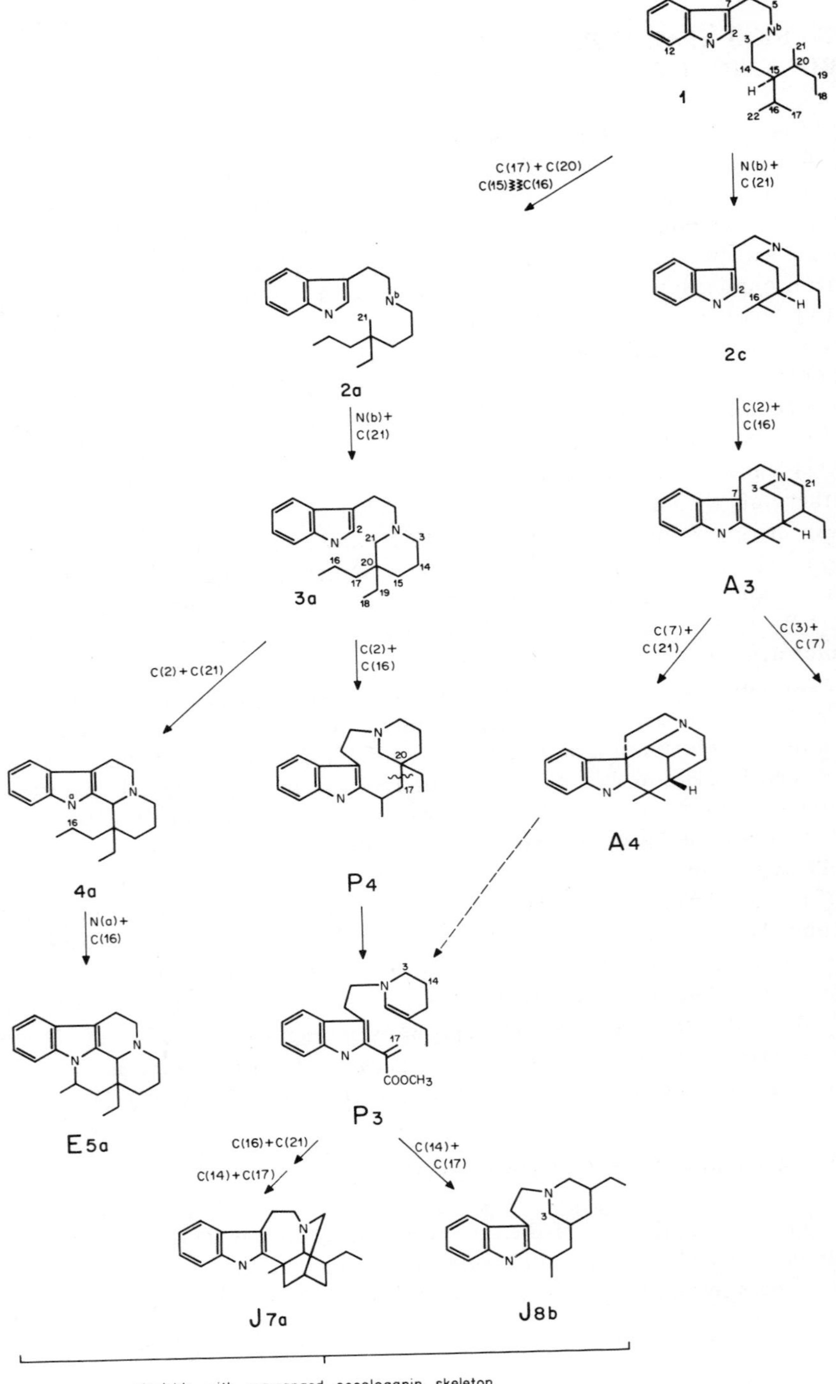

1
C(17) + C(20)
C(15)⋯C(16)
N(b) +
C(21)
2a
2c
N(b) +
C(21)
C(2) +
C(16)
3a
A3
C(2) + C(21)
C(2) +
C(16)
C(7) +
C(21)
C(3) +
C(7)
4a
P4
A4
N(a) +
C(16)
E5a
P3
COOCH3
C(16) + C(21)
C(14) + C(17)
C(14) +
C(17)
J7a
J8b
Alkaloids with rearranged secologanin skeleton
(≈ 1040 isolations)

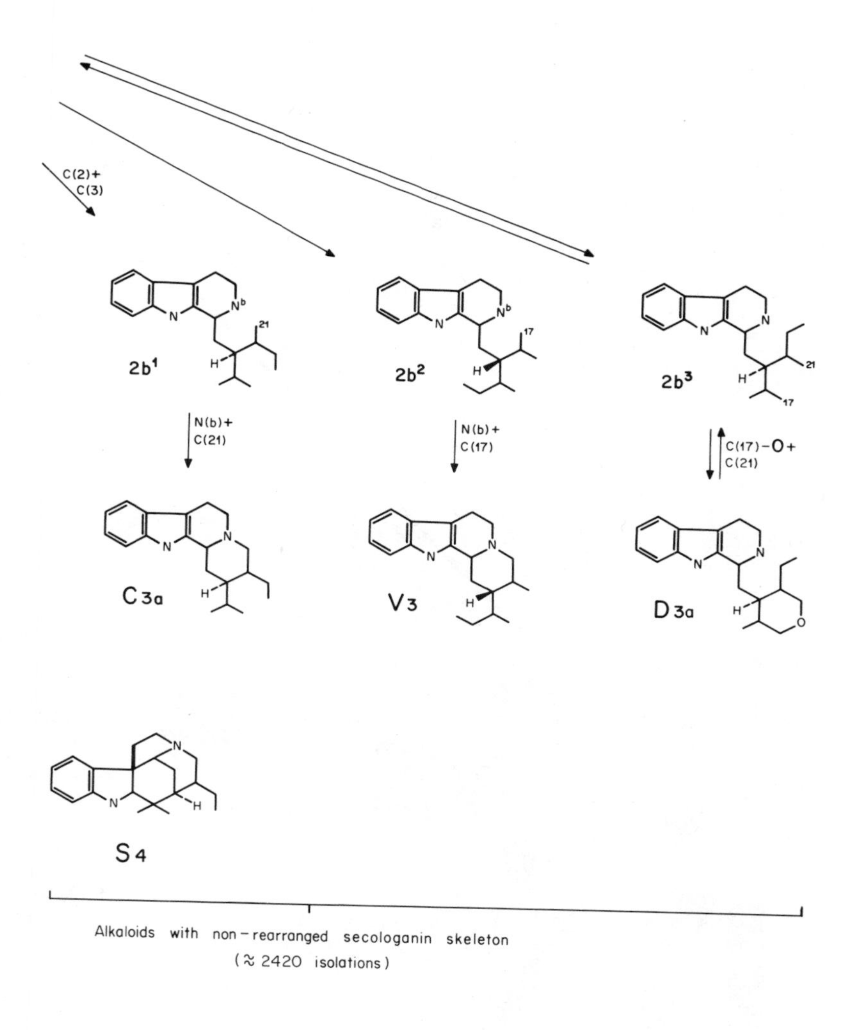

secologanin

Figure 6. Biogenetic relationships of the eight main skeletal types (C, D, V, S, A, E, P, and J) of indole alkaloids with a C_9- or C_{10}-monoterpene moiety.

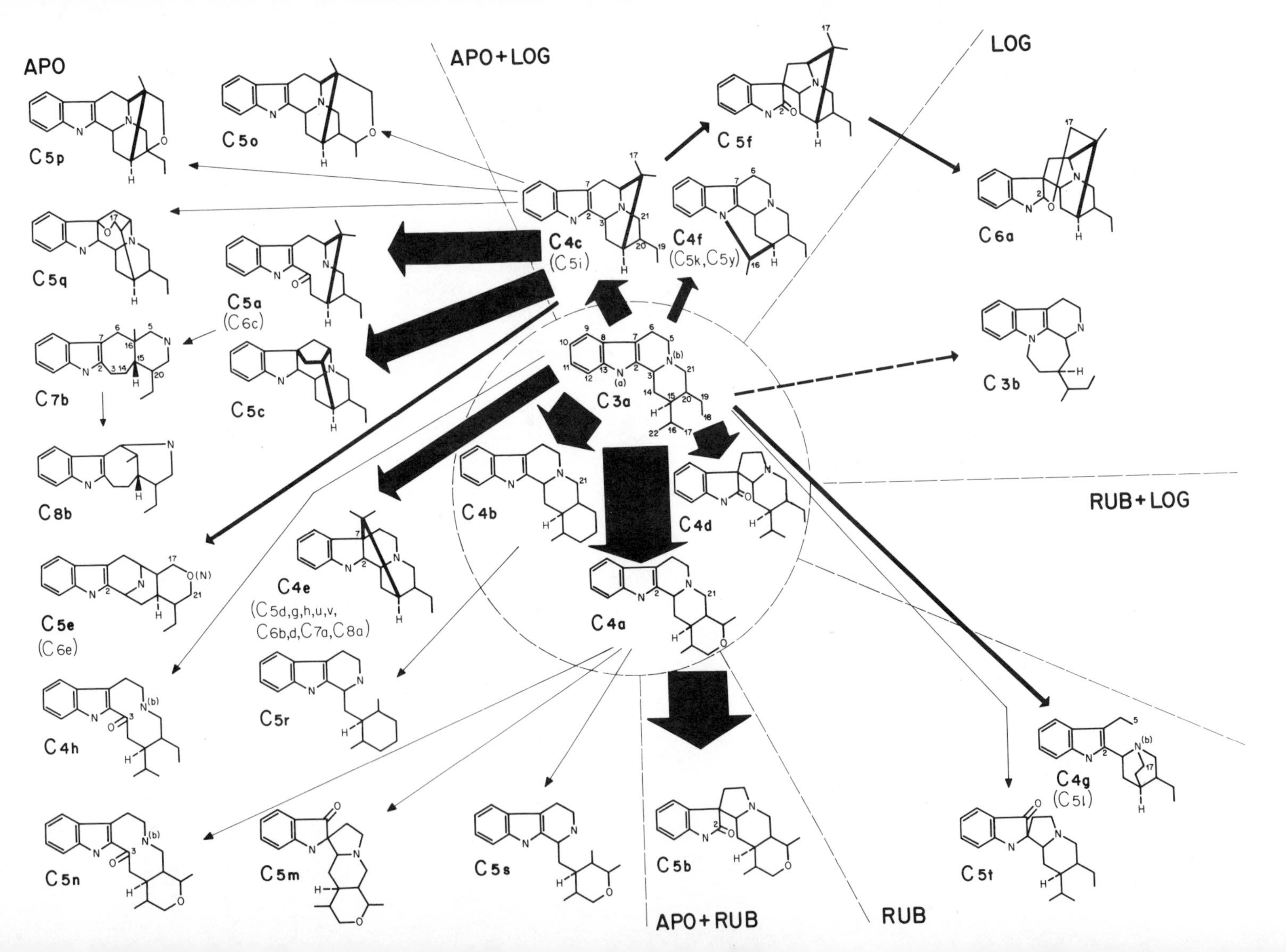

APO
APO+LOG
LOG
RUB+LOG
APO+RUB
RUB
C 5p
C 5o
C 5f
C 6a
C 5q
C 4c
(C 5i)
C 4f
(C 5k, C 5y)
C 5a
(C 6c)
C 7b
C 5c
C 3a
C 3b
C 8b
C 4b
C 4d
C 5e
(C 6e)
C 4e
(C 5d,g,h,u,v,
C 6b,d, C 7a, C 8a)
C 4a
C 4h
C 5r
C 4g
(C 5l)
C 5n
C 5m
C 5s
C 5b
C 5t

Figure 7. Distribution of alkaloids of the C type in the **Apocynaceae**, **Loganiaceae**, and **Rubiaceae**. In this figure, the skeletal types are given according their occurrence in the three plant families. In the middle of the figure (circle), skeletons are given for those alkaloids that have been detected in all three plant families: Apocynaceae, Loganiaceae, and Rubiaceae. Outside the circle in separate regions, those skeletal types are included that occur in only one or two plant families. The thickness of the arrows was chosen to be proportional to the abundance in plants of the particular skeletal type placed at their heads (compare Appendix 9). The direction of arrows is in agreement with biogenetic considerations. The numbers in brackets indicate that certain types include further derivable variations (e.g., **C4e** includes also **C5g**, which is not given in the figure for purpose of simplification. For these further variations, see Appendix 1 and Table 2).

Table 2. Number of Indole Alkaloids of the Corynanthean (C) Type Isolated From Different Plant Species (See Appendix 9)

	Number of Species	Skeletal Types													
		C3a	C4a	C4b	C4c	C4d	C4e	C4f	C5a	C5b	C5c	C5d	C5g	C5h	Other C types
Apocynaceae															
Plumerioideae															
CARISSEAE															
Hunteria	4	4	1	2	2		1	3				2	3		**C4h**:1, **C5p**:1, **C6b**:7, **C7a**:1, **C8a**:5
Melodinus	4			1	3						1				
Picralima	1		2		1		2					2		2	**C5n**:1
Pleiocarpa	3	3			5			5	1						**C5e**:1, **C5o**:1
TABERNAEMONTANEAE															
Callichilia	1								1						**C6d**:1
Stemmadenia	2		1						1						
Tabernaemontana	33	2	2		13		1		106		5				**C6c**:3, **C7b**:4, **C8b**:1
Voacanga	7		1	4	7				23						**C5f**:10, **C5i**:1, **C5o**:1
PLUMERIEAE															
Alstonia	17		3	13	8		7	13			16	6	12	1	**C5e**:20, **C5r**:1, **C5s**:1, **C5u**:1, **C5v**:1, **C6d**:3, **C6e**:1
Amsonia	3	3	2	4	2			1							
Aspidosperma	16	24	11	16	5					2	1	3			**C5m**:1, **C6d**:2
Catharanthus	6	6	11	7	2		1		4	1				1	
Diplorhynchus	1			2	1										
Geissospermum	2	2			3										
Gonioma	2				1			2							**C5k**:1
Rhazya	1	1			2		4								
Tonduzia	1		1	3							2				
Vinca	6	1	16		12		5		1	20	18	5	2	3	**C5y**:1

RAUVOLFIEAE															
Cabucala	5		16	3	1		2		1	6	4		3	2	
Ochrosia	16	17	20	6		1			3	2					
Rauvolfia	47	15	132	136	22		3		2	5	83	3			**C5m**:1, **C5n**:2 **C5e**:2. **C5m**:1, **C5n**:1, **C5q**:1
Vallesia	1			1	1			1							
Loganiaceae															
GELSEMIEAE															
Gelsemium	2			2											
Mostuea	2			2											
STRYCHNEAE															
Gardneria	3				1										
Strychnos	19	20	5	2	14	4	13								**C5f**:9, **C5i**:2, **C6a**:12 **C3b**:12, **C5k**:6
Rubiaceae															
Cinchonoideae															
NAUCLEEAE															
Cephalanthus	1	2				4									
Mitragyna	10	18	18			63				20					
Neonauclea	1	1													
Pertusadina	1	2	1	1											
Uncaria	43	31	42	8		92				256					
CINCHONEAE															**C5t**:3
Cinchona	3		1					2							
Corynanthe	3	3	6	7		2									**C5l**:5
Pausinystalia	3	3	1	12											
Remijia	1							1							
MUSSANDEAE															
Insertia	1														
Guettardoideae															**C5l**:2
GUETTARDEAE															
Guettarda	1		1												

been depicted to simplify the presentation of structural features. Fifteen variations that were omitted in Scheme 4 have been distributed, according to their structural relations, among 26 skeletons occurring in the same plant families. Those skeletal types that include further variations omitted in the scheme are given in brackets underneath the main type. Skeletal types are given according to their occurrence in the three plant families. In the middle of the figure (circle), skeletons are given for those alkaloids that have been detected in all three plant families: **Apocynaceae**, **Loganiaceae**, and **Rubiaceae**. Outside the circle in separate regions, those skeletal types are included that occur in only one or two plant families. The thickness of the arrows was chosen to be proportional to the abundance of the particular skeletal types placed at their heads (see Appendix 9). The direction of the arrows is in agreement with biogenetic considerations. Numbers in brackets indicate that certain types include further derivable variations (e.g., **C4e** also includes **C5g**, which is not given in the figure for simplification; for these further variations, see Appendix 1 and Table 2).

Table 2 offers an outline of all the skeletal types including examples of variations and all the corresponding alkaloids represented by a specific skeletal type. The biogenetic relationships of C skeletal types can be explained as follows. Starting with **C3a** (e.g., corynantheine*), several ring closures can take place, resulting in formation of the main variations. A new bond between C(17) and C(18) leads to **C4b** (e.g., yohimbine). Reaction of C(17)—*O*H with C(19) forms **C4a** (e.g., ajmalicine). Combination of C(16) with N(a) results in **C4f** (e.g., pleiocarpamine). C(16) + C(5) → **C4c** (e.g., normacusine B); C(16) + C(7) → **C4e** (e.g., akuammiline). These main skeletal types can undergo further reactions. Starting with **C3a**, ring closure [C(17) + N(b)] followed by ring opening [N(b)—C(5)] gives **C4g** (e.g., cinchonamine); from **C4c** ring closure [C(17)—*O*H + C(21)] followed by ring opening [N(b)—C(21)] yields **C5e** (e.g., alstophylline). For the corynanthean group of skeletal types, oxidative processes seem to be characteristic. They involve particularly carbon atoms 2, 3, and 7.

The C(2) oxidations with skeletal rearrangement produce alkaloids containing the oxindol chromophore = **C3a** → **C4d** (e.g., rhynchophylline), **C4a** → **C5b** (e.g., mitraphylline), **C4c** → **C5f** (e.g., chitosenine), and **C5e** → **C6e** (e.g., alstonisine). Oxidation at C(3) produces **C5a** (e.g., vobasine) from **C4c**; **C4h** (e.g., burnamicine) from **C3a**; **C5n** (e.g, picraphylline) from **C4a**; and **C5g** (e.g., deformocorymine) from **C4e**. In order to obtain **C5g** from **C4e**, a ring closure [C(2)—N(b)] takes place in addition to the cleavage of the bond N(b)—C(3). In all of the cases given for C(3) oxidations, the bond N(b)—C(3) is to be recognized as cleaved only if N(b) is tertiary so that a transanullar interaction with the C(3)=O group [or C(3)—OH group] in the 10-membered ring can be neglected. Oxidation at C(7), comparable with that of C(2), leads also to a rearrangement of the skeleton and the following pseudoindoxyl chromophores are formed: **C5m** (e.g., isoreserpilinepseudoindoxyl) from **C4a**; **C5k** (e.g., fluorocarpamine) from **C4f**, and **C5t** (e.g., dihydrocorynantheinepseudoin-

*Specific alkaloid examples of skeletal types are illustrated in Appendix 1.

doxyl) from **C3a**. Starting with the initially formed skeletal types, others can be derived through additional ring closures: **C5c** (e.g., ajmaline) by C(7) + C(17) in **C4c**; **C5q** (e.g., rauwolfinine) by C(7) + C(17)—*O*H in **C4c**; **C6c** (e.g., anhydrovobasinediol) by C(17)—*O*H + C(3)—*O*H in **C5a***; **C6a** (e.g., gardneramine) by C(17)—*O*H + C(2)=O in **C5f**; **C5l** (e.g., quinamine) by C(2) + C(5)-*O*H in **C4g**; **C5h** (e.g., pseudoakuammigine) by C(17)—*O*H + C(2) in **C4e**; **C6b** (e.g., corymine) by C(17)—*O*H + C(3)—*O*H in **C5g**; and **C8a** (e.g., erinine) by C(22)—*O*H + C(20) in **C7a**. Oxidative ring closures offer an explanation for the formation of the following skeletal types from **C4c**: **C5p** (e.g., eburnaphylline) by C(17)—*O*H + C(20); **C5i** (e.g., dehydrovoachalotine) by C(17)—*O*H + C(6); **C5o** (e.g., voacoline) by C(17)—*O*H + C(19); **C5d** (e.g., picraline) by C(5)—*O*H + C(2); and **C5u** (e.g., quaternoline) by C(22)—*O*H + C(20) from **C4e**.

Finally, for the formation of some skeletal types, additional ring openings are necessary. The cleavage of the bond C(21)—N(b) yields **C5r** (e.g., alstonilidine) from **C4a**; **C7a** (e.g., eripine) from **C6b**; and **C5s** (e.g., alstonidine) from **C4a**. Intermediate **C6d** (e.g., aspidodasycarpine) results from the cleavage of C(5)—N(b) in **C5d**, and **C5y** (e.g., vinoxine) results from the cleavage of C(6)—C(7) in **C4f**. Type **C5v** (e.g., nareline) can be derived from **C4e** by involvement of atoms C(5), C(6), C(21) and N(b). Alternative to the formation of **C3a** from **2b**^{1} (Fig. 6) is ring closure between N(a) and C(17), which leads to skeleton **C3b** (e.g., akagerine). A more complex reaction, given in Fig. 8, is necessary to explain the formation of alkaloids of type **C7b** (e.g., ervatamine) and **C8b** (e.g., ervitsine) [20, 21].

Alternative possibilities for the formation of some skeletal types discussed here exist.

The established occurrences of indole alkaloids of the C type are given in Table 2. As shown, alkaloids of this type have been detected in all three plant families. A complete summary of their occurrences is given in Table 3 (for a discussion of the occurrences, see Section 4).

2.2. Alkaloids of the Vincosan (D) Type

This relatively small group of alkaloids, being fairly rich in skeletal variations, apparently may arise via two biogenetic pathways. Rotation about bond C(15)-C(20) in **1** (Fig. 6) followed by the ring formations C(2) + C(3) and C(17)-*O*H + C(21) leads to **D3a** (e.g., vincoside). This structure may be accepted as the main skeleton of the D type, from which other skeletal types seem to be derivable (Fig. 9). On the other hand, **D3a** can be formed by condensation of tryptamine (**III**) or tryptophan (**IV**)—with secologanin (**V**) or a direct equivalent of it (Fig. 1). Skeletal types **D6c** (e.g., peraksine) and **D6b** (e.g., perakine) probably belong to the former group of compounds.† Reaction of **D3a** with NH_3 leads to **D3b**,

*Skeletal type **C6c** can also be derived directly from **C4c**.

†Alkaloids of type **C5c** (Fig. 7) frequently contain a hydroxyl group at C(21) and are amino hemiacetals. In such compounds, it is possible that an isomerization to **D6b** with participation of the double bond C(19)=C(20) can take place.

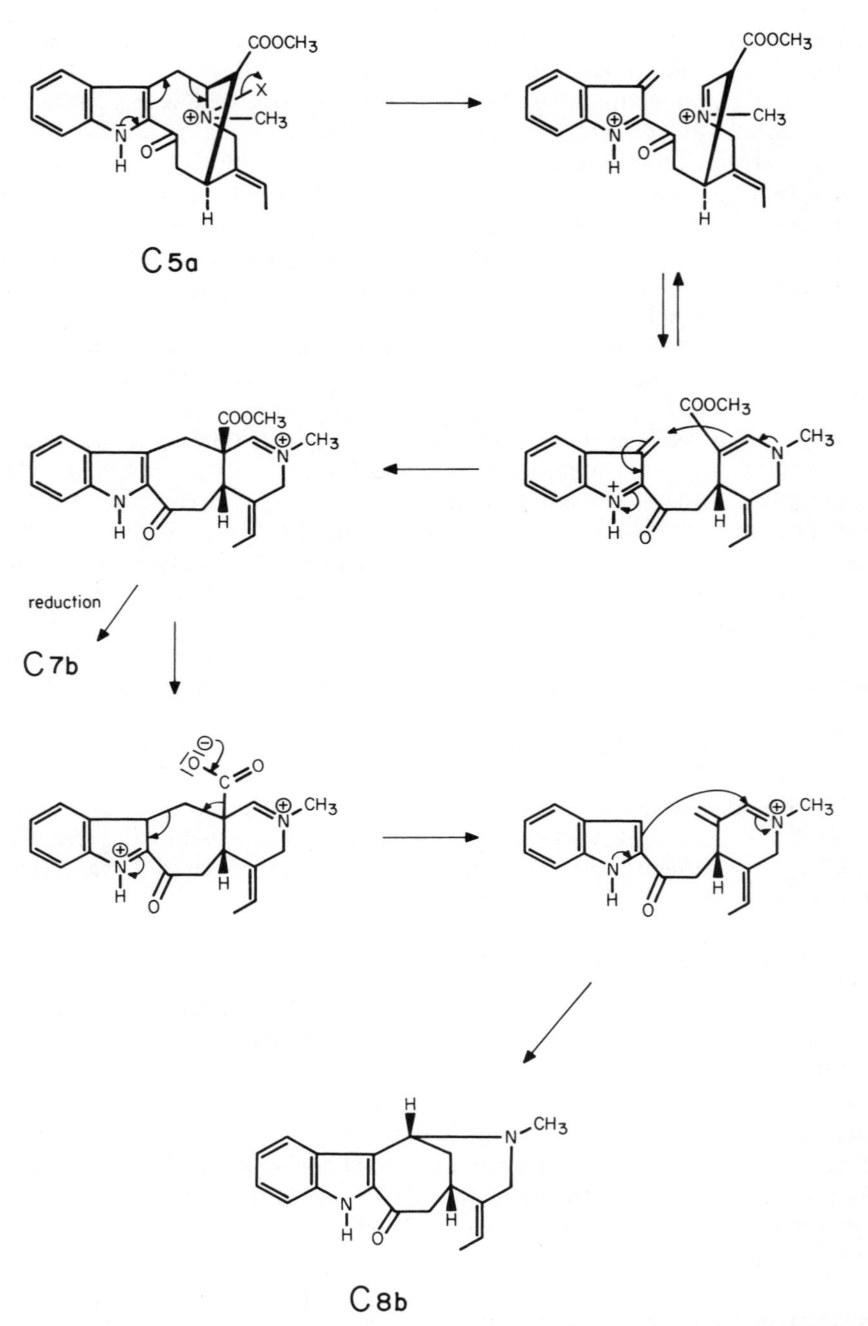

Figure 8. Formation of skeletal types **C7a** and **C8b** from those of type **C5a** [20, 21].

Table 3. Distribution of Alkaloids of the Corynanthean (C) Type in the Three Plant Families (Summarized From Table 2)

Family	Loganiaceae		Apocynaceae				Rubiaceae					
Subfamily			Plumerioideae				Cinchonoideae			Rubioideae		Guettardoideae
Tribe	GEL	STR	CAR	TAB	PLU	RAU	NAU	CIN	MUS	PSY	URO	GUE
Number of genera	2	2	4	4	10	4	5	4	1	—	—	1
Number of species	4	22	12	43	55	69	56	10	1	—	—	1

Number of isolated alkaloids	1789
With absolute configuration	663 = 37%
Number of structurally different alkaloids (bisindole alkaloids are counted twice)	513

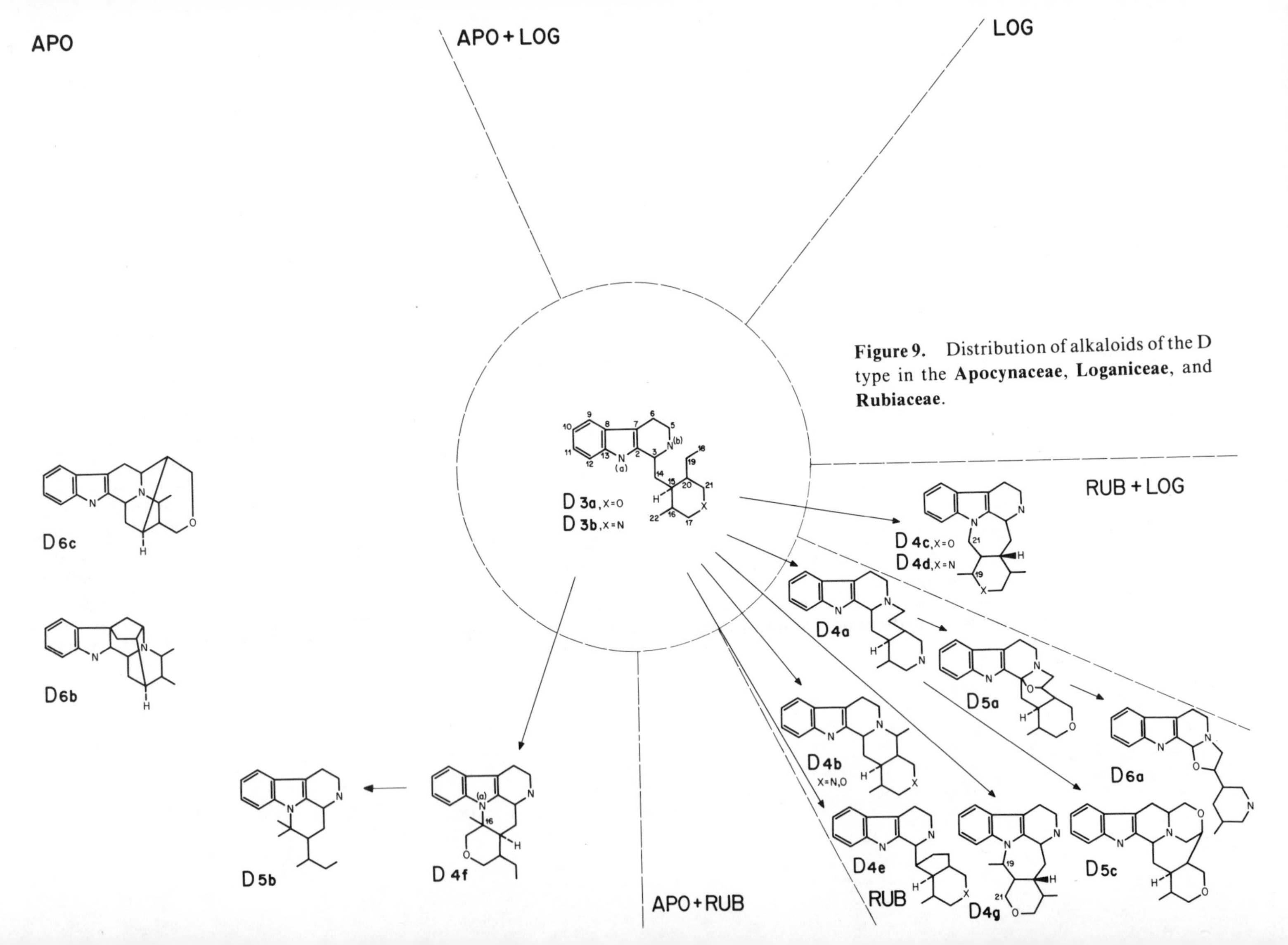

Figure 9. Distribution of alkaloids of the D type in the **Apocynaceae**, **Loganiceae**, and **Rubiaceae**.

(e.g., lyadine). From **D3a** are formed **D4b** (e.g., cadamine), by ring-closure reactions between N(b) and C(19); **D4f**(e.g., talbotine) by closure between N(a) and C(16); **D4g** (e.g., adifoline) by closure between N(a) and C(19); **D4e** (e.g., pauridianthinine) by closure between C(14) and C(18); **D4a** (e.g., 3α-dihydrocadambine) by closure between N(b) and C(18); and **D4c** (e.g., desoxycordifoline lactam) by closure between N(a) and C(22) (see Fig. 9). The ammonia reaction product of **D4c** is **D4d** (e.g., camptoneurine). Formation of **D5a** (e.g., cadambine) and **D5b** (e.g., deformyltalbotinic acid methyl ester) from **D4a** and **D4f**, respectively, is obvious. Reductive ring cleavage of **D5a** can form **D6a** (e.g., α-naucleonine). Skeleton **D5c** (e.g., rubenine) is a special derivative of **D4a** in which a 5-carboxy group forms an additional ring. Examples of structures and a complete list of alkaloids of the D type are given in Appendix 2. In Table 4 the plant sources are given, and Table 5 summarizes the results. Again, alkaloids of D type occur in all three plant families.

2.3. Alkaloids of the Vallesiachotaman (V) Type

As shown in Fig. 6, bases of the vallesiachotaman type can also be derived from **2b**, the common precursor of the corynanthean and vincosan groups. Cyclization between C(17)—*O*H and N(b) in **2b** leads to **V3** (e.g., antirhine), which can be transformed into **V4** (e.g., angustine) by ring closure [C(22)—*O*H or C(17) + C(21); Fig. 10]. Hitherto, only these two skeletal types have been detected. They are distributed in all three plant families; see Appendix 3 and Tables 6 and 7).

2.4. Alkaloids of the Strychnan (S) Type

This group of alkaloids, one of the most populous classes of indole alkaloids, is distributed among the 11 skeletal types given in Fig. 11. The S and A types have a common precursor: **A3** (Fig. 6). From the simplest skeletal type **S4** (e.g., akuammicine), ring closure between C(17)-*O*H and C(19) leads to **S5e** (e.g., spermostrychnine), and closure between C(17)-*O*H and C(18) to **S5b** (e.g., diaboline) (see Appendix 4). By incorporation of an additional C_2 unit on N(a), followed by a C—C linkage with C(17), **S5f** (e.g., isostrychnine) is obtained from **S4**, and **S6a** (e.g., strychnine) from **S5b**. An ether linkage between C(17) and the C_2 unit on N(a) (e.g., acetate) forms **S5c** (e.g., tsilanine). Instead of a C_2 unit, a C_1 unit can also be built onto N(a), giving rise to **S5a** (e.g., geissospermine) from **S4**. In this case, however, it should be emphasized that the C_1 unit in question does not originate from formaldehyde but rather comes from the HC(17)=O group of another *"monomeric"* indole alkaloid. Fairly simple oxidizibility at C(3), leading to the cleavage of the bond C(3)—N(b), is a remarkable characteristic of S-type alkaloids: **S6a** → **S7** (e.g., icajine); **S5c** → **S6c** (e.g., rindline); **S4** → **S5d** (e.g., strychnosilidine); and **S5e** → **S6b** (e.g., strychnobrasiline). A complete list of

Table 4. Number of Indole Alkaloids of the Vincosan (D) Type Isolated From Different Plant Species (See Appendix 9)

	Number of Species	Skeletal Types														
		D3a	**D3b**	**D4a**	**D4b**	**D4c**	**D4d**	**D4e**	**D4f**	**D4g**	**D5a**	**D5b**	**D5c**	**D6a**	**D6b**	**D6c**
Apocynaceae																
Plumerioideae																
CARISSEAE																
Pleiocarpa	1								3			1				
TABERNAEMONTANEAE																
Voacanga	1														1	
PLUMERIEAE																
Alstonia	1															1
Catharanthus	1	2														
Rhazya	1	2														
RAUVOLFIEAE																
Rauvolfia	6														3	6
Loganiaceae																
STRYCHNEAE																
Strychnos	2	1					1									
Rubiaceae																
Rubioideae																
PSYCHOTRIEAE																
Palicourea	1	1														
UROPHYLLEAE																
Pauridiantha	2	2	6					1								
Cinchonoideae																
NAUCLEEAE																
Anthocephalus	1	1		2	4						1					
Haldina	1	2								2						
Nauclea	2			3							1			4		
Pertusadina	1	3		1		1							1			

Table 5. Distribution of Alkaloids of the Vincosan (D) Type in the Three Plant Families (Summarized From Table 4)

Family	**Loganiaceae**		**Apocynaceae**				**Rubiaceae**					
Subfamily			Plumerioideae				Cinchonoideae			Rubioideae		Guettardoideae
Tribe	GEL	STR	CAR	TAB	PLU	RAU	NAU	CIN	MUS	PSY	URO	GUE
Number of genera	—	1	1	1	3	1	4	—	—	1	1	—
Number of species	—	2	1	1	3	6	5	—	—	1	2	—

Number of isolated alkaloids	58
With absolute configuration	17 = 30%
Number of structurally different alkaloids	44

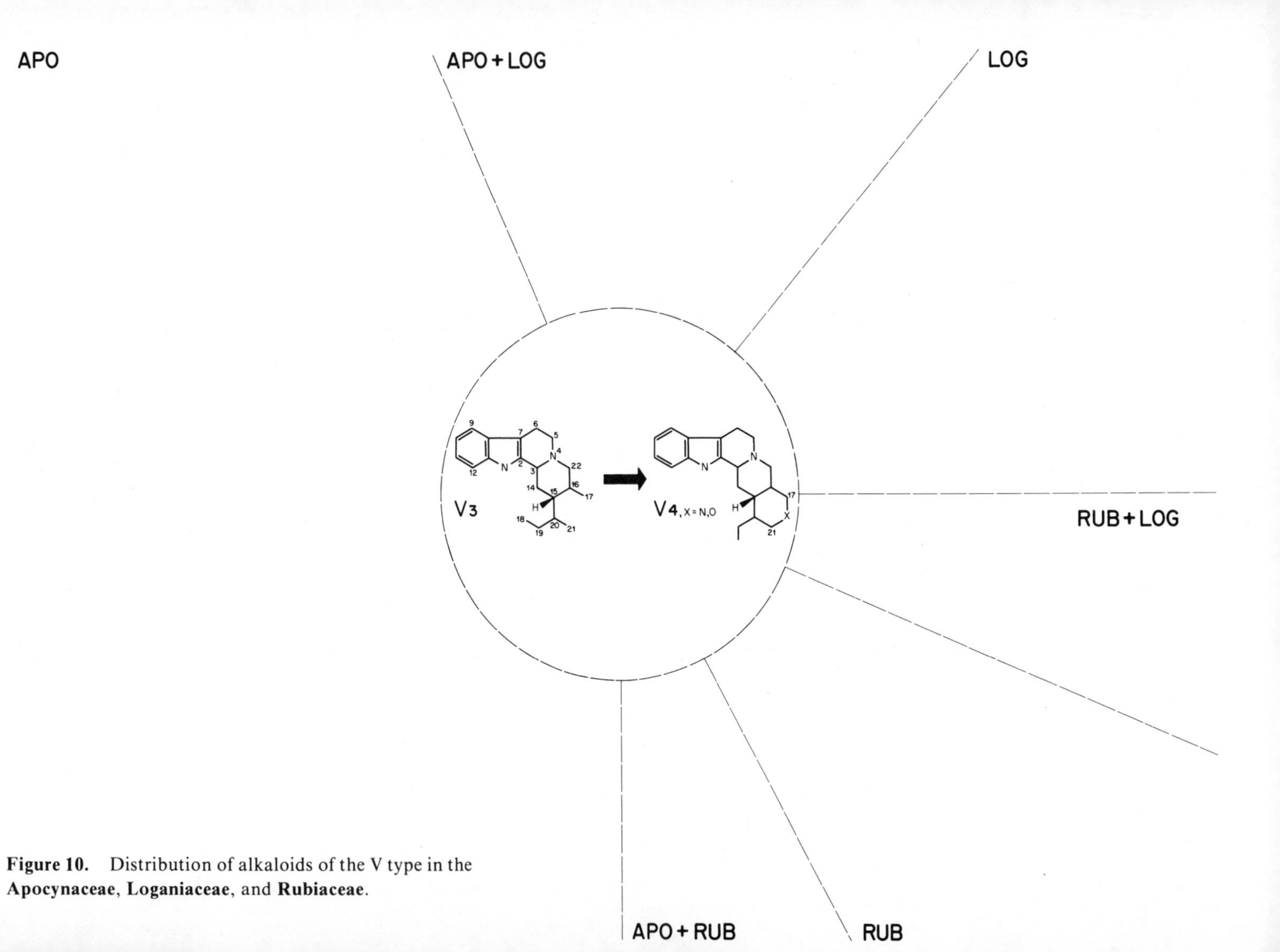

Figure 10. Distribution of alkaloids of the V type in the **Apocynaceae**, **Loganiaceae**, and **Rubiaceae**.

Table 6. Number of Indole Alkaloids of the Vallesiachotaman (V) Type Isolated From Different Plant Species (See Appendix 9)

	Number of Species	Skeletal Types	
		V3	**V4**
Apocynaceae			
Plumerioideae			
CARISSEAE			
Hunteria	2	5	
Pleiocarpa	1	2	
PLUMERIEAE			
Amsonia	2	2	1
Catharanthus	1	1	
Rhazya	1	1	1
RAUVOLFIEAE			
Ochrosia	1	1	
Vallesia	1	1	
Loganiaceae			
STRYCHNEAE			
Strychnos	20	2	47
Rubiaceae			
Cinchonoideae			
NAUCLEEAE			
Mitragyna	2		2
Nauclea	3		8
Pertusadina	1		4
Uncaria	3		8
Guettardoideae			
GUETTARDEAE			
Antirhea	1	1	

alkaloids of the S type with corresponding examples of structures is given in Appendix 4.

It is a remarkable fact that alkaloids of the S type so far, have not been detected in the Rubiaceae. Other than the special type **S5a**, only alkaloids of skeletal type **S4** are produced in the **Apocynaceae** (see Tables 8 and 9); other types are typical of the **Loganiaceae**.

2.5. Alkaloids of the Aspidospermatan (A) Type

This group of natural compounds is distributed among six types of skeletons (Fig. 12). Starting with the common precursor of the aspidospermatan and strychnan types, compound **A3** (e.g., stemmadenine), cyclization between C(7) and C(21) yield **A4b** (e.g., precondylocarpine), and decarboxylation leads to the main group of this type of alkaloids, **A4a** (e.g., tubotaiwine). Oxidation of **A4a** at

Table 7. Distribution of Alkaloids of the Vallesiachotaman (V) Type in the Three Plant Families (Summarized From Table 6)

Family	**Loganiaceae**		**Apocynaceae**				**Rubiaceae**					
Subfamily				Plumerioideae			Cinchonoideae			Rubioideae		Guettardoideae
Tribe	GEL	STR	CAR	TAB	PLU	RAU	NAU	CIN	MUS	PSY	URO	GUE
Number of genera	—	1	2	—	3	2	4	—	—	—	—	1
Number of species	—	20	3	—	4	2	9	—	—	—	—	1

Number of isolated alkaloids	87
With absolute configuration	77 = 89%
Number of structurally different alkaloids	19

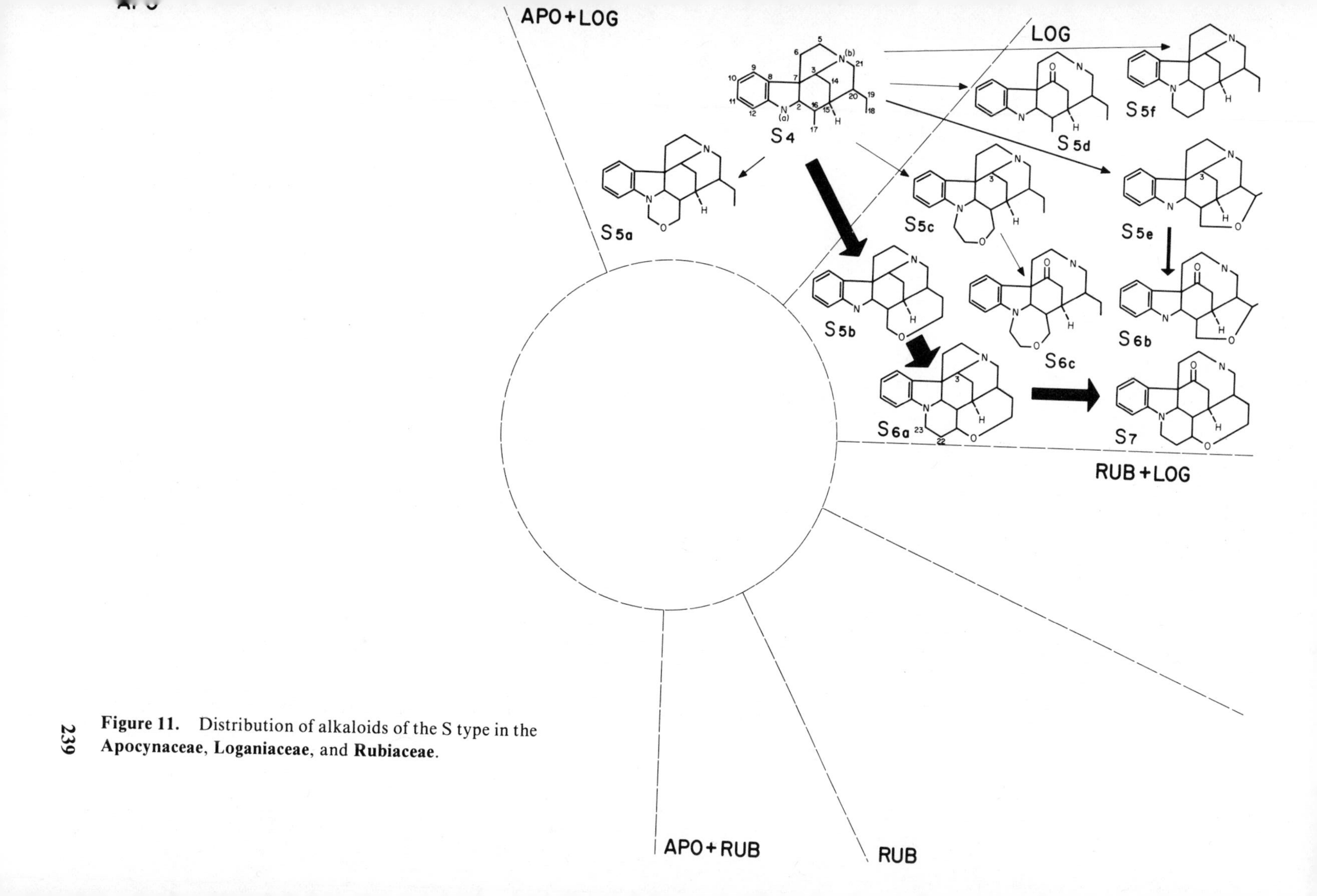

Figure 11. Distribution of alkaloids of the S type in the **Apocynaceae**, **Loganiaceae**, and **Rubiaceae**.

Table 8. Number of Indole Alkaloids of the Strychnan (S) Type Isolated From Different Plant Species (See Appendix 9)

	Number of Species	Skeletal Types										
		S4	S5a	S5b	S5c	S5d	S5e	S5f	S6a	S6b	S6c	S7
Apocynaceae												
Plumerioideae												
CARISSEAE												
Hunteria	2	2										
Picralima	1	2										
Pleiocarpa	1	3										
TABERNAEMONTANEAE												
Tabernaemontana	4	4										
PLUMERIEAE												
Alstonia	5	12										
Amsonia	2	2										
Aspidosperma	1	1										
Catharanthus	3	7										
Diplorhynchus	1	2										
Geissospermum	2	1	3									
Rhazya	1	2										
Vinca	3	9										
RAUVOLFIEAE												
Cabucala	1	1										
Loganiaceae												
STRYCHNEAE												
Strychnos	49	136	3	65	4	4	14	1	55	7	3	64

Table 9. Distribution of Alkaloids of Strychnan (S) Type in the Three Plant Families (Summarized from Table 8)

Family	**Loganiaceae**		**Apocynaceae**				**Rubiaceae**					
Subfamily			Plumerioideae				Cinchonoideae			Rubioideae		Guettardoideae
Tribe	GEL	STR	CAR	TAB	PLU	RAU	NAU	CIN	MUS	PSY	URO	GUE
Number of genera	—	1	3	1	8	1	—	—	—	—	—	—
Number of species	—	49	4	4	18	1	—	—	—	—	—	—

Number of isolated alkaloids	407
With absolute configuration	139 = 34%
Number of structurally different alkaloids (bisindole alkaloids are counted twice)	150

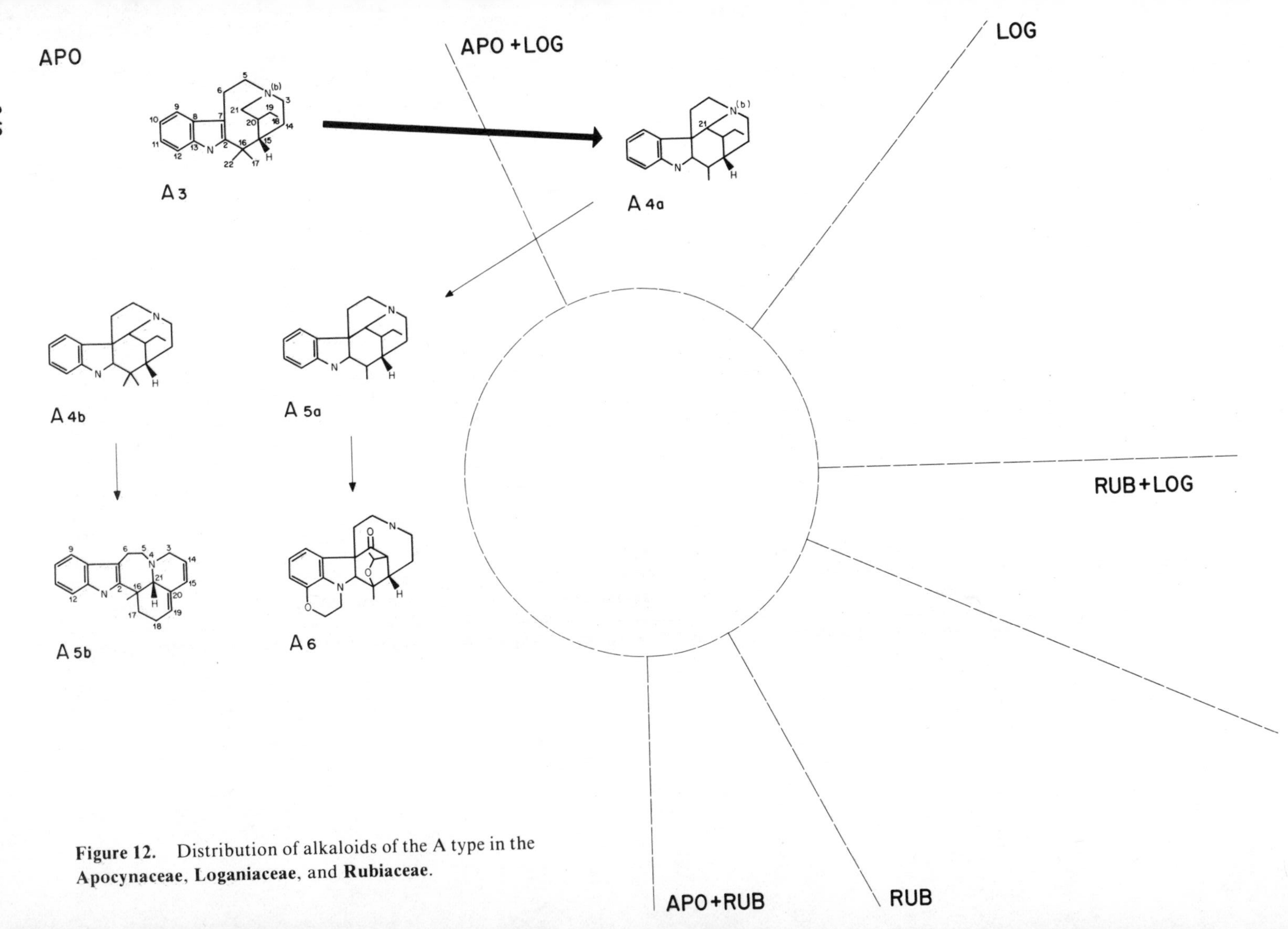

Figure 12. Distribution of alkaloids of the A type in the **Apocynaceae**, **Loganiaceae**, and **Rubiaceae**.

Table 10. Number of Indole Alkaloids of the Aspidospermatan (A) Type Isolated From Different Plant Species (See Appendix 9)

	Number of Species	Skeletal Types					
		A3	**A4a**	**A4b**	**A5a**	**A5b**	**A6**
Apocynaceae							
Plumerioideae							
CARISSEAE							
Melodinus	2	1	3				
Pleiocarpa	1		1				
TABERNAEMONTANEAE							
Stemmadenia	3	3					
Tabernaemontana	7	2	8				
PLUMERIEAE							
Alstonia	2		2				
Amsonia	1		1				
Aspidosperma	9	1	18			2	
Catharanthus	1	1					
Craspidospermum	2	2	2			2	
Diplorhynchus	1	1	1				
Geissospermum	1				1		
RAUVOLFIEAE							
Vallesia	1		3	1			2
Loganiaceae							
STRYCHNEAE							
Strychnos	1		1				

C(21) followed by cleavage of the bond C(21)—N(b) produces **A5a** (e.g., geissovelline), from which by two successive cyclizations and incorporation of an external C_2 unit **A6** (e.g., dichotine) results. Skeletal type **A5b** (e.g., andranginine), of which only two representatives are known, can be derived from **A4b**, represented mainly by precondylocarpine [22]. In contrast to the other skeletal types of the aspidospermatan group, andranginine is a racemate that favors an achiral intermediate between **A4b** and **A5b** [22]. Examples of skeletal types of the A group and the names of all isolated alkaloids are given in Appendix 5. The plants from which the alkaloids were isolated are listed in Table 10, which is summarized in Table 11.

2.6. Alkaloids of the Eburnan (E) Type

This and the two following alkaloid types P and J are derived from the common precursors **2a** and **3a** (Fig. 6) and have a rearranged secologanin skeleton. At present, the eburnan group (Fig. 13) is composed of five skeletal varieties. Cyclization between C(16) and N(a) in precursor **4a** (derived from **3a** by ring closure) forms **E5a** (e.g., eburnamine). An additional cyclization between C(16)

Table 11. Distribution of Alkaloids of the Aspidospermatan (A) Type in the Three Plant Families (Summarized From Table 10)

Family	**Loganiaceae**		**Apocynaceae**				**Rubiaceae**					
Subfamily				Plumerioideae			Cinchonoideae			Rubioideae		Guettardoideae
Tribe	GEL	STR	CAR	TAB	PLU	RAU	NAU	CIN	MUS	PSY	URO	GUE
Number of genera	—	1	2	5	7	1	—	—	—	—	—	—
Number of species	—	1	3	10	17	1	—	—	—	—	—	—
Number of isolated alkaloids			59									
With absolute configuration			39 = 65%									
Number of structurally different alkaloids			24									

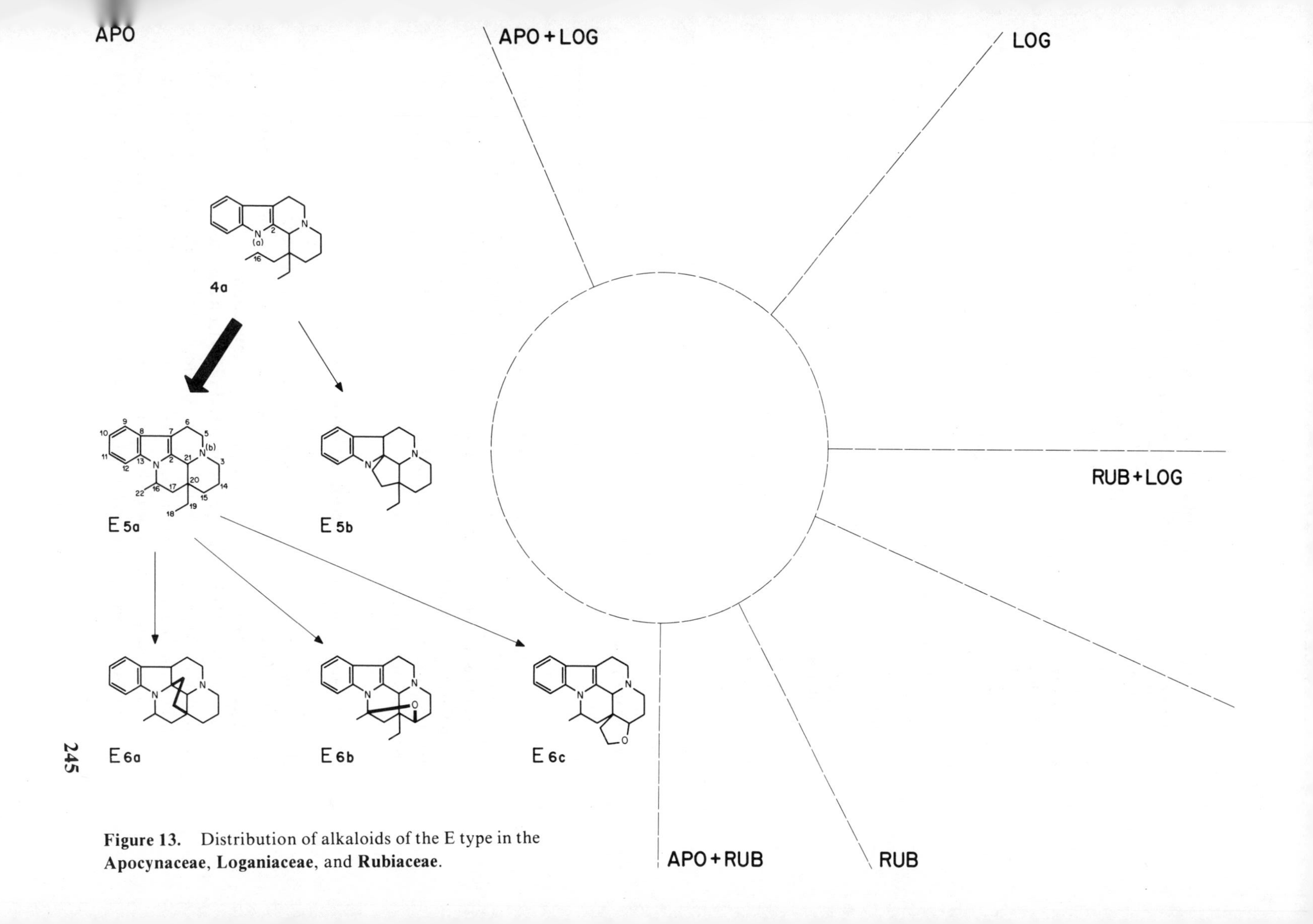

Figure 13. Distribution of alkaloids of the E type in the **Apocynaceae**, **Loganiaceae**, and **Rubiaceae**.

Table 12. Number of Indole Alkaloids of the Eburnan (E) Type Isolated From Different Plant Species (See Appendix 9)

	Number of Species	Skeletal Types				
		E5a	**E5b**	**E6a**	**E6b**	**E6c**
Apocynaceae						
Plumerioideae						
CARISSEAE						
Hunteria	3	7				
Melodinus	3	4				
Pleiocarpa	3	6				
TABERNAEMONTANEAE						
Crioceras	1	5	2		1	
Schizozygia	1		1	6		
Tabernaemontana	2	6				
Voacanga	1	1				2
PLUMERIEAE						
Amsonia	3	7				
Aspidosperma	3	4				
Catharanthus	1				1	
Craspidospermum	2	2	1		1	
Gonioma	1	1				
Haplophyton	1	3				
Rhazya	1	3				
Vinca	4	18				
RAUVOLFIEAE						
Vallesia	1		1			

and C(2) in the same precursor yields **E5b** (e.g., andrangine). Starting with **E5a**, ring closure between C(2) and C(18) leads to **E6a** (e.g., schizogamine); ether linkage between C(15) and C(16) to **E6b** (e.g., craspidospermine), and formation of the bond between C(15) and C(18)—*O*H to **E6c** (e.g., cuanzine). For structural examples and corresponding alkaloids, see Appendix 6. The distribution is given in Tables 12 and 13.

2.7. Alkaloids of the Plumeran (P) Type

This second largest group of indole alkaloids has already been the subject of a detailed chemotaxonomic study [4]. We therefore consider the results of investigations carried out since the previous publication.

Starting with **P4** (e.g., quebrachamine) (Fig. 14), on the one hand, skeletons of the ibogan (J) type (see Section 2.8) can be derived from **P3** (e.g., tetrahydrosecodine), obtained by the cleavage of the ring between C(17) and C(20) in **P4**. On the other hand, various ring-closure reactions lead to the main skeletal types of the plumeran group. Formation of a new bond between C(7) and C(21) in **P4**

Table 13. Distribution of Alkaloids of the Eburnan (E) Type in the Three Plant Families (Summarized From Table 12)

Family	Loganiaceae		Apocynaceae				Rubiaceae					
Subfamily			Plumerioideae				Cinchonoideae			Rubioideae		Guettardoideae
Tribe	GEL	STR	CAR	TAB	PLU	RAU	NAU	CIN	MUS	PSY	URO	GUE
Number of genera	—	—	3	4	8	1	—	—	—	—	—	—
Number of species	—	—	9	5	16	1	—	—	—	—	—	—

Number of isolated alkaloids	83
With absolute configuration	40 = 48%
Number of structurally different alkaloids	42

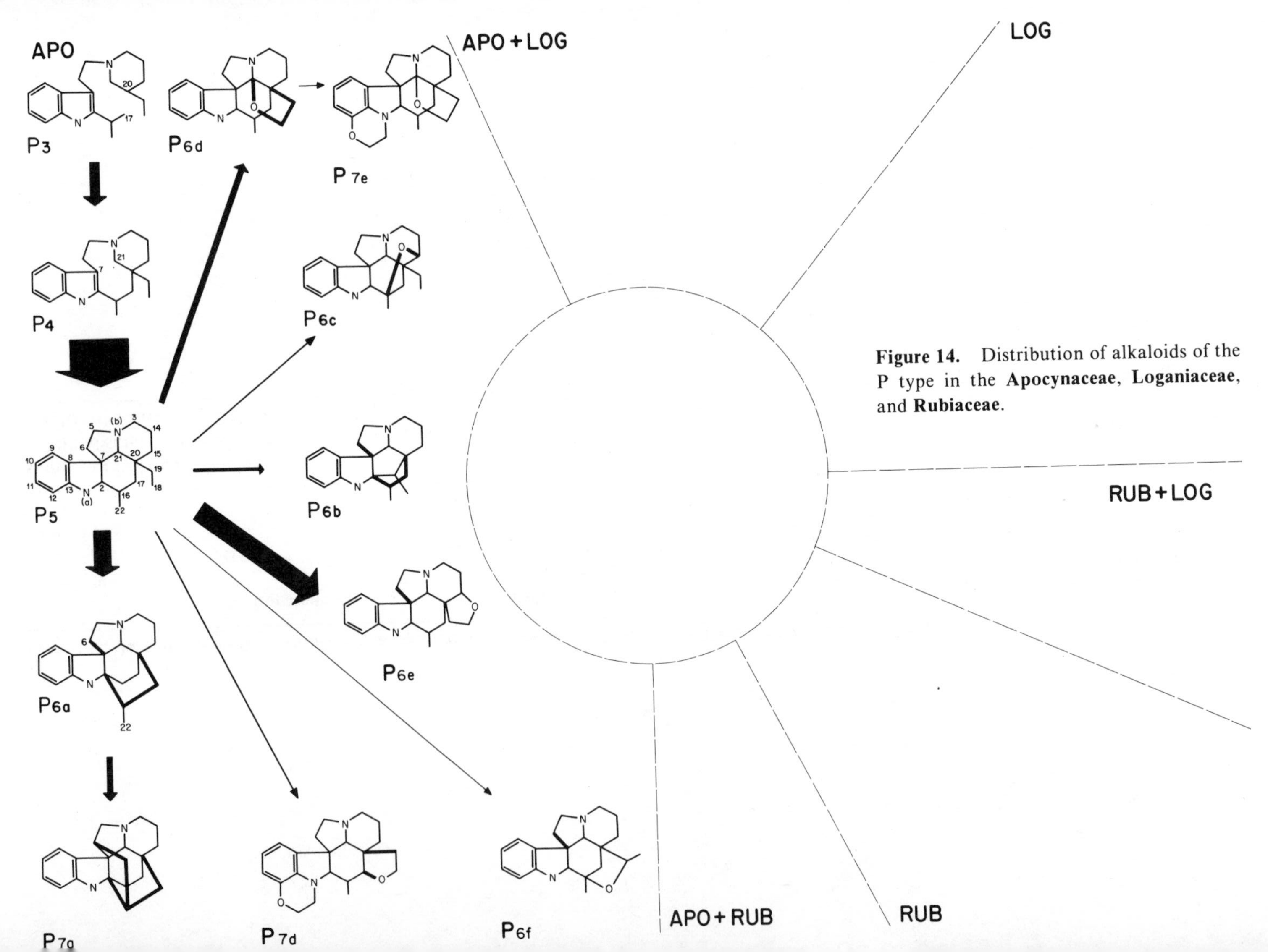

Figure 14. Distribution of alkaloids of the P type in the **Apocynaceae**, **Loganiaceae**, and **Rubiaceae**.

Table 14. Numbers of Indole Alkaloids of the Plumeran (P) Type Isolated From Different Plant Species (See Appendix 9 and Ref. [2])

	Number of Species	Skeletal Types								
		P3	P4	P5	P6a	P6b	P6d	P6e	P7a	Other **P**-Species
Apocynaceae										
Plumerioideae										
CARISSEAE										
Hunteria	2	2	1	5	8					
Melodinus	5		2	26	12	6				**P6f**:1
Pleiocarpa	3		1	2	27	3			3	
TABERNAEMONTANEAE										
Callichilia	2		1					18		
Crioceras	2		2	2				2		
Stemmadenia	3		1	3						
Tabernaemontana	17	1	6	25				8		
Tabernanthe	1		1							
Voacanga	8		3	7				52		
PLUMIERIEAE										
Alstonia	1			6	3					
Amsonia	3	14	7	14						
Aspidosperma	38		8	93	15		17		20	**P7d**:5, **P7e**:1
Catharanthus	5			54	4	5				**P6c**:5, **P6f**:1
Craspidospermum	2			2	2					
Geissospermum	1			4						
Gonioma	2		2	2	1	1				
Haplophyton	1						7			
Rhazya	2	19	2	5						
Vinca	6		7	23	8	2				**P6f**:1
RAUVOLFIEAE										
Cabucala	2			3						
Kopsia	4				3				4	**P7b**:1, **P7c**:2
Vallesia	2		1	8			5			

Table 15. Distribution of Alkaloids of the Plumeran (P) Type in the Three Plant Families (Summarized From Table 14)

Family	**Loganiaceae**		**Apocynaceae**				**Rubiaceae**					
Subfamily			Plumerioideae				Cinchonoideae			Rubioideae		Guettardoideae
Tribe	GEL	STR	CAR	TAB	PLU	RAU	NAU	CIN	MUS	PSY	URO	GUE
Number of genera	—	—	3	6	10	3	—	—	—	—	—	—
Number of species	—	—	10	33	61	8	—	—	—	—	—	—

Number of isolated alkaloids	618
With absolute configuration	415 = 67%
Number of structurally different alkaloids (bisindole alkaloids are counted twice)	283

leads to **P5** (e.g., tabersonine). Various cyclizations in **P5** provide the following variations: C(2) + C(18) → **P6a** (e.g., venalstonine); C(2) + C(19) → **P6b** (e.g., vindolinine); C(15)—*O*H + C(16) → **P6c** (e.g., cathovaline); C(18) + C(21) → **P6d** (e.g., haplophytine); C(16) + C(19)—*O*H → **P6f** (e.g., melobaline); and C(15)—*O*H + C(18) → **P6e** (e.g., beninine). Furthermore, from **P5**, the type **P7d** (e.g., neblinine) can be derived by cyclization between C(17) and C(18)—*O*H followed by an additional ring closure and the inclusion of an external C_2 unit (e.g., acetate) between N(a) and C(12)—*O*H. A similar variation of **P6d** is **P7e** (e.g., alalakine). Type **P7a** (e.g., kopsanole) can be obtained by the formation of a new ring between C(6) and C(22) in **P6a** (Fig. 14). Rearrangement of **P7a** can lead to **P7b** (e.g., decarbomethoxyisokopsine) and **P7c** (e.g., fruticosamine). The absolute configuration demonstrated for types **P6** and **P7** in Fig. 14 corresponds to that of the natural bases so far isolated from this group. The known P alkaloids together with one structural example are given in Appendix 7. The exclusive occurrence of these alkaloids in the subfamily Plumerioideae of the **Apocynaceae** provides a reason for the naming of this group. (Compare Table 14 with Table 15.)

2.8. Alkaloids of the Ibogan (J) Type

The main skeletal types is **J7a** (e.g., conopharyngine) (Fig. 15). Oxidation at C(7) or C(2) in **J7a** leads to **J8a** (e.g., iboluteine) and **J8d** (e.g., crassanine), respectively. On the other hand, an oxidative cyclization between C(3)—*O*H and C(6) yields **J8c** (e.g., eglandine). By cleavage of the C(16)—C(21) bond, **J8b** (e.g., 7*R*-dihydrocleavamine) is formed, which can also be derived from **P3** as demonstrated in Fig. 6.

In **J8b**, cleavage of bonds C(3)—C(14) or C(20)—C(21) leads to the formation of **J9b** (e.g., upper part of catharine) and **J9c** (e.g., upper part of catharinine), respectively. The latter can undergo additional cyclization [C(19)—C(21)] following formation of a new bond [C(3)—C(7)] to produce skeleton **J10b** (e.g., iboxyphylline). Also in **J9c**, an *N*-deformylation of C(21) followed by new bonding of N(b)-C(20) provides skeletal type **J10c** (e.g., ibophyllidine). From **J8b**, by formation of a new bond C(3)—C(7), **J9a** (e.g., pandoline) can be produced. Type **J9a** by an additional reaction step [formation of the bond C(17)—C(21)] can form **J10a** (e.g., pandine). By similar complex reactions, the formation of **J9d** (e.g., upper part of vincathicine) and **J10d** (e.g., lower part of ervafoline) can be explained. For some of the mentioned reaction sequences, other alternatives are conceivable. Representative examples of the J group with the names of the isolated alkaloids belonging to each of skeletal types are given in Appendix 8.

Botanical sources of the ibogan alkaloids are confined to the subfamily Plumerioideae of the **Apocynaceae** as illustrated in Tables 16 and 17.

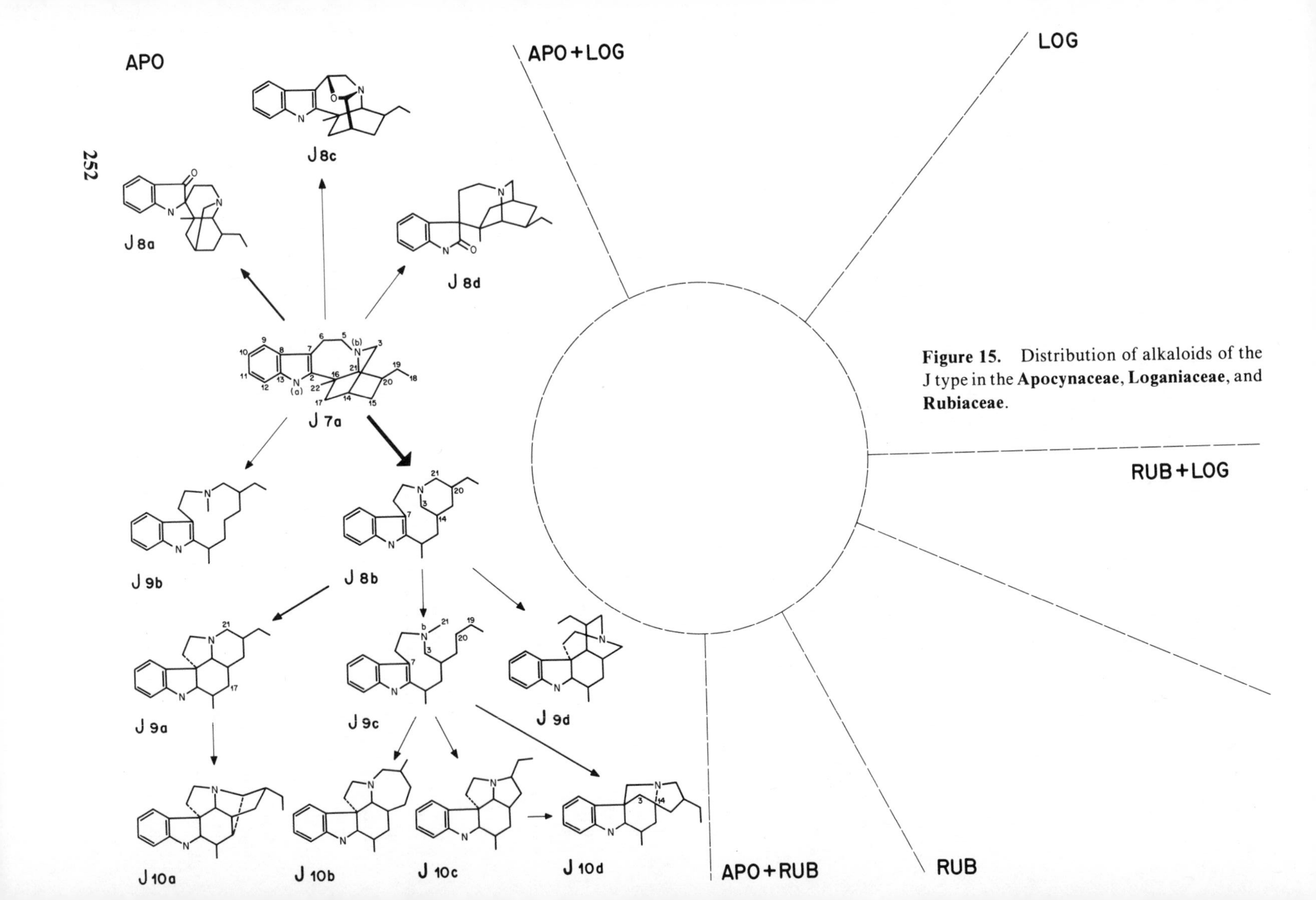

Figure 15. Distribution of alkaloids of the J type in the **Apocynaceae**, **Loganiaceae**, and **Rubiaceae**.

Table 16. Number of Indole Alkaloids of the Ibogan (J) Type Isolated From Different Plant Species (See Appendix 9).

	Number of	Skeletal Types												
	Species	**J7a**	**J8a**	**J8b**	**J8c**	**J8d**	**J9a**	**J9b**	**J9c**	**J9d**	**J10a**	**J10b**	**J10c**	**J10d**
Apocynaceae														
Plumerioideae														
CARISSEAE														
Melodinus	2	2					2							
TABERNAEMONTANEAE														
Callichilia	1	2												
Stemmadenia	4	9												
Tabernaemontana	45	171	9	6	3	1	13				2			4
Tabernanthe	2	12	2			1						2	2	
Voacanga	7	35	3											
PLUMERIEAE														
Alstonia	1	1												
Catharanthus	4	6		15				4	3	1				

Table 17. Distribution of Alkaloids of the Ibogan (J) Type in the Three Plant Families (Summarized From Table 16)

Family	Loganiaceae		Apocynaceae			Rubiaceae						
Subfamily				Plumerioideae			Cinchonoideae			Rubioideae		Guettardoideae
Tribe	GEL	STR	CAR	TAB	PLU	RAU	NAU	CIN	MUS	PSY	URO	GUE
Number of genera	—	—	1	5	2	—	—	—	—	—	—	—
Number of species	—	—	2	59	5	—	—	—	—	—	—	—

Number of isolated alkaloids	311
With absolute configuration	184 = 59%
Number of structurally difference alkaloids (bisindole alkaloids are counted twice)	109

3. ABSOLUTE CONFIGURATION OF INDOLE ALKALOIDS CONTAINING A C_9- OR C_{10}-MONOTERPENE MOIETY

Absolute configuration could be assigned to only 46% of the indole alkaloids containing a C_9- or C_{10}- monoterpene moiety. The absolute configurations of certain remaining alkaloids were not known, or if they were, the information necessary to define the absolute configuration was missing in the original papers. With respect to the expressions used for absolute configuration in this paper, namely α and β, reference should be made to Fig. 5. As shown in Fig. 6, the main skeletal types of indole alkaloids can be divided biogenetically into two main groups: the C, D, V, S, and A types containing a skeleton with a *nonrearranged* secologanin moiety and the E, P, and J types with a *rearranged* secologanin moiety. This classification is confirmed, in addition to the common structural features, by the fact that all of the C-, D-, V-, S- and A-type alkaloids—with known absolute configuration—show the same absolute configuration at C(15) as secologanin (**V**)* at C(7), Fig. 16. Being based on 935 isolations of alkaloids with known absolute configuration and belonging to the five groups of skeletal types (C, D, V, S, A), this statement can be regarded as a valid generalization.

However, while searching the literature for alkaloid isolations of C, D, V, S, and A types, we were confronted with four exceptions, in each of which the configuration appeared to be the opposite of that in (−)-secologanin (**V**):

1. In the first case, (−)-quebrachidine (**XIV**, type **C5c**) isolated from *Cabucala erythrocarpa* (Vatke) Mgf. var. *erythrocarpa* [23] turned out to be a misprint and is in fact (+)-quebrachidine as expected [24] (Fig. 16).
2. The second exception, which unfortunately could not be checked, is represented by a reported isolation of (−)-yohimbine (**XV**, type **C4b**) from *Rauvolfia canescens* L. [25]; yohimbine usually shows an $[\alpha]_D$ value of +80–100° in pyridine. However, since *R. canescens* L. is botanically identical with *R. hirsuta* Jacq., *R. heterophylla* Roem. et Schult., *R. tetraphylla* L., and since also a (+)-yohimbine has been isolated from *R. heterophylla* L., it is likely that only (+)-yohimbine occurs naturally, and that the reported negative value of $[\alpha]_D$ is in error (Fig. 16).
3. The last two cases consist of S-type alkaloids. (+)-2-Epilochneridine (**XVI**, type **S4**) isolated from *Tabernaemontana pandacaqui* Poir. (= *Ervatamia pandacaqui* (Poir.) Pichon) [26] should be assigned to the β group of alkaloids considering its absolute configuration. And pseudo-akuammicine (= (±)-akuammicine) isolated from *Picralima nitida* (Stapf) Th. et H. Dur. (= *P. klaineana* Pierre) [27, 28] with $[\alpha]_D$ = 0 represents a mixture of (+)- and (−)-akuammicine (**XVII**, type **S4**) with the α as well as the β configuration (Fig. 16). The alkaloids (+)-20-epilocheridine (**XVI**)

* C(7) of secologanin corresponds to C(15) of the indole alkaloids being discussed. Natural (−)-secologanin has the same absolute configuration at this carbon atom as the indole alkaloids (Fig. 5).

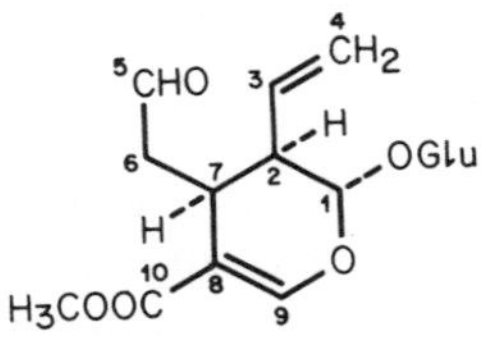

V V(−)-secologanin

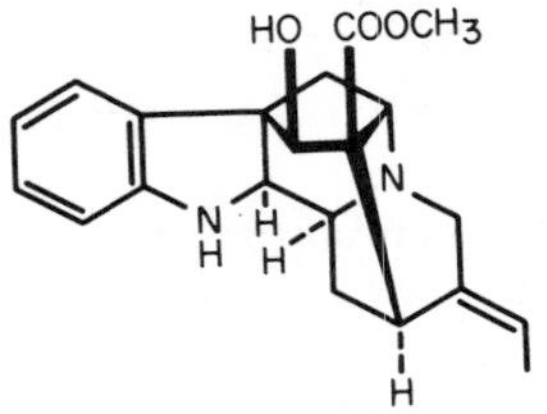

XIV Λ(+)-quebrachidine

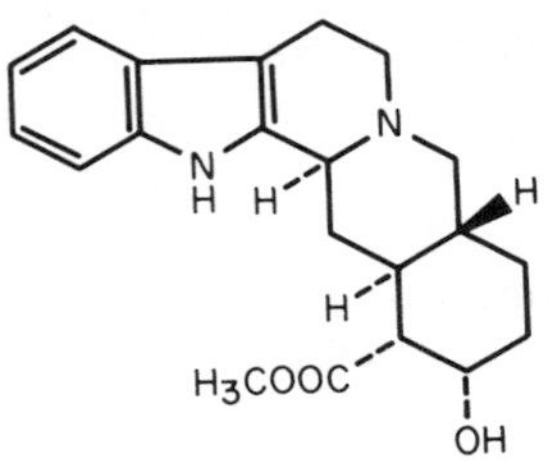

XV V(+)-yohimbine

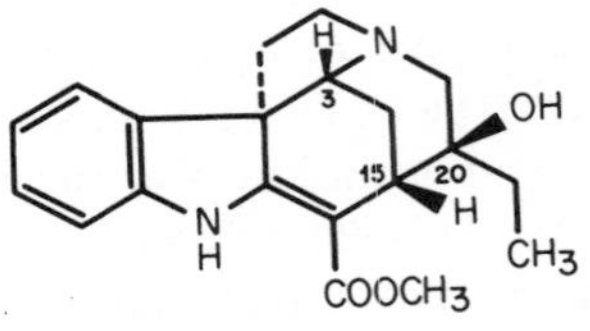

XVI Λ(+)-20-epi-lochneridine

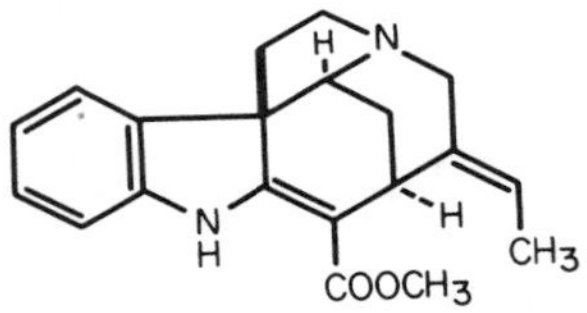

XVII V(−)-akuammicine

Figure 16.

and (−)-akuammicine (**XVII**) each contains a β-anilinoacryl ester chromophor as a common structural feature. Recently, it was shown by a model experiment that exhaustive heating (95° C/50 h) of (−)-19,20-dihydroakuammicine (**XVIII**) in absolute CH_3OH yields a mixture of diastereomers (−)-19,20-dihydroakuammicine (**XVIII**) and (+)-20-epi-19,20-dihydroakuammicine (**XIX**) (ratio ~ 4:1) [29]. According to the mechanism proposed for this isomerization (Fig. 17), chiral centers 3, 7, and 15 can be epimerized. The ratio of **XVIII** to **XIX** is influenced by center 20 in **XX**, which is also chiral (Fig. 17).

If no additional center of chirality is present in the starting molecule, a racemate is formed after equilibrium has been achieved, e.g., (−)-akuammicine (**XVII**). On the other hand, the presence of another chiral center—for example C(20) in (−)-19, 20-dihydroakuammicine (**XVIII**), 20-epilochneridine (**XVI**)—leads to a mixture of diastereomers, the composition of which can not be predicted a priori. Even if the reaction under the previously mentioned

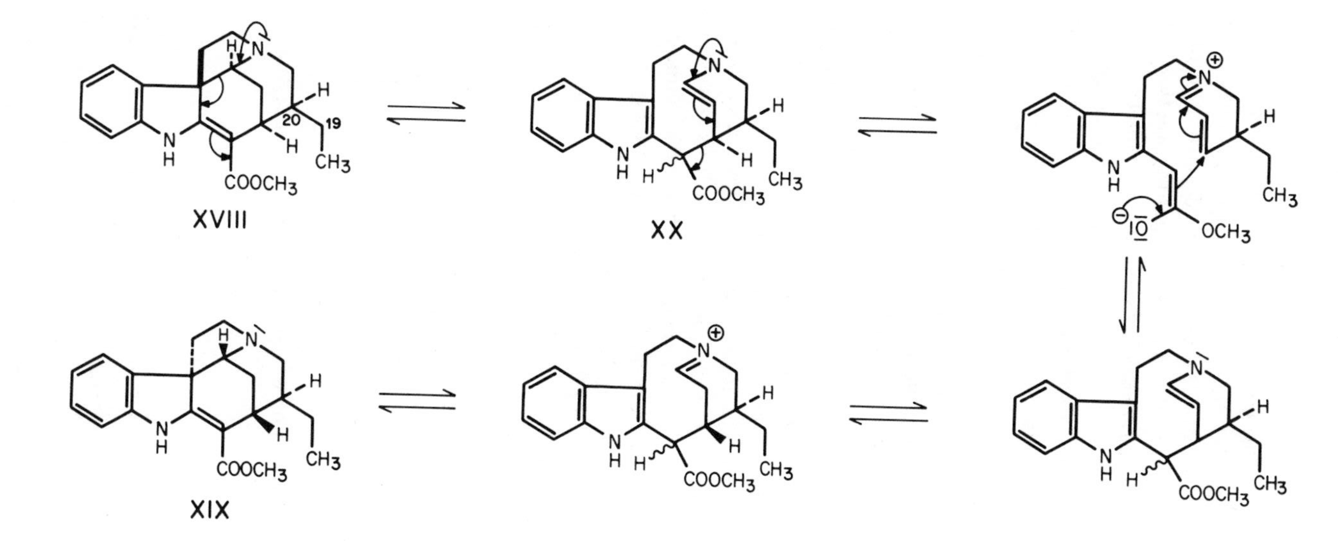

Figure 17. Isomerization of (−)-19,20-dihydroakuammicine (**XVIII**).

conditions definitely cannot take place *in vivo*, there are some hints indicating the possibility of racemization during the extraction or in the plant under modified conditions. Hence it seems plausible that the β configuration of (+)-20-epilochneridine (**XVI**) and (−)-akuammicine (**XVII**) does not demand explanation in terms of the different steric course of a primary biogenetic step. The original configuration in both of these cases is very probably also α as in all of the other alkaloids with a nonrearranged secologanin moiety.

Alkaloids containing a *rearranged* secologanin moiety (E, P, and J types) do not present a uniform characteristic with respect to their configurations, as was the case for alkaloids with a nonrearranged secologanin moiety.

The P alkaloids discussed recently [4] in a separate report exhibit α as well as β configuration (ratio $\alpha:\beta = 1:1.5$). Skeletal types **P4** and **P5**, to which the majority of alkaloids belong, occur as antipodal skeletons, and their configuration therefore is α as well as β. Among the skeletal types derivable from **P5**, **P6a**, **P6b**, **P6e**, and **P6f**, the natural configuration is always α, where **P6d** and **P7d** have only the β configuration (Fig. 14). Skeletal type **P7a**, which can be derived from **P6a**, also occurs with α configuration, as does its precursor. Obviously, the plant enzymes essential for formation of the alkaloids belonging to skeletal types **P6** and **P7** are so specific that only representatives of one particular configuration are synthesized.

Rearrangement of the secologanin moiety does not seem to show any specific characteristics with respect to the configuration at C(7) in **V** (Fig. 16). Alkaloids of the α- and β-plumeran group occur together in the same plant species.

Other alkaloids with a rearranged secologanin moiety, namely those of the E and J types, also fail to show uniform absolute configurations. Alkaloids of the E skeleton (Fig. 13) possess an α configuration if they have a methoxycarbonyl

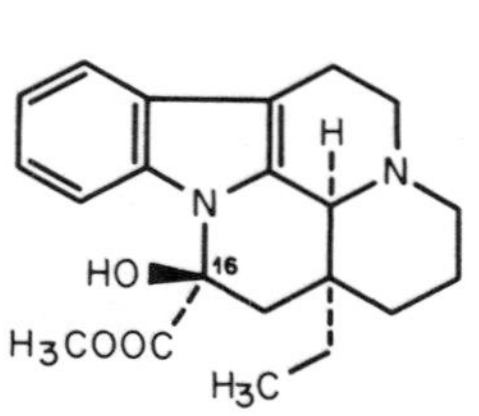

XXI V(+)-vincamine

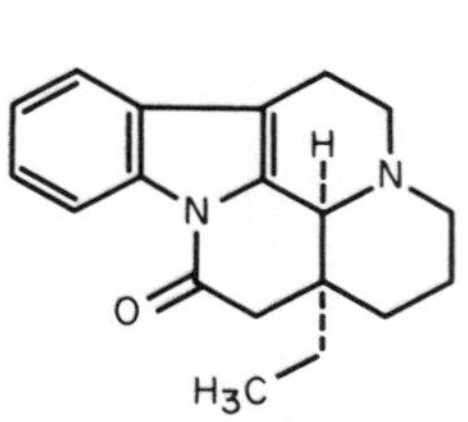

XXII Λ(−)-eburnamonine

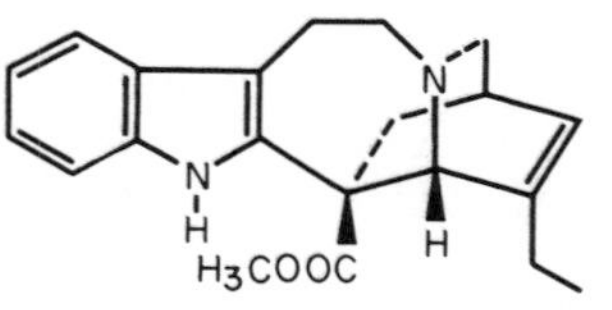

XXIII V(+)-catharanthine

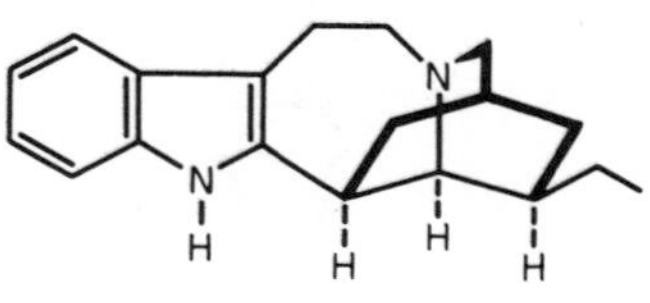

XXIV Λ(−)-ibogamine

Figure 18.

Table 18. The Number of Detected Indole Alkaloids With a C_9- or C_{10}-Monoterpene Moiety and Their Absolute Configuration

	Skeletal Types							
	With Nonrearranged Secologanin Part					With Rearranged Secologanin Part		
Plant Families	C	D	V	S	A	E	P	J
APO	1078	19	15	51	58	83	616	311
Absolute configuration	α	α	α	α	α	$\alpha + \beta$	$\alpha + \beta$	$\alpha + \beta$
LOG	104	2	49	356	1	—	—	—
Absolute configuration	α	α	α	α	α			
RUB	608	36	23	—	—	—	—	—
Absolute configuration	α	α	α					

group at C(16), whereas the β configuration emerges if the same center C(16) is substituted either with two hydrogen atoms or one oxygen in the form of a keto or hydroxy group. However, this observed regularity is contradicted by the exceptions (±)-vincamine and (±)-16-epivincamine (**XXI**, type **E5a**) isolated from *Tabernaemontana rigida* (Miers) Leeuwenberg comb. nov. [=*Anacampta rigida* (Miers) Mgf.] and (−)-and (±)-eburnamonine (**XXII**, type **E5a**) isolated from *Vinca minor* L. (Fig. 18). So far, a reasonable explanation for these exceptions has not been forthcoming.

In the J group, for all of the members of which the β configuration has been established, we also encountered two exceptions of skeletal type **J7a**:(+)-catharanthine (**XXIII**, α configuration, isolated from *Catharanthus* species) and (±)-ibogamine (**XXIV**) isolated from *Tabernaemontana retusa* (Lam.) Pichon [= *Pandaca retusa* (Lam.) Mgf.] (Fig. 18). Only rearranged secologanin alkaloid members with basic skeletons **E5a**, **P4**, **P5**, and **J7a** are known to occur in nature as racemates or with antipodal skeletons. The more complex alkaloids of each type possess only one absolute configuration, α or β depending on the particular skeletal type.

The oçcurrence of alkaloids of all skeletal types with their corresponding absolute configurations is given in Table 18; the uniformity can clearly be recognized again. Alkaloids with a nonrearranged secologanin moiety possess the α configuration, whereas those with a rearranged secologanin moiety may have either configuration.

4. RESULTS AND DISCUSSION

4.1. Differentiation of the Apocynaceae, Loganiaceae, and Rubiaceae With Respect to Structural Types of Indole Alkaloids

More than 99.8% of the isolated indole alkaloids classified in eight skeletal types—five of them with a nonrearranged secologanin moiety and three with a rearranged one—are entirely distributed among three plant families: the

Apocynaceae, the **Loganiaceae**, and the **Rubiaceae**. Reports of the detection and identification of indole alkaloids in some 50 genera and 407 species of these three families suggested to us that it would be worthwhile to attempt a chemotaxonomic study. The objective was to compare the chemotaxonomic data with the botanical classification in order to test the correlation between them and perhaps, eventually, to modify the latter. The occurrence of alkaloids belonging to the skeletal types discussed in the three relevant plant families is summarized in Table 19.

Only alkaloids of the C, D, and V types have been detected in *all* three families. In particular, the number of isolated corynanthean alkaloids reveals an interesting chemotaxonomic distribution (**Apocynaceae** = 1078, **Loganiaceae** = 104, **Rubiaceae** = 608). Also, with respect to absolute configuration, which is exclusively α in case of C-, D-, and V-type alkaloids, it is reasonable to suggest that the corresponding skeletons should be the archetypes of all indole alkaloids with a C_9- or C_{10}-monoterpene moiety. However, the qualitative rather than the quantitative picture is decisive: the relatively lower abundance of D and V alkaloids can be explained in terms of the higher reactivity of these alkaloids, which leads to the generation of C-type alkaloids in the plant.

The other two skeletal types with a nonrearranged secologanin part, S and A alkaloids, are both restricted to the Apocynaceae and the Loganiaceae.

With respect to the skeletal types with a rearranged secologanin moiety, namely the E, P, and J types, the differences between the three plant families are remarkably greater: only the Apocynaceae seem to have the ability to rearrange the secologanin moiety. So far, no alkaloids with a rearranged secologanin moiety have been identified in the Loganiaceae and Rubiaceae. Also alkaloids of A type occur mainly in the Apocynaceae; only one alkaloid has been isolated from a loganiaceous species. In the genera containing A alkaloids, alkaloids with a rearranged secologanin moiety also have been detected (Table 19). This observation allows us to assume a very close biogenetic relationship between the nonrearranged A skeleton and the rearranged skeletal types (E, P and J types). In Fig. 19, this relationship is demonstrated: the decisive step seems to be the isomerization **XXIV** → **XXV**, in which the C(3)=N double bond is transformed to the C(21)=N double bond. In this way, the precursor of the A type (**XXVI**), which is clearly close to a skeleton with a rearranged secologanin moiety (e.g., P type, Fig. 14) is derived from the precursor of the S type (**XXIV**). It can be concluded, with respect to the occurrence of indole alkaloids with a C_9- or C_{10}-monoterpene moiety, that the Apocynaceae, producing *all* of the skeletal variations, must be regarded as the most highly evolved of the three plant families under discussion.

For a further differentiation between the **Loganiaceae** and the **Rubiaceae**, another class of alkaloids, also containing a nonrearranged secologanin moiety, provides a convincing criterion. Besides *Melodinus* (**Apocynaceae**, **Plumerioideae**, **Carisseae**) the **Rubiaceae** are the only plants with the ability to rearrange the indole chromophor into a quinoline structure [e.g., quinine (**XXX**, Fig. 20)]. Furthermore, they can condense secologanin with tyramine instead of tryptamine to produce isoquinoline alkaloids of the emetine type [e.g., emetine (**XXXI**); Fig. 20]. Finally, alkaloids of the tubulosine type (e.g., tubulosine, **XXXII**), in which secologanin is condensed with tryptamine as well as with tyramine, contain a nonrearranged secologanin moiety and have been detected only in the Rubiaceae. Therefore, the Rubiaceae are comparable with the Apocynaceae with respect to their evolutionary development, and both are more developed than the Loganiaceae.

4.2. Analysis of the Tribes of the Apocynaceae [30, 31]

Indole alkaloids containing a C_9- or C_{10}- monoterpene moiety occur only in one subfamily of the **Apocynaceae**, namely, in the Plumerioideae, and not in the other two subfamilies, Cerberoideae and Echitoideae. Of the seven tribes of the Plumerioideae, only four, namely, Carisseae, Tabernaemontaneae, Plumerieae, and Rauvolfieae, contain the indole alkaloids under discussion. So far, there have been no investigations of the other three tribes.

4.2.1. Carisseae. In the Carrisseae, all of the *main* skeletal types occur. Although only 16 species have been investigated, their number, 208 alkaloids, is remarkable. Also the number of subskeletal types, 35 out of the total of 105 skeletal types, is fairly interesting. *Pleiocarpa* seems to be especially typical for Carisseae. Although *Picralima* does not, the three other genera have the ability to form skeletal types with a rearranged secologanin moiety. Only species of *Melodinus* are able to bring about the second rearrangement, leading to the J type.

From the group of C alkaloids, one example provides a particularly good illustration of the chemotaxonomic approach. The structurally simpler alkaloids are more widely distributed than those derived from the simpler precursors by cyclizations, oxidations, etc. In Fig. 21 such a "biogenetic row" is demonstrated. Although alkaloids with the **C3a** skeleton have been detected in all of the plant families under discussion, alkaloids of the **C4e** type derived from **C3a** by the formation of one additional C—C bond occur only in the **Apocynaceae**. The increase in variation of the molecular structure is accompanied by a decrease in the frequency of occurrence. Alkaloids of the most complicated skeletal type **C7a** have been detected only in one species of *Hunteria*. Similar "biogenetic rows" can also be established among other skeletal groups.

4.2.2. Tabernaemontaneae. The Tabernaemontaneae represent a rather uniform tribe by comparison with the Plumerioideae. Indole alkaloids with a C_9- or C_{10}-monoterpene moiety have been detected in 8 genera and 75 species. The formation of J skeletal types seems to be suitable for characterization of Tabernaemontaneae species, because alkaloids of the J type occur—with the exception of *Alstonia* and *Catharanthus* (Plumerieae) and *Melodinus* (Carrisseae)—only in the Tabernaemontaneae. Only two of the eight mentioned genera, namely, *Crioceras* and *Schizozygia*, do not contain alkaloids of the J type.

Table 19. Occurrence of Indole Alkaloids in the Apocynaceae, Loganiaceae, and Rubiaceae (Summary of Tables 2, 4, 6, 8, 10, 12, 14, and 16 and Ref. [2].

	Number of Investigated Species	With Nonrearranged Secologanin Part					With Rearranged Secologanin Part			Total
		C	D	V	S	A	E	P	J	
Apocynaceae										
Plumerioideae										
CARISSEAE	16	64	4	7	7	5	17	100	4	208
Hunteria	4	33		5	2		7	16		
Melodinus	6	5				4	4	47	4	
Picralima	1	10			2					
Pleiocarpa	5	16	4	2	3	1	6	37		
TABERNAEMONTANEAE	70	188	1		4	13	24	129	277	636
Callichilia	2	2						19	2	
Crioceras	1						8	6		
Schizozygia	1						7			
Stemmadenia	4	2				3		4	9	
Tabernaemontana	51	137			4	10	6	40	209	
Tabernanthe	2							1	19	
Voacanga	8	47	1				3	59	38	
PLUMERIEAE	89	326	5	6	39	34	41	360	30	841
Alstonia	17	107	1		12	2		9	1	
Amsonia	3	12		3	2	1	7	35		
Aspidosperma	44	65			1	21	4	159		

Catharanthus	6	33	2	1	7	1	1	69	29	
Craspidospermum	2					6	4	4		
Diplorhynchus	1	3			2	2				
Geissospermum	3	5			4	1		4		
Gonioma	2	4					1	6		
Haplophyton	1						3	7		
Rhazya	2	7	2	2	2		3	26		
Tonduzia	1	6								
Vinca	7	84			9		18	41		
RAUVOLFIEAE	74	500	9	2	1	6	1	27		546
Cabucala	5	38			1			3		
Kopsia	4							10		
Ochrosia	16	52		1						
Rauvolfia	47	407	9							
Vallesia	2	3		1		6	1	14	14	
Apocynaceae/Total	249	1078	19	15	51	58	83	616	311	2231
Loganiaceae										
GELSEMIEAE	4	4								4
Gelsemium	2	2								
Mostuea	2	2								
STRYCHNEAE	71	100	2	49	356	1				508
Gardneria	3	24								
Strychnos	68	76	2	49	356	1				
Loganiaceae/Total	75	104	2	49	356	1				512

Table 19. Cont'd

	Number of Investigated Species	With Nonrearranged Secologanin Part					With Rearranged Secologanin Part			Total
		C	D	V	S	A	E	P	J	
Rubiaceae										
Cinchonoideae										
NAUCLEEAE	62	562	26	22						610
Anthocephalus	1		8							
Cephalanthus	1	6								
Haldina	1		4							
Mitragyna	10	119		2						
Nauclea	4		8	8						
Neonauclea	1	1								
Pertusadina	1	4	6	4						
Uncaria	43	432		8						
CINCHONEAE	10	43								43
Cinchona	3	8								
Corynanthe	3	18								

Pausinystalia	3	16			
Remijia	1	1			
MUSSAENDEAE					2
Isertia	1	2			
Rubioideae PSYCHOTRIEAE					1
Palicourea	1		1		
UROPHYLLEAE					9
Pauridiantha	2		9		
Guettardoideae GUETTARDEAE	2	1		1	2
Antirhea	1			1	
Guettarda	1	1			
Rubiaceae/Total	78	608	36	23	667

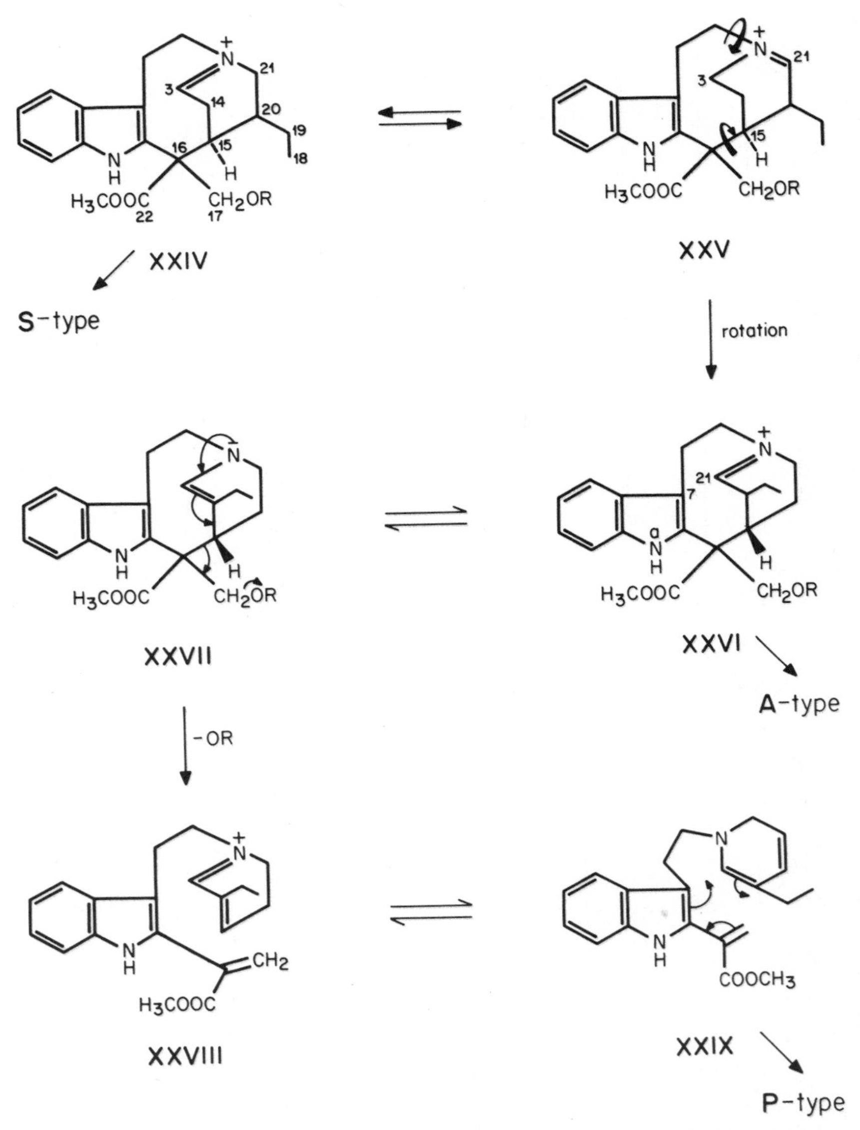

Figure 19. Biogenetic relationship of A skeleton with rearranged skeletal types.

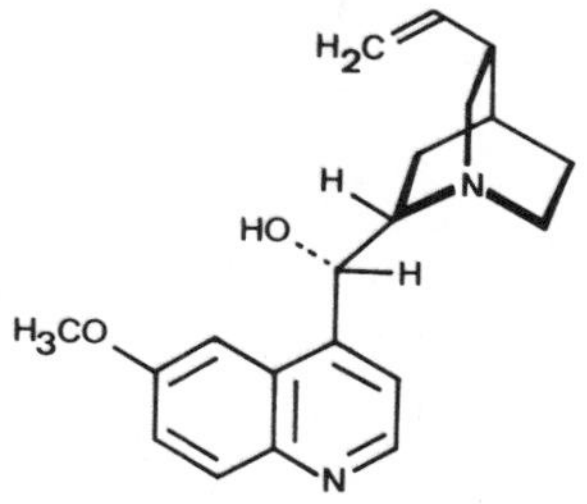

XXX
quinine

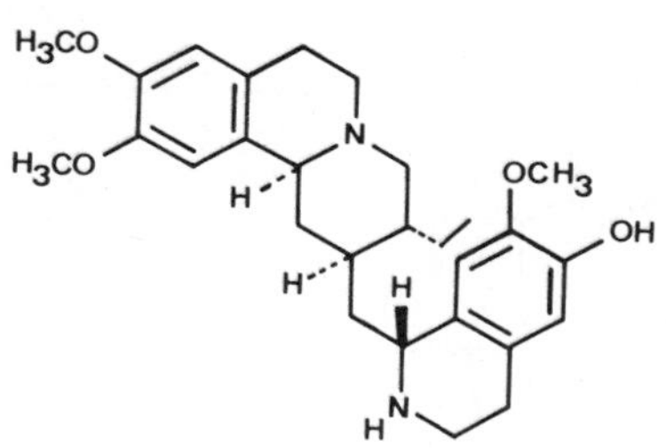

XXXI
emetine

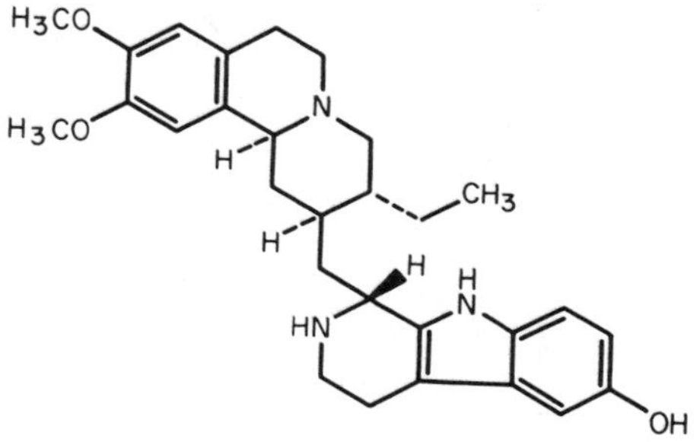

XXXII
tubulosine

Figure 20.

Numerous other representatives of alkaloid types containing a rearranged secologanin moiety (E and P types) have been detected in the Tabernaemontaneae. Few alkaloids with a nonrearranged secologanin moiety (D, S, and A types) (Table 19) were detected; V-type alkaloids were not found.

The Tabernaemontaneae also show uniformity in the case of P alkaloids, (see Ref. [4]). Only skeletal types **P3**, **P4**, **P5**, and **P6e** are formed (Fig. 14 and Table 14). Alkaloids of the special skeletal type **P6e**, which can be derived from **P5** by oxidation, have been detected only in the Tabernaemontaneae. Skeletal types—isolated from *Tabernaemontana* (*Pandaca* and *Capuronetta*) species—that were indicated as **4z**, **5z**, and **6z** types in a previous study of P alkaloids [4] are considered (from the biogenetic point of view) as J alkaloids in this study.

Finally, among 70 investigated species of the Tabernaemontaneae, *Schizozygia coffaeoides* Baill. deserves a special position, as from this species, *only* alkaloids of a specific E skeleton type **E6a** have been isolated.

Generally, it can be ascertained that among all of the skeletal types, those derived from the basic type given in Fig. 2 by oxidative processes were the most frequently isolated. In this context, two skeletal types especially should be

skeletal type	APO CAR	APO TAB	APO PLU	APO RAU	LOG STR	RUB NAU	RUB CIN
	Occurrence in						
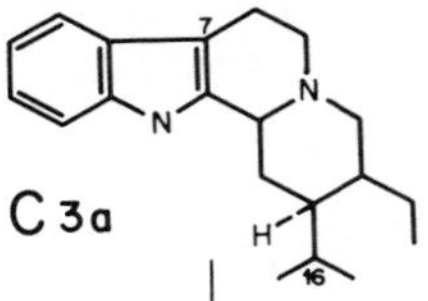 C 3a	+	+	+	+	+	+	+
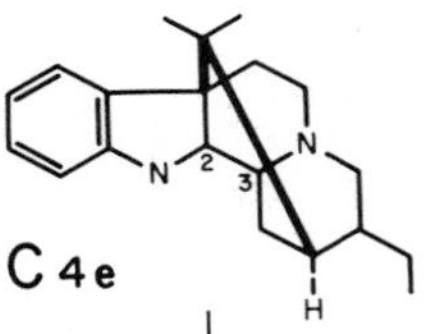 C 4e	+	+	+	+			
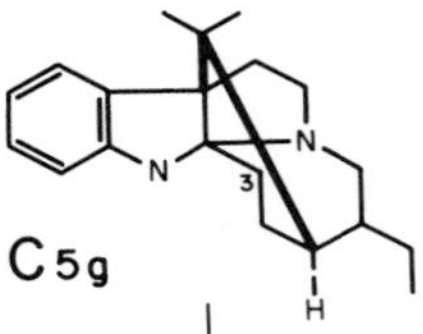 C 5g	+		+	+			
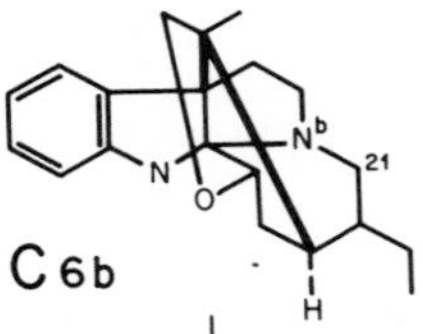 C 6b	+ only Hunteria 4 species						
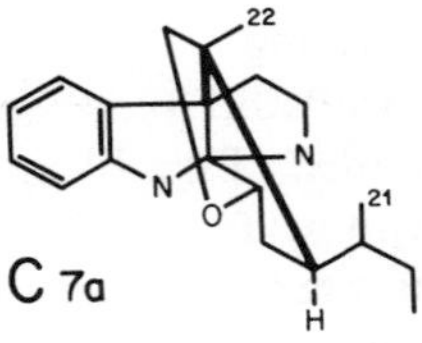 C 7a	+ Hunteria 1 species						

Figure 21. Example of a "biogenetic row."

mentioned: **P6e** (derived from **P5**), with 80 isolated alkaloids, and **C5a** (including **C6c**, derived from **C4c**), with 134 (Tables 14 and 2). On the other hand, skeletal type **C4b** which occurs commonly in the other three tribes, is nearly absent in species of the Tabernaemontaneae (only present in *Voacanga*).

4.2.3. Plumerieae. With 841 isolated alkaloids from 89 species of 12 genera, the Plumerieae form one of the most intensively investigated tribes of the Apocynaceae. With respect to the main alkaloid types, the Plumerieae are considerably less uniform than the Tabernaemontaneae, as alkaloids of *all* the main skeletal types have been isolated. However, C and P alkaloids are most common. With respect to these two main skeletal types, the Plumerieae present the largest variety. With the exception of *Craspidospermum* and *Haplophyton*, the C skeletal types occur in all of the investigated genera. In the Plumerieae, alkaloids with a J skeleton have been isolated from *Alstonia* and *Catharanthus* species. With respect to the content of J alkaloids, the genera *Catharanthus* and *Vinca* differ clearly from each other, based on the evidence that *Vinca* species do not contain J alkaloids. As already noted in a previous chemotaxonomic study [4], there is another chemotaxonomically remarkable difference between these two genera, as bisindole alkaloids were isolated from the species of *Catharanthus*, whereas *Vinca* species contain only "monomeric" bases. Although based upon morphological evidence, *Catharanthus* and *Vinca*, should be counted among the Plumerieae and *not* the Tabernaemontaneae, there are some indications that they should be placed in the Tabernaemontaneae: the occurrence of J skeletal types, which are distinctly characteristic of the Tabernaemontaneae, in *Catharanthus*, and the occurrence of the skeletal type **C5a** in the Tabernaemontaneae and in *Catharanthus* and *Vinca*.

4.2.4. Rauvolfieae. To date, 74 species from 5 genera of Rauvolfieae have been investigated.

With the exception of the J type, *all* of the main skeletal types occur in the Rauvolfieae. However, aside from C alkaloids (500 isolated alkaloids), the number of other types is low. The P type, with 27 alkaloid isolations, is a little more abundant (Table 19).

The genus *Kopsia*, which contains *only* plumeran alkaloids of the "most complicated" skeletal types **P7b / P7c** [4], deserves a special position. *Rauvolfia* contains only alkaloids of the C and D types, and *Ochrosia* only those of the C and V types.

4.3. Analysis of the Tribes of the Rubiaceae

The **Rubiaceae**, with approximately 500 accepted genera and about 6000 species, is one of the three plant families containing indole alkaloids with a C_9- or C_{10}-monoterpene moiety. Also botanically, the **Rubiaceae** provide sufficient evidence to indicate a close relationship with the **Apocynaceae** and the **Loganiaceae**. The latest detailed botanical classification of the genera of the

Rubiaceae originates from K. Schumann [32]. More recently, C. Bremekamp [33], B. Verdcourt [34], and C. E. Ridsdale [35, 36] have made decisive contributions to the classification of this plant family. With respect to the chemotaxonomy of indole alkaloids, we will base our discussion of the **Rubiaceae** on two recent classification systems, those of Leeuwenberg [14] and Ehrendorfer [37]. (In the foregoing tables, **Rubiaceae** are classified according to Leeuwenberg.)

4.3.1. Classification of Leeuwenberg. According to the classification of Leeuwenberg [14], the **Rubiaceae** are divided into four subfamilies, the Rubioideae, Cinchonoideae, Guettardoideae, and Hillioideae (Table 1). In the Rubiaceae, and also in the Gelsemieae of the **Loganiaceae**, only alkaloids containing a nonrearranged secologanin moiety (C, D, and V types) have been detected, of which the C alkaloids, with 608 isolated alkaloids from the species of this family, are the most abundant (Table 19).

On the basis of the classification of Leeuwenberg [14], the occurrence of indole alkaloids of the C, D, and V types is remarkably concentrated in the subfamily Cinchonoideae (the tribes Naucleeae and Cinchoneae). The following are exceptions: *Palicourea* and *Pauridiantha* of the Rubioideae contain exclusively D-type alkaloids; *Isertia* of Mussandeae contains compounds of **C4g** type; and *Antirhea* and *Guettarda* of Guettardoideae contain one alkaloid each of the V and C types.

In the subfamily Cinchonoideae, the tribe Cinchoneae is the only one that contains indole alkaloids with a nonrearranged secologanin moiety as well as quinoline and isoquinoline alkaloids. These are the three main alkaloid types of the **Rubiaceae**. Therefore it seems that this tribe has an important place in the "genealogical tree" of the **Rubiaceae**.

In the Cinchoneae, although C-type alkaloids have been detected in species of *Cinchona*, *Remijia*, *Corynanthe*, and *Pausinystalia*, *Remijia*, *Capirona*, and *Ferdinandusa* species contain isoquinoline alkaloids of the emetine type, and *Cinchona*, *Ladenbergia*, *Remijia* and *Coutarea* species contain quinoline alkaloids of the quinine type that occur only in Cinchoneae. The alkaloidal content of the directly connected tribe Naucleeae is remarkably restricted to indole alkaloids of the C, D, and V types (Tables 2, 4, and 6). The absence of quinoline and isoquinoline alkaloids and the occurrence of indole alkaloids in an increased variety in the Naucleeae compared with the Cinchoneae provide a chemotaxonomic basis for differentiating between these two tribes. Within the Naucleeae, the genera *Mitragyna* and *Uncaria* attract special notice by containing, exclusively, alkaloids of the same C and V types. On the other hand, these two genera are characterized by the occurrence of oxindole alkaloids of type **C5b** (Fig. 6), a fact that supports Ehrendorfer's proposal [37] that *Mitragyna* and *Uncaria* should be recognized as constituting a distinct tribe: the Mitragyneae. *Cephalanthus* and *Anthocephalus* were also considered by Ehrendorfer as a separate tribe (Cephalantheae) and contain only alkaloids of C and D types. Compounds of the **C3** and **C4d** types have been detected in

Cephalanthus, whereas the genus *Anthocephalus* contains alkaloids of the **C4a**, **D4a**, **D4d**, and **D5a** types. The last two types have a restricted occurrence in this genus. In the Mussaendeae, alkaloids of the type **C5l**—which is characteristic of *Cinchona* species—have been detected in *Isertia* species. From species of *Pogonopus* (Rondeletieae) and *Tocoyena* (Gardenieae), only isoquinoline alkaloids of the emetine type have been isolated so far.

In the Rubioideae, other than *Palicourea* (Psychotrieae) and *Pauridiantha* (Urophylleae), which containing exclusively D alkaloids, only isoquinoline alkaloids of the emetine type have so far been detected in *Psychotria* and *Cephaelis* of the Psychotrieae, *Bothriospora* of the Hamelieae, *Borreria*, *Spermacoce*, and *Richardsonia* of the Spermacoceae, and *Manettia* of the Hedyotideae.

Within the fourth subfamily of the Rubiaceae, the Hillioideae, *Hillia* (Hillieae) species contain only emetine-type alkaloids.

4.3.2. Classification of Ehrendorfer [37]. The genera that are of interest for the purposes of this study, and their classification, are given in Table 1. The botanical relationships of the tribes of the subfamily Cinchonoideae as accepted by Ehrendorfer [25] are illustrated in Fig. 23. According to the proposal of Ehrendorfer, only species of Cinchonoideae contain indole alkaloids with a C_9- or C_{10}-monoterpene moiety. In comparison with the classification system of Leeuwenberg (Section 4.3.1), there is a noteworthy difference: Ehrendorfer subdivides the Naucleeae into three tribes, Mitragyneae, Naucleeae, and Cephalantheae. This subdivision can be supported by the following chemotaxonomic arguments. From species of the Mitragyneae, only C and V alkaloids have been isolated. Both genera of the Mitragyneae, *Mitragyna* and *Uncaria*, are clearly characterized by the exclusive occurrence of oxindole alkaloids of type **C5b**. The Naucleeae also contain alkaloids of the D type in addition to C and V alkaloids. In this respect, segregation of the Naucleeae from the Mitragyneae by Ehrendorfer can be corroborated by the detection of D alkaloids and the absence of oxindole alkaloids. Oxindole alkaloids seem to be characteristic of the Mitragyneae. In the Cephalantheae, alkaloids of the C and D types have been detected (**C3** and **C4d** in *Cephalanthus* and **D4a**, **D4b**, and **D5a** in *Anthocephalus*). Alkaloids of type **D4b** and **D5a** have been isolated *only* from this tribe.

4.4. Analysis of the Tribes of the Loganiaceae [38]

The **Loganiaceae**, in contrast with the **Apocynaceae** and **Rubiaceae**, have not been divided into subfamilies. Only two of the tribes, namely, the Gelsemieae and Strychneae, contain indole alkaloids with a C_9- or C_{10}-monoterpene moiety. *Only* C alkaloids have been isolated from *Gelsemium* and *Mostuea* of the Gelsemieae. However, the limited number of alkaloids isolated (only four) does not allow further chemotaxonomic argument. Of the other tribe, Strychneae,

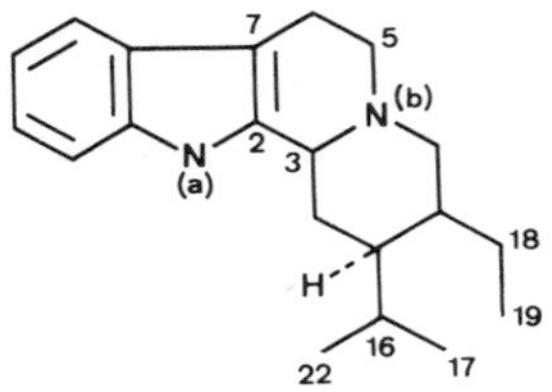

Figure 22. Differentiation of the **Apocynaceae**, **Loganiaceae**, and **Rubiaceae** by means of the main skeletal variations of type **C3a**, oxidation at carbons 2, 3, and 7 as well as new bond formations starting from carbon atoms 16 and 17.

Gardneria species contain only C alkaloids, whereas alkaloids of C, D, V, S, and A types have been isolated from species of *Strychnos*. A peculiarity of the S skeletal types lies in the fact that they contain two additional carbon atoms as skeletal members (Fig. 11).

The most abundant alkaloids in the Loganiaceae are of the S type (11 skeletal variations and 356 isolated alkaloids). In addition to the basic skeletal type **S4** (e.g., akuammicine) (Fig. 11), C(17) is the locus of two additional cyclization possibilities yielding ether rings, e.g., C(17) + C(18)—*O*H → **S5b** (e.g., diabo-line) and C(17) + C(19)—*O*H → **S5e** (e.g., spermostrychnine). Additionally, two further carbon atoms can be inserted between N(a) and C(16) (e.g., **S6a**, strychnine). Most of these skeletons show additional variations in which C(3) is oxidized [e.g., **S4** → **S5d** (e.g., strychnosilidine), **S5e** → **S6b** (e.g, strychno-brasiline), **S6a** → **S7** (e.g., icajine)].

Within the Strychneae, the genera *Gardneria* (only C alkaloids) and *Strychnos* can clearly be differentiated with respect to their alkaloidal content. Recently, *Strychnos* has been investigated chemotaxonomically [39, 40].

Alkaloids of the C type occur in all three plant families. As seen in Fig. 7, this type can be subdivided into a vast number of skeletal variations. It can be seen from the same scheme that within the families some preferred skeletal types exist. It is necessary to examine these types more precisely. The summary of this examination is given in Fig. 22.

The C type **C3a** was chosen as "starting material" for all the other C types. In this sense the main variations result from **C3a** by oxidation and new bond formation. In the case of oxidation, carbon atoms 2 (leading to the skeletal types **C4d, C5b, C5d, C5f, C5h, C6a, C6d, C6e**), 3 (→ **C4h, C5a, C5g, C5n, C6b, C6c, C7a, C7b, C8a, C8b**), and 7 (→ **C5k, C5m, C5t**) are involved (see Appendix 1).

Additional bond formation can start at either of two carbon atoms: C(16) or C(17). The counterparts in the case of C(16) are N(a) (leading to **C4f, C5k, C5y**), C(5) (→ **C4c, C5a, C5c, C5e, C5f, C5i, C5o, C5p, C5q, C6a, C6c, C6e, C7b, C8b**), and C(7) (→ **C4e, C5d, C5g, C5h, C5u, C5v, C6b, C6d, C7a, C8a**). Carbon atom 17 can be connected with N(a) (→ **C3b**), N(b) (→ **C4g, C5l**), C(7) (→ **C5c**), C(18) (→ **C4b, C5r**), or *O*—C(19) (→ **C4a, C5b, C5m, C5n, C5s**). With respect to this compilation, there are great differences between the three plant families.

All the C skeletal types that occur in the **Rubiaceae** contain additional bonds starting from C(17). In the **Apocynaceae** and **Loganiaceae**, both C(16) and C(17) are starting points for new bonds. The converse of the **Rubiaceae** are the

		Oxidation				New Bond Formation										Ratio of C(16)/C(17)			
						New Bonds Starting from C(16) to				New Bonds Starting from C(17) to						in Tribes		in Families	
	Isolated Alkaloids of Type C	at C(2)	at C(3)	at C(7)	% of Oxidized to All Isolated Alkaloids	$N_{(a)}$	C(5)	C(7)	Total	$N_{(a)}$	$N_{(b)}$	C(18)	*O*—C(19)	C(7)	Total	% C(16)	% C(17)	% C(16)	% C(17)
Apocynaceae	1078	86	179	5	25	36	425	94	555	—	—	199	264	132	595			48	52
CARISSEAE	64	2	4	—	9	8	16	25	49	—	—	3	4	1	8	86	14		
TABERNAE-MONTANEAE	188	10	147	—	84	—	188	1	189	—	—	4	4	7	15	93	7		
PLUMER-IEAE	326	55	19	2	23	27	99	55	181	—	—	46	69	37	152	54	46		
RAUVOL-FIEAE	500	19	9	3	6	1	122	13	136	—	—	146	187	87	420	24	76		
Loganiaceae	104	13	—	6	18	19	38	—	57	12	—	6	5	—	23			71	29
Rubiaceae	608	437		3	72	—	—	—	0	—	10	28	346	—	381			—	100
NAUCLEEAE	562	435	—	3	78	—	—	—	0	—	—	9	337	—	346	—	100		
CINCHO-NEAE	43	2	—	—	5	—	—	—	0	—	8	19	8	—	35	—	100		

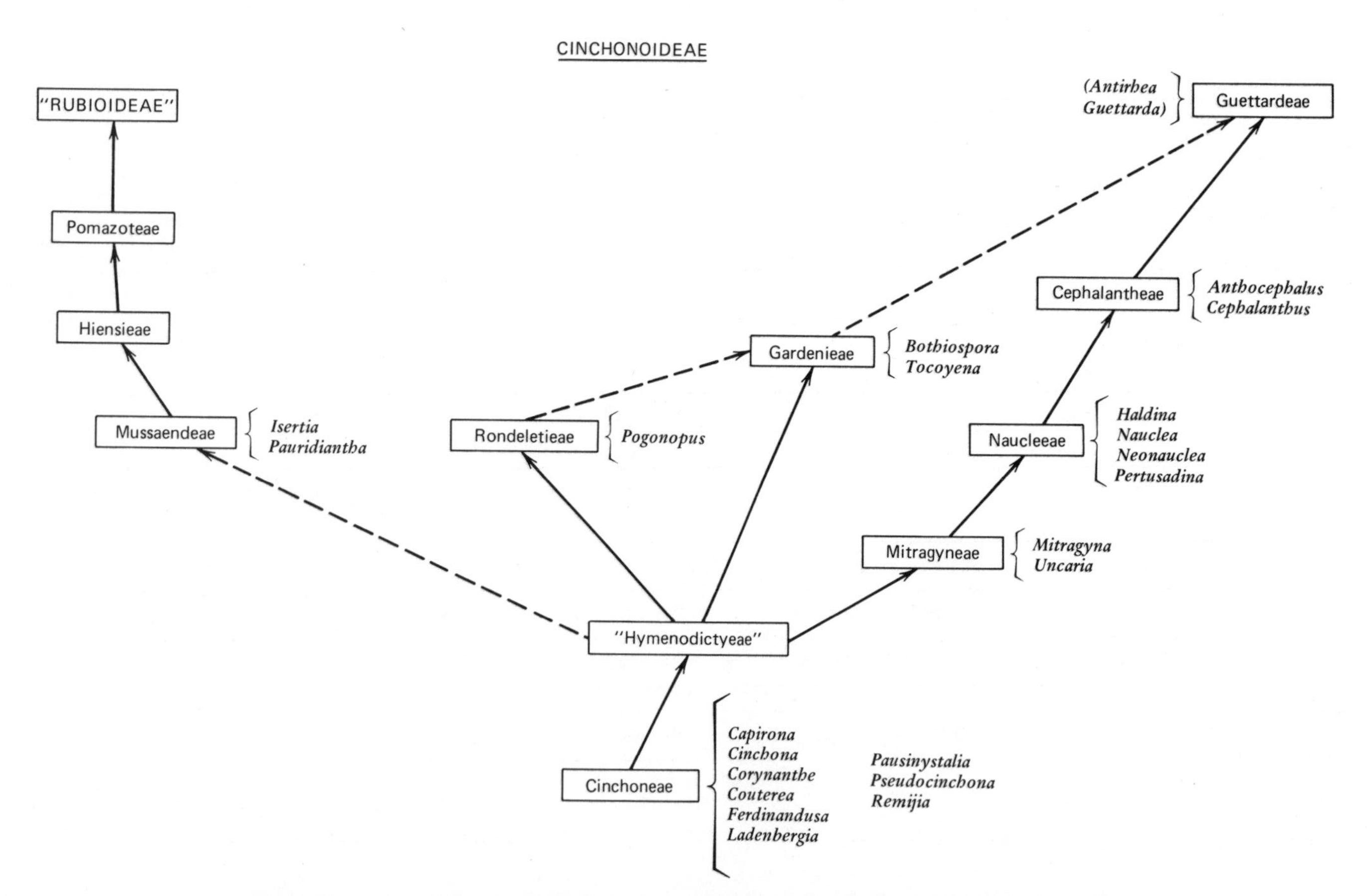

Figure 23. Classification of Cinchonoideae (**Rubiaceae**) according to Ehrendorfer [37].

Tabernaemontaneae (**Apocynaceae**), in which only a few alkaloids occur, but with respect to **C3a** they have bonds starting from C(17). The other families or tribes respectively are placed between these two extremes. Also with respect to the oxidation state of carbons 2, 3, and 7, similar differentations can be established. The tribe Naucleeae of **Rubiaceae** produces most of the alkaloids that are oxidized at C(2). In the Rauvolfieae (**Apocynaceae**) only 6% of all alkaloids are oxidized at C(2).

The values given in Fig. 22 permit the following comments. The **Rubiaceae** are clearly differentiated from the **Loganiaceae** and **Apocynaceae** by the exclusive occurrence of **C3a** derivatives with additional bonds starting from C(17). Within this family, the Naucleeae can also be separated from the Cinchoneae by the fact that 78% of the Cinchoneae alkaloids are oxidized at C(2), whereas only 5% of Naucleeae alkaloids have been detected in an oxidized state at this specific center. Furthermore, the occurrence of alkaloids with a C(17)—N(b) bond in the Cinchoneae suggests that the Cinchoneae should be associated with the Loganiaceae. It also follows from these examinations that the formation of alkaloids of the quinine type can be connected with the ability of plants to oxidize the skeleton at C(2).

Differentation between the three plant families—and in case of the **Apocynaceae** and **Rubiaceae** even between their subfamilies—is possible. The oxidation levels of carbon atoms 2, 3, and 7 and additional bond formation starting from C(16) or C(17) as shown in Fig. 22 give indications that supplement the remarks about the distribution of skeletal types made at the beginning of this chapter.

The common occurrence of different skeletal types and the variations in the C-type skeleton both testify to the central position of Loganiaceae. It can readily be inferred that the **Loganiaceae** should be placed closer to the **Apocynaceae** than to the **Rubiaceae**. Furthermore, the Carisseae should be directly joined with the Loganiaceae. Then the Tabernaemontaneae, the Plumerieae, and finally the Rauvolfieae follow. On the other side, the **Rubiaceae** are placed in connection with the Cinchonoideae according to the classification of Ehrendorfer. This statement proceeds from Fig. 24.

As mentioned before [2], the genus *Holarrhena* (Plumerieae, Plumerioideae, Apocynaceae) seems to be wrongly placed based on chemotaxonomic considerations. This genus exhibits taxonomic characteristics of members of the subfamily Echitoideae. Instead of indole alkaloids, steroidal alkaloids have been isolated from species of this genus [41]. Steroidal alkaloids are typical of Echitoideae, and so we suggest that the genus *Holarrhena* should be included. Furthermore, it is necessary to consider another change. Because of its special indole alkaloid content, the tribe Tabernaemontaneae has to be considered as a subfamily (Tabernaemontanoideae), which should be near to the Plumerioideae (Carisseae, Plumerieae, Rauvolfieae). Unfortunately, not all genera of the Plumerioideae are chemically analyzed, and this prevents a definite classification from the chemotaxonomic point of view.

An attempt to differentiate the genera more precisely with respect to the

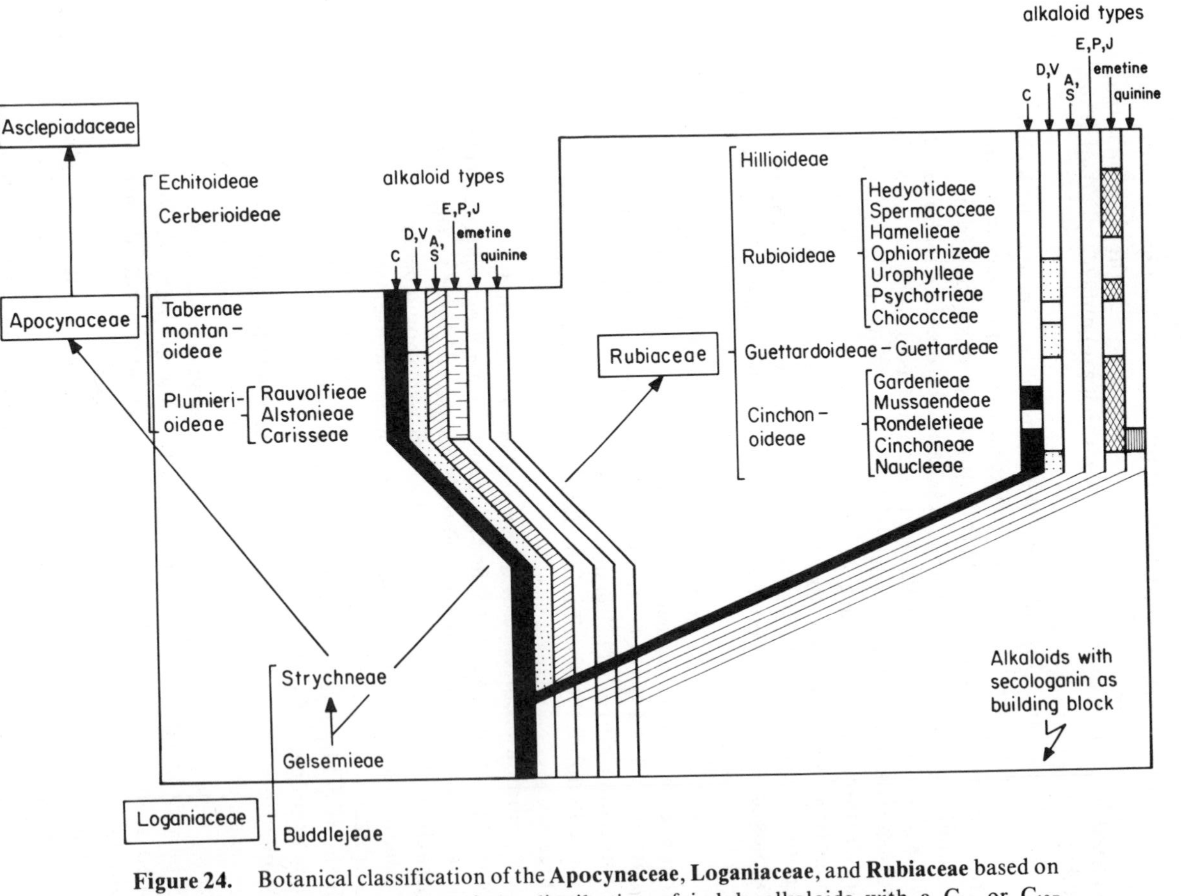

Figure 24. Botanical classification of the **Apocynaceae**, **Loganiaceae**, and **Rubiaceae** based on the chemotaxonomic criterion of the distribution of indole alkaloids with a C_9- or C_{10}-monoterpene moiety.

occurrence of the alkaloids has been successful only to a certain extent (some examples have already been mentioned), because, besides the genera investigated intensively (e.g., *Catharanthus*, *Rauvolfia*, *Uncaria*), there are also some genera that have not been examined closely. There are species from which many alkaloids have been isolated; on the other hand there are also species in which so far only one alkaloid has been detected. Hitherto it has not been possible to use as a quantitative criterion a statement that only *one* alkaloid and *no other* can be isolated. However, it would appear possible to use as a chemotaxonomic criterion quantitative information from the plant in question, provided that due regard is given to seasonal, geographic, and anatomic factors.

LIST OF SYNONYMOUS ALKALOID NAMES

(The preferred version is written in italics)

burnamine = *desacetylpicraline*
caracurine VII = *Wieland-Gumlich aldehyde*
condensamine = *11-methoxyhenningsamine*
desacetyldiaboline = *Wieland-Gumlich aldehyde*
dimethoxypicrinine = *quaternine*
elegantine = *isomajdine*
elliptamine = *reserpiline*
3-epi-α-yohimbine = *isorauhimbine*
19,20-α-epoxy-12,15-dihydroxy-*N*-methyl-*sec*-pseudostrychnine = *19,20-α-epoxy-15-hydroxyvomicine*
19,20-α-epoxy-15-hydroxynovacine = *19,20-α-epoxy-15-hydroxy-10,11-dimethoxy-N-methyl-sec-pseudostrychnine*
grandifoline = *amataine*
henningsamine = *O-acetyldiaboline*
10-hydroxyakuammicine = *sewarine*
3-hydroxy-α-colubrine = *pseudo-α-colubrine*
3-hydroxy-β-colubrine = *pseudo-β-colubrine*
hydroxyindolenine-conopharyngine = *jollyanine*
isovincoside = *strictosidine*
kopsinoline = *kopsinine N-oxide*
11-methoxycompactinervine = *alstovine*
11-methoxy-epi-3α-yohimbine = *quaternatine*
montanine = *voacristine-pseudoindoxyl*
reserpine *N*-oxide = *renoxydine*
rhazinine = *antirhine*

subsesseline = *amataine*
taberpsychine = *anhydrovobasinediol*
tombozine = *normacusine B*
uncarine A = *isoformosanine*
uncarine B = *formosanine*
uncarine C = *pteropodine*
uncarine D = *rauvoxinine*
uncarine E = *isopteropodine*
vinamidine = *catharinine*
vincamidine = *strictamine*
vincovine = *vinorine*
vincristine = *leurocristine*
voacangarine = *voacristine*
voaphylline = *conoflorine*
vomifoline = *peraksine*

LIST OF ABBREVIATIONS

A	Aspidospermatan
APO	Apocynaceae
C	Corynanthean
CAR	Carisseae
CIN	Cinchoneae
D	Vincosan
E	Eburnan
GEL	Gelsemieae
GUE	Guettardeae
J	Ibogan
LOG	Loganiaceae
MUS	Mussaendeae
NAU	Naucleeae
P	Plumeran
PLU	Plumerieae
PSY	Psychotrieae
RAU	Rauvolfieae
RUB	Rubiaceae
S	Strychnan
STR	Strychneae

TAB	Tabernaemontaneae
URO	Urophylleae
V	Vallesiachotaman

ACKNOWLEDGMENTS

This study was supported by the Schweizerischer Nationalfonds zur Förderung der wissenschaftlichen Forschung, to whom we express our sincere appreciation. We are grateful to Professor Dr. F. Markgraf, Universität Zürich and to Professor Dr. F. Ehrendorfer, Universitat Wien, for information about the **Rubiaceae**. Dr. D. Ganzinger, Universität Wien, is gratefully acknowledged for some literature work.

We wish also to express our warm thanks to Mrs. M. Kalt, who showed great patience in typing and retyping the manuscript, and to Dr. D. Crout, University of Exeter, for linguistic corrections.

We acknowledge permission from Academic Press, Inc. (London) Ltd. to reproduce Schemes 3, 4, 6, 8, and 9 from our Chapter 2 in *Indole and Biogenetically Related Alkaloids*, J. D. Phillipson and M. H. Zenk, Eds., Academic Press, London, 1980.

Appendixes to Chapter Five

GENERAL REMARKS TO APPENDIXES 1–8

The sources for each individual alkaloid can be determined from the list given in Appendix 9 together with the references. ° = bisindole alkaloids; given twice if necessary.

APPENDIX 1. CORYNANTHEAN (C) TYPE: EXAMPLES OF STRUCTURES AND THE LIST OF ALL ALKALOIDS DETECTED IN NATURE

Type: C3a

Example:

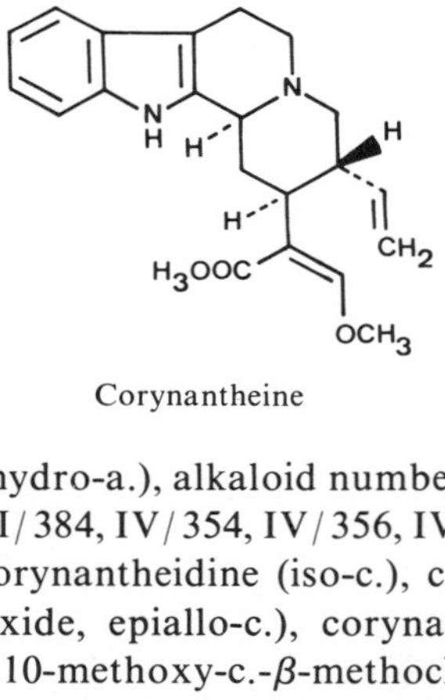

Corynantheine

Alkaloids: Adirubine (anhydro-a.), alkaloid numbers : II/296, II/298, II/328, III/352, III/354, III/382, III/384, IV/354, IV/356, IV/384, IV/386; angusteine, aspexcine, corynantheal, corynantheidine (iso-c.), corynantheine (c.-aldehyd, dihydro-c., dihydro-c.-*N*-oxide, epiallo-c.), corynantheol (18,19-dihydro-c., dihydro-c.-methochloride, 10-methoxy-c.-β-methochloride, 10-methoxydihydro-c.), diploceline, flavocarpine, gambirine, geissoschizal, geissoschizine (decarbomethoxy-g., g.-methyl ether), geissoschizol (10-hydroxy-g., 10-methoxy-g.), geissospermine°, hervine, hirsuteine, hirsutine (h, *N*-oxide), huntrabrine methochloride, isostrychnopentamine A, isostrychnopentamine B, methoxygeissoschizoline, mitraciliatine, ochrolifuanine A (o.- B, o.-*N*-oxide, 3-dehydro-o., hydroxy-o.), ochromianine, ochropposinine, ochrosandwine, paynantheine (iso-p.), serpentinine°, sitsirikine (dihydro-s., 16-epi-s., iso-s.,), speciociliatine, speciogynine, strychnobaridine, strychnopentamine, tschibangensine, usambarensine (3′, 4′-dihydro-u., $N_{(b)}$-methyl-3′, 4′-dihydro-u., *N*′-methyl-u.), usambaridine (u.-Br, u.-Vi, 18,19-dihydro-u.-Br, 18,19-dihydro-u. Vi), usambarine (18,19-dihydro-u.).

Type: C3b

Example:

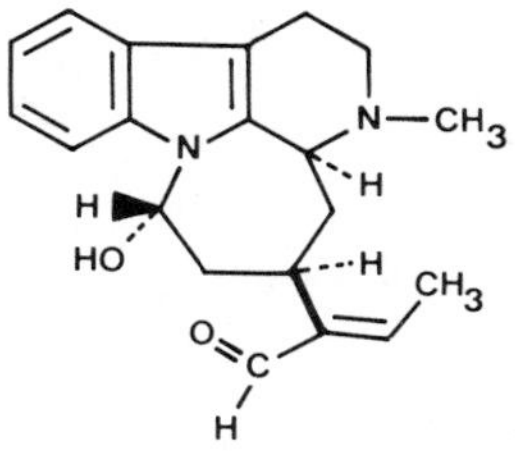

Akagerine

Alkaloids: Akagerine (17-*O*-methyl-a.), kribine (epi-21-*O*-methyl-k., 21-*O*-methyl-k.).

Type: C4a

Example:

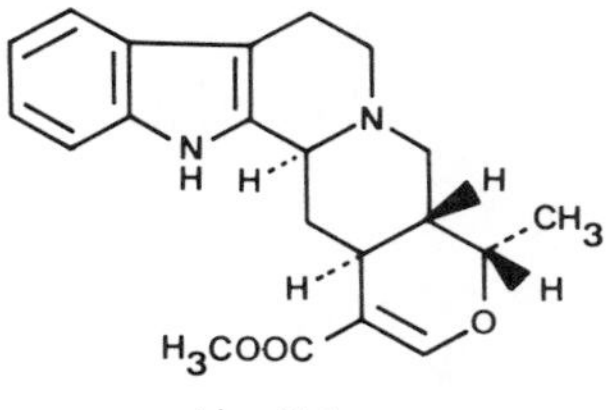

Ajmalicine

Alkaloids: Ajmalicine (10,11-dimethoxy-a., 19-epi-a., 3-iso-a., 3-iso-19-epi-a.), ajmalicinine (10,11-dimethoxy-a.), akuammigine (4*R*-a.-*N*-oxide, 4*S*-a.-*N*-oxide), alkaloid numbers: I/352, I/382a, I/382b, alstonine, aricine, bleekerine, cabucine, cabucinine, cathenamine, ervine, herbaceine, herbaine, holeinine, melionine A (m.-B), mitrajavine, raufloridine, raumitorine, rauniticine (3-iso-r.), raunitidine, rauvanine, reserpiline (19,20-dehydro-r., iso-r., r.-methosalt), reserpinine (iso-r.), roxburghine (r.- A, r.- B, r.- C, r.- D, r.- E), serpentine (19-epi-s., 10-methoxy-s.), serpentinine°, tetrahydroalstonine (5α-carboxy-t., 4*R*-t.-*N*-oxide), tetraphylline, tetraphyllinine.

Type: C4b

Example:

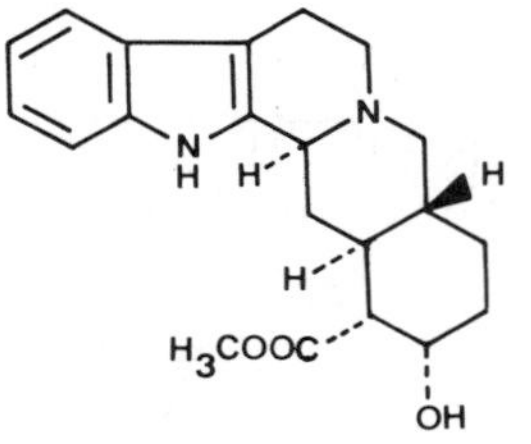

Yohimbine

Alkaloids: Alloyohimbine, alstoniline, anhydroalstonatine, canembine, corynanthine (5α-carboxy-c.), 3,4-dehydroalstovenine, deserpideine, deserpidine, excelsinine, gambirtannine (decarbomethoxydihydro-g., dihydro-g., neooxy-g., oxo-g.), isorauhimbine, melinonine E, methyl deserpidate, methyl reserpate, ourouparine, poweridine, pseudoreserpine (iso-p.), pseudoyohimbine (11-methoxy-p.), quaternatine, raugustine, raujemidine, raunescine (iso-r.), renoxydine, rescidine, rescinnamine, reserpine (iso-r.), sempervirine, seredine, venenatine (iso-v.), veneserpine, venoxidine, yohimbine (*O*-acetyl-y., 19,20-dehydro-y., 18-hydroxy-y., 11-methoxy-y.), 3,4,5,6-tetradehydro-β-yohimbinium chloride, α-yohimbine, β-yohimbine, yohimbol methochloride.

Type: C4c

Example:

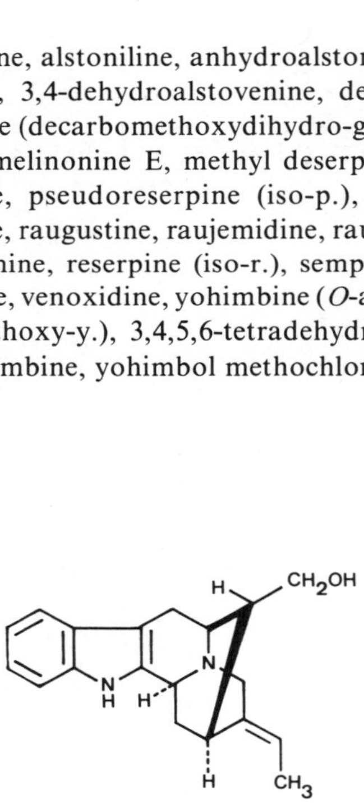

Normacusine B

Alkaloids: Affinisine, akuammidine (a.-methosalt, deformo-a., $N_{(b)}$-methyl-a.-chloride, $N_{(a)}$-methyl-10-methoxy-a., $N_{(b)}$-methyl-10-methoxy-a.), *O*-benzoyltombozine, ervincidine, gardnerine, geissolosimine°, lochneram, lochnerine, lochvinerine, macralstonidine°, macusine A, macusine B (*O*-methyldihydro-m., *O*-methyl-m., *O*-methyl-nor-m., *O*-methyl-nor-m.-B-*N*-oxide, nor-m.), macusine C, majvinine, pericyclivine, polyneuridine (*O*-acetyl-p.), sarpagine ($N_{(a)}$, $N_{(b)}$-

dimethyl-s. chloride, $N_{(a)}$-methyl-s., neo-s.), spegatrine, vellosimine (10-methoxy-v.), voachalotine (17-*O*-acetyl-19,20-dihydro-v., 3-hydroxy-v., 21-hydroxy-v.).

Type: C4d

Example:

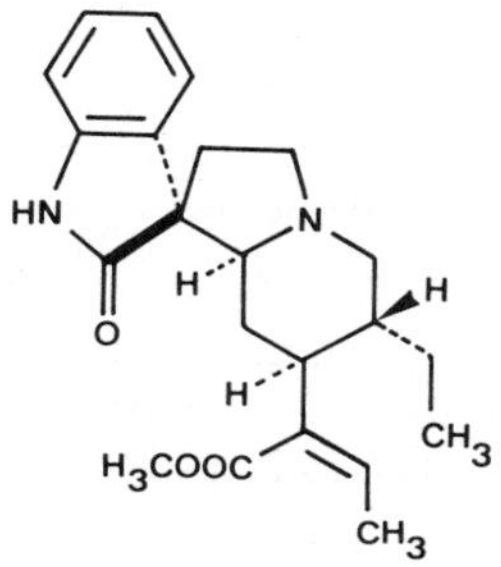

Rhynchophylline

Alkaloids: Ciliaphylline (c.-*N*-oxide), corynoxeine (iso-c.), corynoxine A (c.-B), mitrafoline (iso-m.), mitragynine (m.-oxindol A, m.-oxindol B), ochromianoxine, rhynchociline (r.-*N*-oxide), rhynchophylline (*anti*-iso-r.-*N*-oxide, iso-r., iso-r.-*N*-oxide, r.-*N*-oxide), rontundifoleine (iso-r.), rotundifoline (*anti*-r.-*N*- oxide, 3-epiiso-r., iso-r.), speciofoline (iso-s.), specionoxeine (iso-s.), strychnofoline (iso-s.), strychnophylline (iso-s.).

Type: C4e

Example:

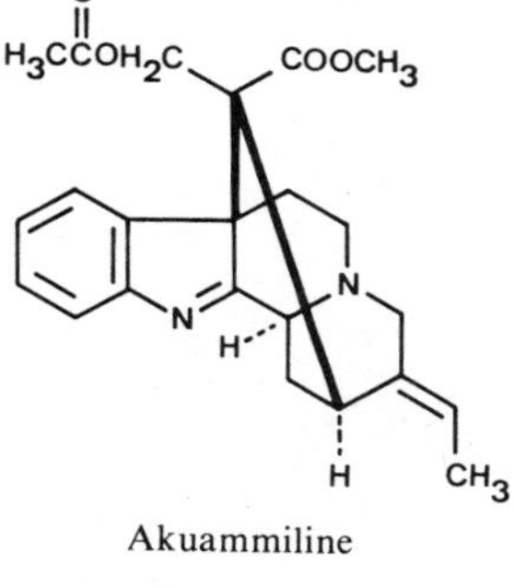

Akuammiline

Alkaloids: Akuammiline (desacetyl-a., desacetyldeformo-a., 10-methoxydesacetyl-a.), cabucraline, cathafoline, pleiocraline°, quaternoxine, raufloricine, rhazinaline, strictalamine, strictamine, umbellamine°.

Type: C4f

Example:

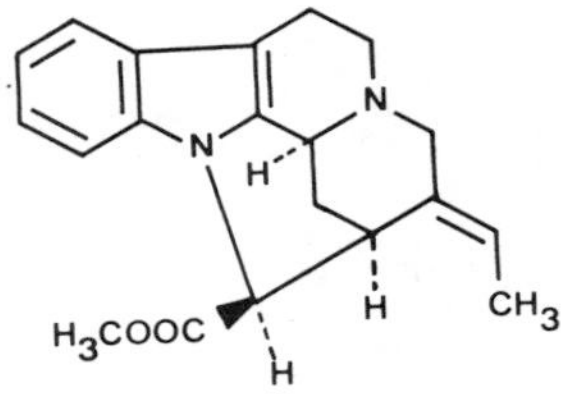

Pleiocarpamine

Alkaloids: Macralstonine°, macrocarpamine°, C-mavacurine, pleiocarpamine (2,7-dihydro-p., p.-methochloride), pleiocorine°, pleiocraline°, pleiomutinine°, C-profluorocurine, pycnanthine° (14′, 15′-dihydro-p.°), pycnanthinine°, villalstonine°.

Type: C4g

Example:

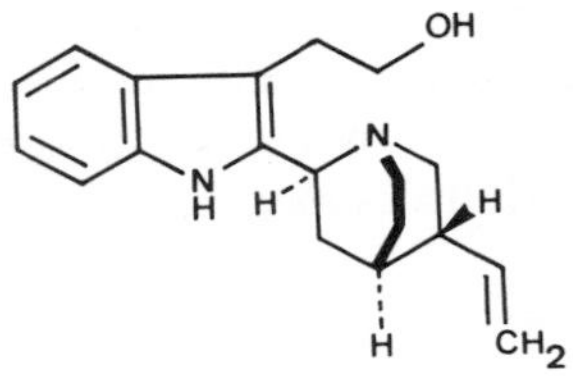

Cinchonamine

Alkaloids: Cinchonamine, cinchophyllamine (iso-c.).

Type: C4h

Example:

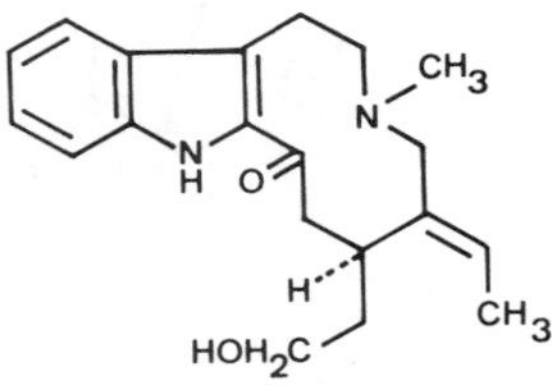

Burnamicine

Alkaloid: Burnamicine.

Type: C5a

Example:

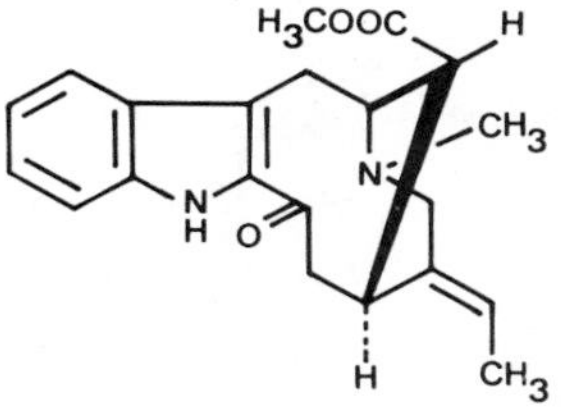

Vobasine

Alkaloids: Accedine (*N*-demethyl-16-epi-a.), accedinine°, accedinisine°, affinine, capuvosidine°, capuvosine° (15-dehydroxy-c., *N*-demethyl-c.). conoduramine° (19,20-epoxy-c.), conodurine° (19-oxo-c.), dregamine, 16-epiaffinine, 16-epivobasinic acid, gabunamine°, gabunine°, *N*-methyl-16-epiaffinine, ochropamine, ochropine, pelirine, periformyline, perivine, tabernaelegantine A° (t.-B, t.-C, t.-D), tabernaelegantinine A° (t.-B), voacarpine, tabernaemontanine, vincadiffine, voacamidine°, voacamine° (16-decarbomethoxy/dihydro-v., 16-decarbomethoxy-20′-epidihydro-v., 18′-decarbomethoxy-v., *N*-demethyl-v., v.-*N*-oxide), voacorine° (19-epi-v.), vobasine (16-decarbomethoxy-19,20-dihydro-v.).

Type: C5b

Example:

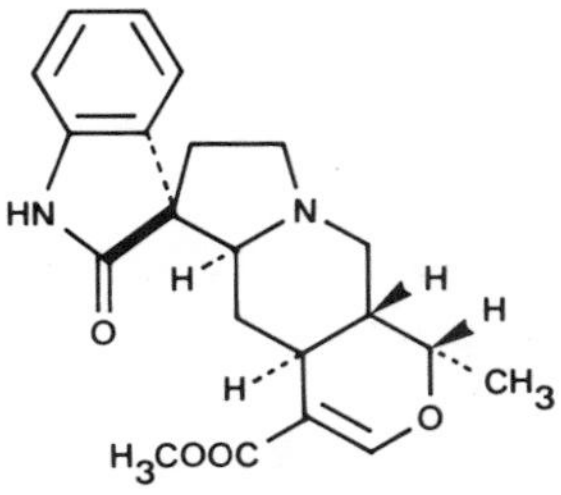

Mitraphylline

Alkaloids: Alkaloid V, caboxine A (iso-c.-A), carapanaubine (c.-*N*-oxide, iso-c.), elegantissine (iso-e.), ericinine, formosanine (iso-f.), gambirdine (iso-g.), majdine (iso-m.), herbaline, herboxine, isocaboxine B, javaphylline, mitraphylline (10,11-dimethoxyiso-m., iso-m.-*N*-oxide, m.-*N*-oxide), rauvoxine, rauvoxinine, speciophylline (s.-*N*-oxide), uncarine F (u.-F-*N*-oxide), vineridine (v.-*N*-oxide), vinerine (*N*-acetyl-v., v.-*N*-oxide), vinerinine, pteropodine (iso-p., iso-p.-*N*-oxide, p.-*N*-oxide).

Type: C5c

Example:

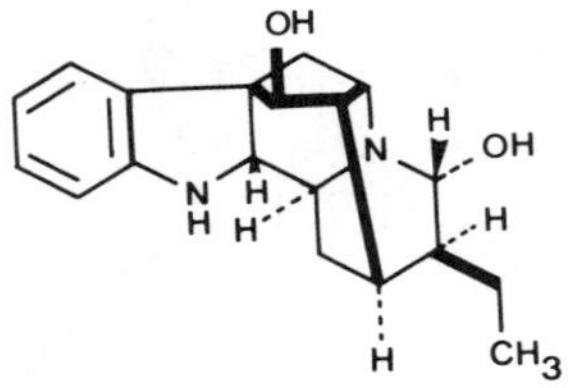

Ajmaline

Alkaloids: Ajmalidine, ajmaline (17-acetyl-a., 1-demethyl-17-*O*-acetyl-21-desoxy-a.-diene(1,19), diacetyl-a., iso-a., 12-methoxy-a., nor-a., 17-*O*-3′, 4′, 5′-trimethoxybenzoyl-a.), alstonisidine°, endolobine, herbadine, herbamine, indolenine RG, majoridine, mauiensine, mitoridine (nor-m.), purpeline ($N_{(a)}$-demethyldihydro-p., $N_{(a)}$-demethyl-p., dihydronor-p., nor-p.), quebrachidine, raucaffricine, rauflorine, rauvomitine ($N_{(a)}$-demethyl-r.), reflexine, sandwicine (iso-s.), seredamine ($N_{(a)}$-demethyl-s., nor-s., 17-*O*-trimethoxybenzoyl-s.), tetraphyllicine (dimethoxy-t., 10-hydroxynor-t., monomethoxy-t., nor-t., trimethoxy-t.), vincamajine (*O*-benzoyl-v., 10-methoxy-v., *O*-3,4,5-trimethoxybenzoylhydroxy-v., *O*-3,4,5-trimethoxybenzoyl-v., *O*-3,4,5-trimethoxycinnamoyl-10-hydroxy-v., *O*-3,4,5-trimethoxycinnamoyl-10-methoxy-v., *O*-3,4,5-trimethoxycinnamoyl-v.), vinçamajoreine, vincamedine, vincarine, vinorine (10-methoxy-v.), vomalidine, vomilenine.

Type: C5d

Example:

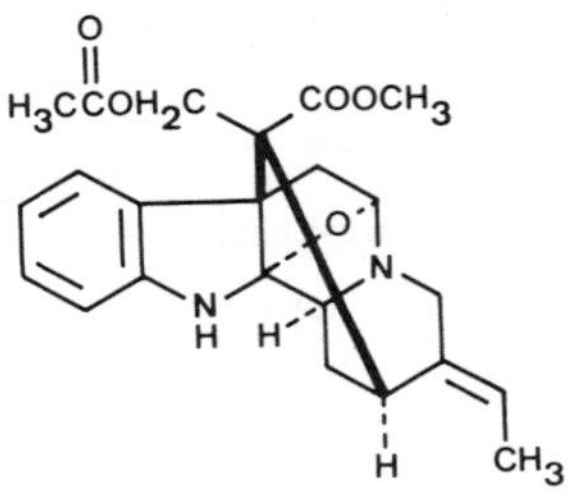

Picraline

Alkaloids: Picralinal, picraline (desacetyl-p.), picralstonine, picrinine, quaternidine, quaternine, vincaricine, vincaridine, vincarinine.

Type: C5e

Example:

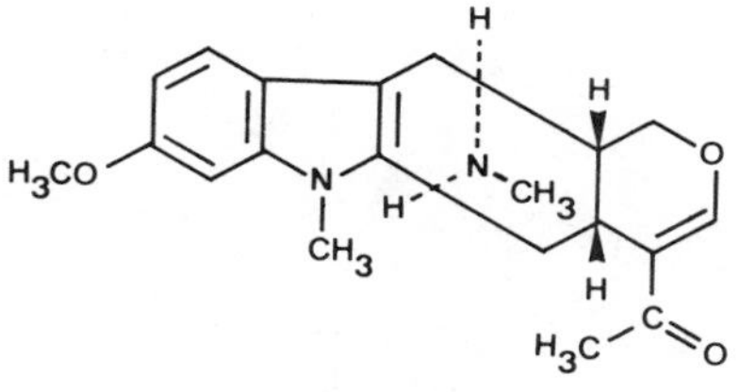

Alstophylline

Alkaloids: Alstonerine, alstonisidine°, alstophylline, macralstonidine°, macralstonine° ($N'_{(a)}$-demethylanhydro-m.), macrocarpamine°, suaveoline, talcarpine, villalstonine°.

Type: C5f

Example:

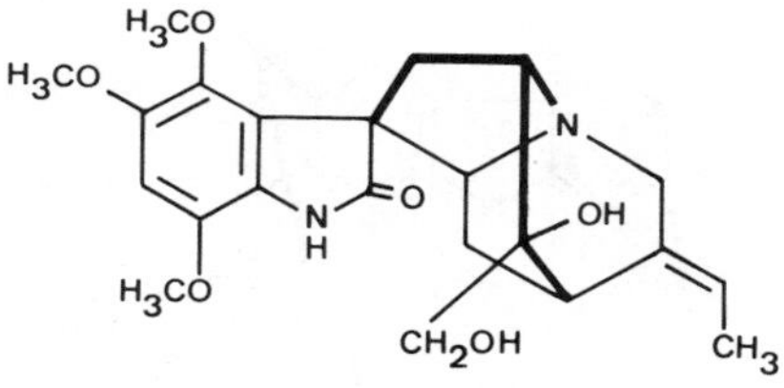

Chitosenine

Alkaloids: Alkaloid I (a.-J, a.-L, a.-M, a.-N), chitosenine, gardmultine°, voachalotineoxindole.

Type: C5g

Example:

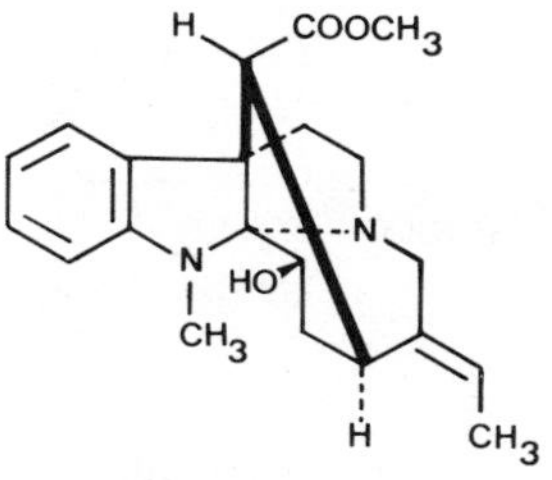

Deformocorymine

Alkaloids: Pleiocorine°, cabuamine (deformo-c.), deformocorymine, echitamine (nor-e.), 3-epi-3,17-dihydrocorymine, vincoridine, vincorine.

Type: **C5h**

Example:

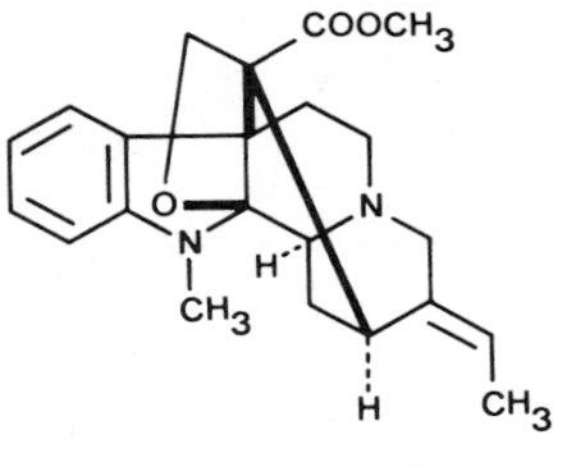

Pseudoakuammigine

Alkaloids: Akuammine (*O*-methyl-a.), pseudoakuammigine.

Type: **C5i**

Example:

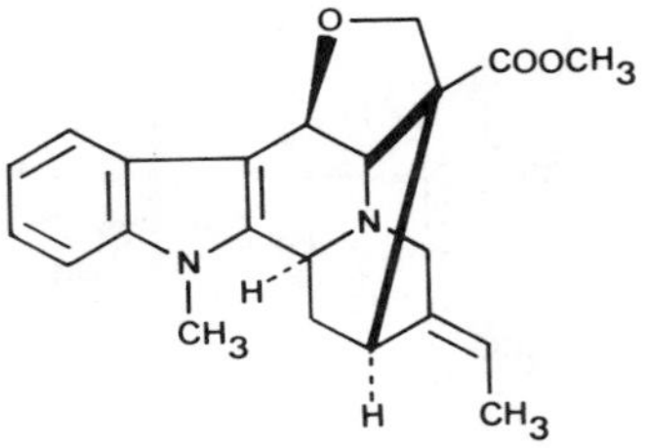

Dehydrovoachalotine

Alkaloids: Dehydrovoachalotine, gardnutine (18-hydroxy-g.).

Type: **C5k**

Example:

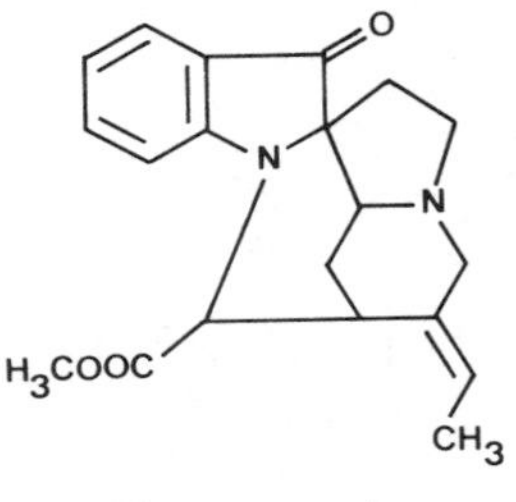

Fluorocarpamine

Alkaloids: Fluorocarpamine, C-fluorocurine.

Type: C5l

Example:

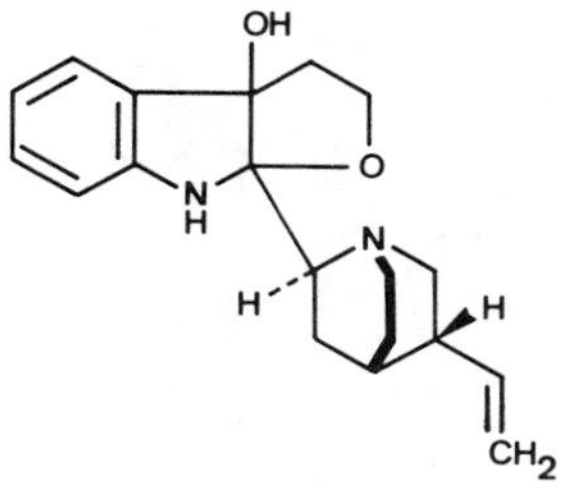

Quinamine

Alkaloids: Conquinamine, quinamine.

Type: C5m

Example:

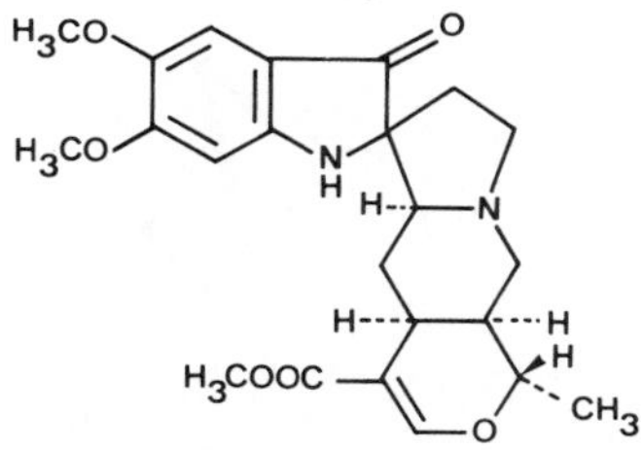

Isoreserpiline-pseudoindoxyl

Alkaloid: Isoreserpiline-pseudoindoxyl.

Type: C5n

Example:

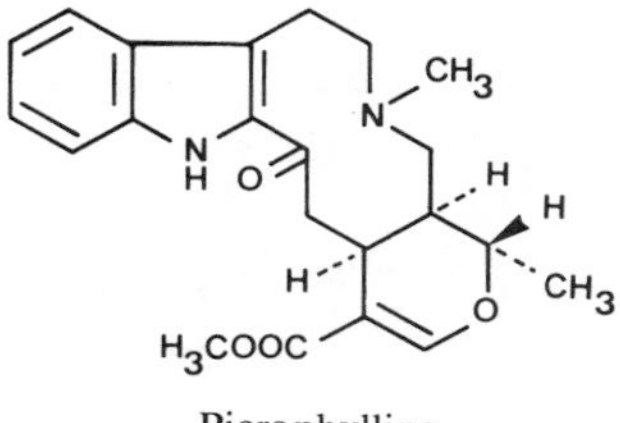

Picraphylline

Alkaloids: Picraphylline (10,11-dimethoxy-p., 20-epi-p.).

Type: C5o

Example:

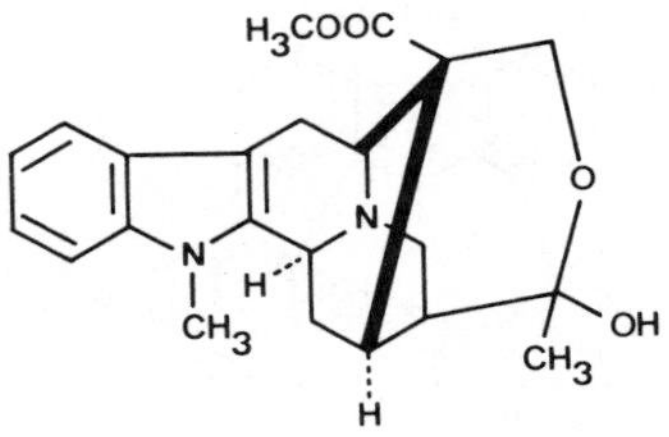

Voacoline

Alkaloids: Talpinine, voacoline.

Type: C5p

Example:

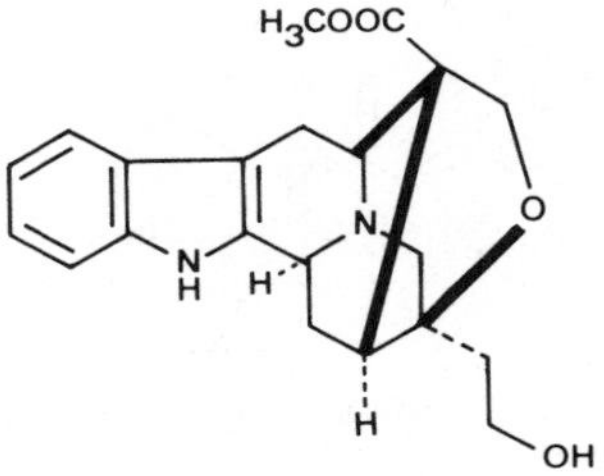

Eburnaphylline

Alkaloid: Eburnaphylline.

Type: C5q

Example:

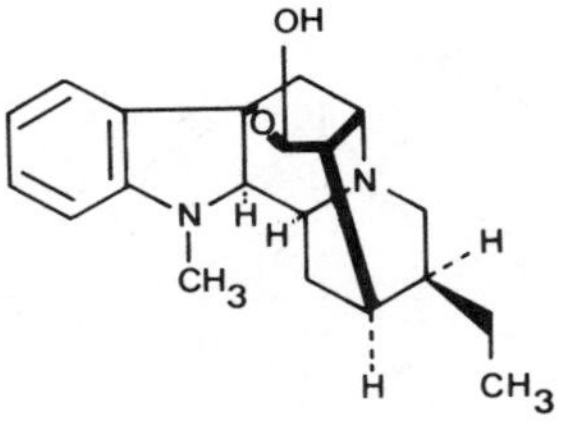

Rauwolfinine

Alkaloid: Rauwolfinine.

Type: C5r

Example:

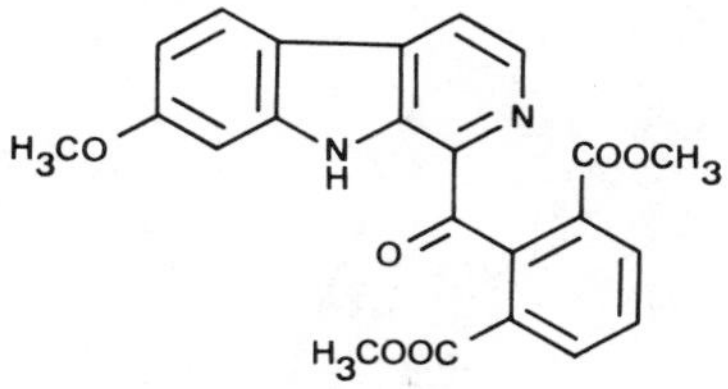

Alstonilidine

Alkaloid: Alstonilidine.

Type: C5s

Example:

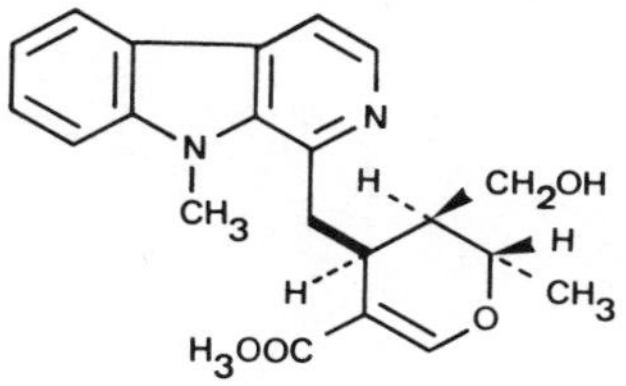

Alstonidine

Alkaloid: Alstonidine.

Type: C5t

Example:

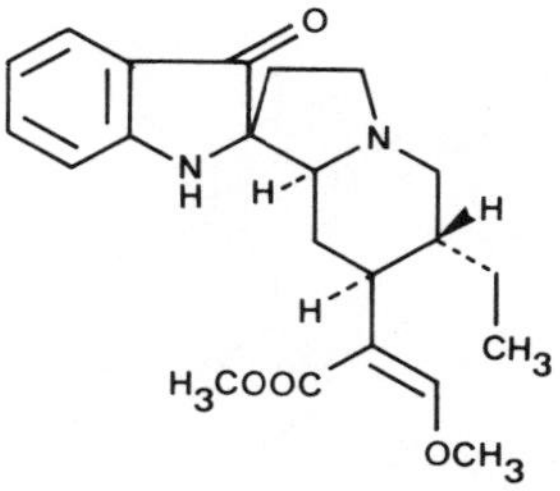

Dihydrocorynantheine-pseudoindoxyl

Alkaloids: Dihydrocorynantheine-pseudoindoxyl.

Type: C5u

Example:

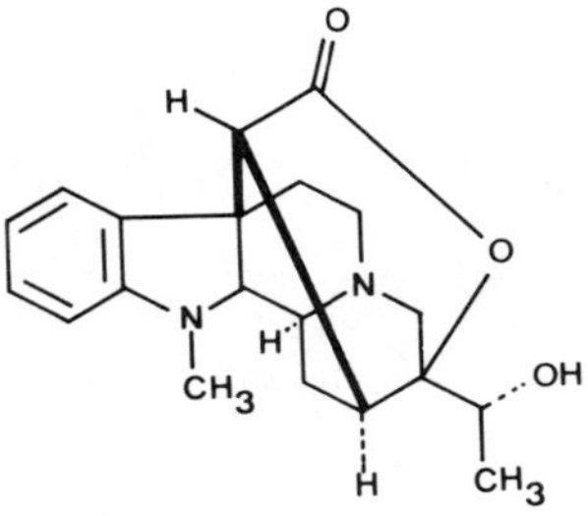

Quaternoline

Alkaloid: Quaternoline.

Type: C5v

Example:

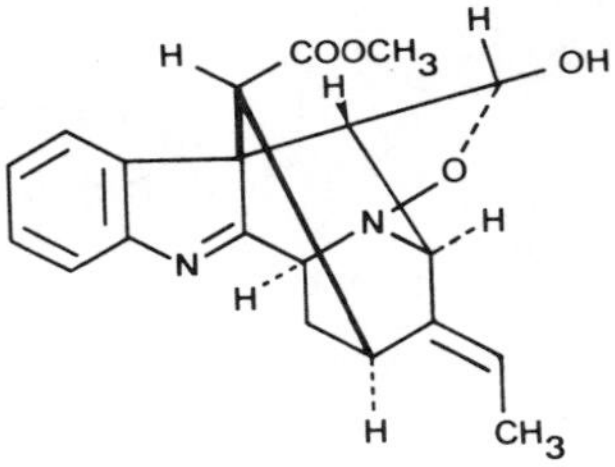

Nareline

Alkaloid: Nareline.

Type: C5y

Example:

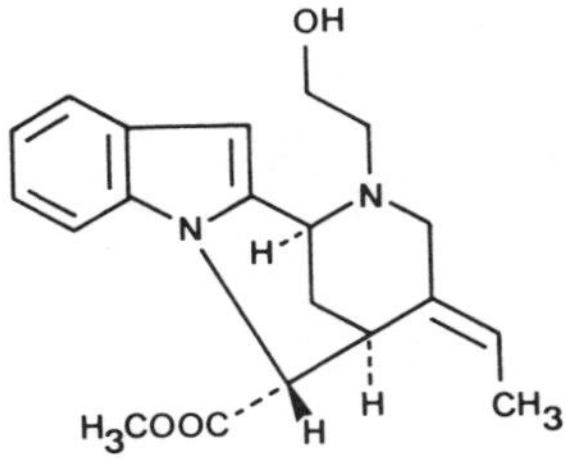

Vinoxine

Alkaloid: Vinoxine.

Type: C6a

Example:

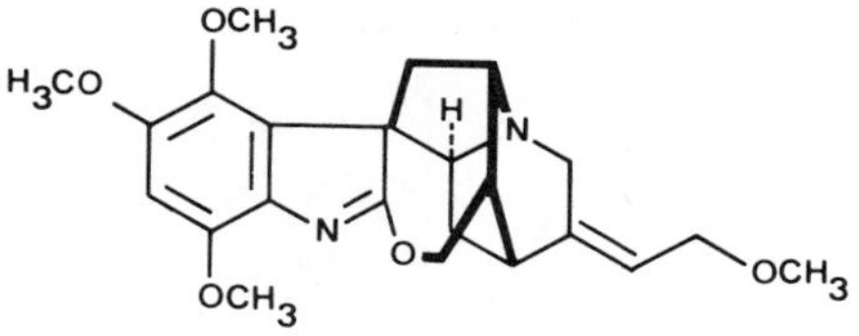

Gardneramine

Alkaloids: Gardfloramine (18-demethox-g.), gardmultine°, gardneramine (18-demethoxy-g., 18-demethyl-g., g.-*N*-oxide).

Type: C6b

Example:

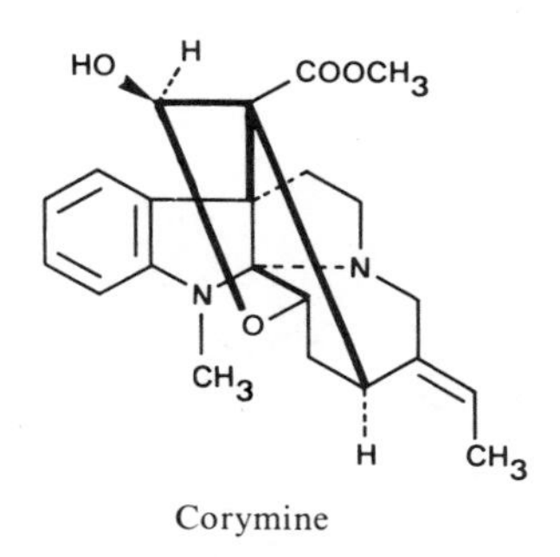

Corymine

Alkaloids: Corymine (acetyl-c.).

Type: C6c

Example:

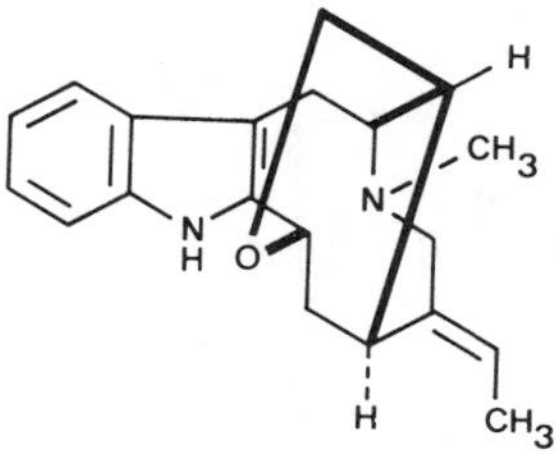

Anhydrovobasinediol

Alkaloid: Anhydrovobasinediol.

Type: **C6d**

Example:

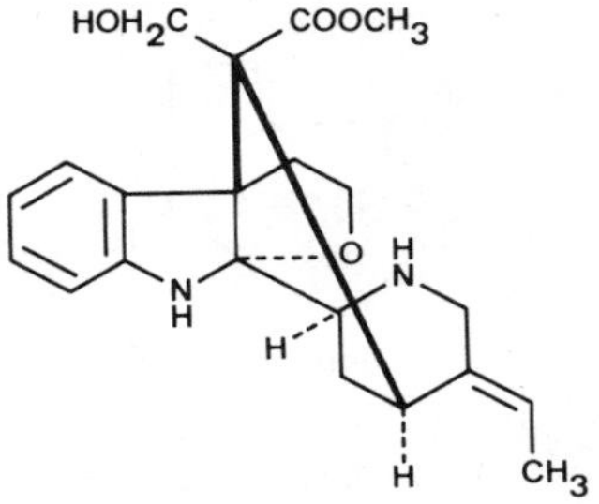

Aspidodasycarpine

Alkaloids: Aspidodasycarpine, lanciferine (10-hydroxy-l., 10-methoxy-l.), lonicerine.

Type: **C6e**

Example:

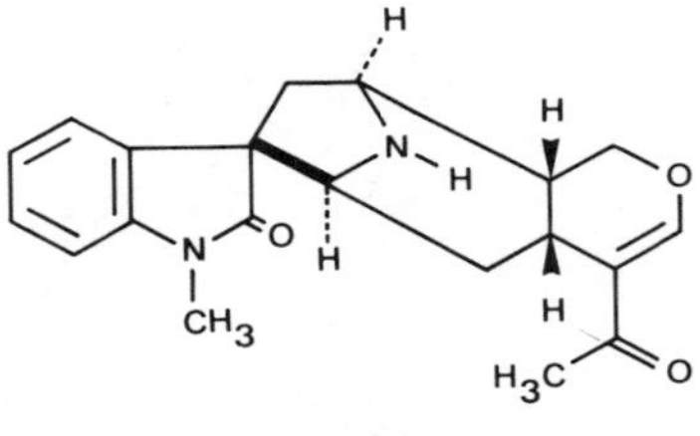

Alstonisine

Alkaloid: Alstonisine.

Type: **C7a**

Example:

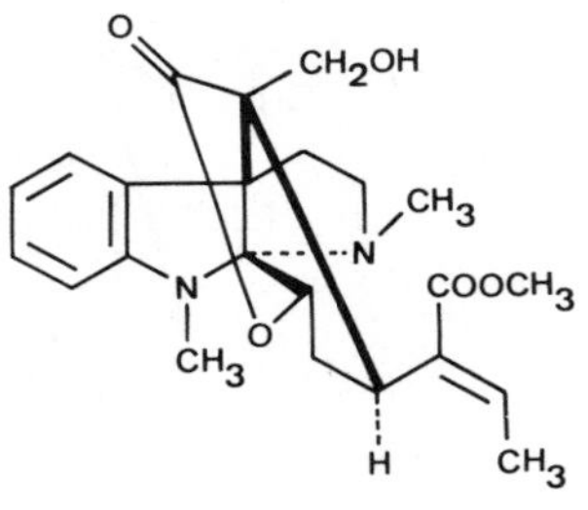

Eripine

Alkaloid: Eripine.

Type: C7b

Example:

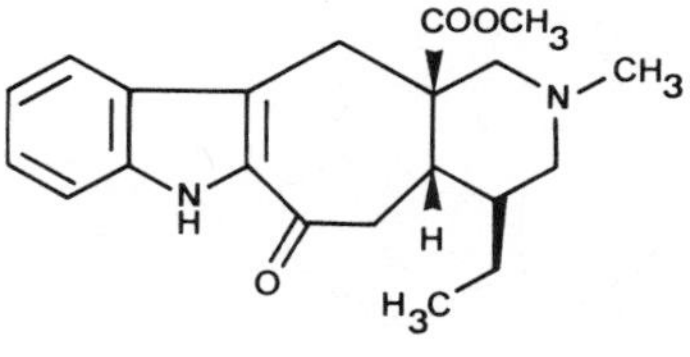

Ervatamine

Alkaloids: Ervatamine (19-dehydro-e., 20-epi-e.), methuenine.

Type: C8a

Example:

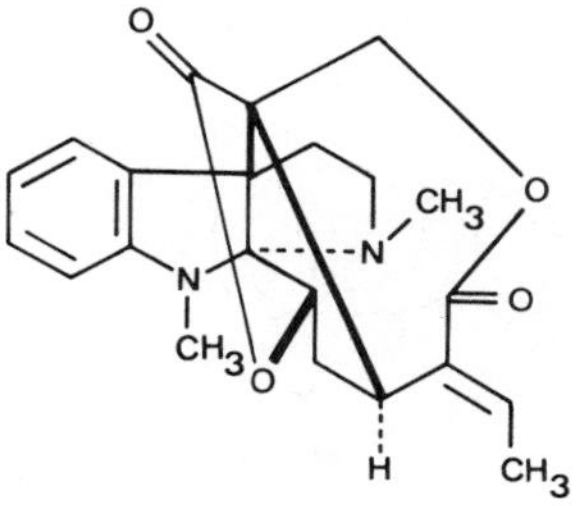

Erinine

Alkaloids: Erinicine, erinine, isocorymine.

Type: C8b

Example:

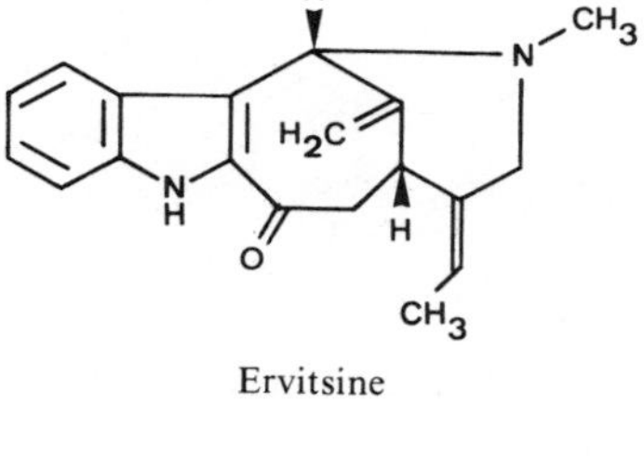

Ervitsine

Alkaloid: Ervitsine.

APPENDIX 2. VINCOSAN (D) TYPE : EXAMPLES OF STRUCTURES AND THE LIST OF ALL ALKALOIDS DETECTED IN NATURE

Type: D3a

Example:

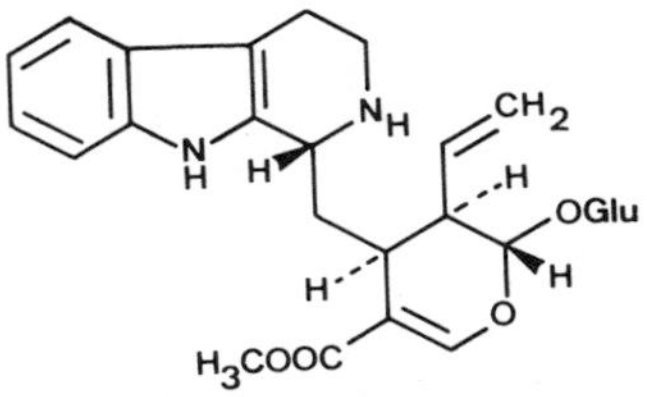

Vincoside

Alkaloids: Cordifoline (10-desoxy-c.), dolichantoside, lyaloside, palinine, pauridianthoside, strictosidine (5α-carboxy-s., 5-oxo-s.), vincoside.

Type: D3b

Example:

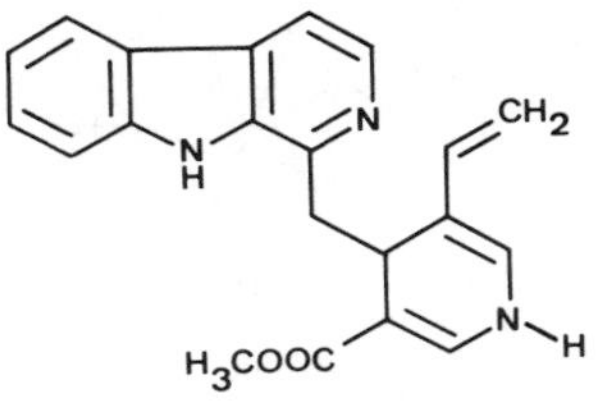

Lyadine

Alkaloids: Lyalidine (19-hydroxy-l.), lyadine, lyaline, pauridianthine, pauridianthinol.

Type: D4a

Example:

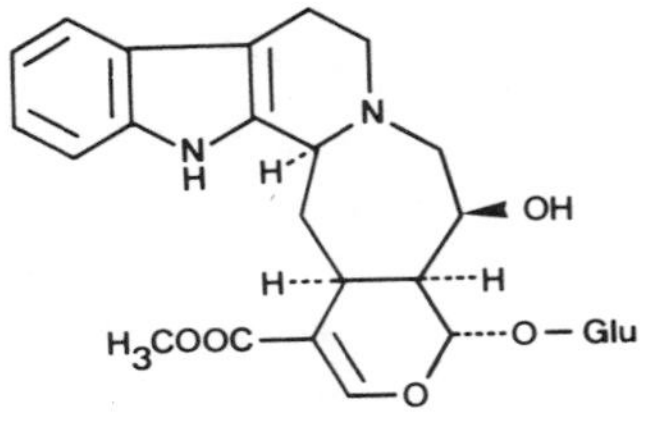

3α-Dihydrocadambine

Alkaloids: Decarbomethoxynauclechine, 3α-dihydrocadambine (3β-dihydro-c.), macrolidine, naufoline.

Type: D4b

Example:

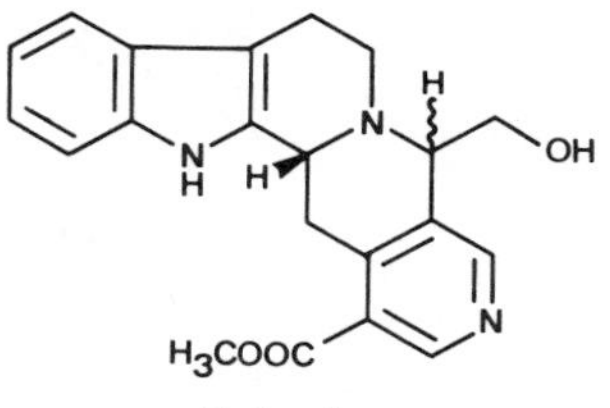

Cadamine

Alkaloids: Cadamine (iso-c.), isodihydrocadambine, 3β-isodihydrocadambine.

Type: D4c

Example:

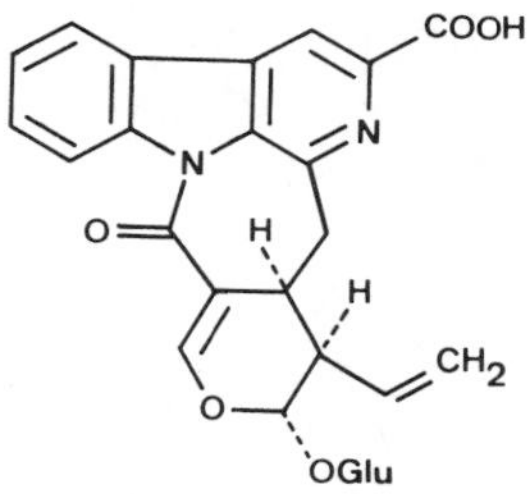

Desoxycordifolinelactam

Alkaloid: Desoxycordifolinelactam.

Type: D4d

Example:

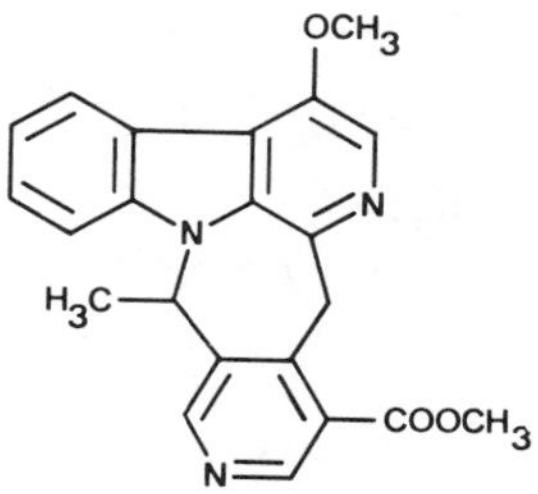

Camptoneurine

Alkaloid: Camptoneurine.

Type: D4e

Example:

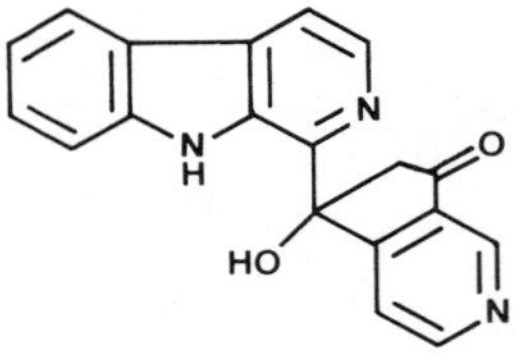

Pauridianthinine

Alkaloid: Pauridianthinine.

Type: D4f

Example:

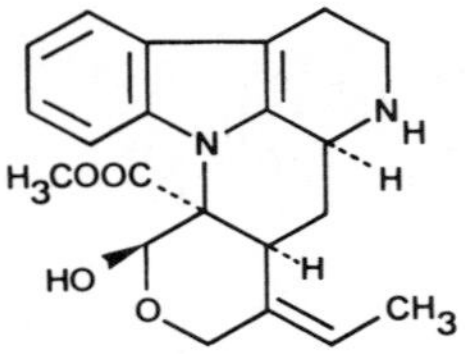

Talbotine

Alkaloids: Talbotine (3,4-dehydro-t., 3,4,5,6-tetradehydro-t.).

Type: D4g

Example:

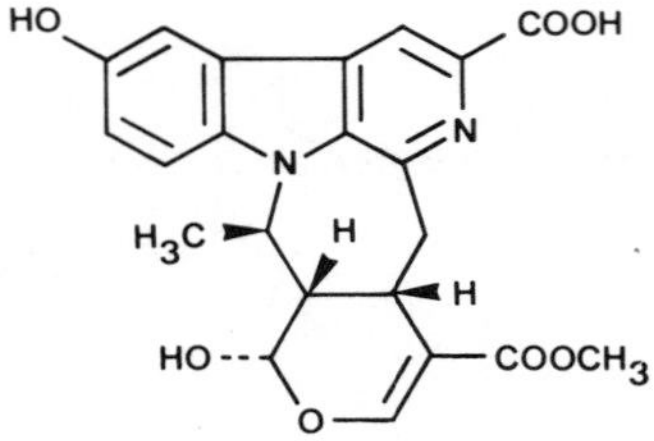

Adifoline

Alkaloids: Adifoline (10-desoxy-a.).

Type: D5a

Example:

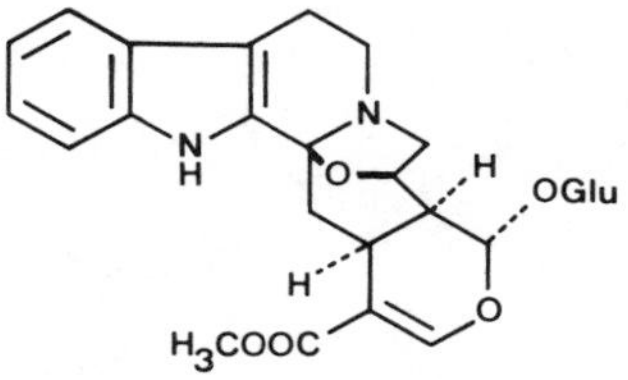

Cadambine

Alkaloid: Cadambine.

Type: D5b

Example:

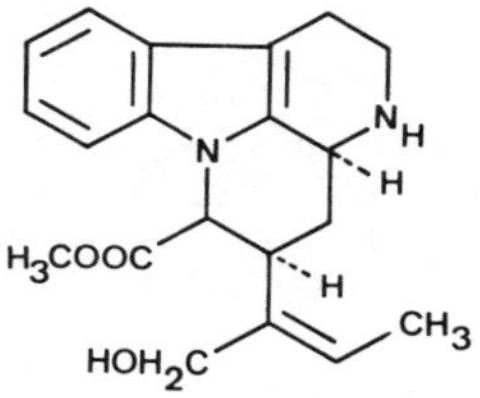

Deformyltalbotinic acid methyl ester

Alkaloid: Deformyltalbotinic acid methyl ester.

Type: D5c

Example:

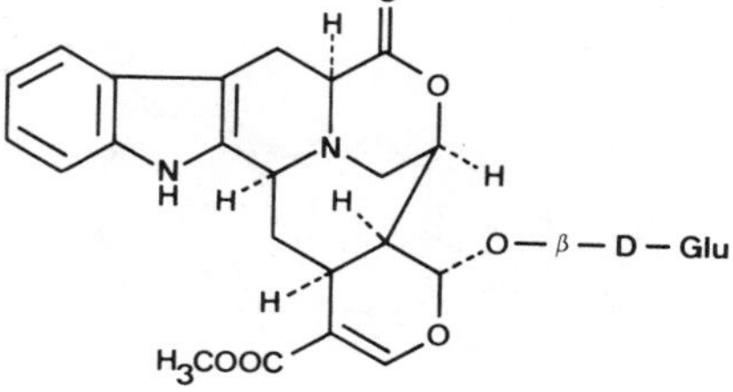

Rubenine

Alkaloid: Rubenine.

Type: D6a

Example:

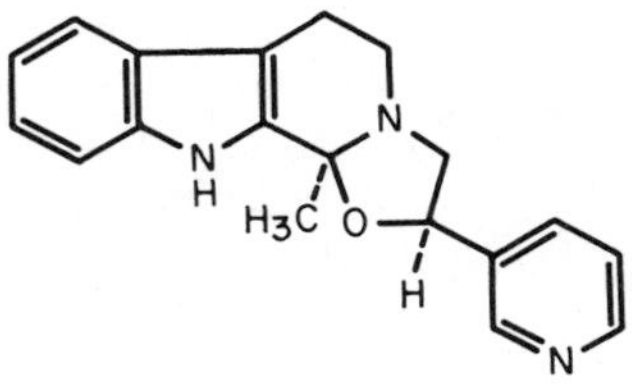

α-Naucleonine

Alkaloids: α-Naucleonidine, β-naucleonidine, α-naucleonine, β-naucleonine.

Type: D6b

Example:

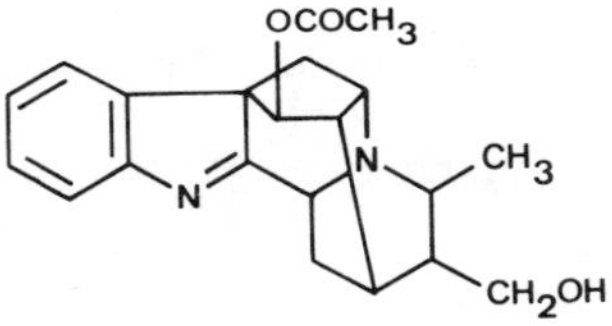

Perakine

Alkaloids: Perakine, raucaffrinoline.

Type: D6c

Example:

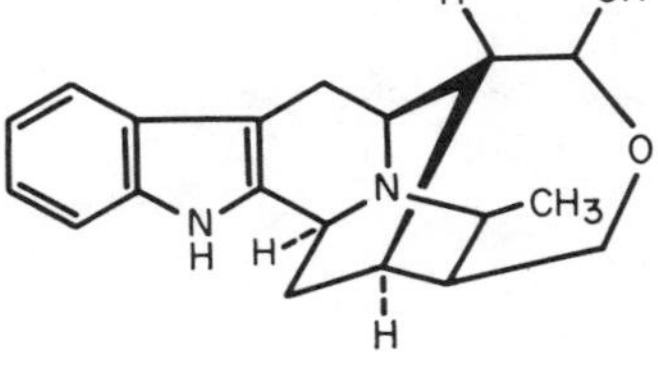

Peraksine

Alkaloids: Macrosalhine, peraksine.

APPENDIX 3. VALLESIACHOTAMAN (V) TYPE : EXAMPLES OF STRUCTURES AND THE LIST OF ALL ALKALOIDS DETECTED IN NATURE

Type: V3

Example:

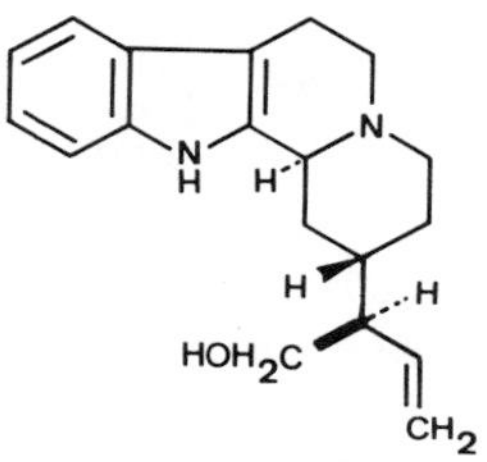

Antirhine

Alkaloids: Antirhine (a.-methochloride, 18,19-dihydro-a.-β-methochloride), hunterburnine-α-methochloride, hunterburnine-β-methochloride, vallesiachotamine.

Type: V4

Example:

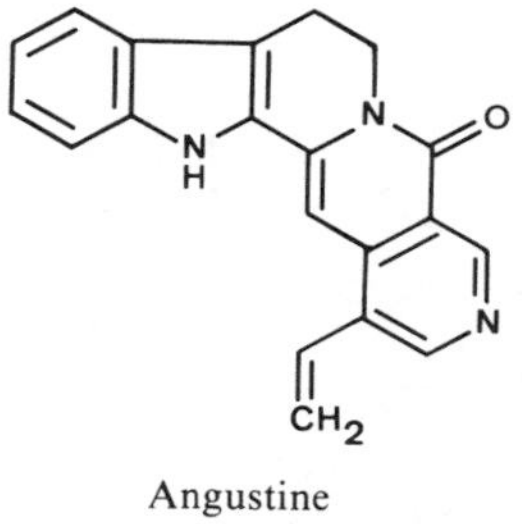

Angustine

Alkaloids: Angustidine, angustine, angustoline, glucoalkaloid, 10-β-D-glucosyloxyvincosidelactam, nauclefine, naucletine, parvine, rubescine, strictosamide, strictosidinelactam, 3α,5α-tetrahydrodesoxycordifolinelactam, 3β, 5α-tetrahydrodesoxycordifolinelactam.

APPENDIX 4. STRYCHNAN (S) TYPE : EXAMPLES OF STRUCTURES AND THE LIST OF ALL ALKALOIDS DETECTED IN NATURE

Type: S4

Example:

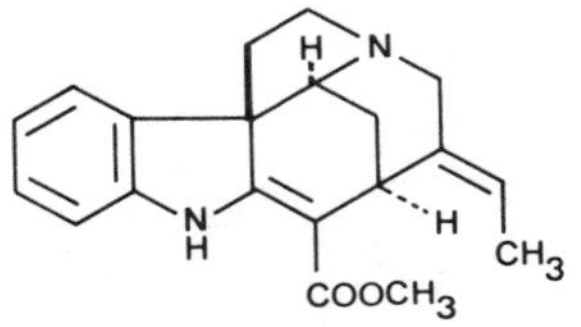

Akuammicine

Alkaloids: *N*-Acetyl-2,16-dihydro-2β-16α-akuammicinal, afrocurarine°, akuammicine (a.-methochloride, a.-*N*-oxide, 19,20-dihydro-a., 2,16-dihydroakuammicine, 18- or 19-hydroxy-19,20-dihydro-a., 11-methoxya., $N_{(a)}$-methyl-2,16-dihydro-a., pre-a., pseudo-a.), C-alkaloid-A° (a.-D°, bisnor-a.-D°, a.-E°, a.-F°, a.-G°, a.-H°, bisnor-a.-H°), alstovine, angustimycine, C-calebassine°, compactinervine, C-curarine I° (bisnor-C-c.°), 18-desoxy-Wieland-Gumlich aldehyde, C-dihydrotoxiferine° (bisnor-d.°, bisnor-d.-di-*N*-oxide°, bisnor-d.-*N*-oxide°, nor-d.°), echitamidine, C-fluorocurarine (nor-f.), geissoschizoline, lochneridine (20-epi-l.), mossambine, retuline ($N_{(a)}$-desacetyl-17-*O*-acetyl-18-hydroxy-iso-r., $N_{(a)}$-desacetyl-18-hydroxy-iso-r., $N_{(a)}$-desacetyl-iso-r., desacetyl-r., 18-hydroxy-iso-r., iso-r., r.-*N*-oxide), sewarine, strychnobiline° (12′-hydroxy-iso-s.°, iso-s.°), C-toxiferine I°, tsilanimbine, tubifolidine, tubifoline, vincanicine, vincanidine, vinervine, vinervinine.

Type: S5a

Example:

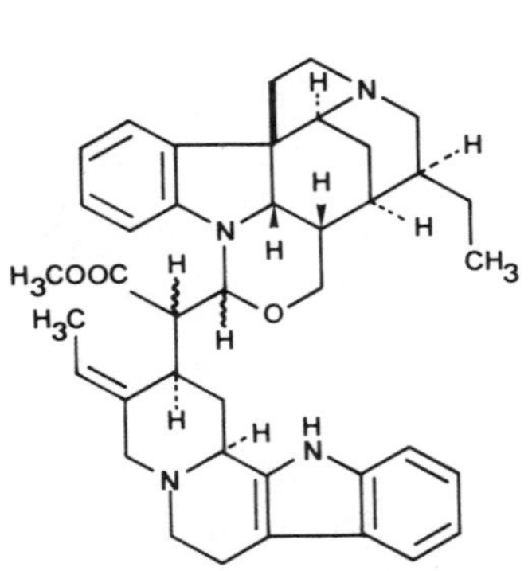

Geissospermine

Alkaloids: Geissospermine°, geissolosimine°, strychnobiline° (12′-hydroxy-iso-s.°, iso-s.°).

Type: S5b

Example:

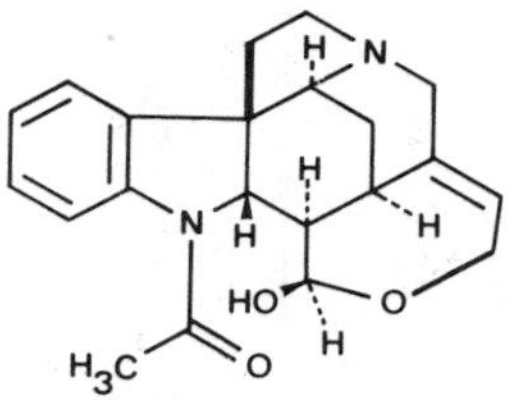

Diaboline

Alkaloids: Caracurine II° (c.-II methosalt°), curacurine V° (c.-V-*N*-oxide°, C-V di-*N*-oxide°), diaboline (*O*-acetyl-d., 2,16-dehydro-d., 11-methoxy-2,16-dehydro-d., 11-methoxy-d.), henningsoline (*O*-acetyl-h.), jobertine, 11-methoxy-henningsamine, Wieland-Gumlich aldehyde ($N_{(b)}$-W.-chlormethylate).

Type: S5c

Example:

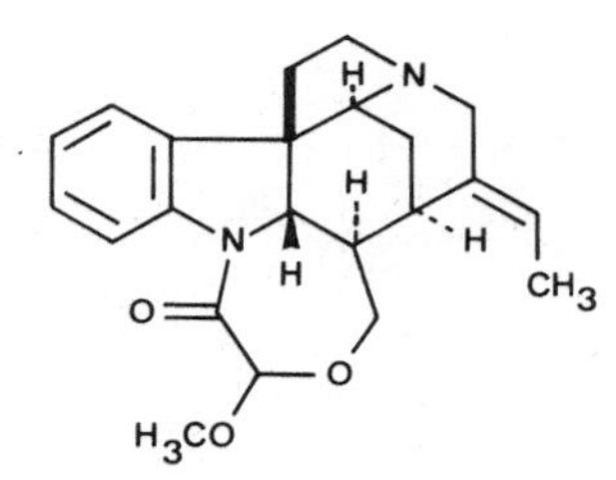

Tsilanine

Alkaloids: Tsilanine (*O*-demethyl-t., 10-methoxy-*O*-demethyl-t., 10-methoxy-t.).

Type: S5d

Example:

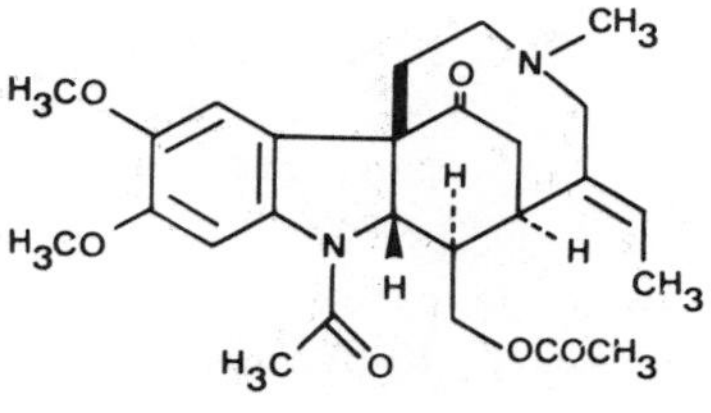

Strychnosilidine

Alkaloids: Strychnosilidine, strychnosiline, tabascanine.

Type: **S5e**

Example:

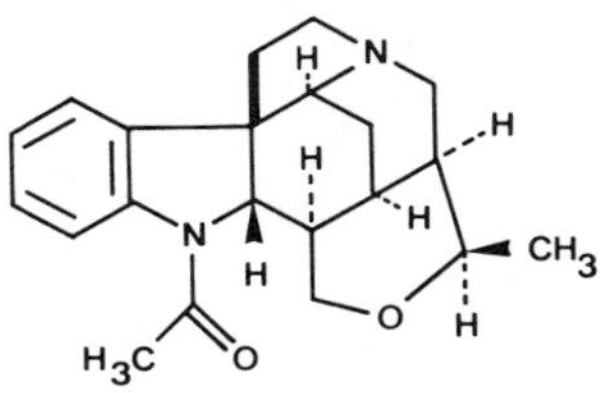

Spermostrychnine

Alkaloids: Spermostrychnine (12-hydroxy-11-methoxy-s.), splendoline (iso-s.), strychnosplendine (*N*-acetyl-iso-s., $N_{(a)}$-acetyl-s., 12-hydroxy-11-methoxy-$N_{(a)}$-acetyl-s., iso-s.), strychnospermine (desacetyl-s.).

Type: **S5f**

Example:

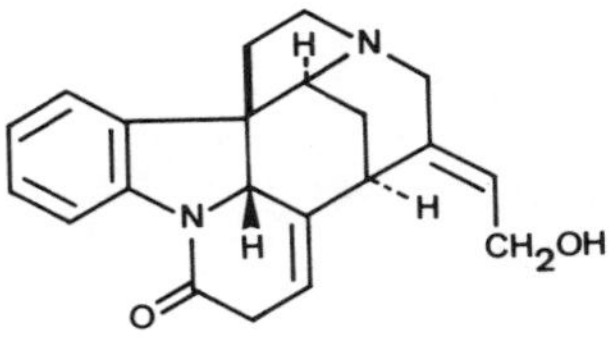

Isostrychnine

Alkaloid: Isostrychnine.

Type: **S6a**

Example:

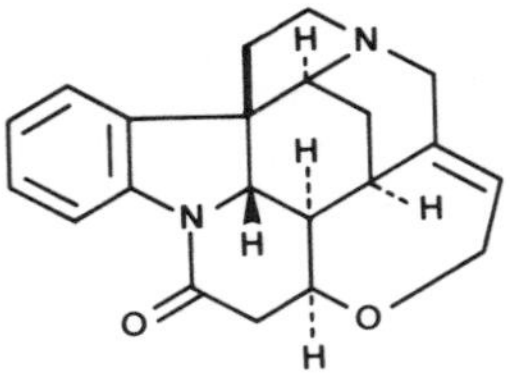

Strychnine

Alkaloids: Brucine (b.-*N*-oxide; pseudo-b.), α-colubrine, β-colubrine, strychnine (12-hydroxy-11-methoxy-s., 12-hydroxy-s., 15-hydroxy-s., 3-methoxy-s., pseudo-s., s.-*N*-oxide).

Type: S6b

Example:

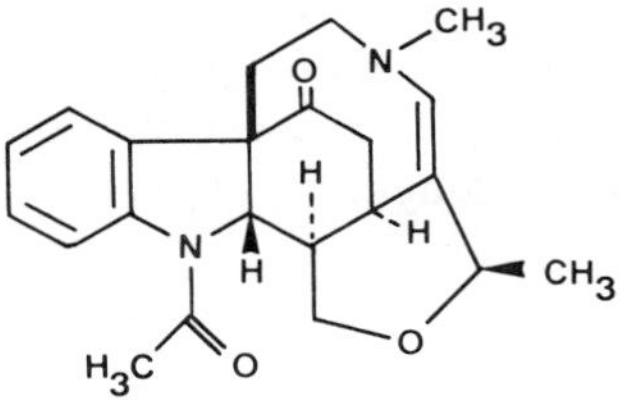

Strychnobrasiline

Alkaloids: $N_{(a)}$-Acetyl-*O*-methylstrychnosplendine, isosplendine, strychnobrasiline (10,11-dimethoxy-s., 12-hydroxy-11-methoxy-s., 10-methoxy-s.), strychnofendlerine (12-hydroxy-11-methoxy-s.).

Type: S6c

Example:

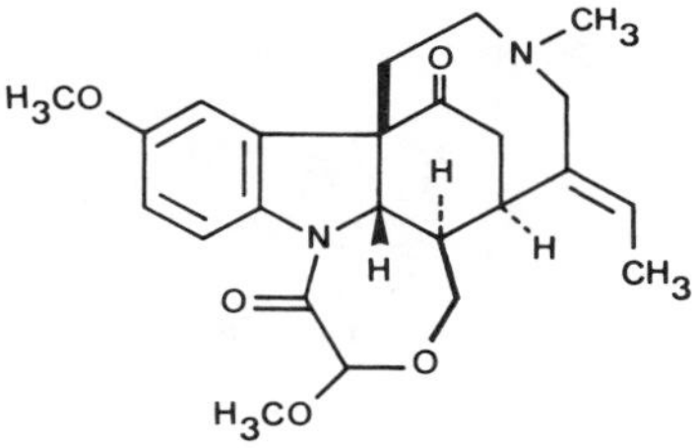

Rindline

Alkaloids: Holstiine, holstiline, rindline.

Type: S7

Example:

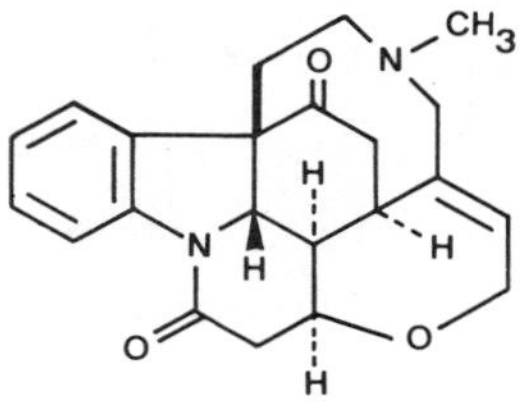

Icajine

Alkaloids: *N*-Cyano-*sec*-pseudobrucine, *N*-Cyano-*sec*-pseudocolubrine, *N*-cyano-*sec*-pseudostrychnine, 12-hydroxy-11-methoxypseudostrychnine, icajine

(15-hydroxy-i., i.-*N*-oxide), *N*-methyl-*sec*-pseudobrucine, *N*-methyl-*sec*-pseudostrychnine (19,20-α-epoxy-11, 12-dimethoxy-*N*-m., 19,20-α-epoxy-10, 11-dimethoxy-*N*-m., 19, 20-α-epoxy-15-hydroxy-*N*-m., 19, 20-α-epoxy-15-hydroxy-12-methoxy-*N*-m., 19, 20-α-epoxy-12-methoxy-*N*-m., 12-hydroxy-*N*-m., 10- or 11-methoxy-*N*-m.), novacine (19, 20-α-epoxy-n., 15-hydroxy-n.), pseudo-α-colubrine, pseudo-β-colubrine (19, 20-α-epoxy-*N*-methyl-*sec*-p., *N*-methyl-*sec*-p.), vomicine (19, 20-α-epoxy-15-hydroxy-v., 19, 20-α-epoxy-11-methoxy-v., 19, 20-α-epoxy-v.).

APPENDIX 5. ASPIDOSPERMATAN (A) TYPE : EXAMPLES OF STRUCTURES AND THE LIST OF ALL ALKALOIDS DETECTED IN NATURE

Type: A3

Example:

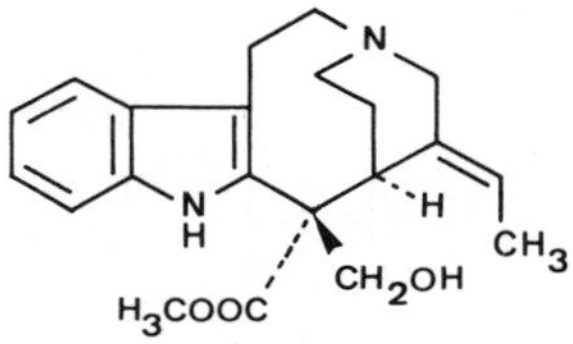

Stemmadenine

Alkaloids: Stemmadenine (deformo-s.).

Type: A4a

Example:

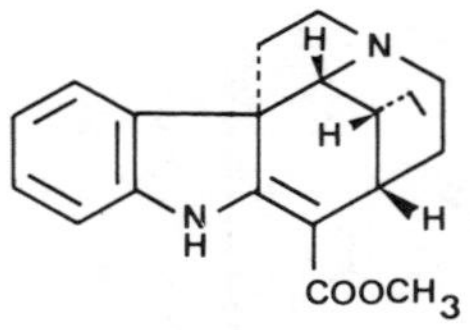

Tubotaiwine

Alkaloids: Aspidospermatidine ($N_{(a)}$-acetyl-a., 11,12-dihydroxy-*N*-acetyl-a., $N_{(a)}$-methyl-a.), aspidospermatine (*N*-acetyl-11-hydroxy-a., desacetyl-a., 19, 20-dihydro-a.), condylocarpine (19, 20-dihydro-c., 11-methoxy-19, 20-dihydro-c.), limatine (11-methoxy-l.), limatinine (11-methoxy-l.), tubotaiwine (t.-*N*-oxide).

Type: **A4b**

Example:

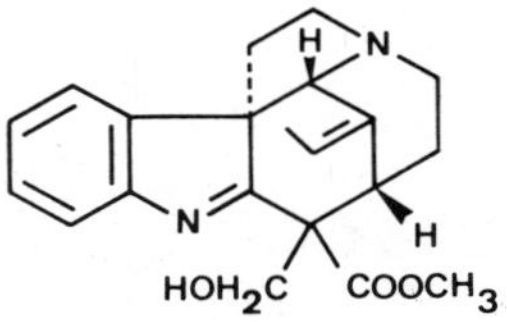

Precondylocarpine

Alkaloid: Precondylocarpine.

Type: **A5a**

Example:

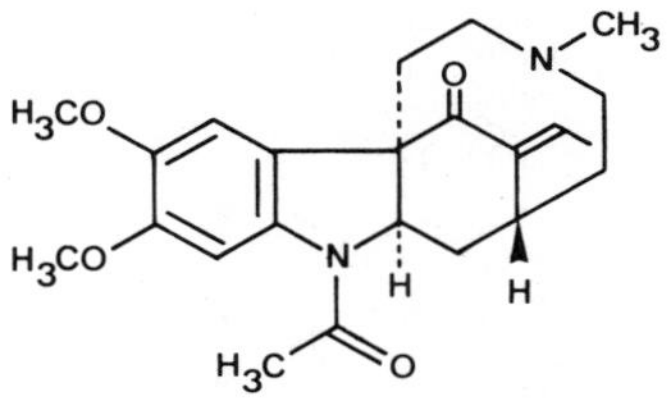

Geissovelline

Alkaloid: Geissovelline.

Type: **A5b**

Example:

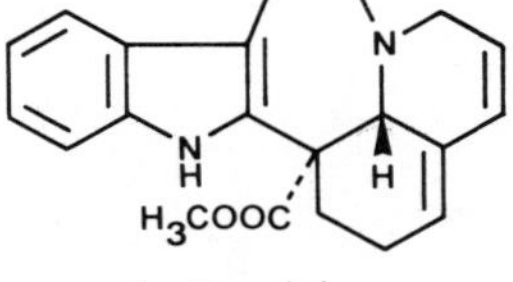

Andranginine

Alkaloid: Andranginine.

Type: A6

Example:

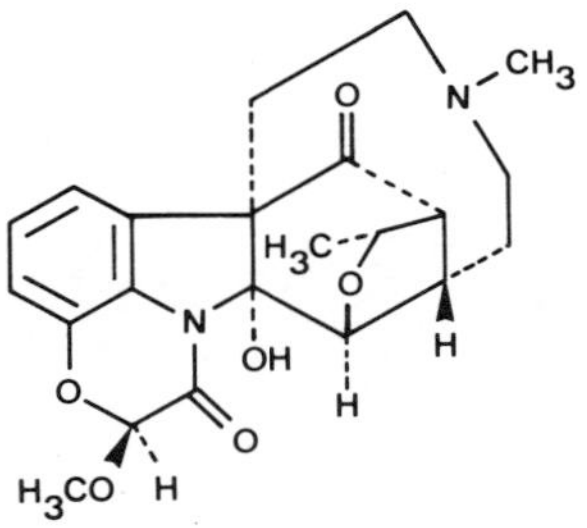

Dichotine

Alkaloids: Dichotine (11-methoxy-d.).

APPENDIX 6. EBURNAN (E) TYPE : EXAMPLES OF STRUCTURES AND THE LIST OF ALL ALKALOIDS DETECTED IN NATURE

Type: E5a

Example:

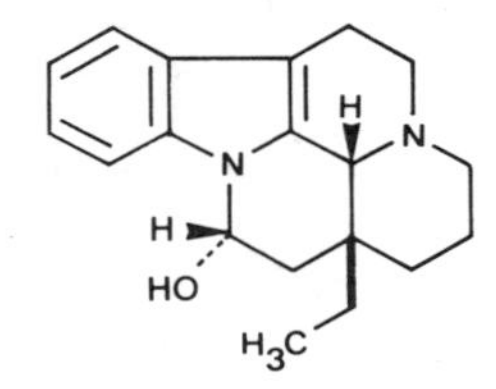

Eburnamine

Alkaloids: Eburnamenine, eburnamine (14,15-dehydro-e., iso-e., *O*-methyl-e.), eburnamonine (11-methoxy-e.), paucivenine°, pleiomutine°, umbellamine°, vincamine (apo-14,15-dehydro-v., apo-v., 14,15-dehydro-v., 16-epi-14,15-dehydro-v., 16-epi-v., 12-methoxy-14,15-dehydro-v.), vincaminine, vincine (14,15-dehydro-epi-v., 14,15-dehydro-v., 16-epi-14,15-dehydro-v.), vincinine.

Type: E5b

Example:

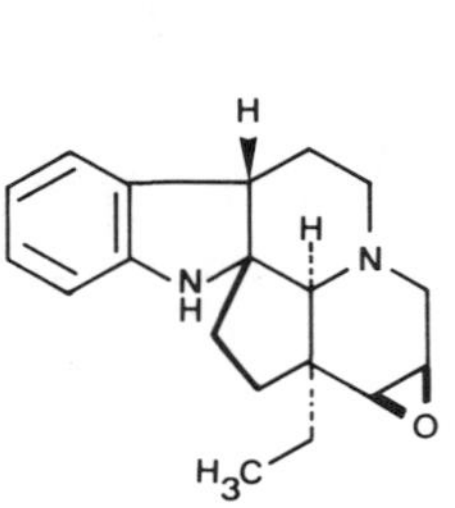

Andrangine

Alkaloids: Andrangine, criophylline°, schizophylline, vallesamidine.

Type: E6a

Example:

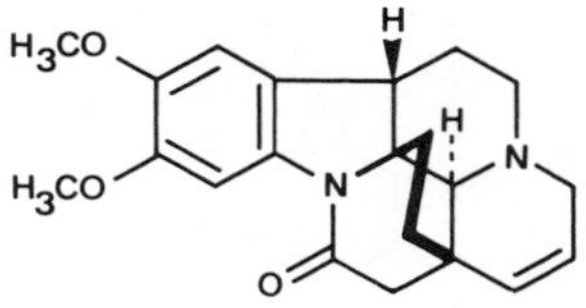

Schizogamine

Alkaloids: Schizogaline (iso-s.), schizogamine (iso-s.), schizozygine, α-schizozygol.

Type: E6b

Example:

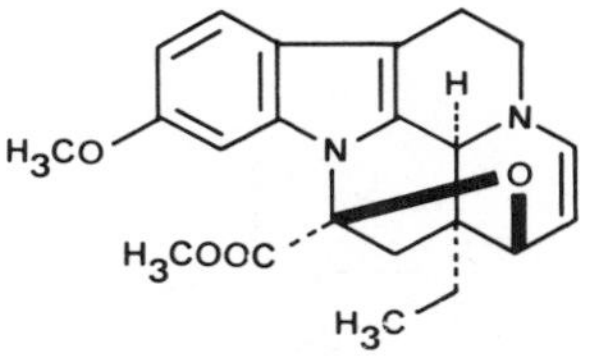

Craspidospermine

Alkaloids: Craspidospermine, criocerine, vincarodine.

Type: E6c

Example:

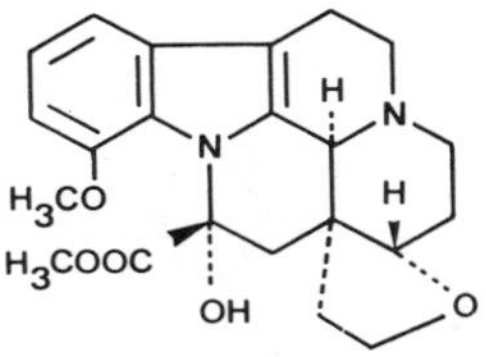

Cuanzine

Alkaloids: Cuanzine, decarbomethoxyapocuanzine.

APPENDIX 7. PLUMERAN (P) TYPE : EXAMPLES OF STRUCTURES AND THE LIST OF ALL ALKALOIDS DETECTED IN NATURE

Type: P3

Example:

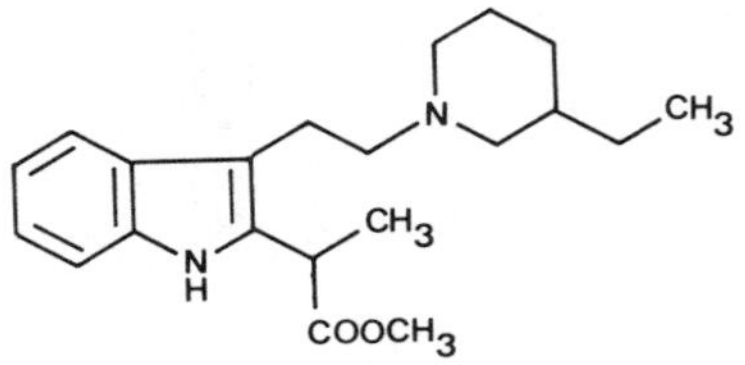

Tetrahydrosecodine

Alkaloids: Decarbomethoxytetrahydrosecodine (16,17-dihydro-s., 16,17-dihydro-x.17-ole), presecamine° (dihydro-p.°, 15,20,15′,20′-tetrahydro-p.°), secamine° (decarbomethoxytetrahydro-s.°, dihydro-s.°, 15,20,15′,20′-tetrahydro-s.°), tetrahydrosecodine (tetrahydro-s.-17-ole).

Type: P4

Example:

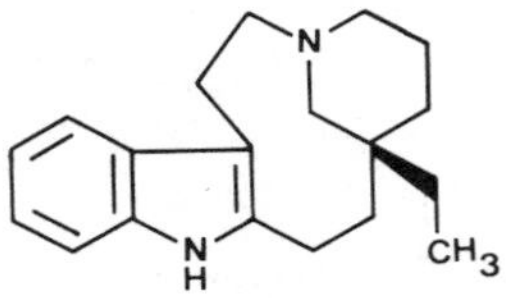

Quebrachamine

Alkaloids: Conoflorine (c.-diole, c.-7-hydroxyindolenine, 12-methoxy-c.), ervinidine, quebrachamine (12-methoxy-qu., *N*-methyl-qu.), rhazidigenine (r.-*N*-oxide), vincadine (14,15-dehydro-epi-v., 14,15-dehydro-v., epi-v.), vincaminoreine, vicaminoridine, vincaminorine.

Type: P5

Example:

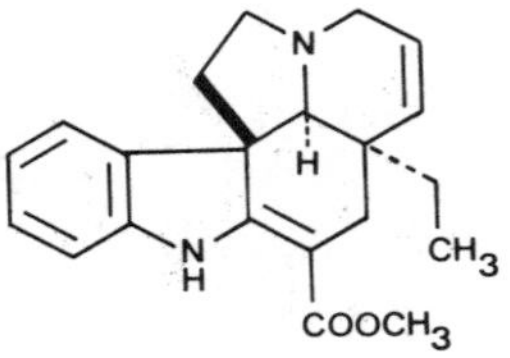

Tabersonine

Alkaloids: *N*-Acetyl-*N*-depropionylaspidoalbinole, aspidocarpine (*O*-demethyl-a.), aspidodispermine (desoxy-a.), aspidolimidinol, aspidolimine, aspidospermidine (*N*-acetyl-a., 16-carbomethoxy-16-hydroxy-14, 15-epoxy-3-oxo-1,2-dehydro-a., 1,2-dehydro-a., 20,21-epi-a., 20,21-epi-a.-*N*-oxide, *N*-formyl-a., $N_{(a)}$-methyl-a.), aspidospermine (demethoxy-a., demethyl-a., desacetyl-a., *N*-methyldesacetyl-a.), baloxine, cathaphylline, catharine°, catharinine°, catharosine, cathovalinine, criophylline°, cylindrocarine (*N*-benzoyl-c., 12-demethoxy-*N*-acetyl-c., *N*-formyl-c., 19-hydroxy-*N*-acetyl-c., 19-hydroxy-*N*-benzoyl-c., 19-hydroxy-*N*-cinnamoyl-c., 19-hydroxy-c., 19-hydroxy-*N*-dihydrocinnamoyl-c., 19-hydroxy-*N*-formyl-c., methyl-c.), cylindrocarpidine (homo-c., 5-oxo-c.), cylindrocarpine (*N*-acetyl-c.-ole, *N*-formyl-c.-ole), 11-demethylaspidolimidine, dimer III°, eburcine, eburenine, eburine, echitoserpidine, echitoserpine, echitovenaldine, echitovenidine, echitovenine, ervafolidine° (iso-e.°), ervafoline°, ervamine, ervincinine, ervinidinine, fendlispermine, folicangine°, hazuntine, hazuntinine, hoerhammericine, hoerhammerinine, leurocolombine°, leurocristine°, leurosidine°, leurosine° (iso-l.°), limapodine (11-methoxy-l.), limaspermine (11-methoxy-l.), lochnericine, lochnerinine, meloceline, melocelinine, minovincine (11-methoxy-m., 3-oxo-m., 5-oxo-m.), minovincinine (14,15-dehydro-19-epi-m., 19-epi-m., 11-methoxy-m.), minovine, palosine (demethoxy-p., *O*-demethyl-p.), paucivenine°, pleurosine°, pseudovincaleukoblastinediol°, pycnanthinine°, pyrifolidine (1,2-dehydrodesacetyl-p., desacetyl-p.), scandomeline (epi-s.), scandomelonine (epi-s.), spegazzinidine, spegazzinine, tabernamine°, tabersonine (19-epi-19-hydroxy-t., 11-hydroxy-t., 19-hydroxy-t., 19*R*-hydroxy-t., 19*S*-hydroxy-t., 11-methoxy-t., 3-oxo-t., t.-$N_{(b)}$-oxide), vallesine, vinblastine° (desacetoxy-v.°), vincadifformine (14, 15-epoxy-3-oxo-v., 11-methoxy-v.), vincathicine°, vincovalicine°, vincovaline°, vincovalinine°, vindoline (demethoxy-v., desacetyl-v.), vindorosine, voafolidine° (iso-v.°), voafoline° (iso-v.°).

Type: P6a

Example:

Venalstonine

Alkaloids: Alkaloid Ld 65 (a. Ld 85), aspidofiline, aspidofractine, aspidofractinine (11,12-dimethoxy-a., *N*-formyl-11,12-dimethoxy-a., *N*-formyl-12-methoxy-a., 12-methoxy-a.), kopsaporine, kopsingine, kopsinic acid methochloride, kopsinilam, kopsinine (14, 15-dehydro-5-oxo-k., epoxy-k., 15-hydroxy-k., 19-hydroxy-k., k.-*N*-oxide), maidinine, pleiocarpine (5, 6-dioxo-p.), pleiocarpinilam, pleiocarpinine, pleiocarpoline, pleiocarpolinine, pleiomutine°, pyrifoline, refractidine, refractine, venalstonidine, venalstonine.

Type: P6b

Example:

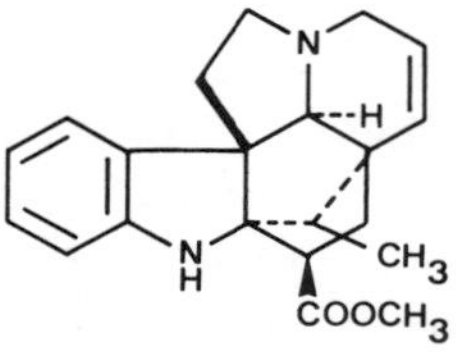

Vindolinine

Alkaloids: Pleiomutinine°, pseudokopsinine (p.-*N*-oxide), pycnanthine° (14′, 15′-dihydro-p.°), tuboxenine (epi-t.), vindolinine (19-epi-v., 19-epi-v. *N*-oxide, v.-*N*-oxide).

Type: P6c

Example:

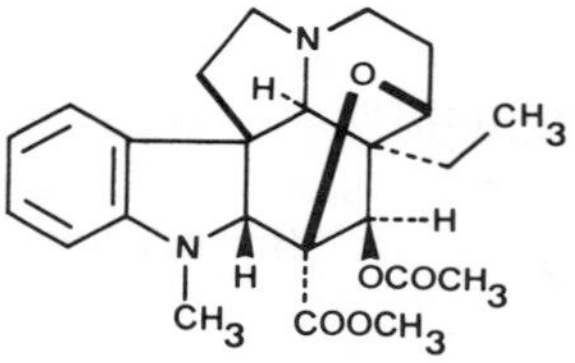

Cathovaline

Alkaloids: Cathanneine, cathovaline (desacetyl-c., 14-hydroxy-c.).

Type: P6d

Example:

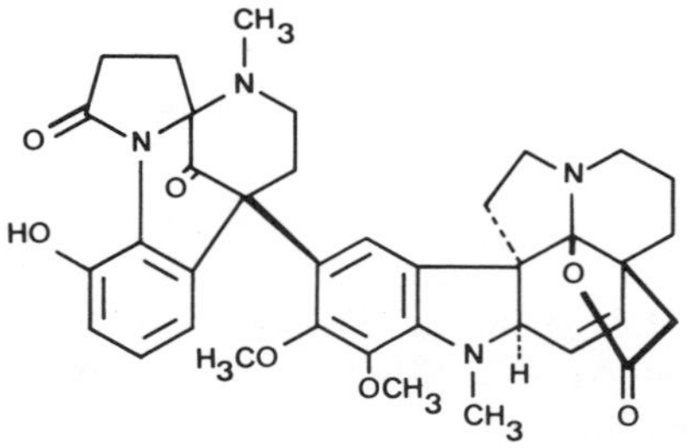

Haplophytine

Alkaloids: Aspidoalbidine (*N*-acetyl-a., *N*-acetyl-12-methoxy-a., *N*-formyl-12-methoxy-a.), aspidoalbine (*N*-acetyl-*N*-depropionyl-a., 18-oxo-a., 18-oxo-*O*-methyl-a.), aspidofendlerine, aspidolimidine (*N*-propionyl-*N*-desacetyl-a.), cimicidine, cimicine, cimiciphytine (nor-c.), dichotamine, fendlerine, haplocidine, haplocine, haplophytine.

Type: P6e

Example:

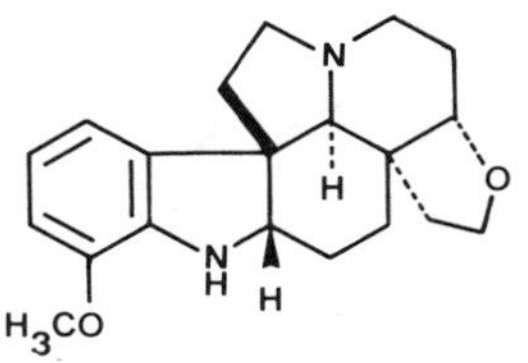

Beninine

Alkaloids: Amataine°, apodine (desoxo-a.), beninine (1,2-dehydro-b), callichiline°, folicangine°, goziline°, hedrantherine (12-hydroxy-h.), owerreine°, quimbeline°, subsessilinelactone°, voafolidine° (iso-v.°), voafoline° (iso-v.°), vobtusine° (12-*O*-demethyl-v.°, desoxy-v.°; 2-desoxy-v.-lactam°, desoxy-v.-lactone°, v.-3-lactam°, v.-3-lactam-*N*-oxide°, v.-lactone°).

Type: P6f

Example:

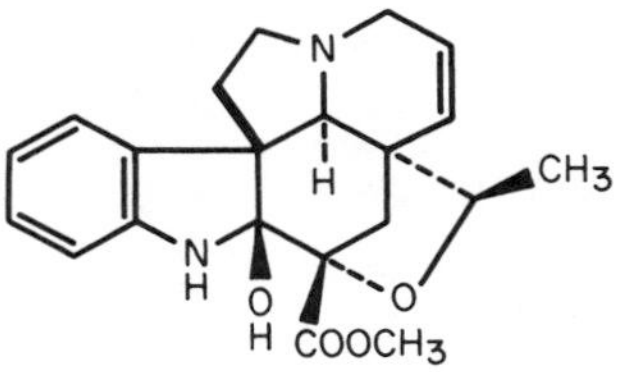

Melobaline

Alkaloids: Melobaline, vincoline.

Type: P7a

Example:

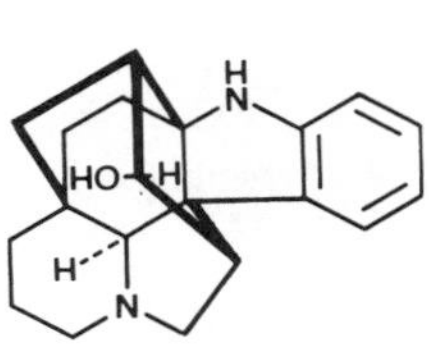

Kopsanole

Alkaloids: *N*-Carbomethoxy-5,22-dioxokopsan, 5,22-dioxokopsan, kopsanol (22-epi-k, *N*-formyl-k., 5-oxo-16-epi-k), kopsanone, kopsine (decarbomethoxy-k., *N*-methyl-5,22-dioxo-k).

Type: P7b

Example:

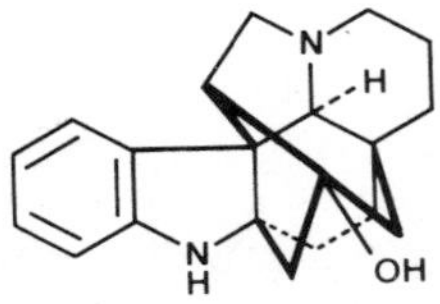

Decarbomethoxyisokopsine

Alkaloid: Decarbomethoxyisokopsine.

Type: P7c

Example:

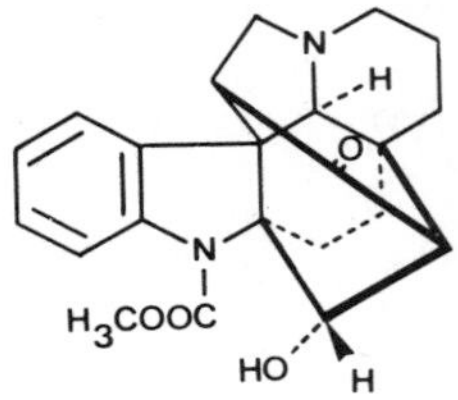

Fruticosamine

Alkaloids: Fruticosamine, fruticosine.

Type: P7d

Example:

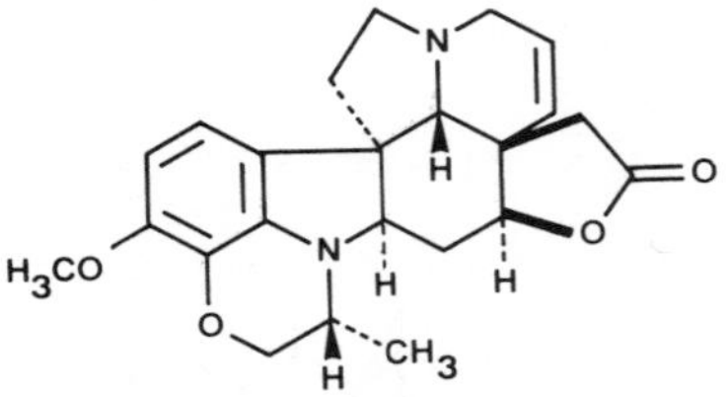

Neblinine

Alkaloids: Neblinine, obscurinervidine (dihydro-o.), obscurinervine (dihydro-o.).

Type: **P7e**

Example:

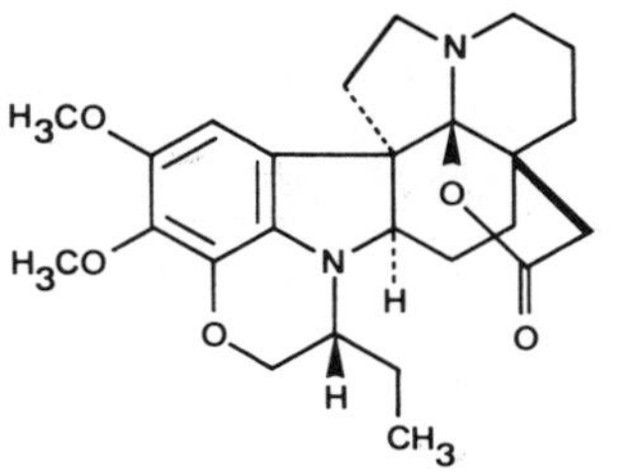

Alalakine

Alkaloid: Alalakine.

APPENDIX 8. IBOGAN (J) TYPE : EXAMPLES OF STRUCTURES AND THE LIST OF ALL ALKALOIDS DETECTED IN NATURE

Type: **J7a**

Example:

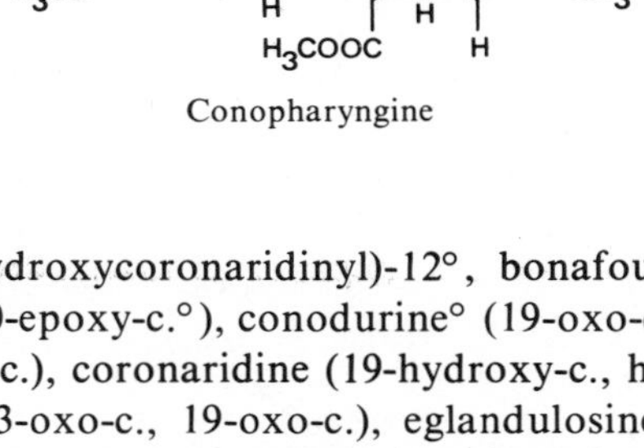

Conopharyngine

Alkaloids: Bis-(11-hydroxycoronaridinyl)-12°, bonafousine, catharanthine, conoduramine° (19, 20-epoxy-c.°), conodurine° (19-oxo-c.°), conopharyngine (19-hydroxy-c., 3-oxo-c.), coronaridine (19-hydroxy-c., hydroxyindolenine-c., 19-hydroxy-3-oxo-c., 3-oxo-c., 19-oxo-c.), eglandulosine, 19-epiiboxygaline, gabunamine°, gabunine°, heyneanine (19-epi-h.), ibogaine (hydroxyindolenine-i.), ibogaline, ibogamine (20-epi-i., hydroxyindolenine-i.), iboxygaine (19-epi-i., hydroxyindolenine-i.), jollyanine, tabernaelegantine A° (t.-B°, t.-C°, t.-D°), tabernaelegantinine A° (t.-B°), tabernamine°, tabernanthine, voacamidine°, voacamine° (16-decarbomethoxydihydro-v.°, 16-decarbomethoxy-20′-epidihydro-v.°, 18′-decarbomethoxy-v.°, *N*-demethyl-v.°, v.-*N*-oxide°), voacangine (hydroxyindolenine-v., 3-hydroxyiso-v., 6-hydroxy-3-oxoiso-v., iso-v., 3-oxo-v., v.-lactam), voacorine° (19-epi-v.°), voacristine (19-epi-v., hydroxyindolenine-v., iso-v.), voacryptine.

Type: J8a

Example:

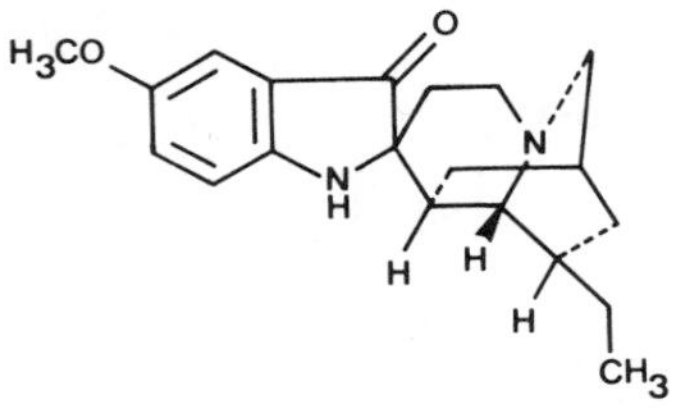

Iboluteine

Alkaloids: Conopharyngine-pseudoindoxyl, coronaridine-pseudoindoxyl, iboluteine (demethoxy-i.), rupicoline, voacangine-pseudoindoxyl, voacristine-pseudoindoxyl, voaluteine.

Type: J8b

Example:

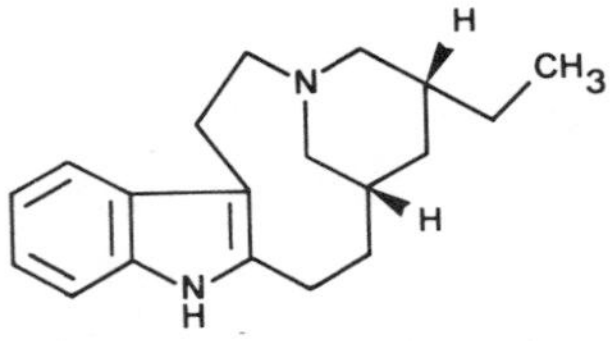

7*R*-Dihydrocleavamine

Alkaloids: Capuvosine° (15-dehydroxy-c.°, *N*-demethyl-c.°), capuronine, 20*R*-dihydrocleavamine (20*S*-dihydro-c.), leurocolombine°, leurocristine°, leurosidine°, leurosine° (iso-l.°), pleurosine°, pseudovincaleukoblastinediol°, vinblastine° (desacetoxy-v.°), vincovalicine°, vincovaline°, vincovalinine°.

Type: J8c

Example:

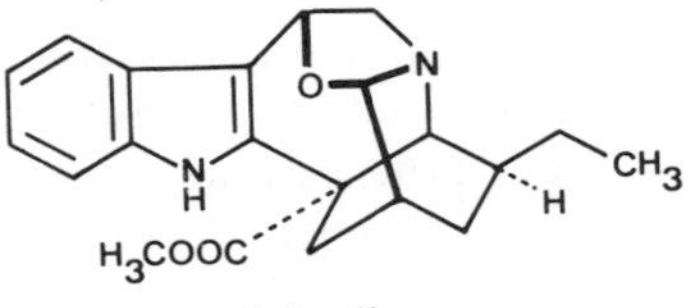

Eglandine

Alkaloids: Eglandine, 3, 6-oxidoisovoacangine.

Type: J8d

Example:

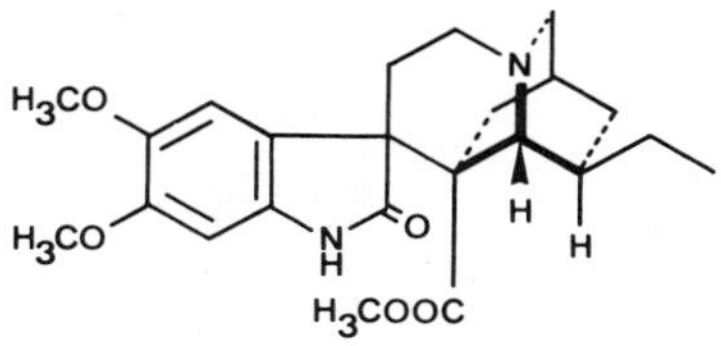

Crassanine

Alkaloids: Crassanine, kisantine.

Type: J9a

Example:

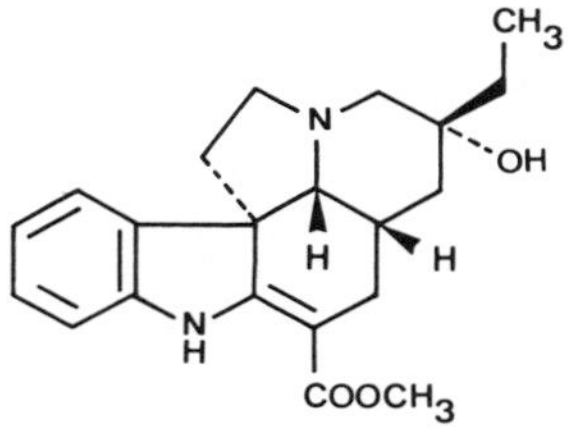

Pandoline

Alkaloids: Capuvosidine°, capuronidine (14, 15-anhydro-c., 14, 15-anhydro-1, 2-dihydro-c.), 20*S*-1, 2-dehydro-pseudoaspidospermidine, pandoline (20-epi-p.), pseudotabersonine, 20*R*-pseudovincadifformine.

Type: J9b

Example:

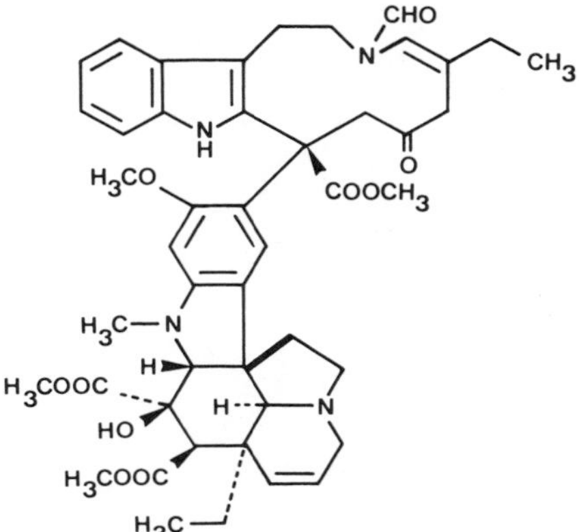

Catharine

Alkaloid: Catharine°.

Type: J9c

Example:

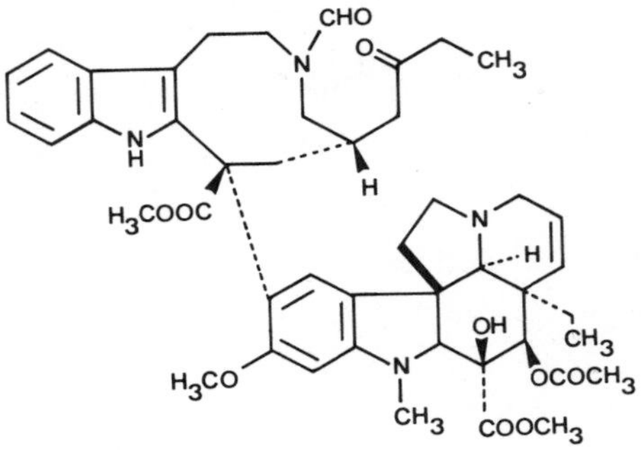

Catharinine

Alkaloid: Catharinine°.

Type: J9d

Example:

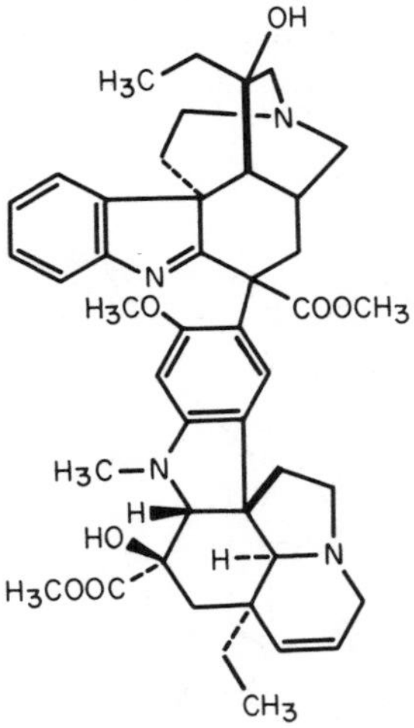

Vincathicine

Alkaloid: Vincathicine°.

Type: J10a

Example:

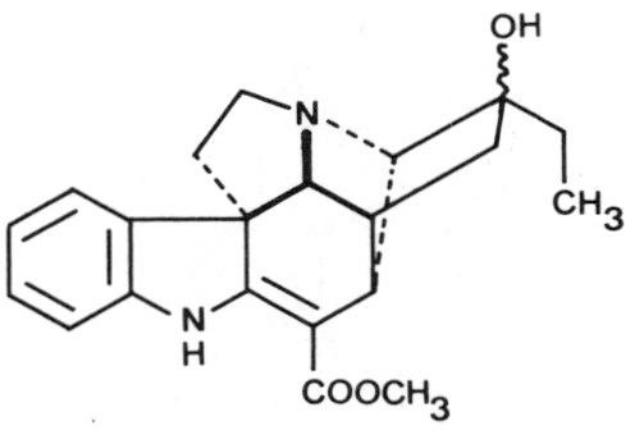

Pandine

Alkaloid: Pandine.

Type: J10b

Example:

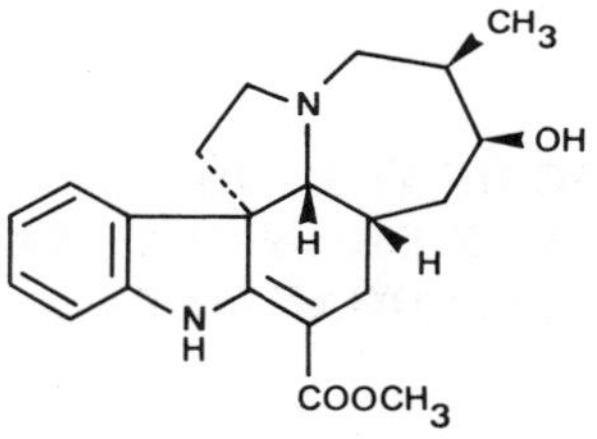

Iboxyphylline

Alkaloid: Iboxyphylline.

Type: J10c

Example:

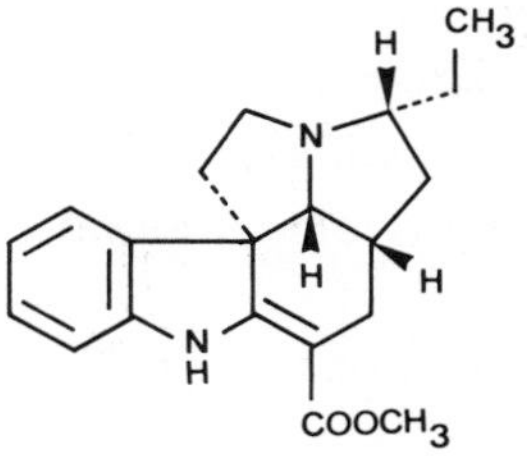

Ibophyllidine

Alkaloid: Ibophyllidine.

Type: J10d

Example:

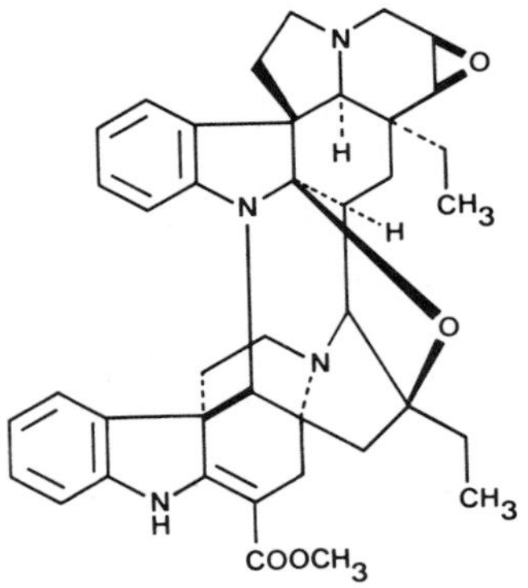

Ervafoline

Alkaloids: Ervafolidine° (iso-e.°), ervafoline.

APPENDIX 9. CATALOG OF THE INDOLE ALKALOIDS WITH C_9- OR C_{10}-MONOTERPENE MOIETY OCCURRING IN THE PLANT FAMILIES APOCYNACEAE, LOGANIACEAE, AND RUBIACEAE

In this table the families, genera, species, and skeletal types (A, C, D, V, etc.) are listed alphabetically. The tribes to which the listed genera belong are indicated by corresponding abbreviations in parentheses. Both skeletal types are given for the bisindole alkaloids (e.g., **C4f-C5e**: (+)-villalstonine). If known, the absolute configuration of the isolated alkaloids is also illustrated as α or β, as mentioned in Fig. 5. The following symbols are used:

(+)-Alkaloid (α): the alkaloid has the absolute configuration α, $[\alpha]_D = +$.

($\pm$)-Alkaloid ($\alpha + \beta$): racemate.

(0)-Alkaloid (α): the specific rotation is 0, but the compound is not a racemate.

( )-Alkaloid (β): the absolute configuration is known, but not the specific rotation.

Alkaloid*: C(15) is not a center of chirality.

We have attempted to cite all the references up to the end of 1978. In the case of plumeran alkaloids, only those isolations are listed that were either not considered in Ref. [4] or have been isolated since then.

Apocynaceae (APO)

Alstonia (PLU). *A. actinophylla* (Cunn) K. Schum. (= *A. verticillosa* F. v. Müll.): **C5g**: (−)-echitamine (α; [8]). *A. angustiloba* Miq.: **C5g**: (−)-echitamine (α; [741]). *A. boonei* De Wild.: **C4c**: akuammidine [742]; **C5g**: echitamine [743]; **J7a**: voacangine [742]. *A. congensis* Engl. (= *A. gilletii* De Wild.): **C5g**: (−)-echitamine (α; [741]); **S4**: (−)-echitamidine (α; [744]). *A. constricta* F. v. Mull.: **C4a**: (+)-alstonine (α; [745, 746]), tetrahydroalstonine [746]; **C4b**: alstoniline* [747], (−)-reserpine (α; [748]), α-yohimbine [746]; **C5c**: (−)-*O*-3,4,5-trimethoxybenzoylvincamajine (α; [749]), (−)-*O*-3,4,5-trimethoxycinnamoylvincamajine (α; [749]), (−)-vincamajine (α; [749]; **C5r**: alstonilidine* [749]; **C5s**: (−)-alstonidine [749]. *A. deplanchei* van Heurck et Müll. Arg.: **C4e**: cabucraline [750]; **C4e-C4f**: (+)-pleiocraline [751]; **C4f-C5g**: (+)-pleiocorine [751, 752]; **C5g**: vincorine [750]. *A. gilletii* De Wild. (see *A. congensis*). *A. glabriflora* Mgf.: **C4f**: (+)-pleiocarpamine (α; [753]); **C4f-C5e**: (+)-villalstonine [753]; **C5e**: (−)-alstophylline [753]; **C5e-C5e**: (+)-macralstonine [753]. *A. lanceolifera S. Moore (see A. lenormandii* var. *lanceolifera*). *A. lenormandii* von Heurck et Müll. Agr. var. *lanceolifera* (S. Moore) Monach. (= *A. lanceolifera* S. Moore): **C4c**: (+)-$N_{(a)}$-methyl-10-methoxyakuammidine [754], $N_{(b)}$-methyl-10-methoxyakuammidine [42]; **C5c**: (−)-10-methoxyvincamajine (α; [754]), (−)-*O*-3,4,5-trimethoxybenzoylhydroxyvincamajine (α; [754]), (−)-*O*-3,4,5-trimethoxycinnamoylvincamajine (α; [754]), (−)-*O*-3,4,5-trimethoxycinnamoyl-10-methoxyvincamajine (α; [754]), (−)-*O*-3,4,5-trimethoxycinnamoyl-10-hydroxyvincamajine (α; [754]); **C6d**: (−)-10-hydroxylanciferine (α; [42]), (−)-lanciferine (α; [42]), (−)-10-methoxylanciferine (α; [42]). *A. macrophylla* Wall. ex G. Don: **C4c**: (+)-affinisine (α; [43]); **C4c-C5e**: (+)-macralstonidine [44]; **C4f**: (+)-pleiocarpamine (α; [44]); **C4f-C5e**: (−)-macrocarpamine [45], (+)-villalstonine [46]; **C5c**: (−)-*O*-benzoylvincamajine (α; [47]), quebrachidine [48]; **C5d**: (−)-picralstonine [43], (−)-picrinine (α; [43]); **C5e**: (−)-alstophylline [49]; **C5e-C5e**: (+)-macralstonine [50]; **D6c**: (+)-macrosalhine [51]; **S4**: $N_{(a)}$-methyl-2,16-dihydroakuammicine [44]. *A. muelleriana* Domin: **C4f**: 2,7-dihydropleiocarpamine [52], pleiocarpamine [52]; **C4f-C5e**: (+)-villalstonine [52, 53]; **C5c**: quebrachidine [52]; **C5c-C5e**: (−)-alstonisidine [52, 54]; **C5e**: (−)-alstonerine [52, 55]; **C5e-C5e**: (+)-macralstonine [52, 57], (+)-$N'_{(a)}$-demethylanhydromacralstonine [52]; **C6e**: (+)-alstonisine [52, 55]; **S4**: 11-methoxyakuammicine [58], (−)-vinervinine [52]. *A. neriifolia* D. Don.: **C5g**: echitamine [59]. *A. quaternata* van Heurck et Müll. Arg.: **A4a**: (+)-tubotaiwine (α; [60]); **C4b**: (+)-pseudoyohimbine (α; [60]), (+)-quaternatine (α; [60]), (+)-yohimbine (α; [60]); **C4e**: ( )-cathafoline (α; [60]), (−)-quaternoxine [60]; **C5c**: (−)-vincamajine (α; [60]); **C5d**: (−)-quaternidine [60], (−)-quaternine (α; [60]); **C5u**: quaternoline [60]. *A. scholaris* (L.) R. Br.: **A4a**: tubotaiwine [61, 62]; **C4a**: tetrahydroalstonine [63]; **C4c**: akuammidine [64, 65]; **C4e**: strictamine [63]; **C5d**: (−)-picralinal (α; [64]); (−)-picrinine (α; [61, 64]); **C5g**: (−)-echitamine (α; [59, 61, 62]), norechitamine [61, 62]; **C5h**: pseudoakuammigine [61, 62]; **C5v**: (−)-nareline (α; [66]); **S4**: akuammicine [61, 62], akuammicine methiodide [61, 62], akuammicine *N*-oxide [61, 62], (−)-

echitamidine (α; [62, 744]); 18- or 19-hydroxy-19,20-dihydroakuammicine [61], hydroxy-19,20-dihydroakuammicine [61, 62]. *A. somersetensis* Bailey (see *A. spectabilis*). *A. spatulata* Bl.: **C5g**: (−)-echitamine (α; [8]). *A. spectabilis* R. Br. (= *A. somersetensis* Bailey, *A. villosa* Bl.): **C4c**: (+)-$N_{(a)}$-methylsarpagine (α; [753]); **C4c-C5e**: (+)-macralstonidine [753]; **C4f**: (+)-pleiocarpamine (α; [753]); **C4f-C5e**: (+)-villalstonine [67, 753]; **C5c**: (+)-quebrachidine (α; [753]), (−)-vincamajine (α; [753]); **C5g**: echitamine [68]. *A. venenata* R. Br.: **C4b**: 3,4-dehydroalstovenine [69], anhydroalstonatine* [70], (+)-isovenenatine (α; [71]), reserpine [71], (−)-venenatine (α; [71]), (−)-veneserpine (α; [72]), (−)-venoxidine (α; [73]). *A. villosa* Bl. (see *A. spectabilis*). *A. vitiensis* Seem. var. *novo-ebudica* Monach.: **C4e**: (−)-cabucraline (α; [750]), quaternoxine [750]; **C4f**: pleiocarpamine [750]; **C5g**: (−)-vincorine (α; [750]); **S4**: (−)-alstovine [750].

Ammocallis (PLU). *A. rosea* Small (see *Catharanthus roseus*).

Amsonia (PLU). *A. angustifolia* (Ait.) Michx. (see *A. ciliata*). *A. ciliata* Walt. [= *A. angustifolia* (Ait.) Michx.]: **C3a**: angusteine [74]; **C4b**: β-yohimbine [9]; **C4c**: akuammidine [74]; **E5a**: (−)-eburnamenine (α; [74]), (+)-eburnamonine (β; [74]); **P5**: (+)-vincadifformine (β; [74]); **S4**: angustimycine [74]; **V4**: angustine* [74]. *A. elliptica* (Thunb.) Roem. et Schult.: **C3a**: 10-hydroxygeissoschizol [75, 76]; **C4a**: tetrahydroalstonine [75, 77]; **C4b**: 3,4,5,6-tetradehydro-β-yohimbinium chloride [75, 76], (+)-yohimbine (α; [75, 76]), (−)-β-yohimbine (α; [75, 76]); **C4f**: (+)-pleiocarpamine (α; [75, 76]); **E5a**: 14,15-dehydrovincamine [75, 77], 16-epi-14,15-dehydrovincamine [75, 77]; **P3-P3**: presecamine [75], secamine [75], tetrahydropresecamine [75]; **P5**: 16-carbomethoxy-16-hydroxy-14,15-epoxy-3-oxo-1,2-dehydroaspidospermidine [77], 14,15-epoxy-3-oxo-vincadifformine [75, 77], (−)-3-oxo-tabersonine [75, 77], (−)-tabersonine (α; [75, 77]), tabersonine *N*-oxide [77]; **V3**: (+)-antirhine (α; [75, 76]), antirhine α-methochloride [75, 76]. *A. tabernaemontana* Walt.: **A4a**: (+)-19,20-dihydrocondylocarpine (α; [78]); **C3a**: (−)-18,19-dihydrocorynantheol (α; [78]); **C4a**: (−)-tetrahydroalstonine (α; [78, 79]); **C4c**: (+)-akuammidine (α; [78]); **E5a**: (−)-eburnamine (β; [79, 80]), (+)-eburnamonine (β; [79]), isoeburnamine [80]; **S4**: (−)-norfluorocurarine (α; [79]).

Anacampta (TAB). *A. rigida* (Miers) Mgf. (see *Tabernaemontana rigida*). *A. rupicola* (Benth.) Mgf. (see *Tabernaemontana rupicola*).

Aspidosperma (PLU). *A. album* (Vahl.) R. Ben. ex Pichon (= *A. desmanthum* Benth. ex Müll. Arg.): **A4a**: (+)-condylocarpine (α; [83]), (+)-tubotaiwine (α; [83]); **A5b**: (±)-andranginine ($\alpha + \beta$; [83]); **C3a**: (+)-16-episitsirikine (α; [83]), (−)-isositsirikine (α; [83]), (+)-sitsirikine (α; [83]); **E5a**: (±)-vincamine ($\alpha + \beta$; [83]); **P5**: (−)-*N*-acetylaspidospermidine (α; [83]), (+)-aspidospermidine [83], (+)-limaspermine (β; [83]), (+)-11-methoxylimaspermine (β; [83]), (+)-vincadifformine (β; [83]); **P6d**: aspidoalbine [88], (+)-fendlerine (β; [83]; **P7e**: (−)-alalakine [83]. *A. auriculatum* Mgf.: **C3a**: (−)-dihydrocorynantheol (α; [84]);

C4a: reserpinine [84]. *A. carapanauba* Pichon: **C5b**: (−)-carapanaubine (α; [85]). *A. compactinervum* Kuhlm. (see *A. eburneum*). *A. cuspa* (H.B.K.) S.F. Blake: **C3a**: (−)-16-episitsirikine (α; [86]); **C5d**: (−)-desacetylpicraline (α; [87]); **C6d**: (−)-aspidodasycarpine (α; [86, 87]); **P7a**: (−)-22-epikopsanol (α; [86]), kopsanol [86], (−)-kopsanone (α; [86]). *A. dasycarpon* A. DC. (see *A. tomentosum*). *A. desmanthum* Benth. ex Müll. Arg. (see *A. album*). *A. discolor* A. DC.: **C3a**: (−)-10-methoxydihydrocorynantheol (α; [89]), (−)-10-methoxygeissoschizol (α; [89]); **C4a**: (−)-isoreserpiline (α; [89]), (−)-reserpiline (α; [89]); **C4b**: (+)-yohimbine (α; [89]), (−)-β-yohimbine (α; [89]); **C5m**: (−)-isoreserpiline-pseudoindoxyl (α; [89]). *A. eburneum* F. Allem. ex Sald. (= *A. compactinervum* Kuhlm.): **A4a**: *N*-acetyl-11-hydroxyaspidospermatidine [84]; **C4b**: (−)-β-yohimbine (α; [90]); **S4**: (−)-compactinervine (*a*; [91]). *A. exalatum* Monach.: **P5**: *N*-acetylaspidospermidine [92], aspidospermine [92], demethoxypalosine [92], *O*-demethylpalosine [92], limaspermine [92]; **P6d**: cimicine [92], fendlerine [92]. *A. excelsum* Benth.: **C3a**: (−)-aspexcine [9], methoxygeissoschizoline [9]; **C4b**: ( )-*O*-acetylyohimbine (α; [93]), (−)-excelsinine (α; [93]), (+)-yohimbine (α; [93]), (−)-α-yohimbine (α; [94]). *A. formosanum* A.P. Duarte: **P5**: (+)-aspidocarpine (β; [88]). *A. laxiflorum* Kuhlm. (see *A. rigidum*). *A. limae* Woods.: **A4a**: (+)-limatine (α; [95]), (+)-limatinine (α; [95]), (+)-11-methoxylimatine (α; [95]), (+)-11-methoxylimatinine (α; [95]), (+)-tubotaiwine (α; [95]). *A. marcgravianum* Woods.: **C3a**: (−)-dihydrocorynantheol (α; [96]); **C4a**: aricine [97], reserpiline [97]. *A. neblinae* Monach.: **E5a**: eburnamonine [98]. *A. nitidum* Benth. ex Müll. Arg.: **C3a**: (−)-10-methoxydihydrocorynantheol (α; [97]). *A. oblongum* A. DC.: **C3a**: alkaloid II/296 [99], alkaloid II/298 [99], alkaloid II/328 [99], alkaloid III/352 [99], alkaloid III/354 [99], alkaloid III/382 [99], alkaloid III/384 [99], alkaloid IV/354 [99], alkaloid IV/356 [99], alkaloid IV/384 [99], alkaloid IV/386 [99], (−)-10-methoxygeissoschizol (α; [84]); **C4a**: alkaloid I/352 [99], alkaloid I/382a [99], alkaloid I/382b [99], isoreserpiline [99], reserpiline [99]; **C4b**: 11-methoxyyohimbine [100], (+)-pseudoyohimbine (α; [101]), yohimbine [101], (−)-β-yohimbine (α; [100]). *A. peroba* F. Allem. ex Sald. (see *A. polyneuron*). *A. polyneuron* Müll. Arg. (= *A. peroba* F. Allem. ex Sald.): **A4a**: aspidospermatine [102]; **C4b**: yohimbine [102]; **C4c**: (+)-macusine B (α; [103]), (+)-normacusine B (α; [90]), (−)-polyneuridine (α; [90]). *A. populifolium* A. DC. (see *A. pyrifolium*). *A. pyricollum* Müll. Arg.: **A3**: (+)-stemmadenine (α; [97]); **C4b**: (+)-19,20-dehydroyohimbine (α; [104]), yohimbine [104], β-yohimbine [104]. *A. pyrifolium* Mart. (= *A. populifolium A.* DC.): **A4a**: (+)-11-methoxy-19,20-dihydrocondylocarpine [84]. *A. quebracho-blanco* Schlecht.: **A4a**: $N_{(a)}$-acetylaspidospermatidine [105], aspidospermatidine [105], (−)-aspidospermatine (α; [105]), desacetylaspidospermatine [105], 19,20-dihydroaspidospermatine [105], $N_{(a)}$-methylaspidospermatidine [105]; **C4b**: yohimbine [105]; **C4c**: akuammidine [106]; **C5c**: (+)-quebrachidine (α; [107]); **E5a**: eburnamenine [108]. *A. rigidum* Rusby (= *A. laxiflorum* Kuhlm.): **C4a**: reserpiline [97]; **C5b**: (−)-carapanaubine (α; [97]); **C5d**: desacetylpicraline [97], picraline [97]. *A. spegazzini* Molfino ex T. Meyer (see *Rauvolfia schuelii*). *A.* species No. 9610: **A4a**(+)-11,12-dihydroxy-*N*-acetylaspidospermatidine [109]. *A.* species No. RJB 119070: **C3a**: 10-methoxydihydrocorynantheol [84]. *A. tomentosum* Mart. (= *A. dasycarpon* A. DC.): **A4a**: (+)-limatinine (α; [97]); **C4c**: polyneuridine [110]; **C6d**: (−)-aspidodasycarpine (α; [110]).

Bleekeria (RAU). *B. coccinea* (Miq.) Koidz. (see *Ochrosia coccinea*). *B. compta* (K. Schum.) Wilbur. (see *Ochrosia compta*). *B. elliptica* (Labill.) Koidz. (see *Ochrosia elliptica*). *B. vitiensis* (Mgf.) A.C. Smith (see *Ochrosia vitiensis*).

Bonafousia (TAB). *B. tetrastachya* (H.B.K.) Mgf. (see *Tabernaemontana siphilitica*).

Cabucala (RAU). *C. erythrocarpa* (Vatke) Mgf. var. *erythrocarpa*: **C4a**: (−)-aricine (α; [23]), (−)-cabucine (α; [23]), (0)-cabucinine (α; [23]); **C4b**: reserpine [23]; **C4c**: (+)-vellosimine (α; [23]); **C4e**: (−)-cabucraline (α; [23]); **C5a**: (−)-ochropamine [23]; **C5c**: (+)-quebrachidine (α; [23, 24]); **C5g**: (−)-cabuamine (α; [23, 114]), (−)-deformocabuamine (α; [23]), vincorine [750]; **C5h**: (−)-akuammine (α; [23]); **S4**: (−)-akuammicine (α; [23]). *C. fasciculata* Pichon: **C4a**: (−)-cabucine (α; [115]), (0)-cabucinine (α; [115]); **C5b**: (−)-caboxine A (α; [115, 116]), (−)-carapanaubine (α; [115]), (+)-10,11-dimethoxyisomitraphylline [115], (+)-isocaboxine A (α; [115, 116]), (+)-isocaboxine B (α; [115, 116]), (+)-rauvoxinine (α; [115]); **C5h**: (−)-*O*-methylakuammine (α; [115]). *C. glauca* Pichon (see *C. madagascariensis*). *C. madagascariensis* (A. DC.) Pichon (= *C. glauca* Pichon): **C4a**: (−)-cabucine (α; [117]), (0)-cabucinine (α; [117]); **C4b**: (−)-reserpine (α; [117]). *C. striolata* Pichon: **C4a**: (−)-ajmalicine (α; [118]), (0)-ajmalicinine (α [118]), (−)-cabucine (α; [118]), (0)-cabucinine (α; [118]), (−)-10,11-dimethoxyajmalicine (α; [118]), (−)-10,11-dimethoxyajmalicinine (α; [118]); **C4b**: (−)-reserpine (α; [118]); **C5c**: (+)-quebrachidine (α; [118]). *C. torulosa* Pichon: **C4a**: (−)-aricine (α; [119]), (−)-cabucine (α; [119]), (−)-tetrahydroalstonine (α; [119]); **C4e**: (−)-cabucraline (α; [119]); **C5c**: (+)-quebrachidine (α; [119]), (−)-vincamajine (α; [119]).

Callichilia (TAB). *C. barteri* (Hook. f.) Stapf [= *Hedranthera barteri* (Hook. f.) Pichon]: **C5a-J7a**: (−)-voacamine (α-β; [219]); **C6d**: (−)-lonicerine (α; [220]); **J7a**: (−)-voacangine (β; [219]). *C. subsessilis* (Benth.) Stapf (= *Tabernaemontana subsessilis* Benth.): **P6e-P6e**: (−)-amataine (α-α; [120]).

Capuronetta (TAB). *C. elegans* Mgf. (see *Tabernaemontana capuronii*).

Carpodinus (CAR). *C. umbellatus* K. Schum. (see *Hunteria umbellata*).

Catharanthus (PLU). *C. lanceus* (Boj. ex A. DC.) Pichon [= *Vinca lancea* (Boj. ex A. DC.) K. Schum., *Lochnera lancea* (Boj. ex A. DC.) K. Schum.]: **C4a**: (−)-ajmalicine (α; [123]), (−)-tetrahydroalstonine (α; [123]); **C4b**: (+)-yohimbine (α; [123]); **C4c**: (+)-pericyclivine (α; [124]); **C5a**: ( )-periformyline (α; [125]), ( )-perivine (α; [126]); **P5**: (−)-14,15-dehydro-19-epiminovincinine (α; [127]);

J7a: catharanthine [128]; **J8b-P5**: (+)-leurosine (α-α; [129]); **J9b-P5**: (−)-catharine (α-α; [130]). *C. longifolius* (Pichon) Pichon: **C4a**: ajmalicine [131]; **C4c**: akuammidine [131], normacusine B [131], pericyclivine [131]; **C4e**: (−)-cabucraline (α; [750]); **C5a**: perivine [131]; **J7a**: ( )-catharanthine (α; [131]); **J8b-P5**: ( )-leurosine (α-α; [131]), vinblastine [131]; **J9b-P5**: catharine [131]; **J9c-P5**: (−)-catharinine (α-α; [132]); **P5**: desacetylvindoline [133]; **P6a**: 15-hydroxykopsinine [133]; **P6c**: cathovaline [133]; **S4**: akuammicine [131]; 2,16-dihydroakuammicine [131], $N_{(a)}$-methyl-2,16-dihydroakuammicine [131]; **V3**: antirhine [131]. *C. ovalis* Mgf.: **C4a**: (+)-serpentine (α; [134]); **J7a**: (+)-catharanthine (α; [134]), (−)-coronaridine (β; [134]); **J8b-P5**: vincovalicine [135], (−)-vincovaline (β-α; [135]), vincovalinine [135]; **J9b-P5**: (−)-catharine (α-α; [136]). **J9c-P5**: (−)-catharinine (α-α; [132]); **P5**: (−)-cathovalinine (α; [137]), ( )-19-epi-19-hydroxytabersonine (α; [138]), ( )-19-hydroxytabersonine (α; [138]). *C. pusillus* (Murr.) G. Don (= *Vinca pusilla* Murr.): **C4a**: (−)-ajmalicine (α; [139]); **C4b**: α-yohimbine [140]. *C. roseus* (L.) G. Don [= *Vinca rosea* (L.) Reichb., *Ammocallis rosea* (L.) Small, *Lochnera rosea* (L.) Reichb.]: **A3**: stemmadenine [141]; **C3a**: corynantheine [141], corynantheinealdehyde [141], (−)-dihydrositsirikine (α; [142]), geissoschizine [141], (−)-isositsirikine (α; [142]), (+)-sitsirikine (α; [83, 142]); **C4a**: (−)-ajmalicine (α; [143]), alstonine [9], (+)-serpentine (α; [143]), (−)-tetrahydroalstonine (α; [143]); **C4b**: (−)-reserpine (α; [143]); **C4c**: (+)-lochnerine (α; [144]); **C5a**: (−)-perivine (α; [145]); **C5b**: (−)-mitraphylline (α; [143]); **C5h**: (−)-akuammine (α; [143]); **D3a**: (−)-strictosidine (α; [15]), (−)-vincoside (α; [15, 141]); **E6b**: (−)-vincarodine (α; [146, 147]); **J7a**: (+)-catharanthine (α; [145]), coronaridine [141]; **J8b-P5**: (+)-desacetoxyvinblastine (α-α; [148]), (+)-isoleurosine (α-α; [149, 150]), leurocolombine [151], (+)-leurocristine (α-α; [152]), (+)-leurosidine (α-α; [153]), (+)-leurosine (α-α; [145, 150]), (+)-pleurosine (α-α; [154]), pseudovincaleukoblastinediol [151], (+)-vinblastine (α-α; [145]); **J9b-P5**: (−)-catharine (α-α; [136, 149]); **J9c-P5**: catharinine (α-α; [132]); **J9d-P5**: vincathicine [155]; **P6a**: venalstonine [156]; **P6f**: ( )-vincoline (α; [138, 157]); **S4**: (−)-akuammicine (α; [141, 143]), (−)-lochneridine (α; [91]), ( )-preakuammicine (α; [141]). *C. trichophyllus* (Bak.) Pichon: **C4a**: (−)-ajmalicine (α; [158]), (−)-tetrahydroalstonine (α; [158]); **C4b**: (+)-pseudoyohimbine (α; [159]); **P5**: (−)-cathaphylline (α; [159]), (−)-echitovenine (α; [159]), (−)-hoerhammericine (α; [159]), (−)-lochnericine (α; [159]), (−)-minovincine (α; [159]), (−)-minovincinine (α; [159]), vindorosine [159]; **P6b**: (−)-vindolinine (α; [158]); **S4**: akuammicine [159].

Conopharyngia (TAB). *C. brachyantha* (Stapf) Stapf (see *Tabernaemontana brachyantha*). *C. contorta* (Stapf) Stapf (see *Tabernaemontana contorta*). *C. crassa* (Benth.) Stapf (see *Tabernaemontana crassa*). *C. cumminsii* Stapf (see *Tabernaemontana pachysiphon*). *C. durissima* (Stapf) Stapf (see *Tabernaemontana crassa*). *C. elegans* (Stapf) Stapf (see *Tabernaemontana elegans*). *C. holstii* (K. Schum.) Stapf (see *Tabernaemontana holstii*). *C. johnstonii* Stapf (see *Tabernaemontana johnstonii*). *C. jollyana* Stapf (see *Tabernaemontana crassa*). *C. longiflora* (Benth.) Stapf (see *Tabernaemontana longiflora*). *C.*

pachysiphon (Stapf) Stapf (see *Tabernaemontana pachysiphon*). *C. penduliflora* (K. Schum.) Stapf (see *Tabernaemontana penduliflora*). *C. retusa* (Lam.) G. Don (see *Tabernaemontana retusa*).

Craspidospermum (PLU). *C. verticillatum* Boj. ex A. DC.: **A4a**: (+)-tubotaiwine (α; [178]); **E5a**: ( )-14,15-dehydro-16-epivincine (α; [179, 180], (+)-14,15-dehydrovincine (α; [179, 180]); **E6b**: (−)-craspidospermine (α; [180]); **P5**: (−)-11-hydroxytabersonine [180]. *C. verticillatum* Boj. ex A. DC. var. *petiolare* A. DC.: **A3**: (+)-deformostemmadenine [181], stemmadenine [181]; **A4a**: condylocarpine [181]; **A5b**: (±)-andranginine ($\alpha + \beta$; [22, 181]); **E5b**: (−)-andrangine (α; [181]).

Crioceras (TAB). *C. dipladeniiflorus* (Stapf) K. Schum. (= *C. longiflorus* Pierre): **E5a**: (+)-apo-14,15-dehydrovincamine (α; [182, 183]), (+)-14,15-dehydrovincamine (α; [182, 183, 187]), (+)-16-epi-14,15-dehydrovincamine (α; [182, 183, 187]), (+)-12-methoxy-14,15-dehydrovincamine (α; [182, 183, 184]); **E5b**: (−)-andrangine (α; [185]); **E5b-P5**: (−)-criophylline [185, 186]; **E6b**: (−)-criocerine (α; [183]); **P6e-P6e**: (−)-vobtusine (α-α; [187]). *C. longiflorus* Pierre (see *C. dipladeniiflorus*).

Diplorhynchus (PLU). *D. condylocarpon* (Müll. Arg.) Pichon (= *D. condylocarpon*) sp. *mossambicensis* (Benth.) Duvign.): **A3**: (+)-stemmadenine (α; [188, 189]); **A4a**: (+)-condylocarpine (α; [189]); **C4b**: (+)-yohimbine (α; [189]), (−)-β-yohimbine (α; [189]); **C4c**: (+)-normacusine B (α; [189]); **S4**: (−)-mossambine (α; [189]), (−)-norfluorocurarine (α; [189]).

Ervatamia (TAB). *E. coronaria* (Jacq.) Stapf (see *Tabernaemontana divaricata*). *E. dichotoma* (Roxb.) Blatter (see *Tabernaemontana dichotoma*). *E. divaricata* (L.) Burkill (see *Tabernaemontana divaricata*). *E. macrocarpa* (Jack) Merr. (see *Tabernaemontana macrocarpa*). *E. mucronata* (Merr.) Mgf. (see *Tabernaemontana mucronata*. *E. obtusiuscula* Mgf. (see *Tabernaemontana orientalis*). *E. orientalis* (R. Br.) Domin (see *Tabernaemontana orientalis*). *E. pandacaqui* (Poir.) Pichon (see *Tabernaemontana pandacaqui*).

Excavatia (RAU). *E. balansae* Guill. (see *Ochrosia balansae*). *E. coccinea* (Teysm. et Binn.) Mgf. (see *Ochrosia coccinea*). *E. elliptica* (Labill.) Mgf. (see *Ochrosia elliptica*). *E. vitiensis* Mgf. (see *Ochrosia vitiensis*).

Gabunia (TAB). *G. eglandulosa* (Stapf) Stapf (see *Tabernaemontana eglandulosa*). *G. longiflora* Stapf (see *Tabernaemontana eglandulosa*). *G. odoratissima* Stapf (see *Tabernaemontana odoratissima*).

Geissospermum (PLU). *G. argenteum* Woods: **P5**: (+)-aspidocarpine (β; [205]), (−)-aspidospermine (β; [205]), (−)-demethoxyaspidospermine (α; [205]), (+)-demethylaspidospermine [205]. *G. laeve* (Vell.) Miers. (= *G. vellosii* F. Allem.,

Tabernaemontana laevis Vell.): **A5a**: (−)-geissovelline [207]; **C3a-S5a**: (−)-geissospermine (α-α; [208]); **C4c**: (+)-normacusine B (α; [134]), (+)-vellosimine (α; [208]); **C4c-S5a**: (+)-geissolosimine (α-α; [209]); **S4**: (+)-geissoschizoline (α; [210]). *G. sericeum* Benth. et Hook f.: **C3a-S5a**: (−)-geissospermine (α-α; [206]). *G. vellosii* F. Allem. (see *G. laeve*).

Gonioma (PLU). *G. kamassi* E. Mey.: **C4c**: (+)-akuammidine (α; [211]); **C4f**: pleiocarpamine [211]; **C5k**: fluorocarpamine [211]; **E5a**: (−)-eburnamine (β; [211]). *G. malagasy* Mgf. et Boiteau: **C4f-P6b**: (+)-14′15′-dihydropycnanthine (α--; [212]).

Haplophyton (PLU). *H. cimicidum* A. DC.: **E5a**: eburnamine [213], isoeburnamine [213], *O*-methyleburnamine [213]; **P6d**: (−)-cimiciphytine [214], (−)-norcimiciphytine [214].

Hazunta (TAB). *H. angustifolia* Pichon (see *Tabernaemontana coffeoides*). *H. coffeoides* (Boj. ex A. DC.) Pichon (see *Tabernaemontana coffeoides*). *H. costata* Mgf. (see *Tabernaemontana coffeoides*). *H. membranacea* (A. DC.) Pichon (see *Tabernaemontana coffeoides*). *H. membranacea* (A. DC.) Pichon forma *pilifera* Mgf. (see *Tabernaemontana coffeoides*). *H. modesta* (Bak.) Pichon (see *Tabernaemontana coffeoides*). *H. modesta* (Bak.) Pichon var. *methuenii* (Stapf et M. L. Green) Pichon subvar. *methuenii* (see *Tabernaemontana coffeoides*). *H. modesta* (Bak.) Pichon var. *methuenii* (Stapf et M. L. Green) Pichon subvar. *velutina* (Pichon) Mgf. (see *Tabernaemontana coffeoides*). *H. modesta* (Bak.) Pichon var. *modesta* subvar. *modesta* (see *Tabernaemontana coffeoides*). *H. silicicola* Pichon (see *Tabernaemontana coffeoides*). *H. velutina* Pichon (see *Tabernaemontana coffeoides*).

Hedranthera (TAB). *H. barteri* (Hook. f.) Pichon (see *Callichilia barteri*).

Hunteria (CAR). *H. corymbosa* Roxb. (see *H. zeylanica*). *H. eburnea* Pichon: **C3a**: (+)-dihydrocorynantheol methochloride (α; [221]), (−)-geissoschizol (α; [222]), (+)-huntrabrine methochloride [221]; **C4b**: (+)-yohimbol methochloride (α; [221]); **C4f**: (+)-pleiocarpamine (α; [223]), (+)-pleiocarpamine methochloride (α; [224]); **C4h**: (−)-burnamicine (α; [223]); **C5d**: (−)-desacetylpicraline (α; [223]); **C5g**: (−)-deformocorymine (α; [222]); **C5p**: (+)-eburnaphylline (α; [222, 225, 226]); **C6b**: (−)-acetylcorymine (α; [222]), (+)-corymine (α; [222]); **C8a**: (−)-erinicine (α; [222]), (−)-erinine (α; [222]); **E5a**: (+)-eburnamenine (β; [223, 227]), (−)-eburnamine (β; 223, 227]), (+)-eburnamonine (β; [223, 227]), (+)-isoeburnamine (β; [223, 227]); **S4**: (−)-akuammicine methochloride (α; [221]); **V3**: (+)-antirhine β-methochloride (α; [224]), (+)-18,19-dihydroantirhine β-methochloride (α; [224]), ( )-hunterburnine α-methochloride (α; [221]), (+)-hunterburnine β-methochloride (α; [221]). *H. elliottii* (Stapf) Pichon: **C4a**: huntrabrine methochloride [228]; **C4a**: tetrahydroalstonine [229]; **C4b**: yohimbol methochloride [228]; **C4c**: akuammidine [228]; **C4f**: pleiocarpamine [228];

C5d: desacetylpicraline [228]; **C5g**: ( )-deformocorymine (α; [229]), (−)-3-epi-3,17-dihydrocorymine (α; [229]); **C6b**: acetylcorymine [229], corymine [229]; **E5a**: eburnamine [229], (−)-12-methoxy-14,15-dehydrovincamine (β; [229]); **P3-P3**: tetrahydropresecamine [229]; **P4**: quebrachamine [229]; **P5**: (−)-1,2-dehydroaspidospermidine (α; [229]), (−)-vincadifformine (α; [229]); **P6a**: (−)-kopsinine (α; [229]), pleiocarpine [228]; **S4**: (−)-akuammicine (α; [229]); **V3**: hunterburnine α-methochloride [229]. *H. umbellata* (K. Schum.) Hall. f. [= *Carpodinus umbellatus* K. Schum., *Polyadoa umbellata* (K. Schum.) Stapf, *Picralima umbellata* (K. Schum.) Stapf]: **C4e-E5a**: (−)-umbellamine (--β; [230]); **C6b**: (+)-acetylcorymine (α; [231]), (+)-corymine (α; [231, 232]; **C7a**: (−)-eripine (α; [233]); **C8a**: (−)-erinicine (α; [232, 234]), (−)-erinine (α; [232, 234]), (−)-isocorymine (α; [235]). *H. zeylanica* (Retz.) Gardn. ex Thw. (= *H. corymbosa* Roxb.): **C4c**: (0)-akuammidine (α; [236]); **C6b**: (+)-corymine (α; [237]).

Lochnera (PLU). *L. lancea* (Boj. ex A. DC.) K. Schum. (see *Catharanthus lanceus*). *L. rosea* (L.) Reichb. (see *Catharanthus roseus*).

Melodinus (CAR). *M. aeneus* Baill.: **A4a**: (+)-tubotaiwine (α; [238]), (+)-tubotaiwine *N*-oxide (α; [238]); **E5a**: (+)-16-epi-14,15-dehydrovincamine (α; [238]), (+)-16-epi-14,15-dehydrovincine (α; [238]); **J7a**: (+)-20-epiibogamine [238], (−)-ibogamine (β; [238]); **P5**: (+)-20,21-epiaspidospermidine [239], (+)-20,21-epiaspidospermidine *N*-oxide [239], meloceline [239], melocelinine [239], (−)-11-methoxytabersonine (α; [238]), (−)-lochnericine (α; [238]), (−)-lochnerinine (α; [238]), (−)-tabersonine (α; [238]), (+)-vincadifformine (β; [238]). *M. australis* (F. v. Müll.) Pierre: **A3**: (+)-stemmadenine (α; [240]); **A4a**: (+)-condylocarpine (α; [240]); **C4c**: (+)-akuammidine (α; [240]). *M. balansae* Baill. var. *paucivenosus* (S. Moore) Boiteau: **C4b**: (+)-renoxydine [241]; **C5c**: (+)-ajmaline (α; [241]); **E5a-P5**: paucivenine [241]; **P6a**: (−)-venalstonidine (α; [241]), (−)-venalstonine (α; [241]); **P6f**: (−)-melobaline (α; [242]). *M. buxifolius* Baill. (see *M. celastroides*). *M. celastroides* Baill. (= *M. buxifolius* Baill.): **C4c**: (+)-akuammidine (α; [243]); **E5a**: 14,15-dehydroeburnamine [244]; **P5**: ( )-19*R*-hydroxytabersonine (α; [243]), ( )-19*S*-hydroxytabersonine (α; [243]), (−)-tabersonine (α; [243]); **P6a**: (−)-venalstonine (α; [243]); **P6b**: (−)-19-epivindolinine (α; [243]), (−)-vindolinine (α; [243]). *M. polyadenus* (Baill.) Boiteau: **J9a**: epipandoline (β; [198]), (+)-pandoline (β; [198]). *M. scandens* Forst.: **C4c**: (+)-akuammidine (α; [245]); **P5**: (−)-episcandomeline [246], (+)-episcandomelonine [246], (−)-scandomeline [246], (−)-scandomelonine [246].

Muntafara (TAB). *M. sessilifolia* (Bak.) Pichon (see *Tabernaemontana sessilifolia*).

Neiosperma (RAU). *N. glomerata* (Bl.) Fosb. et Sach. (see *Ochrosia glomerata*). *N. lifuana* (Guill.) Boiteau (see *Ochrosia lifuana*). *N. miana* (Baill. ex White) Boiteau (see *Ochrosia miana*). *N. nakaiana* (Koidz.) Fosb. et Sach. (see *Ochrosia nakaiana*). *N. oppositifolia* (Lam.) Fosb. et Sach. (see *Ochrosia oppositifolia*). *N. poweri* (F. M. Bailey) Fosb. et Sach. (see *Ochrosia poweri*).

Ochrosia (RAU). *O. balansae* (Guill.) Guill. (= *Excavatia balansae* Guill.): **C4a**: (−)-aricine (α; [259]), (−)-isoreserpiline (α; [259]), (−)-reserpiline (α; [259]); **C5n**: (−)-10,11-dimethoxypicraphylline (α; [259, 260]). *O. borbonica* Gmel. (= *O. maculata* Jacq.): **C4b**: (−)-reserpine (α; [261]). *O. coccinea* (Teysm. et Binn.) Miq. [= *Bleekeria coccinea* (Miq.) Koidz., *Excavatia coccinea* (Teysm. et Binn.) Mgf.]: **C4a**: reserpiline [248]. *O. compta* K. Schum. [= *O. sandwicensis* auct. non A. DC., *Bleekeria compta* (K. Schum.) Wilbur.]: **C3a**: (+)-ochrosandwine (α; [262]); **C4a**: (−)-holeinine (α; [263]); **V3**: (+)-hunterburnine α-methochloride (α; [262]). *O. confusa* Pichon: **C3a**: (−)-10-methoxydihydrocorynantheol (α; [264]), (+)-ochrolifuanine (α; [264]); **C5b**: (+)-carapanaubine (α; [264]), (+)-rauvoxine (α; [264]). *O. elliptica* Labill. [= *Bleekeria elliptica* (Labill.) Koidz., *Excavatia elliptica* (Labill.) Mgf.]: **C4a**: (−)-isoreserpiline (α; [265]), reserpiline [257]. *O. glomerata* (Bl.) Valeton [= *Neiosperma glomerata* (Bl.) Fosb. et Sach.]: **C4a**: reserpiline [248]. *O. lifuana* Guill. [= *Neiosperma lifuana* (Guill.) Boiteau]: **C3a**: 3-dehydroochrolifuanine [249], ochrolifuanine A [249], ochrolifuanine B [249], ochrolifuanine *N*-oxide [249]; **C4b**: (−)-decarbomethoxydihydrogambirtannine [250]. *O. maculata* Jacq. (see *O. borbonica*). *O. miana* Baill. [ex Guill., nomen and *O. m.* Baill. ex White] [= *Neiosperma miana* (Baill. ex White) Boiteau]: **C3a**: hydroxyochrolifuanine [251], ochrolifuanine A [251], ochrolifuanine B [251], (−)-ochromianine (α; [252]); **C4b**: (−)-decarbomethoxydihydrogambirtannine [250]; **C4d**: (+)-ochromianoxine (α; [252]). *O. moorei* (F. v. Mull.) F. v. Müll. ex Benth.: **C4a**: reserpiline [248]. *O. mulsantii* Montrouz (= *O. vieillardii* Guill.): **C3a**: (−)-10-methoxydihydrocorynantheol (α; [266]), (−)-ochropposinine [255]; **C4a**: (−)-isoreserpiline (α; [266]), (−)-reserpiline (α; [255]); **C5n**: (−)-10,11-dimethoxypicraphylline (α; [255]). *O. nakaiana* (Koidz.) Koidz. ex Hara [= *Neiosperma nakaiana* (Koidz.) Fosb. et Sach.]: **C3a**: (+)-10-methoxycorynantheol β-methosalt [75, 253]; **C4a**: 10-methoxyserpentine [75], (+)-reserpiline (α; [75, 253]), serpentine [75, 253]; **C4c**: (+)-akuammidine (α; [75, 253]); **C5a**: (−)-vobasine (α; [75,253]). *O. oppositifolia* (Lam.) K. Schum. [= *Neiosperma oppositifolia* (Lam.) Fosb. et Sach.]: **C3a**: (−)-10-methoxydihydrocorynantheol (α; [254]), (+)-ochrolifuanine (α; [254]), ochropposinine [255]; **C4a**: (−)-isoreserpiline (α; [254]), (−)-isoreserpinine (α; [254]), (−)-reserpiline (α; [254]), (−)-reserpinine (α; [254]). *O. poweri* F. M. Bailey [= *Neiosperma poweri* (F. M. Bailey) Fosb. et Sach.]: **C4a**: (−)-isoreserpiline (α; [256]), reserpiline [257]; **C4b**: (−)-poweridine [256], reserpine [256]; **C5a**: (−)-ochropamine [258], (−)-ochropine [258]. *O. sandwicensis* auct. non A. DC. (see *O. compta*). *O. silvatica* Daen.: **C4a**: isoreserpiline [267]. *O. vieillardii* Guill. (see *O. mulsantii*). *O. vitiensis* (Mgf.) Pichon [= *Excavatia vitiensis* Mgf., *Bleekeria vitiensis* (Mgf.) A. C. Smith]: **C4a**: (+)-bleekerine [268], holeinine [268], isoreserpiline [268]; **C5m**: isoreserpiline-pseudoindoxyl [268].

Pagiantha (TAB). *P. cerifera* (Panch. et Sebert) Mgf. (see *Tabernaemontana cerifera*). *P. heyneana* (Wall.) Mgf. (see *Tabernaemontana heyneana*). *P. macrocarpa* (Jack) Mgf. (see *Tabernaemontana macrocarpa*). *P. sphaerocarpa* (Bl.) Mgf. (see *Tabernaemontana sphaerocarpa*).

Pandaca (TAB). *P. boiteaui* Mgf. (see *Tabernaemontana mocquerysii* for type; paratype Boiteau 2121 is *T. callosa*). *P. caducifolia* Mgf. (see *Tabernaemontana calcarea*). *P. calcarea* (Pichon) Mgf. (see *Tabernaemontana calcarea*). *P. crassifolia* (Pichon) Mgf. (see *Tabernaemontana crassifolia*). *P. debrayi* Mgf. (see *Tabernaemontana debrayi*). *P. eusepala* (A. DC.) Mgf. (see *Tabernaemontana eusepala*). *P. mauritiana* (Poir.) Mgf. et Boiteau (see *Tabernaemontana mauritiana*). *P. minutiflora* (Pichon) Mgf. (see *Tabernaemontana minutiflora*). *P. mocquerysii* (A. DC.) Mgf. var. *pendula* Mgf. (see *Tabernaemontana mocquerysii*). *P. ochrascens* (Pichon) Mgf. (see *Tabernaemontana humblotii*). *P.retusa* (Lam.) Mgf. (see *Tabernaemontana retusa*). *P. speciosa* Mgf. (see *Tabernaemontana humblotii*). *P. stellata* (Pichon) Mgf. (see *Tabernaemontana stellata*).

Peschiera (TAB). *P. accedens* (Müll. Arg.) ined. (see *Tabernaemontana accedens*). *P. affinis* (Müll. Arg.) Miers (see *Tabernaemontana affinis*). *P. australis* (Müll. Arg.) Miers (see *Tabernaemontana australis*). *P. fuchsiifolia* (A. DC.) Miers (see *Tabernaemontana fuchsiifolia*). *P. laeta* (Mart.) Miers (see *Tabernaemontana laeta*). *P. lundii* (A. DC.) Miers (see *Tabernaemontana lundii*). *P. psychotriifolia* (H. B. K.) Miers (see *Tabernaemontana psychotriifolia*).

Phrissocarpus (TAB). *P. rigidus* Miers (see *Tabernaemontana rigida*).

Picralima (CAR.) *P. klaineana* Pierre (see *P. nitida*). *P. nitida* (Stapf) Th. et H. Dur. (= *P. klaineana* Pierre): **C4a**: (−)-akuammigine (α; [298]), (−)-melinonine A (α; [299]); **C4c**: (+)-akuammidine (α; [28, 300]); **C4e**: (+)-akuammiline (α; [301]), (+)-desacetylakuammiline (α; [302]); **C5d**: (−)-desacetylpicraline (α; [303]), (−)-picraline (α; [302]); **C5h**: akuammine [302], (−)-pseudoakuammigine (α; [302]); **C5n**: (−)-picraphylline (α; [304]); **S4**: (−)-akuammicine (α; [28]), (±)-pseudoakuammicine (α + β; [27]). *P. umbellata* (K. Schum.) Stapf (see *Hunteria umbellata*).

Pleiocarpa (CAR). *P. mutica* Benth.: **C3a**: flavocarpine* [305], (+)-huntrabrine methochloride [306]; **C4c**: (+)-$N_{(a)}$, $N_{(b)}$-dimethylsarpagine chloride (α; [306]); **C4f**: (+)-pleiocarpamine (α; [307]), (+)-pleiocarpamine methochloride (α; [306]); **C4f-P6b**: (+)-pleiomutinine [307]; **E5a**: eburnamenine [307], eburnamine [307]; **E5a-P6a**: (−)-pleiomutine (α--; [308]); **V3**: hunterburnine α-methochloride [306], (−)-hunterburnine β-methochloride (α; [306]). *P. pycnantha* (K. Schum.) Stapf (= *P. pycnantha* var. *pycnantha*): **A4a**: (+)-tubotaiwine (α; [313]); **C3a**: (+)-huntrabrine methochloride [313]; **C4c**: (+)-$N_{(a)}$, $N_{(b)}$-dimethylsarpagine chloride (α; [313]), (+)-macusine B (α; [313, 309]); **C4f**: (+)-pleiocarpamine (α; [309]); **C4f-P5**: (+)-pycnanthinine (α-α; [309, 314]); **C4f-P6b**: (+)-pycnanthine (α--; [309]); **E5a**: (−)-eburnamine (β; [309, 314]); **E5a-P6a**: (−)-pleiomutine (α--; [313]); **S4**: (−)-19, 20-dihydroakuammicine (α; [313]), (−)-tubifolidine (α; [313]), (−)-tubifoline (α; [313]). *P. talbotii* Wernh.: **C4c**: (+)-normacusine B (α; [310]);

C5a: (−)-16-epiaffinine (α; [311]); **C5e**: talcarpine [311]; **C5o**: (−)-talpinine [311]; **D4f**: (−)-3,4-dehydrotalbotine (α; [310]), (−)-talbotine (α; [312]), (−)-3,4,5, 6-tetradehydrotalbotine (α; [312]); **D5b**: (+)-deformyltalbotinic acid methyl ester (α; [310]). *P. tubicina* Stapf (see *P. pycnantha*).

Plumeria (PLU). *P. retusa* Lam. (see *Tabernaemontana retusa*).

Polyadoa (CAR). *P. umbellata* (K. Schum.) Stapf (see *Hunteria umbellata*).

Rauvolfia (RAU). *R. affinis* Müll. Arg. (see *R. grandiflora*). *R. amsoniifolia* A. DC.: **C4a**: ajmalicine [315], aricine [315]; **C4b**: deserpidine [315], rescinnamine [315], reserpine [315], methyl reserpate [315], yohimbine [315]. *R. bahiensis* A. DC.: **C4a**: reserpiline [316]; **C4b**: rescinnamine [316], reserpine [316]. *R. beddomei* Hook. f.: **C4a**: (−)-ajmalicine (α; [8]); **C4c**: (+)-sarpagine (α; [8]). *R. boliviana* Mgf. (see *R. schuelii*). *R. caffra* Sond. (= *R. inebrians* K. Schum., *R. natalensis* Sond., *R. welwitschii* Stapf): **C4a**: ajmalicine [316, 317], reserpiline [316, 317], serpentine [317, 318], tetrahydroalstonine [318]; **C4b**: rescinnamine [316, 317], reserpine [316, 317, 319]; **C4c**: sarpagine [318]; **C5c**: (+)-ajmaline (α; [317, 318, 319]), raucaffricine [320]; **D6b**: raucaffrinoline [321], (+)-perakine [318]; **D6c**: peraksine [318]. *R. cambodiana* Pierre ex Pitard: **C4a**: aricine [322], (−)-isoreserpiline (α; [323]), reserpiline [322]; **C4b**: (−)-reserpine (α; [323]); **C5a**: pelirine [322]; **C5c**: ajmaline [322]. *R. canescens* L. (see *R. tetraphylla*). *R. capuronii* Mgf.: **C4b**: (+)-11-methoxypseudoyohimbine (α; [324]), (+)-11-methoxyyohimbine (α; [324]). *R. chinensis* (Hance) Hemsl. (see *R. verticillata*). *R. concolor* Pichon (in chemical literature published as *R. discolor* Pichon): **C4a**: (−)-isoreserpiline (α; [89, 336]), (−)-reserpiline (α; [337]), tetraphylline [336, 338], (−)-tetraphyllinine [336]; **C4b**: (−)-reserpine (α; [337]); **C5a**: (−)-16-decarbomethoxy-19,20-dihydrovobasine (α; [339]), (−)-tabernaemontanine (α; [336, 340]); **C5c**: (+)-quebrachidine (α; [336]). *R. confertiflora* Pichon: **C4a**: isoreserpiline [325], (+)-raufloridine (α; [325]), reserpiline [325], ( )-reserpiline methosalt (α; [326]), tetraphylline [325], tetraphyllinine [325]; **C4b**: reserpine [325]; **C4e**: (+)-raufloricine [327]; **C5c**: ajmaline [325], mauiensine [325], quebrachidine [325], (+)-rauflorine [325], tetraphyllicine [325]. *R. cubana* A. DC.: **C4a**: reserpiline [316]; **C4b**: deserpidine [316], rescinnamine [316], reserpine [316]. *R. cumminsii* Stapf: **C3a**: ( )-corynantheal (α; [328, 329]), ( )-corynantheol (α; [328, 329, 330]); **C4a**: ajmalicine [328, 331], ajmalicinine [328, 330], aricine [328, 331], 10, 11-dimethoxyajmalicine [328, 330], 19-episerpentine [328, 331], serpentine [328, 331]; **C4b**: 18-hydroxyyohimbine [328, 330], rescinnamine [328, 331], reserpine [328, 331], yohimbine [328, 331], α-yohimbine [328, 331]; **C4c**: *O*-methylnormacusine B [328, 329, 330], *O*-methylnormacusine B *N*-oxide [328, 330], normacusine B [328, 330], pericyclivine [328], sarpagine [328]; **C5c**: ajmaline [328], (0)-N_a-demethylpurpeline (α; [332]), N_a-demethyldihydropurpeline (α; [332]), diacetylajmaline [328], dihydronorpurpeline [328, 330], endolobine [328, 329], normitoridine [328, 329], norpurpeline [328, 330], norseredamine [328], nortetraphyllicine [328, 330], purpeline [328, 330], seredamine [328, 330], tetraphyllicine [328], 17-*O*-trimethoxybenzoylseredamine [328], vomalidine [328]; **C5d**: ( )-desacetylpicraline (α; [330]), (−)-picrinine (α; [329]); **C5n**: 20-epipicraphylline [328]; **D6c**: peraksine [330]. *R. decurva* Hook. f.: **C4a**: (−)-isoreserpiline (α; [333]), (−)-reserpiline (α; [333]), reserpinine [333]; **C4b**: rescinnamine [333], (−)-reserpine (α; [333]); **C4c**: (+)-sarpagine (α; [333]). *R. degeneri* Sherff: **C3a-C4a**: serpentinine [334]; **C4a**: tetraphylline [334]; **C5c**: (+)-ajmaline (α; [334]), tetraphyllicine [334]. *R. densiflora* (Wall.) Benth. et Hook. f.: **C4a**: reserpinine [8]; **C4b**: reserpine [335]; **C4c**: sarpagine [8]; **C5c**: ajmaline [335]. *R. discolor* Pichon (see *R. concolor*). *R. fruticosa* Burck: **C4a**: ajmalicine [8], aricine [8], serpentine [8]; **C4b**: yohimbine [8], α-yohimbine [8]; **C5c**: ajmaline [8]. *R. grandiflora* Mart. ex. A. DC. (= *R. affinis* Müll. Arg.): **C4a**: reserpiline [316], reserpinine [316]; **C4b**: deserpidine [316], pseudoreserpine [8], raugustine [8], raunescine [8], rescinnamine [316], reserpine [144]. *R. heterophylla* Roem. et Schult. (see *R. tetraphylla*). *R. hirsuta* Jacq. (see *R. tetraphylla*). *R. indecora* Woods. (see *R. ligustrina*). *R. inebrians* K. Schum. (see *R. caffra*). *R. lamarkii* A. DC. (see *R. viridis*). *R. ligustrina* Roem. et Schult. (= *R. indecora* Woods., *R. ternifolia* H. B. K.): **C3a-C4a**: serpentinine [341]; **C4a**: ajmalicine [341], aricine [341], isoreserpiline [341], isoreserpinine [341], reserpiline [316], reserpinine [316], serpentine [341], tetrahydroalstonine [341]; **C4b**: deserpidine [316], isopseudoreserpine [9], isoraunescine [341], isoreserpine [341], pseudoreserpine [341], (−)-raugustine (α; [341]), raunescine [341], renoxydine [341], rescinnamine [316], reserpine [316], yohimbine [341], α-yohimbine [341]; **C4c**: sarpagine [341]; **C5c**: ajmaline [341]; **C5m**: (−)-isoreserpiline-pseudoindoxyl (α; [8]). *R. littoralis* Rusby (= *R. macrocarpa* Standl.): **C4a**: reserpiline [316]; **C4b**: reserpine [316]. *R. longiacuminata* De Wild et Dur.: **C4b**: reserpine [342]. *R. longifolia* A. DC. (see *Tonduzia longifolia*). *R. macrocarpa* Standl. (see *R. littoralis*). *R. macrophylla* Stapf: **C4a**: ajmalicine [343], serpentine [343]; **C4b**: rescinnamine [343], reserpine [343], yohimbine [343], **C5c**: ajmaline [343], norajmaline [343]. *R. mannii* Stapf: **C4b**: reserpine [8]; **C5c**: (−)-vincamajine (α; [344]). *R. mattfeldiana* Mgf.: **C4a**: reserpiline [8]; **C4b**: pseudoreserpine [8], raugustine [8]. *R. mauiensis* Sherff: **C3a-C4a**: serpentinine [334]; **C5c**: ajmalidine [345], (+)-mauiensine (α; [334]), (+)-sandwicine (α; [334]), tetraphyllicine [334]. *R. micrantha* Hook. f.: **C4a**: ajmalicine [346], reserpiline [8]; **C4b**: reserpine [346]; **C4c**: neosarpagine [8]; **C5c**: ajmaline [8]. *R. mombasiana* Stapf: **C3a**: corynantheol [347]; **C4a**: (−)-ajmalicine (α; [347]), reserpiline [316]; **C4b**: rescinnamine [316], reserpine [316], yohimbine [347], **C4c**: normacusine B [347]; **C5c**: ajmaline [8], endolobine [347], norpurpeline [347], (+)-purpeline (α; [347]); **D6c**: peraksine [347]. *R. nana* Bruce: **C4b**: reserpine [8]. *R. natalensis* Sond. (see *R. caffra*). *R. nitida* Jacq.: **C4a**: ajmalicine [348], aricine [348], isoreserpiline [348], (−)-isoreserpinine (α; [348]), (−)-rauniticine (α; [348]), (−)-raunitidine (α; [348]), reserpiline [316], reserpinine [348]; **C4b**: (−)-deserpideine (α; [349]), (−)-deserpidine (α; [349]), rescinnamine [316], reserpine [316]. *R. obscura* K. Schum.: **C3a**: 10-methoxygeissoschizol [350]; **C4a**: (+)-alstonine (α; [342, 351, 352]), tetrahydroalstonine [351]; **C4b**: 19, 20-dehydroyohimbine [353], deserpidine [351], methyl deserpidate [351], methyl reserpate [351], rescinnamine

[351], reserpine [351], α-yohimbine [350, 351]; **C5c**: ajmaline [351], 12-methoxyajmaline [351], norajmaline [351], rauvomitine [351], tetraphyllicine [351], 17-*O*-3′, 4′, 5′-trimethoxybenzoylajmaline [351], vomalidine [351]. *R. oreogiton* Mgf. (see *R. volkensii*). *R. paraensis* Ducke: **C4a**: reserpiline [316]; **C4b**: rescinnamine [316], reserpine [316]. *R. pentaphylla* (Hub.) Ducke: **C4a**: ajmalicine [316], reserpiline [316], reserpinine [316]; **C4b**: deserpidine [316], rescinnamine [316], reserpine [316]. *R. perakensis* King et Gamble: **C4a**: aricine [355], (−)-isoreserpiline (α; [355]); **C4b**: reserpine [355], α-yohimbine [356]; **C4c**: normacusine B [356], (+)-sarpagine (α; [355]); **C5a**: (−)-pelirine [355]; **C5c**: (+)-ajmaline (α; [355]), 1-demethyl-17-*O*-acetyldesoxyajmalinediene-1,19 [356]; **D6b**: (+)-perakine (α; [355]); **D6c**: (+)-peraksine (α; [356]). *R. reflexa* Teijsm. et Binn.: **C5c**: purpeline [357], (+)-reflexine (α; [357]). *R. rosea* K. Schum.: **C4a**: ajmalicine [316], deserpidine [316], rescinnamine [316], reserpiline [316]; **C4b**: reserpine [316]. *R. salicifolia* Griseb.: **C4a**: reserpiline [316]; **C4b**: deserpidine [316], rescinnamine [316], reserpine [316]. *R. sandwicensis* A. DC.: **C3a-C4a**: serpentinine [334]; **C4a**: reserpiline [316], tetraphylline [334]; **C4b**: reserpine [316]; **C5c**: (+)-sandwicine (α; [334]), tetraphyllicine [334]. *R. sarapiquensis* Woods.: **C4b**: reserpine [358]. *R. schuelii* Speg. (= *Aspidosperma spegazzini* Molfino ex T. Meyer, *R. boliviana* Mgf.): **C4a**: (−)-aricine (α; [359]), (−)-isoreserpiline (α; [359]), (−)-reserpiline (α; [359]); **C4b**: (−)-reserpine (α; [359]); **C4c**: (+)-$N_{(b)}$-methylakuammidinechloride (α; [360]), (+)-spegatrine (α; [360]); **C5c**: (+)-ajmaline (α; [360]). *R. sellowii* Müll. Arg.: **C4a**: (−)-ajmalicine (α; [361]), (−)-aricine (α; [361]), serpentine [8], (−)-tetrahydroalstonine (α; [361]); **C4b**: (−)-reserpine (α; [361]); **C5c**: ajmalidine [361], (+)-ajmaline (α; [361]), (+)-tetraphyllicine (α; [361]). *R. serpentina* (L.) Benth.: **C3a-C4a**: (+)-serpentinine [362]; **C4a**: (−)-ajmalicine (α; [362]), (−)-reserpiline (α; [362]), (−)-reserpinine (α; [362]), (+)-serpentine (α; [362]); **C4b**: (−)-corynanthine (α; [362]), deserpidine [363], (−)-isorauhimbine (α; [364]), (−)-methyl reserpate (α; [362]), raunescine [8], (−)-rescinnamine [362, 363], renoxydine [365], (−)-reserpine (α; [362, 363]), (+)-yohimbine (α; [366]), (−)-α-yohimbine (α; [362]); **C4c**: (+)-sarpagine (α; [362]); **C5c**: (+)-ajmaline (α; [362]), (+)-isoajmaline (α; [362]), tetraphyllicine [362]; **C5q**: (−)-rauvolfinine [367]. *R. sprucei* Müll. Arg.: **C4a**: reserpiline [316]; **C4b**: deserpidine [316], rescinnamine [316], reserpine [316]. *R. suaveolens* S. Moore: **C5e**: (0)-suaveoline [369]. *R. sumatrana* Jack: **C3a-C4a**: serpentinine [368]; **C4a**: ajmalicine [368], aricine [368], reserpiline [316], serpentine [368]; **C4b**: rescinnamine [316], reserpine [316], yohimbine [368], α-yohimbine [368]; **C5c**: ajmaline [368], (+)-$N_{(a)}$-demethylseredamine (α; [368]). *R. ternifolia* H. B. K. (see *R. ligustrina*). *R. tetraphylla* L. (= *R. canescens* L., *R. heterophylla* Roem. et Schult., *R. hirsuta* Jacq.): **C3a-C4a**: serpentinine [370]; **C4a**: (−)-ajmalicine (α; [371]), alstonine [372], (−)-aricine (α; [373, 374]), (−)-isoreserpiline (α; [374]), (−)-isoreserpinine (α; [374]), (−)-reserpiline (α; [373, 374]), (−)-reserpinine (α; [371]), serpentine [375], (−)-tetraphylline [370]; **C4b**: (+)-canembine [8], corynanthine [8], (−)-deserpidine (α; [363]), (−)-isoraunescine [376], (−)-pseudoreserpine [377], (+)-pseudoyohimbine (α; [370]), (−)-raunescine [376], (−)-raujemidine [378], renoxydine [365], (−)-reserpine (α; [373]), (−??

[375])-yohimbine (α; [375]), (−)-α-yohimbine (α; [373, 374]), (−)-β-yohimbine (α; [379]); **C4c**:(+)-sarpagine (α; [371]); **C5c**:(+)-ajmaline (α; [371]), (+)-tetraphyllicine (α; [370]). *R. verticillata* (Lour.) Baill. [= *R. chinensis* (Hance) Hemsl.]: **C4a**:(−)-ajmalicine (α; [8, 380], aricine [9], serpentine [381]; **C4b**:(−)-reserpine (α; [8, 381]), yohimbine [381]; **C4c**:(+)-vellosimine (α; [382]); **C5c**:ajmaline [8, 381]; **D6c**:(+)-peraksine (α; [382]). *R. viridis* Roem. et Schult. (= *R. lamarkii* A. DC.): **C4a**:reserpiline [316], reserpinine [316]; **C4b**:deserpidine [316], rescinnamine [316], reserpine [316]. *R. volkensii* Stapf (= *R. oreogiton* Mgf.): **C4a**:ajmalicine [354], reserpiline [354], serpentine [354]; **C4b**:renoxydine [354], rescinnamine [354], reserpine [354], α-yohimbine [354]; **C5c**:ajmaline [354]. *R. vomitoria* Afzel.: **C3a**:geissoschizol [383], 10-hydroxygeissoschizol [384], 10-methoxygeissoschizol [384]; **C3a-C4a**:serpentinine [8]; **C4a**:ajmalicine [8], (+)-alstonine (α; [352]), (−)-aricine (α; [385, 386]), 19, 20-dehydroreserpiline [384], (−)-isoreserpiline (α; [385, 387]), (+)-raumitorine (α; [388]), (+)-rauvanine (α; [388]), (−)-reserpiline (α; [384, 386, 389]), tetrahydroalstonine [385]; **C4b**:deserpidine [363], 18-hydroxyyohimbine [384], methyl reserpate [384], rescinnamine [386], (−)-renoxydine (α; [365]), (−)-rescidine [8], (−)-reserpine (α; [390]), (−)-seredine [391], (+)-yohimbine (α; [384, 392]), (−)-α-yohimbine (α; [386, 392]); **C4c**:normacusine B [384], (+)-sarpagine (α; [384, 387]); **C4e**:(+)-desacetyldeformoakuammiline (α; [393]); **C5b**:(−)-carapanaubine (α; [384, 385, 386]), carapanaubine *N*-oxide [386], isocarapanaubine [384], (+)-rauvoxine (α; [386, 394]), (+)-rauvoxinine (α; [394]); **C5c**:(+)-17-acetylajmaline (α; [395]), ajmaline [386], $N_{(a)}$-demethylrauvomitine [386], 10-hydroxynortetraphyllicine [384], indolenine RG [384], isoajmaline [8], isosandwicine [386], (+)-mitoridine (α; [386, 396]), norpurpeline [384]), norseredamine [384], nortetraphyllicine [384], (+)-purpeline (α; [384, 386, 396]), (−)-rauvomitine (α; [386, 396]), sandwicine [386], (+)-seredamine (α; [386]), tetraphyllicine [386], (+)-vomalidine (α; [397]), (−)-vomilenine (α; [398]); **C5d**:(−)-picrinine (α; [393]); **C5e**:suaveoline [386]; **C5m**:(−)-isoreserpiline-pseudoindoxyl (α; [384, 386]); **D6c**:(+)-peraksine (α; [383]). *R. welwitschii* Stapf (see *R. caffra*). *R. yunnanensis* Tsiang: **C4a**:ajmalicine [9]; **C4b**:reserpine [9].

Rejoua (TAB). *R. aurantiaca* (Gaud.) Gaud. (see *Tabernaemontana aurantiaca*).

Rhazya (PLU). *R. orientalis* A. DC.: **D3a**:(−)-5α-carboxystrictosidine (α; [400]), (−)-strictosidine (α; [401]). *R. stricta* Decaisne: **C3a**:(+)-geissoschizine (α; [402]); **C4c**:(0)-akuammidine (α; [108]), (0)-akuammidine methosalt (α; [300]); **C4e**:(+)-rhazinaline [402], (−)-strictalamine (α; [403]); (+)-strictamine (α; [403]); **D3a**:(−)-strictosidine (α; [401]); **E5a**:eburnamenine [108], eburnamine [108], eburnamonine [108]; **S4**:(−)-norfluorocurarine (α; [403]), (−)-sewarine (α; [403, 404, 405]); **V3**:(−)-antirhine (α; [406]); **V4**:(−)-strictosamide (α; [406]).

Schizozygia (TAB). *S. coffaeoides* Baill.: **E5b**:(−)-schizophylline [407, 408];

E6a: (−)-isoschizogaline [407, 408], (−)-isoschizogamine [407, 408], (+)-schizogaline [407, 408], (−)-schizogamine [407, 408], (+)-schizozygine [407], (+)-α-schizozygol [407, 408].

Stemmadenia (TAB). *S. donnell-smithii* (Rose) Woods.: **A3**: (+)-stemmadenine (α; [188, 409]); **C5a-J7a**: (−)-voacamine (α-β; [409]); **J7a**: (−)-coronaridine (β; [410]), (−)-isovoacangine (β; [409]), (−)-tabernanthine (β; [409]), (−)-voacangine (β; [409]). *S. galeottiana* (A. Rich.) Miers: **J7a**: (−)-ibogamine (β; [409]). *S. obovata* (Hook. et Arn.) K. Schum.: **A3**: stemmadenine [410]; **C4a**: (−)-ajmalicine (α; [410]); **J7a**: coronaridine [410], voacangine [410]. *S. tomentosa* Greenman var. *palmeri* (Rose) Woods.: **A3**: stemmadenine [410]; **J7a**: coronaridine [410].

Stenosolen (TAB). *S. heterophyllus* (Vahl) Mgf. (see *Tabernaemontana heterophylla*).

Tabernaemontana (TAB). *T. accedens* Müll. Arg. [= *Peschiera accedens* (Müll. Arg.) ined.]: **C4c**: affinisine [287]; **C5a-C5a**: (−)-accedinine (α-α; [287]); (−)-accedinisine (α-α; [287]); **C5a**: (+)-accedine (α; [288]), (+)-*N*-demethyl-16-epiaccedine (α; [289]), (−)-$N_{(a)}$-methyl-16-epiaffinine (α; [288]); **C5a-J7a**: (−)-*N*-demethylvoacamine (α-β; [287]), (−)-voacamidine (α-β; [287]), (−)-voacamine (α-β; [287]), (−)-voacamine *N*-oxide (α-β; [287]). *T. affinis* Müll. Arg. [= *Peschiera affinis* (Müll. Arg.) Miers]: **C4c**: affinisine [290, 291]; **C5a**: (−)-affinine (α; [291]), vobasine [291]; **J7a**: coronaridine [290], (−)-19-epiheyneanine (β; [290]), heyneanine [290]; **J8a**: coronaridine-pseudoindoxyl [290]. *T. alba* Mill.: **J7a**: coronaridine [410]. *T. apoda* Wright (= *T. armeniaca* Areces ex Iglesias et L. Diatta): **P6e**: apodine [412], desoxoapodine [413]. *T. armeniaca* Areces ex Iglesias et L. Diatta, nomen (see *T. apoda*). *T. aurantiaca* Gaud. [= *Rejoua aurantiaca* (Gaud.) Gaud.]: **J7a**: voacangine [399]; **J8a**: iboluteine [399], (−)-voaluteine [399]. *T. australis* Müll. Arg. [= *Peschiera australis* (Müll. Arg.) Miers]: **C5a-J7a**: voacamine [191]; **J7a**: voacangine [191]. *T. barteri* Hook. f. (see *Callichilia barteri*). *T. brachyantha* Stapf [= *Conopharyngia brachyantha* (Stapf) Stapf]: **C4c**: (+)-normacusine B (α; [160]); **C5a-J7a**: (−)-19-epivoacorine (α-β; [160]), (−)-voacorine (α-β; [160]); **C6c**: (−)-anhydrovobasinediol (α; [160]). *T. calcarea* Pichon [= *Pandaca caducifolia* Mgf., *P. calcarea* (Pichon) Mgf.]: **C5a**: (−)-dregamine (α; [275, 277]); **J9a**: (+)-20-epipandoline (β; [275, 276]), (+)-pandoline (β; [275, 276, 277]), (+)-pseudotabersonine (β; [275]), (+)-20*R*-pseudovincadifformine (β; [275]); **J10a**: (+)-pandine [277]. *T. callosa* Pichon (Boiteau 2121, paratype of *Pandaca boiteaui* Mgf.), *T. capuronii* Leeuwenberg, nom. nov. (= *Capuronetta elegans* Mgf. non *T. elegans* Stapf): **C5a-J8b**: (−)-capuvosine (α-β; [121]), (−)-dehydroxycapuvosine [122], (−)-*N*-demethylcapuvosine [122]; **C5a-J9a**: (−)-capuvosidine (α-β; [122]); **J8b**: (+)-capuronine (β; [121]); **J9a**: (−)-14,15-anhydrocapuronidine (β; [122]), (−)-14,15-anhydro-1,2-dihydrocapuronidine (β; [122]), (+)-capuronidine (β; [121]). *T. cerifera* Panch. et Sebert [= *Pagiantha cerifera* (Panch. et Sebert) Mgf.]: **C5a**:

(−)-vobasine (α; [269]); **J7a**:(−)-ibogaine (β; [270]), (−)-voacangine (β; [270]), (−)-hydroxyindoleninevoacangine (β; [270]). *T. chartacea* Pichon (= *T. eglandulosa* Stapf). *T. citrifolia* L. [= *T. oppositifolia* (Spreng.) Urb.]: **C5a-C7a**: voacamine [191]; **J7a**: coronaridine [191], ibogamine [191], voacangine [191]. *T. coffeoides* Boj. ex A. DC. [= *Hazunta angustifolia* Pichon, *H. coffeoides* (Boj. ex A. DC.) Pichon, *H. costata* Mgf., *H. membranacea* (A. DC.) Pichon, *H. membranacea* (A. DC.) Pichon forma *pilifera* Mgf., *H. modesta* (Bak.) Pichon, *H. modesta* (Bak.) Pichon var. *methuenii* (Stapf et M.L. Green) Pichon subvar. *methuenii*, *H. modesta* (Bak.) Pichon var. *methuenii* (Stapf et M. L. Green) Pichon subvar. *velutina* (Pichon) Mgf., *H. modesta* (Bak.) Pichon var. *modesta* subvar. *modesta*, *H. silicola* Pichon, *H. velutina* Pichon, *Tabernaemontana membranacea* A. DC., *T. modesta* Bak.]: **A3**: stemmadenine [215]; **C4a**: isoreserpiline [215], reserpiline [215]; **C4c**: normacusine B [215]; **C5a**: (−)-dregamine (α; [215, 216, 218]), (−)-tabernaemontanine (α; [215, 216, 218]), (+)-voacarpine (α; [218]), (−)-vobasine (α; [215, 218]); **C5a-J7a**: (−)-tabernaelegantine A [217]; **C5c**: dimethoxytetraphyllicine [215], monomethoxytetraphyllicine [215], tetraphyllicine [215], trimethoxytetraphyllicine [215]; **J7a**: (−)-coronaridine (β; [217]), (−)-epiheyneanine (β; [217]), (−)-ibogamine (β; [215, 216]), isovoacangine [215], (−)-voacangine (β; [217]); **S4**: vincanidine [215]. *T. contorta* Stapf [= *Conopharyngia contorta* (Stapf) Stapf]: **J7a**: conopharyngine [161], coronaridine [161], (−)-ibogaine (β; [161]), (−)-voacangine (β; [161]), voacristine [161]. *T. coronaria* (Jacq.) Willd. (see *T. divaricata*). *T. crassa* Benth. [= *Conopharyngia crassa* (Benth.) Stapf, *C. durissima* (Stapf) Stapf, *C. jollyana* Stapf, *Tabernaemontana durissima* Stapf]: **C4c**: *O*-acetylpolyneuridine [174]; **C4e**: (+)-akuammiline (α; [165]); **C5a-J7a**: (−)-conoduramine (α-β; [166]), (−)-conodurine (α-β; [166]); **C6c**: (−)-anhydrovobasinediol (α; [165]); **J7a**: (−)-conopharyngine (β; [162, 167]), (−)-coronaridine (β; [168, 175]), heyneanine [174], (−)-19-hydroxyconopharyngine [162, 174], 19-hydroxycoronaridine [174], (−)-hydroxyindoleninecoronaridine (β; [168]), 19-hydroxy-3-oxocoronaridine [176], (−)-isovoacangine (β; [167]), (−)-jollyanine [174], (−)-3-oxoconopharyngine [176], 3-oxocoronaridine [176], voacristine [174]; **J8d**: (+)-crassanine [162]. *T. crassifolia* Pichon [= *Pandaca crassifolia* (Pichon) Mgf.]: **J7a**: (−)-ibogamine (β; [278]), (−)-tabernanthine (β; [278]). *T. debrayi* (Mgf.) Leeuwenberg, comb. nov. (= *Pandaca debrayi* Mgf.): **C5a**: (−)-dregamine (α; [277, 278]); **J9a**: (+)-pandoline (β; [277]); **J10a**: (+)-pandine [277]. *T. dichotoma* Roxb. [= *Ervatamia dichotoma* (Roxb.) Blatter]: **J7a**: coronaridine [194], (−)-heyneanine (β; [195]), (−)-hydroxyindoleninevoacristine (β; [195]). *T. divaricata* (L.) R. Br. ex Roem. et Schult. [= *Ervatamia coronaria* (Jacq.) Stapf, *E. divaricata* (L.) Burkill, *Tabernaemontana coronaria* (Jacq.) Willd.]: **C5a**: dregamine [190, 191], (−)-tabernaemontanine (α; [190, 191, 196]; **J7a**: (−)-coronaridine (β; [190, 192, 196], (−)-3-oxocoronaridine (β; [192]), (−)-voacangine (β; [193, 196]), voacristine [190]; **P4**: (+)-conoflorine [196]; **P5**: (−)-lochnericine (α; [196]. *T. durissima* Stapf (see *T. crassa*). *T. eglandulosa* Stapf [= *Gabunia eglandulosa* (Stapf) Stapf, *G. longiflora* Stapf, *T. chartacea* Pichon]: **C5a**: perivine [202], vobasine [202]; **C5a-J7a**: voacamine [202]; **J7a**: conopharyngine [161], (−)-coronaridine (β; [161, 202]), (−)-eglandulosine (β; [203]), 19-hydroxycoronaridine [202], (−)-3-hydroxyisovoacangine (β; [202]), (−)-isovoacangine (β; [202]), (−)-voacangine (β; [161]); **J8c** : (−)-eglandine (β; [203]). *T. elegans* Stapf [= *Conopharyngia elegans* (Stapf) Stapf]: **C5a** : (−)-dregamine (α; [169]), (−)-tabernaemontanine (α; [169]); **C5a-J7a** : (−)-conoduramine (α-β; [169, 170]), (−)-tabernaelegantine A [169, 170], (+)-tabernaelegantine B [169, 170], (−)-tabernaelegantine C [169, 170], (+)-tabernaelegantine D [169, 170] *T. eusepala* Aug. DC. [= *Pandaca eusepala* (Aug. DC.) Mgf.]: **A4a** : (+)-19,20-dihydrocondylocarpine (α; [279]); **C5a** : (−)-vobasine (α; [279]); **J7a** : (−)-19-epivoacristine (β; [279]), (−)-ibogaine (β; [279]), (+)-hydroxyindolenineibogaine (β; [279]); **J8b** : (+)-20*R*-dihydrocleavamine (β; [279]), (−)-20*S*-dihydrocleavamine (β; [279]); **J9a** : (+)-20*S*-1,2-dehydropseudoaspidospermidine (β; [279]). *T. fuchsiifolia* A. DC. [= *Peschiera fuchsiifolia* (A. DC.) Miers]: **C4c** : (+)-affinisine (α; [292]), ( )-voachalotine (α; [292]). *T. heterophylla* Vahl [= *Stenosolen heterophyllus* (Vahl) Mgf.]: **P5-J10d** : (+)-ervafoline (α-β; [411]). *T. heyneana* Wall. [= *Pagiantha heyneana* (Wall.) Mgf.]: **J7a** : (−)-coronaridine (β; [271]), (−)-heyneanine (β; [272]), (−)-ibogamine (β; [271]), 3-oxocoronaridine [271], (−)-voacangine (β; [271, 273a]); **J8a** : voacangine-pseudoindoxyl [271]. *T. holstii* K. Schum. [= *Conopharyngia holstii* (K. Schum.) Stapf]: **A4a** : (+)-tubotaiwine *N*-oxide (α; [414]); **C4c** : pericyclivine [415]; **C5a** : perivine [415], vobasine [415]; **C5a-J7a** : conoduramine [415], condodurine [415], gabunine [415], 19-oxoconodurine [415]; **J7a** : coronaridine [415], 19-oxocoronaridine [415]. *T. humblotii* (Baill.) Pichon [= *Pandaca ochrascens* (Pichon) Mgf., *P. speciosa* Mgf., *T. ochrascens* Pichon]: **A4a** : (+)-19,20-dihydrocondylocarpine (α; [283]); **C4c** : (+)-akuammidine (α; [283]); **C5a-J7a** : (−)-decarbomethoxyvoacamine (α-β; [286]); **E5a** : (+)-16-epi-14,15-dehydrovincamine (α; [283]); **J7a** : (−)-19-epiiboxygaline (β; [283]), (−)-ibogaine (β; [283, 286]), (−)-ibogaline (β; [283]), (−)-19-epiiboxygaine (β; [283]), (−)-iboxygaine (β; [286]), (−)-voacristine (β; [286]), (−)-voacangine (β; [286]); **J8a** : iboluteine [283, 286]; **S4** : (−)-akuammicine (α; [283]). *T. johnstonii* (Stapf) Pichon (= *Conopharyngia johnstonii* Stapf): **A4a** : (+)-tubotaiwine (α; [171]), (+)-tubotaiwine *N*-oxide (α; [171]); **C4c** : pericyclivine [172]; **C5a** : perivine [172]; **C5a-J7a** : conoduramine [172], conodurine [172], 19,20-epoxyconoduramine [172], gabunamine [172], gabunine [172], (−)-tabernamine (α-β; [172, 173]); **J7a** : ibogamine [172], isovoacangine [172]. *T. laeta* Mart. [= *Peschiera laeta* (Mart.) Miers]: **C3a** : (−)-geissoschizol (α; [293, 294]); **C4c** : (+)-akuammidine (α; [293]), (+)-normacusine B (α; [293]); **C5a** : (−)-affinine (α; [293]), (−)-vobasine (α; [293]); **C5a-J7a** : (−)-conodurine (α-β; [293]), (−)-voacamine (α-β; [293]). *T. laevis* Vell. (see *Geissospermum laeve*). *T. laurifolia* L.: **J7a** : coronaridine [416], ibogamine [416], iboxygaine [416], isovoacangine [416], (−)-isovoacristine (β; [416]), tabernanthine [416]. *T. longiflora* Benth. [= *Conopharyngia longiflora* (Benth.) Stapf]: **C5a-J7a** : 18′-decarbomethoxyvoacamine [9], voacamine [9], voacorine [9]; **J7a** : conopharyngine [177], voacangine [177], voacristine [9]. *T. lundii* A. DC. [= *Peschiera lundii* (A. DC.) Miers]: **C5a** : vobasine [295]; **J7a** : coronaridine [295], (−)-19-epivoacristine (β; [295]), ibogaine [295], iboxygaine [295], (+)-hydroxyindolenineiboxygaine (β; [295]), voacangine [295], (−)-

voacristine (β; [295]); **J8a**: (−)-voacristine-pseudoindoxyl (β; [295]). *T. macrocarpa* Jack [= *Ervatamia macrocarpa* (Jack) Merr., *Pagiantha macrocarpa* (Jack) Mgf.]: **J7a**: coronaridine [273b], (−)-voacangine (β; [273b]), hydroxyindoleninevoacangine [273b]; **P4**: conoflorine [273b]. *T. macrophylla* Müll. Arg. non Poir. (see *T. rigida*). *T. mauritiana* Poir. [= *Pandaca mauritiana* (Poir.) Mgf. et Boiteau]: **A4a**: tubotaiwine [280]; **C5a**: dregamine [277, 280], vobasine [277, 280]. *T. membranacea* A. DC. (see *T. coffeoides*). *T. minutiflora* Pichon [= *Pandaca minutiflora* (Pichon) Mgf.]: **A3**: (+)-stemmadenine (α; [281]); **A4a**: (+)-condylocarpine (α; [281]), (+)-tubotaiwine (α; [281]); **C5a**: (−)-vobasine (α; [281]); **J7a**: (−)-coronaridine (β; [281]); **P5**: (+)-vincadifformine (β; [281]). *T. mocquerysii* A. DC. [= *Pandaca boiteaui* Mgf. excl. paratype Boiteau 2121, which belongs to *P. callosa*, *P. mocquerysii* (A. DC.) Mgf. var. *pendula* Mgf.]: **J7a**: (−)-coronaridine (β; [282]), (−)-19-epiheyneanine (β; [282]), (−)-heyneanine (β; [282]), (−)-19-epivoacristine (β; [282]), (−)-voacristine (β; [282]), (−)-voacangine (β; [282]). *T. modesta* Bak. (see *T. coffeoides*). *T. mucronata* Merr. [= *Ervatamia mucronata* (Merr.) Mgf.]: **C5a**: tabernaemontanine [197]; **J7a**: coronaridine [197]. *T. noronhiana* Boj. ex A. DC. (see *T. retusa*). *T. ochrascens* Pichon (see *T. humblotii*). *T. odoratissima* (Stapf) Leeuwenberg, comb. nov. (= *Gabunia odoratissima* Stapf): **C4c**: (+)-pericyclivine (α; [204]); **C5a**: (−)-perivine (α; [204]), (−)-vobasine (α; [204]); **C5a-J7a**: (−)-conoduramine (α-β; [204]), (−)-conodurine (α-β; [204]), (−)-gabunine (α-β; [204]); **J7a**: (−)-coronaridine (β; [204]), (−)-ibogamine (β; [204]), (−)-isovoacangine (β; [204]). *T. oppositifolia* (Spreng.) Urb. (see *T. citrifolia*). *T. orientalis* R. Br. [= *Ervatamia obtusiuscula* Mgf., *E. orientalis* (R. Br.) Domin]: **C5a**: (−)-dregamine (α; [199]), (−)-tabernaemontanine (α; [199]), vobasine [199]; **C5a-J7a**: (+)-16-decarbomethoxydihydrovoacamine (α-β; [199]), (+)-16-decarbomethoxy-20′-epidihydrovoacamine (α-β [199]), decarbomethoxyvoacamine [199], voacamine [199]; **C7b**: (+)-19-dehydroervatamine (α; [199]), (−)-20-epiervatamine (α; [199]), (−)-ervatamine (α; [199]); **J7a**: (−)-ibogaine (β; [199]), (−)-iboxygaine (β; [199]), (−)-voacristine (β; [199]); **J9a**: ( )-20-epipandoline (β; [198]), ( )-pandoline (β; [198]). *T. pachysiphon* Stapf [= *Conopharyngia cumminsii* Stapf, *C. pachysiphon* (Stapf) Stapf, *Tabernaemontana pachysiphon* var. *cumminsii* (Stapf) H. Huber]: **C5a**: (−)-affinine (α; [161]); **J7a**: (−)-conopharyngine (β; [161]), coronaridine [161], jollyanine [163], 19-hydroxyconopharyngine [164], (−)-voacangine (β; [161]); **J8a**: conopharyngine-pseudoindoxyl [164]. *T. pandacaqui* Poir. [= *Ervatamia pandacaqui* (Poir.) Pichon]: **C4c**: deformoakuammidine [200]; **C5a**: (−)-tabernaemontanine (α; [200]); **J7a**: (−)-coronaridine (β; [201]); **J10d-P5**: (+)-ervafolidine (α-β; [200]), (+)-ervafoline (α-β; [200]), (+)-isoervafolidine (α-β; [200]); **S4**: (+)-20-epilochneridine (β; [200]). *T. penduliflora* K. Schum. [= *Conopharyngia penduliflora* (K. Schum.) Stapf]: **J7a**: (−)-conopharyngine (β; [161]), coronaridine [161], voacangine [161]. *T. psychotriifolia* H. B. K. [= *Peschiera psychotriifolia* (H. B. K.) Miers]: **C5a**: (−)-affinine (α; [296]), 16-epivobasinic acid [296]; **C5a-J7a**: voacamine [191]; **C6c**: (−)-anhydrovobasinediol (α; [296, 297]); **J7a**: coronaridine [191], voacangine [191]. *T. retusa* (Lam.) Pichon [= *Conopharyngia retusa* (Lam.) G. Don, *Pandaca* *retusa* (Lam.) Mgf., *Plumeria retusa* Lam., *Tabernaemontana noronhiana* Boj. ex A. DC.]: **J7a**: (−)-coronaridine (β; [278, 284]), (−)-heyneanine (β; [278, 284]), (−)-hydroxyindoleninecoronaridine (β; [278]), (±)-ibogamine (α + β; [278]), 3-oxovoacangine [284], (−)-voacangine (β, [284, 285]), voacristine [284]. *T. rigida* (Miers) Leeuwenberg, comb. nov. [= *Anacampta rigida* (Miers) Mgf., *Phrissocarpus rigidus* Miers, *Tabernaemontana macrophylla* Müll. Arg. non Poir.]: **E5a**: (+)-apovincamine (α; [81]), (−)-16-epivincamine (α; [81]), (±)-16-epivincamine (α + β; [81]), (+)-vincamine (α; [81]), (±)-vincamine (α + β; [81]). *T. rupicola* Benth. [= *Anacampta rupicola* (Benth.) Mgf.]: **J8a**: (−)-rupicoline [82], (−)-voacristine-pseudoindoxyl (β; [82]). *T. sessilifolia* Bak. [= *Muntafara sessilifolia* (Bak.) Pichon]: **C5a**: (−)-dregamine (α; [247]), (−)-tabernaemontanine (α; [247]; **J7a**: (−)-coronaridine (β; [247]), (−)-eglandulosine (β; [247]), (−)-6-hydroxy-3-oxoisovoacangine (β; [247]), (−)-isovoacangine (β; [247]); **J8c**: (−)-eglandine (β; [247]), (−)-3,6-oxidoisovoacangine (β; [247]). *T. siphilitica* (L. f.) Leeuwenberg, comb. nov. [= *Bonafusia tetrastachya* (H. B. K.) Mgf., *Echites siphilitica* L. f. *Tabernaemontana tatrastachya* H. B. K.]: **C3a** (+)-geissoschizine (α; [111]); **J7a**: (−)-bonafousine (β; [112]), (−)-coronaridine (β; [113]), (−)-isovoacangine (β; [113]), (−)-voacangine (β; [113]); **J7a-J7a**: (−)-bis-(11-hydroxycoronaridinyl-12) (β-β; [113]). *T. sphaerocarpa* Bl. [= *Pagiantha sphaerocarpa* (Bl.) Mgf.]: **C5a**: dregamine [274], tabernaemontanine [274]. *T. stellata* Pichon [= *Pandaca stellata* (Pichon) Mgf.]: **J7a**: (−)-coronaridine (β; [278]). *T. subsessilis* Benth. (see *Callichilia subsessilis*). *T. tetrastachya* H. B. K. (see *T. siphilitica*). *T. undulata* Vahl: **J7a**: coronaridine [417].

Tabernanthe (TAB). *T. iboga* Baill. [418]: **J7a**: (−)-coronaridine (β; [419]), (+)-hydroxyindolenineibogaine (β; [420]), (+)-hydroxyindolenineibogamine [420], (−)-ibogaine (β; [420]), (−)-ibogaline (β; [418]), (−)-ibogamine (β; [420]), (−)-iboxygaine (β; [420]), (−)-isovoacangine (β; [418]), (−)-tabernanthine (β; [420]), (−)-voacangine (β; [420]); **J8a**: demethoxyiboluteine [420], (−)-iboluteine [420, 421]; **J8d**: (−)-kisantine [420]; **J10b**: (+)-iboxyphylline (β; [422]); **J10c**: (+)-ibophyllidine (β; [422]). *T. subsessilis* Stapf: **J7a**: (−)-ibogamine (β; [422]); **J10b**: (+)-iboxyphylline (β; [422]); **J10c**: (+)-ibophyllidine (β; [422]).

Tonduzia (PLU). *T. longifolia* (A. DC.) Mgf. (= *Rauvolfia longifolia* A. DC.): **C4a**: ajmalicine [316]; **C4b**: deserpidine [423], rescinnamine [316], reserpine [316]; **C5c**: ajmaline [423], (−)-vincamajine (α; [424]).

Vallesia (RAU). *V. dichotoma* Ruiz et Pav. (see *V. glabra*). *V. glabra* (Cav.) Link. (= *V. dichotoma* Ruiz et Pav.): **A4a**: (−)-$N_{(a)}$-acetylaspidospermatidine (α; [425]), condylocarpine [425], (+)-19,20-dihydrocondylocarpine (α; [425]); **A4b**: precondylocarpine [425]; **A6**: (+)-dichotine (α; [426, 427]), (+)-11-methoxydichotine (α; [426, 427]); **C4b**: reserpine [428]; **C4c**: (+)-akuammidine (α; [425]); **C4f**: pleiocarpamine [425]; **E5b**: (−)-vallesamidine (β; [429]); **V3**: (+)-vallesiachotamine (α; [430]).

Vinca (PLU). *V. difformis* Pourr.: **C4c**: (+)-akuammidine (α; [431, 432]), (+)-sarpagine (α; [433]), (+)-vellosimine (α; [431]); **C5a**: (−)-vincadiffine (α; [434]); **C5c**: (−)-vincamajine (α; [431, 435]), vincamedine (α; [431, 435]); **E5a**: (+)-vincamine (α; [436]). *V. elegantissima* Hort. (see *V. major* ssp. *major* cv. *variegata*). *V. erecta* Rgl. et Schmalh.: **C4a**: ajmalicine [8], ervine [8], reserpinine [8]; **C4c**: akuammidine [9], (+)-*O*-benzoyltombozine (α; [439]), (+)-ervincidine [440], normacusine B [441]; **C5b**: (−)-*N*-acetylvinerine (α; [442, 443]), ericinine [444], (−)-majdine (α; [445]), (+)-vineridine (α; [442, 443]), (+)-vineridine *N*-oxide (α; [446]), (+)-vinerine (α; [443]), ( )-vinerine *N*-oxide (α; [447]), (−)-vinerinine [448]; **C5c**: 10-methoxyvinorine [449], (+)-vincarine (α; [450]); **C5d**: (0)-vincaricine [451], (−)-vincaridine [451, 452], (−)-vincarinine [453]; **C5h**: akuammine [9], *O*-methylakuammine [454]; **E5a**: (−)-eburnamonine (α; [455]), (+)-vincamine (α; [436]), vincine [456]; **P6a**: maidinine [447]; **S4**: (−)-akuammicine (α; [457]), (−)-norfluorocurarine (α; [458]), (−)-vincanicine [459, 460], (−)-vincanidine (α; [459, 461]), (−)-vinervine (α; [457, 461]), (−)-vinervinine [441, 459]. *V. herbacea* Waldst. et Kif. (= *V. libanotica* Zucc.): **C3a**: (−)-hervine (α; [462]); **C4a**: (−)-herbaceine (α; [463]), (−)-herbaine [464], isoreserpinine [465], (−)-rauniticine (α; [472]), (−)-reserpinine (α; [9, 472]); **C4e**: (+)-strictamine (α; [472]); **C5b**: (−)-herbaline (α; [466]), herboxine [467], (−)-isomajdine (α; [468, 469]), (−)-majdine (α; [468, 469]); **C5c**: herbadine [470, 473], (−)-herbamine (α; [470, 472, 474]), (+)-quebrachidine (α; [472]), (−)-vincamajine (α; [472]), vincarine [471]; **C5d**: picrinine [472]; **P6f**: (−)-vincoline (α; [472]); **S4**: (−)-norfluorocurarine (α; [468]). *V. lancea* (Boj. ex A. DC.) K. Schum. (see *Catharanthus lanceus*). *V. libanotica* Zucc. (see *V. herbacea*). *V. major* L. (= *V. pubescens* Urv.): **C4a**: ervine [9], (−)-reserpinine (α; [475]), serpentine [8], tetrahydroalstonine [8]; **C4c**: sarpagine [8]; **C5b**: alkaloid V [476], (−)-carapanaubine (α; [477]), (−)-isomajdine (α; [469]), (−)-majdine (α; [469, 475]); **C5c**: (−)-majoridine (α; [478, 479]), (−)-vincamajine (α; [478]), (−)-vincamedine (α; [9]); **C5h**: (−)-akuammine (α; [475]); **E5a**: (+)-vincamine (α; [436]), (+)-vincine (α; [476, 480]); **S4**: akuammicine [9], (−)-norfluorocurarine (α; [9]). *V. major* L. ssp. *major*: **C4a**: (−)-reserpinine (α; [437, 481]); **C4c**: lochnerine [438]; lochvinerine [481], majvinine [482], (+)-10-methoxyvellosimine (α; [481, 483]); **C5b**: (+)-elegantissine (α; [438]), (+)-isoelegantissine (α; [438]), (−)-isomajdine (α; [437, 438]), (−)-majdine (α; [438]); **C5c**: (−)-majoridine (α; [481, 484]), vincamajoreine [438, 481]. *V. minor* L.: **C4b**: reserpine [9]; **C4e**: (+)-akuammiline (α; [485]), desacetylakuammiline [485], (+)-10-methoxydesacetyl-akuammiline (α; [485]), (−)-strictamine (α; [486]); **C5c**: ( )-vinorine (α; [487]); **C5d**: picrinine [488]; **C5g**: (−)-vincoridine (α; [489]), (−)-vincorine (α; [114, 490]); **C5y**: (−)-vinoxine [491]; **E5a**: (+)-eburnamenine (β; [492]), (−)-eburnamine (β; [492]), (−)-eburnamonine (α; [492]), (±)-eburnamonine (α + β; [492]), (−)-16-epivincamine (α; [493]), (+)-isoeburnamine (β; [492]), (−)-11-methoxyeburnamonine [494], (+)-vincamine (α; [152]), (+)-vincaminine (α; [495]), (+)-vincine (α; [496]), (+)-vincinine (α; [495]). *V. pubescens* Urv. (see *V. major*). *V. pusilla* Murr. (see *Catharanthus pusillus*). *V. rosea* (L.) Reichb. (see *Catharanthus roseus*).

Voacanga (TAB). *V. africana* Stapf (= *V. schweinfurthii* Stapf): **C4b**: isorauhimbine [497], pseudoyohimbine [497], reserpine [497], β-yohimbine [497]; **C5a**: vobasine [497]; **C5a-J7a**: 18′-decarbomethoxyvoacamine [497], (−)-voacamidine (α-β; [166]), (−)-voacamine (α-β; [497, 498, 520]), (−)-voacamine *N*-oxide (α-β; [499]), (−)-voacorine (α-β; [489, 521]); **D6b**: perakine [497]; **J7a**: coronaridine [497], ibogaine [497], ibogamine [497], iboxygaine [497], voacangine [9, 497], hydroxyindoleninevoacangine [497, 500], voacanginelactam [497], (−)-voacristine (β; [501]), (+)-voacryptine (β; [502]); **J8a**: iboluteine [497]. *V. bracteata* Stapf (= *V. bracteata* var. *bracteata*): **C5a-J7a**: (−)-19-epivoacorine (α-β; [503, 504]), (−)-voacamine (α-β; [503, 504]), (−)-voacamine *N*-oxide (α-β; [505]), (−)-voacorine (α-β; [503, 504]); **J7a**: (−)-19-epivoacristine (β; [503, 504]), (−)-voacangine (β; [503, 504]); (−)-voacristine (β; [503, 504]). *V. chalotiana* Pierre ex Stapf: **C4a**: (−)-tetrahydroalstonine (α; [506]); **C4c**: (−)-17-*O*-acetyl-19,20-dihydrovoachalotine (α; [507]), (+)-akuammidine (α; [508]), (+)-3-hydroxyvoachalotine [506], 21-hydroxyvoachalotine [508], polyneuridine [508], (−)-voachalotine (α; [509]); **C5a**: (+)-voacarpine (α; [510]); **C5f**: (−)-voachalotine-oxindole (α; [508]); **C5i**: (+)-dehydrovoachalotine (α; [506, 511]); **C5o**: (−)voacoline (α; [512]); **E5a**: (+)-14,15-dehydrovincamine (α; [506]); **E6c**: (−)cuanzine (α; [506, 513]), (−)-decarbomethoxyapocuanzine (α; [506, 514]). *V. dregei* E. Mey (see *V. thouarsii*). *V. globosa* (Blanco) Merr.: **C5a**: tabernaemontanine [156]; **C5a-J7a**: (−)-voacamine (α-β; [515]); **J7a**: (−)-voacangine (β; [515]); **P6e-P6e**: (−)-quimbeline [516]. *V. grandiflora* Miq. (by mistake for *V. grandifolia*). *V. grandifolia* (Miq.) Rolfe: **C4c**: (+)-akuammidine (α; [517]); **C5a-J7a**: (−)-voacamine (α-β; [518]); **J7a**: (−)-voacangine (β; [518]); **P6e-P6e**: (−)-amataine (α-α [120]). *V. megacarpa* Merr.: **C5a-J7a**: voacamine [9]. *V. papuana* (F. v. Mull.) K. Schum.: **C5a-J7a**: voacamine [519]; **J7a**: voacangine [519]. *V. schweinfurthii* Stapf (see *V. africana*). *V. thouarsii* Roem. et Schult. [= *V. thouarsii* var. *dregei* (E. Mey.) Pichon, *V. dregei* E. Mey., *V. thouarsii* Roem. et Schult. var. *obtusa* (K. Schum.) Pichon]: **C5a**: (−)-dregamine (α; [522]), (−)-vobasine (α; [523]); **C5a-J7a**: 18′-decarbomethoxyvoacamine [523], (−)-voacamine (α-β; [523]), voacorine [9]; **J7a**: conopharyngine [9], ibogaine [523, 524], (−)-voacangine (β; [523, 524, 525]), (−)-voacristine (β; [523, 524]); **J8a**: iboluteine [523], (−)-voaluteine [523]; **P6e-P6e**: 12-*O*-demethylvobtusine [526], 2′-desoxyvobtusinelactam [526], (−)-subsessilinelactone (α-α; [527]), (−)-vobtusine-3-lactam *N*-oxide (α-α; [526]), (−)-vobtusine-3-lactam (α-α; [524, 526]).

Loganiaceae

Gardneria (STR). *G. angustifolia* Wall. (= *G. shimadai* Hayata): **C5f**: (−)-chitosenine (α; [528]); **C5f-C6a**: (−)-gardmultine [528]; **C6a**: (−)-18-demethylgardneramine (α; [528]), (−)-gardneramine (α; [528, 529]). *G. insularis* Nakai (see *G. nutans*). *G. multiflora* Makino: **C5f**: alkaloid I [75], alkaloid J [75], alkaloid L [75], alkaloid M [75], alkaloid N [75], (−)-chitosenine (α; [528, 530]); **C5f-C6a**: (−)-gardmultine [528, 531]; **C6a**: 18-demethoxygardfloramine [530], (−)-18-demethoxygardneramine [530], (−)-18-demethylgardneramine (α; [528,

529]), (−)-gardfloramine [530], (−)-gardneramine (α; [528, 529, 532]), gardneramine *N*-oxide [75]. *G. nutans* Sieb. et Zucc. (= *G. insularis* Nakai): **C4c**: (−)-gardnerine (α; [528, 533, 534, 535]); **C5i**: (+)-gardnutine (α; [528, 533, 535]), (+)-18-hydroxygardnutine (α; [528, 533, 535]); **C6a**: (−)-18-demethylgardneramine (α; [536]), (−)-gardneramine (α; [529]). *G. shimadai* Hayata (see *G. angustifolia*).

Gelsemium (GEL). *G. elegans* (Gardn. et Champ.) Benth.: **C4b**: sempervirine* [537]. *G. sempervirens* (L.) J. St. Hil.: **C4b**: sempervirine* [8].

Mostuea (GEL). *M. batesii* Bak. (= *M. stimulans* A. Chev.): **C4b**: sempervirine* [538]. *M. brunonis* Didr. var. *brunonis* (= *M. buchholzii* Engl.): **C4b**: sempervirine* [538]. *M. buchholzii* Engl. (see *M. brunonis*). *M. stimulans* A. Chev. (see *M. batesii*).

Strychnos (STR) [582]. Calebassen-Curare: **C4c**: (+)-lochneram (α; [539]), (+)-lochnerine (α; [539]), (+)-macusine B (α; [540]); **C4f**: (+)-C-mavacurine (α; [75, 541]), C-profluorocurine [8]; **C5k**: (+)-C-fluorocurine (α; [541]); **S4**: (−)-C-fluorocurarine (α; [8]); **S4-S4**: (+)-C-alkaloid A (α-α; [8, 9]), (−)-C-alkaloid D (α-α; [8, 9]), ( )-C-alkaloid E (α-α; [8, 9]), ( )-C-alkaloid F (α-α; [8, 9]), ( )-C-alkaloid G (α-α; [8, 9]), (−)-C-alkaloid H (α-α; [8, 9]), (+)-C-calebassine (α-α; [8, 9]), (+)-C-curarine I (α-α; [8, 9]), (−)-C-dihydrotoxiferine (α-α; [8, 9]), (−)-C-toxiferine (α-α; [8, 9]); **S5b-S5b**: (−)-caracurine II (α-α; [8, 9]). *S. aenea* A. W. Hill (see *S. vanprukii*). *S. afzelii* Gilg: **S4-S4**: bisnor-C-alkaloid H [542], bisnordihydrotoxiferine [542]; **S5b**: Wieland-Gumlich aldehyde [542]; **S5b-S5b**: caracurine V [542]. *S. amazonica* Krukoff: **C4c**: (+)-macusine B (α; [543]); **C4f**: C-mavacurine [544]; **S4**: ( )-18-desoxy-Wieland-Gumlich aldehyde (α; [545]); **S4-S4**: (−)-bisnordihydrotoxiferine (α-α; [545]); **S5b**: (+)-11-methoxydiaboline (α; [543]). *S. angolensis* Gilg: **V4**: angustidine* [546], angustine* [546]. *S. angustiflora* Benth. (= *S. angustifolia* Benth. ex Jackson): **V4**: angustidine* [547], angustine* [547], angustoline* [547]. *S. angustifolia* Benth. ex Jackson (see *S. angustiflora*). *S. axillaris* Colebr. (= *S. malaccensis* Benth., *S. psilosperma* F. v. Müll.): **S5e**: (−)-desacetylstrychnospermine (α; [580]), (+)-spermostrychnine (α; [603]), (+)-strychnospermine (α; [603]). *S. bakanko* Bourquelet et Hérissey (see *S. madagascariensis*). *S. borneensis* Leenhouts: **V4**: angustidine* [546], angustine* [546], angustoline* [546]. *S. brachiata* Ruiz et Pavón: **S5b**: (+)-11-methoxydiaboline (α; [543]), (−)-Wieland-Gumlich aldehyde (α; [543]). *S. brasiliensis* (Spreng.) Mart.: **S5d**: (+)-strychnosilidine (α; [548]), (+)-strychnosiline (α; [548]); **S5e**: (−)-12-hydroxy-11-methoxyspermostrychnine (α; [548]), (+)-spermostrychnine (α; [548]), **S6b**: (+)-10,11-dimethoxystrychnobrasiline (α; [548]), (−)-12-hydroxy-11-methoxystrychnobrasiline (α; [548]), (+)-strychnobrasiline (α; [548]). *S. burtoni* Bak. (see *S. madagascariensis*). *S. camptoneura* Gilg et Busse: **C3b**: akagerine [549]: **C4a**: alstonine [550], serpentine [550]; **D4d**: (+)-camptoneurine [551]; **S4**: retuline [549], ( )-retuline *N*-oxide (α; [551]); **V3**: (+)-antirhine (α; [546, 552]), (+)-anthirine methosalt (α; [546, 552]);

V4: angustine* [546]. *S. castelnaeana* Wedd.: **S4-S4**: C-alkaloid D [553]; **S5b**: diaboline [553], jobertine [553]. *S. chlorantha* Prog.: **S5b**: *O*-acetyldiaboline [554], diaboline [554]. *S. cinnamomifolia* Thwaites (see *S. wallichiana*). *S. colubrina* L. (see *S. wallichiana*). *S. cuspidata* A. W. Hill (see *S. ignatii*). *S. dale* De Wild.: **C3b**: akagerine [555], kribine [555], 17-*O*-methylakagerine [555], (−)-21-*O*-methylkribine [555], (−)-epi-21-*O*-methylkribine [555]. *S. decussata* (Pappe) Gilg: **V4**: (−)-glucoalkaloid [556]. *S. dewevrei* Gilg (see *S. icaja*). *S. diaboli* Sandwith: **S5b**: (+)-diaboline (α; [557]). *S. divaricans* Ducke: **C4f**: C-mavacurine [558]; **S4**: C-fluorocurarine [558]; **S4-S4**: C-calebassine [558], C-curarine I [558]. *S. dolichothyrsa* Gilg ex Onochie et Hepper: **S4**: 18-desoxy-Wieland-Gumlich aldehyde [559]; **S4-S4**: bisnor-C-alkaloid D [559], bisnor-C-curarine [559], bisnordihydrotoxiferine [559], bisnordihydroxiferine di-*N*-oxide [559], bisnordihydrotoxiferine *N*-oxide [559]; **S5b-S5b**: caracurine V [560], caracurine V di-*N*-oxide [560], caracurine V *N*-oxide [560]. *S. dysophylla* Benth. (see *S. madagascariensis*). *S. elaeocarpa* Gilg ex Leeuwenberg: **C3b**: akagerine [555], kribine [555], 17-*O*-methylakagerine [555], (−)-21-*O*-methylkribine [555], (−)-epi-21-*O*-methylkribine [555]. *S. fendleri* Sprague et Sandwith: **S5e**: (+)-N_a-acetylstrychnosplendine (α; [561]), (−)-12-hydroxy-11-methoxy-$N_{(a)}$-acetylstrychnosplendine (α; [561]); **S6b**: (−)-12-hydroxy-11-methoxystrychnofendlerine (α; [561]), (+)-strychnofendlerine (α; [561]). *S. fernandiae* Duvign. ex Denöel (see *S. usambarensis*). *S. floribunda* Gilg: **V4**: angustine* [546]. *S. froesii* Ducke: **C4f**: C-mavacurine [562]; **S4**: ( )-18-desoxy-Wieland-Gumlich aldehyde (α; [545]); **S4-S4**: C-alkaloid E [563], C-curarine I [563], C-dihydrotoxiferine [563], C-toxiferine I [563]; **S5b**: diaboline [545], Wieland-Gumlich aldehyde [545]. *S. gauthierana* Pierre ex Dop (see *S. wallichiana*). *S. gossweileri* Exell: **C3a**: ( )-diploceline (α; [564]); **C4a**: alstonine [565]; **D3a**: ( )-dolichanthoside (α; [566]). *S. henningsii* Gilg [= *S. holstii* forma *condensata* Duvign., *S. holstii* Gilg var. *reticulata* (Burtt-Davy et Honoré) Duvign. forma *condensata* Duvign., *S. procera* Gilg et Busse, *S. reticulata* Burtt-Davy et Honoré]: **S4**: $N_{(a)}$-desacetyl-17-*O*-acetyl-18-hydroxyisoretuline [567], $N_{(a)}$-desacetyl-18-hydroxyisoretuline [567], $N_{(a)}$-desacetylisoretuline [567], 18-hydroxyisoretuline [567], (+)-retuline (α; [567, 568, 569]), tsilanimbine [567]; **S5b**: (−)-*O*-acetyldiaboline (α; [570]), (−)-*O*-acetylhenningsoline (α; [571]), 2, 16-dehydrodiaboline [572], (+)-diaboline (α; [571, 572]), (−)-henningsoline (α; [571]), 11-methoxy-2, 16-dehydrodiaboline [572], 11-methoxydiaboline [572], (−)-11-methoxyhenningsamine [571], Wieland-Gumlich aldehyde [562]; **S5c**: *O*-demethyltsilanine [567, 572], 10-methoxy-*O*-demethyltsilanine [567, 572], (+)-10methoxytsilanine [567, 572], (+)-tsilanine (α; [567, 572]); **S6c**: (+)-holstiine [573], (+)-holstiline [573], (+)-rindline [571, 573, 574]. *S. heterodoxa* Gilg (see *S. potatorum*). *S. holstii* forma *condensata* Duvign. (see *S. henningsii*). *S. holstii* Gilg var. *reticulata* (Burtt-Davy et Honoré) Duvign. forma *condensata* Duvign. (see *S. henningsii*). *S. icaja* Baill. (= *S. dewevrei* Gilg, *S. kipapa* Gilg): **S6a**: brucine [8], (−)-12-hydroxystrychnine [575], (−)-3-methoxystrychnine (α; [576]), (−)-strychnine (α; [575]); **S7**: 19, 20-α-epoxy-11, 12-dimethoxy-*N*-methyl-*sec*-pseudostrychnine [577], (+)-19, 20-α-epoxy-15-hydroxy-10, 11-dimethoxy-*N*-methyl-*sec*-pseudostrychnine (α; [576, 577]), (+)-19,20-α-epoxy-15-hydroxy-*N*-methyl-*sec*-pseudostrychnine (α; [576]), 19,20-α-epoxy-15-hydroxy-12-methoxy-*N*-methyl-*sec*-pseudostrychnine [577], (+)-19,20-α-epoxy-15-hydroxyvomicine (α; [576, 577]), 19,20-α-epoxy-12-methoxy-*N*-methyl-*sec*-pseudostrychnine [577], (+)-19,20-α-epoxy-11-methoxyvomicine (α; [576, 577]), 19,20-α-epoxy-*N*-methyl-*sec*-pseudo-β-colubrine [576], (+)-19, 20-α-epoxynovacine (α; [577, 578]), (+)-19, 20-α-epoxyvomicine (α; [576, 577]), (−)-15-hydroxyicajine (α; [576]), (−)-icajine (α; [577, 579]), 2- or 3-methoxy-*N*-methyl-*sec*-pseudostrychnine [577], novacine [577], pseudostrychnine [580], (+)-vomicine (α; [577, 579]). *S. ignatii* Berg. (= *S. cuspidata* A. W. Hill, *S.* KL 1929, *S. lanceolaris* Miq., *S. ovalifolia* Wall. ex G. Don, *S. tieuté* Lesch.): **S5b**: diaboline [580], Wieland-Gumlich aldehyde [562]; **S6a**: brucine [581], brucine *N*-oxide [581], α-colubrine [582], β-colubrine [582], 12-hydroxystrychnine [582], strychnine [581], strychnine *N*-oxide [581]; **S7**: *N*-cyano-*sec*-pseudocolubrine [583], *N*-cyano-*sec*-pseudostrychnine [583], icajine [582], novacine [582], pseudobrucine [581], pseudostrychnine [581], vomicine [582]. *S. javanica* Hardy ex Rabuteau et Piétri (see *S. wallichiana*). *S. jobertiana* Baill.: **S5b**: *O*-acetyldiaboline [584], diaboline [584], jobertine [584]. *S. kipapa* Gilg (see *S. icaja*). *S.* KL 1929 (see *S. ignatii*). *S. lanceolaris* Miq. (see *S. ignatii*). *S. ledermannii* Gilg et Bened.: **S5b**: *O*-acetyldiaboline [582], (+)-diaboline (α; [585]); **V4**: angustine* [546]. *S. ligustrina* Blume (see *S. lucida*). *S. lucida* R. Br. (= *S. ligustrina* Blume): **C4c**: normacusine B [582], sarpagine [582]; **S5b**: diaboline [582]; **S6a**: brucine [580], brucine-*N*-oxide [582], α-colubrine [582], β-colubrine [586], strychnine [580], strychnine *N*-oxide [582]; **S7**: icajine [582], novacine [582], pseudobrucine [582]. *S. macrophylla* Barb. Rodr.: **C4f**: C-mavacurine [587]; **C5k**: C-fluorocurine [587]. *S. madagascariensis* Poir. (= *S. bakanko* Bourquelet et Hérissey, *S. burtoni* Bak., *S. dysophylla* Benth., *S. quaqua* Gilg, *S. vacacoua* Baill., *S. wakefieldii* Bak.): **S6a**: strychnine [39]. *S. maingayi* C. B. Clarke: **S6a**: brucine [582], strychnine [582]; **S7**: icajine [582], vomicine [582]. *S. malaccensis* Benth. (see *S. axillaris*). *S. malacoclados* C. H. Wright: **S5b**: 11-methoxydiaboline [550]. *S. medeola* Sagot ex Prog.: **C4c**: (+)-normacusine B (α; [588]); **S5b**: 11-methoxydiaboline [588]. *S. melinoniana* Baill.: **C4a**: (−)-melinonine B [589], (−)-melinonine A (α; [590]); **C4b**: melinonine E [591]; **C4f**: (+)-C-mavacurine (α; [591]); **C5k**: (+)-C-fluorocurine (α; [591]). *S. menyanthoides* Duvign. ex Denöel (see *S. usambarensis*). *S. micans* *S. Moore* (see *S. usambarensis*). *S. minor* Dennst.: **S5b**: *O*-acetyldiaboline [582], diaboline [582], methoxydiaboline [582]; **S6a**: brucine [582], brucine *N*-oxide [582], strychnine [582], strychnine *N*-oxide [582]; **S7**: novacine [582]; **V4**: angustidine* [546], angustine* [546], angustoline* [546]. *S. mitscherlichii* Rich. Schomb. (= *S. solerederi* Gilg): **C4f**: C-mavacurine [558]; **C5k**: C-fluorocurine [558]; **S4**: C-fluorocurarine [558]; **S4-S4**: C-alkaloid D [558], C-calebassine [558], C-curarine I [558]; **S5b**: diaboline [562]. *S. nux-blanda* A. W. Hill: **S5b**: diaboline [580]; **S6a**: brucine [582], brucine *N*-oxide [582], strychnine [582], strychnine *N*-oxide [582]; **S7**: icajine [582], novacine [582], pseudobrucine [582], pseudostrychnine [582], vomicine [582]. *S. nux-vomica* L.: **A4a**: condylocarpine [592]; **C3a**: decarbomethoxygeissoschizine [592], geissoschizal [592], geissoschizine [592];

C4c: normacusine B [592]; **C4f**:(+)-C-mavacurine (α; [593]); **S5b**:diaboline [582], Wieland-Gumlich aldehyde [592]; **S5f**:isostrychnine [594]; **S6a**:(−)-brucine (α; [593, 595]), brucine *N*-oxide [596], (−)-α-colubrine (α; [580, 597]), (−)-β-colubrine (α; [596, 597]), 12-hydroxystrychnine [596], (−)-15-hydroxystrychnine (α; [598]), (−)-strychnine (α; [593, 595]), strychnine *N*-oxide [596]; **S7**:(−)-icajine (α; [595, 596]), *N*-methyl-*sec*-pseudo-β-colubrine [596], (−)-novacine (α; [595, 596]), pseudobrucine [599], pseudo-α-colubrine [580, 594], pseudo-β-colubrine [580, 594], (−)-pseudostrychnine (α; [596, 597]), (−)-vomicine (α; [593, 595, 600]). *S. odorata* A. Chev.:**V4**:angustidine* [546], angustine* [546], angustoline* [546]. *S. oleifolia* A. W. Hill:**S5b**:*O*-acetyldiaboline [582], diaboline [582]. *S. ovalifolia* Wall. ex G. Don (see *S. ignatii*). *S. ovata* A. W. Hill:**S6a**:brucine [582], strychnine [582]; **V4**:angustidine* [546], angustine* [546], angustoline* [546]. *S. panamensis* Seem.:**S4**:C-fluorocurarine [562]; **S4-S4**:C-alkaloid F [562], C-alkaloid G [562], C-dihydrotoxiferine [562]; **S5b**:diaboline [562]; **S6a**:brucine [562], strychnine [562]. *S. parvifolia* A. DC.:**C4f**:C-mavacurine [562]. *S. pierreana* A. W. Hill (see *S. wallichiana*). *S. potatorum* L. f. (= *S. heterodoxa* Gilg, *S. stuhlmannii* Gilg):**S5b**:diaboline [601]; **S6a**:brucine [582], strychnine [582]; **S7**:icajine [582], novacine [582]; **V4**:angustine* [546]. *S. procera* Gilg et Busse (see *S. henningsii*). *S. pseudoquina* A. St. Hil.:**S4-S4**:(−)-nordihydrotoxiferine (α; [602]). *S. psilosperma* F. v. Müll. (see *S. axillaris*). *S. quadrangularis* A. W. Hill (see *S. vanprukii*). *S. quaqua* Gilg (see *S. madagascariensis*). *S. reticulata* Burtt-Davy et Honoré (see *S. henningsii*). *S. rheedii* C. B. Clarke (see *S. wallichiana*). *S. romeu-belenii* Krukoff et Barneby:**S5b**:(+)-11-methoxydiaboline (α; [588, 604]). *S. rondeletioides* Spruce ex Benth.):**S5b**:diaboline [584], Wieland-Gumlich aldehyde [584]. *S. rubiginosa* A. DC.:**S4**:C-fluorocurarine [563]; **S4-S4**:C-calebassine [563]. *S. rupicola* Pierre ex Dop:**S6a**:brucine [582], brucine *N*-oxide [582], strychnine [582]; **S7**:icajine [582], *N*-methyl-*sec*-pseudobrucine [582], *N*-methyl-*sec*-pseudostrychnine [582], novacine [582], pseudobrucine [582]; **V4**:angustidine* [582], angustine* [582], angustoline* [582]. *S. samba* Duvign.:**V4**:angustidine* [546], angustine* [546], angustoline* [546]. *S. scheffleri* Gilg (= *S. subaquatica* De Wild.; *S. volubilis* Duvign. ex Denöel):**V4**:angustidine* [546], angustine* [546], angustoline* [546]. *S. solerederi* Gilg (see *S. mitscherlichii*). *S. solimoesana* Krukoff:**S4**:C-fluorocurarine [605]; **S4-S4**:C-alkaloid D [605], C-alkaloid E [605], C-alkaloid F [605], C-alkaloid G [605], C-calebassine [605], C-curarine I [605]. *S. splendens* Gilg:**S5e**:(+)-N_a-acetylisostrychnosplendine (α; [603, 606, 607]), (+)-isosplendoline (α; [603, 606, 607]), (−)-isostrychnosplendine (α; [603, 606, 607]), (+)-splendoline (α; [603, 606]), (−)-strychnosplendine (α; [603, 606, 607]); **S6b**:(+)-isosplendine (α; [603, 607]). *S. stuhlmannii* Gilg (see *S. potatorum*). *S. subaquatica* De Wild (see *S. scheffleri*). *S. subcordata* Spruce ex Benth.:**C4f** C-mavacurine [608]; **C5k**:C-fluorocurine [608]; **S4**:C-fluorocurarine [608]; **S5b**:Wieland-Gumlich aldehyde [608]. *S. tabascana* Sprague et Sandwith: **S5d**:(+)-strychnosilidine (α; [609]), ( )-tabascanine (α; [609]); **S6b**:( )-10-methoxystrychnobrasiline (α; [609]), ( )-N_a-acetyl-*O*-methylstrychnosplendine (α; [609]), (+)-strychnobrasiline (α; [609]). *S. tieuté* Lesch. (see *S. ignatii*). *S.*

tomentosa Benth.: **S4**: C-fluorocurarine [610]; **S4-S4**: C-alkaloid E [610], C-toxiferine I [610]. *S. toxifera* Rich. Schomb. ex Benth.: **C4c**: (−)-macusine A (α; [611]), (+)-macusine B (α; [611]), (−)-macusine C (α; [540]); **C4f**: (+)-C-mavacurine (α; [612]), C-profluorocurine [612]; **C5k**: (+)-C-fluorocurine (α; [612]); **S4-S4**: C-alkaloid A [611], (−)-nordihydrotoxiferine (α; [613]), (−)-C-toxiferine I (α; [611]); **S5b**: (−)-Wieland-Gumlich aldehyde (α; [611]), (−)-N_b-Wieland-Gumlich aldehydechlormethylate (α; [611]); **S5b-S5b**: (−)-caracurine II (α-α; [614]), (−)-caracurine II methosalt (α-α; [615]), (+)-caracurine V (α-α; [612]). *S. trichoneura* Leeuwenberg: **V4**: angustidine* [546], angustine* [546], angustoline* [546]. *S. trinervis* (Vell.) Mart.: **S4**: C-fluorocurarine [616]; **S4-S4**: C-alkaloid H [616], C-calebassine [616], C-curarine I [616], C-dihydro-toxiferine [616]. *S. tchibangensis* Pellegr.: **C3a**: (+)-tschibagensine (α; [617]). *S. umbellata* (Lour.) Merr.: **S5b**: diaboline [582]; **S6a**: brucine [582], strychnine [582]; **S7**: icajine [582]; **V4**: angustidine* [546], angustine* [546], angustoline* [546]. *S. urceolata* Leeuwenberg: **S4-S4**: bisnor-C-alkaloid H [618], bisnordi-hydrotoxiferine [618]; **S5b-S5b**: caracurine V [618]. *S. usambarensis* Gilg (= *S. fernandiae* Duvign. ex Denöel; *S. menyanthoides* Duvign. ex Denöel; *S. micans* S. Moore): **C3a**: 3′,4′-dihydrousambarensine [619, 620], ( )-18, 19-dihydrous-ambaridine Br (α; 621), ( )-18, 19-dihydrousambaridine Vi (α; [621]), 18, 19-dihydrousambarine [621], $N'_{(b)}$-methyl-3′, 4′-dihydrousambarensine [622], ( )-isostrychnopentamine A (α; [621]), ( )-isostrychnopentamine B (α; [621]), $N'_{(b)}$-methylusambarensine [622], ( )-strychnobaridine (α; [621]), ( )-strych-nopentamine (α; [621, 623]), ( )-usambarensine (α; [619, 622]), usambaridine [624], ( )-usambaridine Br (α; [621]), ( )-usambaridine Vi (α; [621], ( )-usambarine (α; [620, 621, 624, 625]); **C3b**: ( )-akagerine (α; [626]); **C4b**: decarbomethoxydihydrogambirtannine [627]; **C4c**: macusine B [628], *O*-meth-yldihydromacusine B [628], *O*-methylmacusine B [628]; **C4d**: ( )-isostrychno-foline (α; [629]), ( )-isostrychnophylline (α; [629]), ( )-strychnofoline (α; [630]), ( )-strychnophylline (α; [629]); **S4-S4**: afrocurarine [622], C-calebassine [622], C-curarine [622], C-dihydrotoxiferine [622]; **V4**: angustine* [546]. *S. vacacoua* Baill. (see *S. madagascariensis*). *S. vanprukii* Craib (= *S. aenea* A. W. Hill; *S. quadrangularis* A. W. Hill): **V4**: angustidine* [546], angustine* [546], angustoline* [546]. *S. variabilis* De Wild.: **S4**: (+)-N_a-acetyl-2, 16-dihydro-2β, 16α-akuammicinal (α; [631]), desacetylisoretuline [632], (−)-desacetylretuline (α; [633]), isoretuline [632], retuline [632]; **S4-S4**: nordihydrotoxiferine [634]; **S4-S5a**: 12′-hydroxyisostrychnobiline [632], isostrychnobiline [634], strychno-biline [634]. *S. volubilis* Duvign. ex Denöel (see *S. scheffleri*). *S. wakefieldii* Bak. (see *S. madagascariensis*). *S. wallichiana* Steud. ex A. DC. (= *S. cinnamomifolia* Thwaites, *S. colubrina* L., *S. gauthierana* Pierre ex Dop, *S. javanica* Hardy ex Rabuteau et Piétri; *S. pierreana* A. W. Hill, *S. rheedii* C. B. Clarke): **S6a**: brucine [635, 636], brucine *N*-oxide [636], α-colubrine [635], β-colubrine [582], (−)-12-hydroxy-11-methoxystrychnine (α; [635]), 12-hydroxystrychnine [635, 636], strychnine [635, 636], strychnine*N*-oxide [636]; **S7**: *N*-cyano-*sec*-pseudobrucine [583, 636], *N*-cyano-*sec*-pseudostrychnine [583, 636], 15-hydroxyicajine [636], (+)-12-hydroxy-11-methoxy-*N*-methyl-*sec*-pseudostrychnine (α; [635]), 12-hydroxy-11-methoxy-pseudostrychnine [582], 15-hydroxynovacine [636], icajine [636], icajine *N*-oxide [636], *N*-methyl-*sec*-pseudo-β-colubrine [636], novacine [635, 636], (−)-pseudobrucine (α; [636, 637]), (−)-pseudostrychnine (α; [583, 636, 637]), vomicine [635]; **V4**: angustidine* [582], angustine* [582], angustoline* [582]. *S. xantha* Leeuwenberg: **V4**: angustidine* [546], angustine* [546], angustoline * [546].

Rubiaceae (RUB)

Adina (NAU). *A. cordifolia* (Roxb.) Hook. f. ex Brandis (see *Haldina cordifolia*). *A. rubescens* Hemsl. (see *Pertusadina eurhyncha*). *A. rubrostipulata* K. Schum. (see *Mitragyna rubrostipulata*).

Anthocephalus (NAU). *A. cadamba* (Roxb.) Miq. (see *A. chinensis*). *A. chinensis* (Lam.) Rich. ex Walp. [= *A. cadamba* (Roxb.) Miq.]: **D3a**: strictosidine [652]; **D4a**: (−)-3α-dihydrocadambine (α; [652, 653]), 3β-dihydrocadambine [652]; **D4b**: (−)-cadamine* [654], isocadamine [654], ( )-isodihydrocadambine (α; [655]), 3β-isodihydrocadambine [652]; **D5a**: (−)-cadambine (α; [653]).

Antirhea (GUE). *A. putaminosa* (F. v. Müll.) Bailey: **V3**: (−)-antirhine (α; [656]).

Cephalanthus (NAU). *C. occidentalis* L.: **C3a**: dihydrocorynantheine [657], hirsutine [657]; **C4d**: antiisorhynchophylline *N*-oxide [657], isorhynchophylline [657], rhynchophylline [657], rhynchophylline *N*-oxide [657].

Cinchona (CIN). *C. calisaya* Wedd. var. *javanica* Moens (see *C. officinalis*). *C. erythrantha* Pav. (see *C. pubescens*). *C. ledgeriana* Moens ex Trimen (see *C. officinalis*). *C. nitida* Ruiz et Pav. (see *C. officinalis*). *C. officinalis* L. (= *C. calisaya* Wedd. var. *javanica* Moens, *C. ledgeriana* Moens ex Trimen, *C. nitida* Ruiz. et Pav.): **C4g**: (+)-cinchophyllamine [658], (+)-isocinchophyllamine (α; [658]); **C5l**: (+)-quinamine (α; [658]). *C. pelletieriana* Wedd. (see *C. pubescens*). *C. pubescens* Vahl (= *C. erythrantha* Pav., *C. pelletieriana* Wedd., *C. succirubra* Pav. ex Klotzsch): **C4a**: aricine [8]; **C5l**: (+)-quinamine (α; [659]), (+)-conquinamine (α; [660]). *C. rosulenta* Howard ex Wedd.: **C5l**: (+)-quinamine (α [659]), (+)-conquinamine (α; [660]). *C. succirubra* Pav. ex Klotzsch (see *C. pubescens*).

Corynanthe (CIN). *C. johimbe* K. Schum. (see *Pausinystalia johimbe*). *C. macroceras* K. Schum. (see *Pausinystalia macroceras*). *C. mayumbensis* (R. Good) N. Hallé [= *Pseudocinchona mayumbensis* (R. Good) Ray-Ham.]: **C4a**: akuammigine [661], (−)-19-epiajmalicine [661], 3-iso-19-epiajmalicine [661], (+)-3-isorauniticine (α; [661]), rauniticine [661], tetrahydroalstonine [661]; **C4b**: (−)-corynanthine (α; [662]). *C. pachyceras* K. Schum. (= *Pseudocinchona africana* A. Chev. ex Perrot): **C3a**: (−)-corynantheidine (α; [663]), (−)-corynantheine (α; [663, 664]), (+)-dihydrocorynantheine (α; [663]); **C4b**:(−)-corynanthine (α; [662]), (−)-α-yohimbine (α; [665]), (−)-β-yohimbine (α; [665]); **C4d**:(−)-corynoxeine (α; [666]), (−)-corynoxine (α; [666]). *C. paniculata* Welw.: **C4b**:alloyohimbine [667], pseudoyohimbine [667], yohimbine [667], β-yohimbine [667]. *C. yohimbe* K. Schum. (see *Pausinystalia johimbe*).

Guettarda (GUE). *G. eximia* Baill.: **C4a**:(−)-cathenamine (α; [668]).

Haldina (NAU). *H. cordifolia* (Roxb.) Ridsd. [= *Adina cordifolia* (Roxb.) Hook. f. ex Brandis]: **D3a**:cordifoline [638], (−)-10-desoxycordifoline (α; [639]); **D4g**:( )-adifoline (α; [640]), (+)-10-desoxyadifoline [383].

Isertia (MUS). *I. hypoleuca* Benth.: **C5l**:(+)-18,19-dihydroquinamine [669, 670], (−)-isodihydroquinamine [670].

Mitragyna (NAU). *M. africana* (Willd.) Korth. (see *M. inermis*). *M. ciliata* Aubr. et Pellegr.: **C3a**:( )-mitraciliatine (α; [671, 672]); **C4d**:(−)-ciliaphylline (α; [671, 672, 673]), isorhynchophylline [671, 672], isorotundifoline [671, 672], (+)-rhynchociline (α; [671, 672, 673]), rhynchophylline [671], rotundifoline [671, 672]. *M. diversifolia* (Wall. ex G. Don) Havil. (= *M. javanica* Koord. et Valeton): **C4a**:(−)-ajmalicine (α; [678]), (−)-mitrajavine [675, 679]; **C5b**:(+)-isomitraphylline (α; [675]), (+)-javaphylline (α; [678]), (−)-mitraphylline (α; [675]), (+)-vineridine (α; [675]); **V4**:angustine* [546]. *M. diversifolia* (auct. non Wall. ex G. Don) Havil. (see *M. rotundifolia*). *M. hirsuta* Havil.: **C3a**: hirsuteine [674], (+)-hirsutine (α; [675]); **C4a**:mitrajavine [674]; **C4d**:(+)-isorhynchophylline (α; [675]), (−)-rhynchophylline (α; [675]); **C5b**:(+)-isomitraphylline (α; [675]), (−)-mitraphylline (α; [675]). *M. inermis* (Willd.) O. Kuntze [= *M. africana* (Willd.) Korth.]: **C3a**:( )-mitraciliatine [671, 676], speciogynine [671, 676]; **C4d**:ciliaphylline [671], isorhynchophylline [671], ( )-isorhynchophylline *N*-oxide (α; [677]), isorotundifoline [671], rhynchociline [671], rhynchophylline [671], ( )-rhynchophylline *N*-oxide (α; [677]), rotundifoline [671]; **C5b**:speciophylline [671], uncarine F [671]. *M. javanica* Koord. et Valeton (see *M. diversifolia*). *M. macrophylla* Hiern (see *M. stipulosa*). *M. parvifolia* (Roxb.) Korth.: **C3a**:dihydrocorynantheine [680], hirsutine [680, 681]; **C4a**:ajmalicine [682], akuammigine [680, 681], 3-isoajmalicine [680], tetrahydroalstonine [682]; **C4d**:ciliaphylline [680, 681], isorhynchophylline [680, 681], isorotundifoleine [683], isorotundifoline [680, 683], rhynchociline [680, 681], rhynchophylline [680, 681], rotundifoleine [683], rotundifoline [680]; **C5b**:isomitraphylline [680], isopteropodine [680, 681], mitraphylline [680], pteropodine [680, 681], speciophylline [680, 681], uncarine F [680, 681]; **V4**:angustine* [546]. *M. rotundifolia* (Roxb.) O. Kuntze [= *M. diversifolia* (auct. non Wall. ex. G. Don) Havil.]: **C4a**:3-isoajmalicine [684]; **C4d**:ciliaphylline [684], corynoxeine [684], ( )-isocorynoxeine (α; [684]), isorhynchophylline [684], ( )-isorhynchophylline *N*-oxide (α; [677]), rhynchociline [684], (−)-rhynchophylline (α; [684, 685]), ( )-rhynchophylline *N*-oxide (α; [677]),

(+)-rotundifoline (α; [685]); **C5b**: isomitraphylline [684], mitraphylline [684]. *M. rubrostipulaceae* Havil. (see *M. rubrostipulata*). *M. rubrostipulata* (K. Schum.) Havil. (= *Adina rubrostipulata* K. Schum., *M. rubrostipulaceae* Havil.): **C3a**: hirsuteine [686], hirsutine [686]; **C4d**: ( )-antirotundifoline *N*-oxide (α; [686, 687]), isorhynchophylline [686], isorotundifoline [686], rhynchophylline [686], rhynchophylline *N*-oxide [686], rotundifoline [686]; **C5b**: isomitraphylline [686], (−)-mitraphylline (α; [686]). *M. speciosa* (Korth.) Havil.: **C3a**: (−)-corynantheidine (α; [688]), isocorynantheidine [689], isopaynantheine [689], mitraciliatine [689], (−)-mitragynine (α; [690]), (−)-paynantheine (α; [688, 690]), (−)-speciociliatine (α; [690]), (+)-speciogynine (α; [688, 690]); **C4a**: (−)-ajmalicine (α; [688, 690]), akuammigine [689], 3-isoajmalicine [689], mitrajavine [689]; **C4d**: ciliaphylline [689, 690], corynoxeine [690], corynoxine A [690], corynoxine B [690], ( )-3-epiisorotundifoline (α; [683]), ( )-isomitrafoline (α; [683]), isorhynchophylline [689, 690], ( )-isorotundifoleine (α; [683]), ( )-isorotundifoline (α; [683]), ( )-isospeciofoline (α; [683, 690]), ( )-isospecionoxeine (α; [673]), ( )-mitrafoline (α; [683, 690]), mitragynine oxindol A [690], mitragynine oxindol B [690], rhynchociline [689, 690], (−)-rhynchophylline (α; [673, 690]), ( )-rotundifoleine (α; [683]), (+)-rotundifoline (α; [691]), ( )-speciofoline (α; [683, 690]), ( )-specionoxeine (α; [673]); **C5b**: (+)-isomitraphylline (α; [688]), javaphylline [689], ( )-mitraphylline (α; [688]), (+)-speciophylline (α; [688]). *M. stipulosa* (DC.) O. Kuntze (= *M. macrophylla* Hiern): **C4d**: (+)-isorhynchophylline (α; [671, 692]), (−)-isorotundifoline (α; [671, 692]), (−)-rhynchophylline (α; [671, 692]), (+)-rotundifoline (α; [671, 692]); **C5b**: isomitraphylline [9], (−)-mitraphylline (α; [671, 692]). *M. tubulosa* (Arn.) Havil.: **C4d**: ciliaphylline *N*-oxide [693], ( )-isorhynchophylline-*N*-oxide (α; [693]), rhynchociline *N*-oxide [693], ( )-rhynchophylline *N*-oxide (α; [693]).

Nauclea (NAU). *N. coadunata* Roxb. ex J. E. Smith (see *N. orientalis*). *N. diderrichii* (De Wild. ex Th. Dur.) Merr.: **D4a**: 3α-dihydrocadambine [694, 695]; **D5a**: cadambine [695]; **D6a**: α-naucleonidine [695, 696], β-naucleonidine [695, 696], α-naucleonine [695, 696], β-naucleonine [695, 696]. *N. latifolia* Smith (= *Sarcocephalus esculentus* Afzel. ex Sabine): **D4a**: decarbomethoxynauclechine* [709], naufoline* [709]; **V4**: angustoline* [710], angustine* [710], nauclefine [710], naucletine [710], strictosamide [711], (−)-strictosidinelactam (α; [712]). *N. orientalis* (L.) L. [= *N. coadunata* Roxb. ex J. E. Smith, *Sarcocephalus coadunata* (Roxb. ex J. E. Smith) Druce]: **V4**: angustine* [546]. *N. parva* Havil.) Merr. (= *Sarcocephalus parvus* Havil.): **V4**: parvine* [697].

Neonauclea (NAU). *N. schlechteri* (Val.) Merr. et Perry: **C3a**: (+)-gambirine (α; [698]).

Ourouparia (NAU). *O. africana* (G. Don.) Baill. (see *Uncaria africana* G. Don ssp. *africana*). *O. gambir* Baill. (see *Uncaria gambir*). *O. guianensis* Aubl. (see *Uncaria guianensis*). *O. rhynchophylla* (Miq.) Matsum. (see *Uncaria rhynchophylla*).

Palicourea (PSY). *P. alpina* (Sw.) A. DC.: **D3a**: (−)-palinine [699].

Pauridiantha (URO). *P. callicarpoides* (Hiern) Bremek.: **D3b**: pauridianthine* [700]; **D4e**: pauridianthinine* [700]. *P. lyalli* (Bak.) Bremek.: **D3a**: (−)-lyaloside [701], (−)-pauridianthoside [702]; **D3b**: 19-hydroxylyalidine [703], (+)-lyadine [704], (0)-lyalidine [703], (+)-lyaline [704], pauridianthinol [705].

Pausinystalia (CIN). *P. angolensis* Wernham: **C4b**: alloyohimbine [667], corynanthine [667], pseudoyohimbine [667], yohimbine [667], β-yohimbine [667]. *P. johimbe* (K. Schum.) Pierre [= *Corynanthe johimbe* (yohimbe) K. Schum.] : **C3a**: (+)-corynantheine (α; [706]), (+)-dihydrocorynantheine (α; [706]), (−)-dihydrositsirikine (α; [142]); **C4a**: (−)-ajmalicine (α; [667, 707]); **C4b**: (−)-alloyohimbine (α; [667, 707]), corynanthine [667], pseudoyohimbine [667], yohimbine [667], (−)-α-yohimbine (α; [667, 707]), (−)-β-yohimbine (α; [707]). *P. macroceras* (K. Schum.) Pierre ex Beille (= *Corynanthe macroceras* K. Schum.): **C4b**: yohimbine [8].

Pertusadina (NAU). *P. eurhyncha* (Miq.) Ridsd. (= *Adina rubescens* Hemsl.): **C3a**: (−)-adirubine (α; [641, 642]), (−)-anhydroadirubine (α; [642]); **C4a**: (−)-5α-carboxytetrahydroalstonine (α; [6]); **C4b**: (−)-5α-carboxycorynanthine (α; [643]); **D3a**: (−)-10-desoxycordifoline (α; [644]), ( )-5-oxostrictosidine (α; [645]), (−)-vincoside (α; [646]); **D4a**: ( )-macrolidine (α; [647]); **D4c**: (−)-desoxycordifolinelactam [648]; **D5c**: (−)-rubenine (α; [649]); **V4**: (−)-10β-D-glucosyloxyvincosidelactam (α; [7]), (−)-rubescine (α; [650]) (−)-3α,5α-tetrahydrodesoxycordifolinelactam (α; [651]), (−)-3β,5α-tetrahydrodesoxycordifolinelactam (α; [651]).

Pseudocinchona (CIN). *P. africana* A. Chev. ex E. Perrot (see *Corynanthe pachyceras*). *P. mayumbensis* (Good) Ray-Ham. (see *Corynanthe mayumbensis*).

Remijia (CIN). *R. purdieana* Wedd.: **C4g**: (+)-cinchonamine (α; [708]).

Sarcocephalus (NAU). *S. coadunata* (Roxb. ex J. E. Smith) Druce (see *Nauclea orientalis*). *S. esculentus* Afzel ex Sabine (see *Nauclea latifolia*). *S. parvus* (see *Nauclea parva*).

Uncaria (NAU). *U. acida* Hook. f. et Thoms. (see *U. elliptica*). *U. acida* (Hunt.) Roxb. (see *U. acida* var. *acida*). *U. acida* (Hunt.) Roxb. var. *acida* [= *U. acida* (Hunt.) Roxb., *U. ovalifolia* Roxb.]: **C4d**: isorhynchophylline [713], rhynchophylline [713], rhynchophylline *N*-oxide [713]. *U. acida* (Hunt.) Roxb. var. *papuana* Val. (= *U. firma* Val.): **C4a**: 3-isoajmalicine [713]; **C4d**: corynoxeine [713], isorhynchophylline [713], isorhynchophylline *N*-oxide [713], rhynchophylline [713], rhynchophylline *N*-oxide [713]. *U. aculeata* Willd. (see *U. guianensis*). *U. africana* G. Don (see *U. africana* ssp. *africana*). *U. africana* G.

Don ssp. *africana* [= *U. africana* G. Don, *Ourouparia africana* (G. Don) Baill.]: **C3a**: dihydrocorynantheine [713]; **C4a**: ajmalicine [713], 19-epiajmalicine [713], 3-isoajmalicine [713], 3-iso-19-epiajmalicine [713], tetrahydroalstonine [713]; **C4b**: ourouparine* [714]; **C4d**: isorhynchophylline [713], (−)-rhynchophylline (α; [8, 713]); **C5b**: isomitraphylline [713], mitraphylline [713], mitraphylline *N*-oxide [713]; **C5t**: dihydrocorynantheine-pseudoindoxyl [713]. *U. africana* G. Don ssp. *angolensis* (Havil.) Ridsd. [= *U. angolensis* (Havil.) Welw. ex Hutch. et Dalz.]: **C4d**: isorhynchophylline [713], isorhynchophylline *N*-oxide [713], rhynchophylline [713], rhynchophylline *N*-oxide [713]. *U. angolensis* (Havil.) Welw. ex Hutch. et Dalz. (see *U. africana* ssp. *angolensis*). *U. appendiculata* Benth. (see *U. lanosa* var. *appendiculata* forma *appendiculata*). *U. attenuata* Korth. [= *U. attenuata* Korth. ssp. *attenuata* Korth., *U. attenuata* Korth. ssp. *bulusanensis* (Elm.) Ridsd., *U. bulusanensis* Elm., *U. canescens* auct. non Korth., *U. gambir* Wall., *U. salaccensis* Bakh. f., nom. provis.]: **C3a**: dihydrocoryantheine [715], epiallocorynantheine [715], hirsuteine [715], hirsutine [715]; **C4a**: akuammigine [715], 3-isoajmalicine [715], 3-iso-19-epiajmalicine [715]; **C4b**: pseudoyohimbine [715]; **C4d**: corynoxeine [715], corynoxine B [715], isocorynoxeine [715], isorhynchophylline [715], isorhynchophylline *N*-oxide [715], isorotundifoline [715], rhynchophylline [715], rhynchophylline *N*-oxide [715], rotundifoline [715], speciofoline [715]; **C5b**: formosanine [715], isoformosanine [715], isomitraphylline [715], isomitraphylline *N*-oxide [715], mitraphylline [715], mitraphylline *N*-oxide [715], speciophylline [715]; **C5t**: dihydrocorynantheine-pseudoindoxyl [715]. *U. attenuata* Korth. ssp. *attenuata* Korth. (see *U. attenuata*). *U. attenuata* Korth. ssp. *bulusanensis* (Elm.) Ridsd. (see *U. attenuata*). *U. avenia* Val. (see *U. callophylla*). *U. bernaysii* F. v. Müll. [= *U. bernaysioides* Merr. et Perry, *U. salomonensis* (Rech.) Merr. et Perry, *U. sclerophylla* Warb.]: **C4a**: ajmalicine [713], akuammigine [713, 716], 3-isoajmalicine [713], tetrahydroalstonine [716]; **C4d**: isorhynchophylline [713], isorhynchophylline *N*-oxide [713], rhynchophylline [713], rhynchophylline *N*-oxide [713]; **C5b**: isomitraphylline [713], (−)-isopteropodine (α; [717]), ( )-isopteropodine *N*-oxide (α; [716]), mitraphylline [713], (−)-pteropodine (α; [717, 718]), ( )-pteropodine *N*-oxide (α; [716]), (+)-speciophylline (α; [717, 718]), ( )-speciophylline *N*-oxide (α; [716]), (+)-uncarine F (α; [717]), ( )-uncarine F *N*-oxide (α; [716]); **V4**: angustine* [546]. *U. bernaysioides* Merr. et Perry (see *U. bernaysii*). *U. brevicarpa* Elm. (see *U. roxburghiana*). *U. brevispina* Maing. ex Hook. f. (see *U. elliptica*). *U. bulusanensis* Elm. (see *U. attenuata*). *U. callophylla* Bl. ex Korth. (= *U. avenia* Val., *U. jasminiflora* Hook. f., *U. luzoniensis* Merr., *U. wrayi* King et Gamble): **C3a**: dihydrocorynantheine [713], gambirine [719]; **C4d**: isorhynchophylline [713], isorotundifoline [713], rhynchophylline [713], rotundifoline [713]; **C5b**: isomitraphylline [713], mitraphylline [713]. *U. callophylla* var. *oligoneura* Korth. ex Miq. (see *U. longiflora* var. *longiflora*). *U. canescens* auct. non Korth. (see *U. attenuata*). *U. canescens* Korth. ssp. *velutina* (Havil.) Ridsd. Phillip. et Heming. (see *U. velutina*). *U. celebica* Koord. (see *U. lanosa var. appendiculata* forma *philippinensis*). *U. cirrhiflora* Roxb. (see *U. cordata* var. *cordata* forma *cordata*). *U. clavisepala* Elm. (see *U. velutina*). *U.*

cordata (Lour.) Merr. (see *U. cordata* var. *cordata* forma *cordata*). *U. cordata* (Lour.) Merr. var. *cordata* forma *cordata* [= *U. cirrhiflora* Roxb., *U. cordata* (Lour.) Merr., *U. eumececarpa* Korth., *U. ferruginea* Kurz, *U. intermedia* Val., *U. multiflora* K. Schum. et Laut., *U. nemorosa* Korth., *U. pedicellata* Roxb., *U. sclerophylla* (Hunt.) Roxb., *U. sclerophylla* Wall., *U. speciosa* Wall. ex G. Don]: **C4d**: corynoxine A [713], corynoxine B [713], isorhynchophylline [713], rhynchophylline [713]. *U. cordata* (Lour.) Merr. var. *ferruginea* (Bl.) Ridsd. forma *ferruginea* (Bl.) Ridsd. [= *U. ferruginea* (Bl.) DC., *U. glaucescens* Craib.]: **C3a**: dihydrocorynantheine [713]. *U. cordata* (Lour.) Merr. var. *ferruginea* (Bl.) Ridsd. forma *insignis* (Bart. in DC.) Ridsd. (= *U. hallii* Korth., *U. insignis* Bart. in DC., *U.sclerophylla* F. Vill., *U. wallichii* Korth.): **C3a**: dihydrocorynantheine [713]; **C4d**: isorhynchophylline [713], rhynchophylline [713]. *U. donisii* Petit: **C5b**: isopteropodine [713], isopteropodine *N*-oxide [713], pteropodine [713], pteropodine *N*-oxide [713], speciophylline [713], speciophylline *N*-oxide [713], uncarine F [713], uncarine F *N*-oxide [713]. *U. dasycarpa* Pierre msc. (see *U. laevigata*). *U. dasyoneura* Korth. (see *U. elliptica*). *U. dasyoneura* Korth. var. *thwaitesii* Hook. f. (see *U. elliptica*). *U. elliptica* R. Br. ex G. Don (= *U. acida* Hook. f. et Thoms., *U. brevispina* Maing. ex Hook. f., *U. dasyoneura* Korth., *U. dasyoneura* Korth. var. *thwaitesii* Hook. f., *U. gambir* Thw., *U. rostrata* Pierre ex Pitard, *U. thwaitesii* Alston): **C3a**: dihydrocorynantheine [713], gambirine [713]; **C4a**: akuammigine [713], 3-isoajmalicine [713], roxburghine A [720], roxburghine B [720], roxburghine C [713, 720], roxburghine D [713, 720], roxburghine E [713, 720]; **C4d**: isorhynchophylline [713], isorotundifoline [713], rhynchophylline [713], rotundifoline [713]. *U. eumececarpa* Korth. (see *U. cordata* var. *cordata* forma *cordata*). *U. ferrea* F. Vill. (see *U. perrottetii*). *U. ferrea* ssp. *appendiculata* Val. (see *U. lanosa* var. *appendiculata* forma *appendiculata*). *U. ferrea* (Bl.) A. DC. (see *U. lanosa* var. *ferrea* forma *ferrea*). *U. ferruginea* (BL.) DC. (see *U. ferruginea* forma *ferruginea*). *U. ferruginea* Kurz. (see *U. cordata* var. *cordata* forma *cordata*). *U. ferruginea* Kurz. (see *U. cordata* var. *ferruginea* forma *ferruginea*. *U. firma* Val. (see *U. acida* var. *papuana*). *U. florida* Vidal, p.p. excl. lectotype (see *U. lanosa* var. *appendiculata* forma *philippinensis*). *U. florida* Vidal (see *U. lanosa* var. *appendiculata* forma *setiloba*). *U. formosana* (Matsum.) Hayata (see *U. hirsuta*). *U. gambir* (Hunt.) Roxb. (= *Ourouparia gambir* Baill.): **C3a**: (+)-dihydrocorynantheine (α; [721, 722]), (+)-gambirine (α; [721, 723a]); **C4a**: (−)-4*R*-akuammigine *N*-oxide (α; [724]), (−)-4*S*-akuammigine *N*-oxide (α; [749]), (−)-roxburghine A [722], (−)-roxburghine B (α; [722, 725]), (−)-roxburghine C (α; [722, 725]), (+)-roxburghine D (α; [722, 725]), (−)-roxburghine E (α; [722, 725]), (−)-tetrahydroalstonine (α; [722]), (+)-4*R*-tetrahydroalstonine *N*-oxide (α; [724]); **C4b**: (−)-dihydrogambirtannine* [723b], gambirtannine* [723b], 11-methoxyyohimbine [8], neooxygambirtannine* [723b], oxogambirtannine* [723b], ourouparine* [726]; **C4d**: isorhynchophylline [721], rhynchophylline [721], rotundifoline [721], **C5b**: formosanine [713], (+)-gambirdine [727], isoformosanine [713], (+)-isogambirdine [727], (−)-mitraphylline (α; [727]). *U. gambir* Thw. (see *U. elliptica*). *U. gambir* Wall. (see *U. attenuata*). *U. glabrata*

(Bl.) A. DC. (see *U. lanosa* var. *glabrata*). *U. glabrata* F. Vill. (see *U. lanosa* var. *appendiculata* forma *philippinensis*). *U. glabrescens* Merr. et Perry (see *U. lanosa* var. *appendiculata* forma *glabrescens*). *U. glaucescens* Craib. (see *U. cordata* var. *ferruginea* forma *ferruginea*). *U. guianensis* (Aubl.) Gmel. (= *Ourouparia guianesis* Aubl., *U. aculeata* Willd., *U. spinosa* Raeusch): **C3a**: dihydrocorynantheine [713], hirsuteine [713], hirsutine [713]; **C4d**: isorhynchophylline [713], isorhynchophylline *N*-oxide [713], rhynchophylline [713], rhynchophylline *N*-oxide [713]; **C5b**: isomitraphylline [713], isomitraphylline *N*-oxide [713], mitraphylline [713], mitraphylline *N*-oxide [713]; **V4**: angustine* [546], angustoline* [713]. *U. hallii* Korth. (see *U. cordata* var. *ferruginea* forma *insignis*). *U. havilandiana* S. Moore (see *U. longiflora var. longiflora*). *U. hirsuta* Havil. [= *U. formosana* (Matsum.) Hayata, *U. uraiensis* Hayata]: **C4a**: 3-isoajmalicine [713]; **C5b**: (+)-formosanine (α; [8, 713]), isoformosanine [713], isomitraphylline [713], isomitraphylline *N*-oxide [713], mitraphylline [713], mitraphylline *N*-oxide [713]. *U. homomalla* Miq. [= *U. lanosa* var. *parviflora* Ridl., *U. parviflora* (Ridl.) Ridl., *U. quadrangularis* Geddes, *U. tonkinensis* Havil.]: **C4a**: 3-isoajmalicine [713]; **C5b**: isomitraphylline [713], isomitraphylline *N*-oxide [713], isopteropodine [713, 728], isopteropodine *N*-oxide [713], mitraphylline [713], mitraphylline *N*-oxide [713], pteropodine [713, 728], pteropodine *N*-oxide [713], speciophylline [713, 728], speciophylline *N*-oxide [713], uncarine F [713, 728], uncarine F *N*-oxide [713]; **V4**: angustidine* [546], angustine* [546], angustoline* [546]. *U. hookeri* Vidal (see *U. perrottetii*). *U. horsfieldiana* Miq. (see *U. lanosa* var. *ferrea* forma *ferrea*). *U. inermis* Val. (see *U. nervosa*). *U. insignis* Bart. in DC. (see *U. cordata* var. *ferruginea* forma *insignis*). *U. intermedia* Val. (see *U. cordata* var. *cordata* forma *cordata*). *U. jasminiflora* Hook. f. (see *U. callophylla*). *U. kawakamii* Hayata (see *U. lanosa* var. *appendiculata* forma *phillippinensis*). *U. korrensis* Kanehira (see *U. lanosa* var. *korrensis*). *U. kunstleri* King: **C3a**: hirsutine [713]; **C4d**: corynoxine A [713], corynoxine B [713], isorhynchophylline [713], isorhynchophylline *N*-oxide [713]. *U. laevifolia* Elm. (see *U. longiflora* var. *pteropoda*). *U. laevigata* Wall. ex G. Don (= *U. dasycarpa* Pierre msc.): **C5b**: formosanine [713], isoformosanine [713], isomitraphylline [713], isomitraphylline *N*-oxide [713], isopteropodine [713], mitraphylline [713], mitraphylline *N*-oxide [713], speciophylline [713]. *U. lancifolia* Hutch.: **C5b**: isomitraphylline [713], mitraphylline [713], mitraphylline *N*-oxide [713]. *U. lanosa* Wall. var. *appendiculata* (Benth.) Ridsd. forma *appendiculata* (Benth.) Ridsd. (= *U. appendiculata* Benth., *U. ferrea* var. *appendiculata* Val.): **C5b**: isomitraphylline [713], isomitraphylline *N*-oxide [713], isopteropodine [713], isopteropodine *N*-oxide [713], mitraphylline [713], pteropodine [713], pteropodine *N*-oxide [713], speciophylline [713], speciophylline *N*-oxide [713], uncarine F [713], uncarine F *N*-oxide [713]. *U. lanosa* Wall. var. *appendiculata* (Benth.) Ridsd. forma *glabrescens* (Merr. et Perr.) Ridsd. (= *U. glabrescens* Merr. et Perry): **C4a**: akuammigine [713]; **C5b**: isopteropodine [713], isopteropodine *N*-oxide [713], pteropodine [713], pteropodine *N*-oxide [713], speciophylline [713], speciophylline *N*-oxide [713], uncarine F [713], uncarine F *N*-oxide [713]. *U. lanosa* Wall. var. *appendiculata* (Benth.) Ridsd.

forma *philippinensis* (Elm.) Ridsd. (= *U. celebica* Koord., *U. florida* Vidal, p.p. excl. lectotype, *U. glabrata* F. Vill., *U. kawakamii* Hayata, *U. philippinensis* Elm., *U. setiloba* Sasaki): **C5b**: (+)-formosanine (α; [75, 729, 730]), (+)-isoformosanine (α; [75, 729, 730]), isopteropodine [713], isopteropodine *N*-oxide [713], (−)-mitraphylline (α; [731]), pteropodine [713], pteropodine *N*-oxide [713], speciophylline [713], speciophylline *N*-oxide [713], uncarine F [713], uncarine F *N*-oxide [713]. *U. lanosa* Wall. var. *appendiculata* (Benth.) Ridsd. forma *setiloba* (Benth.) Ridsd. (= *U. florida* Vidal, *U. setiloba* Benth.): **C5b**: isomitraphylline [713], isomitraphylline *N*-oxide [713], isopteropodine [75], isopteropodine *N*-oxide [713], mitraphylline [713], mitraphylline *N*-oxide [713], (−)-pteropodine (α; [75, 729, 732]), pteropodine *N*-oxide [713], (+)-speciophylline (α; [729, 732]), speciophylline *N*-oxide [713], uncarine F [713], uncarine F *N*-oxide [713]. *U. lanosa* Wall var. *ferrea* (Bl.) Ridsd. forma *ferrea* (Bl.) Ridsd. (= *U. ferrea* (Bl.) A. DC., *U. horsfieldiana* Miq.): **C5b**: isomitraphylline [713], (−)-isopteropodine (α; [717]), isopteropodine *N*-oxide [713], mitraphylline [713], mitraphylline *N*-oxide [713], (−)-pteropodine (α; [717]), pteropodine *N*-oxide [713], (+)-speciophylline (α; [717]), speciophylline *N*-oxide [713], (+)-uncarine F (α; [717]), uncarine F *N*-oxide [713]. *U. lanosa* Wall. var. *glabrata* (Bl.) Ridsd. (= *U. glabrata* (Bl.) A. DC., *U. lobbii* Hook. f.): **C5b**: isopteropodine [713], isopteropodine *N*-oxide [713], pteropodine [713], pteropodine *N*-oxide [713], speciophylline [713], speciophylline *N*-oxide [713], uncarine F [713], uncarine F *N*-oxide [713]. *U. lanosa* Wall. var. *korrensis* (Kanehira) Ridsd. (= *U. korrensis* Kanehira): **C5b**: isomitraphylline [713], isomitraphylline *N*-oxide [713], isopteropodine [713], isopteropodine *N*-oxide [713], mitraphylline [713], mitraphylline *N*-oxide [713], pteropodine [713], pteropodine *N*-oxide [713], speciophylline [713], speciophylline *N*-oxide [713], uncarine F [713], uncarine F *N*-oxide [713]. *U. lanosa* Wall. var. *lanosa*: **C5b**: isopteropodine [713], isopteropodine *N*-oxide [713], pteropodine [713], pteropodine *N*-oxide [713], speciophylline [713], speciophylline *N*-oxide [713], uncarine F [713], uncarine F *N*-oxide [713]. *U. lanosa* var. *parviflora* Ridl. (see *U. homomalla*). *U. lanosa* Wall. var. *toppingii* (Merr.) Ridsd. forma *toppingii* (Merr.) Ridsd. (= *U. toppingii* Merr.): **C5b**: isomitraphylline [713], isomitraphylline *N*-oxide [713], isopteropodine [713], isopteropodine *N*-oxide [713], mitraphylline [713], mitraphylline *N*-oxide [713], pteropodine [713], speciophylline [713], speciophylline *N*-oxide [713], uncarine F [713]. *U. lobbii* Hook. f. (see *U. lanosa* var. *glabrata*). *U. longiflora* (Poir.) Merr. var. *longiflora* (= *U. havilandiana* S. Moore, *U. callophylla* var. *oligoneura* Korth. ex Miq., *U. pachyphylla* Merr., *U. pteropoda* Merr., *U. trinervis* Havil.): **C4d**: corynoxeine or corynoxine [713], corynoxine B [713], isocorynoxeine [713], isorhynchophylline [713], isorhynchophylline *N*-oxide [713], rhynchophylline [713], rhynchophylline *N*-oxide [713]; **C5b**: isomitraphylline [713], isomitraphylline *N*-oxide [713], isopteropodine [713], isopteropodine *N*-oxide [713], mitraphylline [713], mitraphylline *N*-oxide [713], pteropodine [713], pteropodine *N*-oxide [713], speciophylline [713], speciophylline *N*-oxide [713], uncarine F [713], uncarine F *N*-oxide [713]. *U. longiflora* (Poir.) Merr. var. *pteropoda* (Miq.) Ridsd. (= *U. laevifolia* Elm., *U. pteropoda*

Miq.): **C4d**: corynoxeine [713], isocorynoxeine [713], isorhynchophylline [713], rhynchophylline [713]; **C5b**: isomitraphylline [733], (−)-isopteropodine (α; [733, 734]), isopteropodine *N*-oxide [733], mitraphylline [733], (−)-pteropodine (α; [733, 734]), pteropodine *N*-oxide [733], speciophylline [733], speciophylline *N*-oxide [733], uncarine F [733]. *U. luzoniensis* Merr. (see *U. callophylla*). *U. macrophylla* Wall. (= *U. sessifolia* Roxb. ex Kurz.): **C4d**: (−)-corynoxine A (α; [735]), (−)-corynoxine B (α; [735]), isorhynchophylline [735], isorhynchophylline *N*-oxide [713], rhynchophylline [735], rhynchophylline *N*-oxide [713]. *U. membranifolia* How (see *U. sinensis*). *U. multiflora* K. Schum. et Laut. (see *U. cordata* var. *cordata* forma *cordata*). *U. nemorosa* Korth. (see *U. cordata* var. *cordata* forma *cordata*). *U. nervosa* Elm. (= *U. inermis* Val., *U. sclerophylloides* Val., *U. valetoniana* Merr. et Perry): **C3a**: dihydrocorynantheine [713], hirsuteine [713], hirsutine [713]. *U. orientalis* Guill.: **C4a**: ajmalicine [713], akuammigine [713], 3-isoajmalicine [715], 3-iso-19-epiajmalicine [713]; **C5b**: formosanine [713], isoformosanine [713], isomitraphylline [715], isomitraphylline *N*-oxide [715], isopteropodine [715], isopteropodine *N*-oxide [715], mitraphylline [715], mitraphylline *N*-oxide [715], pteropodine [715], pteropodine *N*-oxide [715], speciophylline [715], speciophylline *N*-oxide [715], uncarine F [715], uncarine F *N*-oxide [715]. *U. ovalifolia* Roxb. (see *U. acida* var. *acida*). *U. pachyphylla* Merr. (see *U. longiflora* var. *longiflora*). *U. parviflora* (Ridl.) Ridl. (see *U. homomalla*). *U. pedicellata* Roxb. (see *U. cordata* var. *cordata* forma *cordata*). *U. perrottetii* (A. Rich.) Merr. (= *U. ferrea* F. Vill., *U. hookeri* Vidal.): **C5b**: isomitraphylline [713], isomitraphylline *N*-oxide [713], isopteropodine [713], mitraphylline [713], mitraphylline *N*-oxide [713], (−)-pteropodine (α; [718]), (+)-speciophylline (α; [718]), uncarine F [713]. *U. philippinensis* Elm. (see *U. lanosa* var. *appendiculata* forma *philippinensis*). *U. pilosa* Roxb. (see *U. scandens*). *U. pteropoda* Merr. (see *U. longiflora* var. *longiflora*). *U. pteropoda* Miq. (see *U. longiflora* var. *pteropoda*). *U. quadrangularis* Geddes (see *U. homomalla*). *U. rhynchophylla* (Miq.) Miq. ex. Havil. [= *Ourouparia rhynchophylla* (Miq.) Matsum., *U. rhynchophylloides* How]: **C3a**: corynantheine [75, 732, 736], (+)-dihydrocorynantheine (α; [75, 732, 736]), (+)-geissoschizine methyl ether (α; [75, 737]), (+)-hirsuteine (α; [75, 732, 736]), (+)-hirsutine (α; [75, 732, 736]); **C4a**: (−)-akuammigine (α; [75, 737]); **C4d**: (−)-corynoxeine (α; [75, 732, 736, 738]), isocorynoxeine [75, 732, 736], (+)-isorhynchophylline (α; [75, 729]), (−)-rhynchophylline (α; [75, 729, 736]), rhynchophylline *N*-oxide [713]; **V4**: angustidine* [546], angustine* [546], angustoline* [546]. *U. rhynchophylloides* How (see *U. rhynchophylla*). *U. rostrata* Pierre ex Pitard (see *U. elliptica*). *U. roxburghiana* Korth. (= *U. brevicarpa* Elm.): **C5b**: isopteropodine [713], isopteropodine *N*-oxide [713], pteropodine [713], pteropodine *N*-oxide [713], speciophylline [713], speciophylline *N*-oxide [713], uncarine F [713], uncarine F *N*-oxide [713]. *U. salaccensis* Bakh. f., nom. provis. (see *U. attenuata*). *U. salomonensis* (Rech.) Merr. et Perry (see *U. bernaysii*). *U. scandens* (Smith.) Hutch. (= *U. pilosa* Roxb., *U. wangii* How): **C5b**: isomitraphylline [713], isomitraphylline *N*-oxide [713], isopteropodine [713], isopteropodine *N*-oxide [713], mitraphylline [713], mitraphylline *N*-oxide [713], pteropodine [713], pteropodine *N*-oxide [713], speciophylline [713], speciophylline *N*-oxide [713], uncarine F [713]. *U. sclerophylla* (Hunt.) Roxb. (see *U. cordata* var. *cordata* forma *cordata*). *U. sclerophylla* F. Vill. (see *U. cordata* var. *ferruginea* forma *insignis*). *U. sclerophylla* Wall. (see *U. cordata* var. *cordata* forma *cordata*). *U. sclerophylla* Warb. (see *U. bernaysii*). *U. sclerophylloides* Val. (see *U. nervosa*). *U. sessifolia* Roxb. ex Kurz. (see *U. macrophylla*). *U. sessilifructus* Roxb.: **C3a**: hirsutine [713]; **C4a**: akuammigine [713], 3-isoajmalicine [713], 3-iso-19-epi-ajmalicine [713]; **C4d**: corynoxine A [713], corynoxine B [713], isorhynchophylline [713], rhynchophylline [713]; **C5b**: formosanine [713], isoformosanine [713], isomitraphylline [713], isomitraphylline *N*-oxide [713], mitraphylline [713], mitraphylline *N*-oxide [713], uncarine F [713]. *U. setiloba* Benth. (see *U. lanosa* var. *appendiculata* forma *setiloba*). *U. setiloba* Sasaki (see *U. lanosa* var. *appendiculata* forma *philippinensis*). *U. sinensis* (Oliv.) Havil. (= *U. membranifolia* How): **C4a**: akuammigine [713]; **C5b**: isopteropodine [713], isopteropodine *N*-oxide [713], pteropodine [713], pteropodine *N*-oxide [713], speciophylline [713], speciophylline *N*-oxide [713], uncarine F [713], uncarine F *N*-oxide [713]. *U. speciosa* Wall. ex. G. Don (see *U. cordata* var. *cordata* forma *cordata*). *U. spinosa* Raeusch (see *U. guianensis*). *U. sterrophylla* Merr. et Perry: **C4a**: 3-isoajmalicine [713]; **C4d**: isorhynchophylline [713], rhynchophylline [713]; **C5b**: isomitraphylline [713], isopteropodine [713], mitraphylline [713], pteropodine [713], speciophylline [713], speciophylline *N*-oxide [713], uncarine F [713]. *U. surinamensis* Miq. (see *U. tomentosa*). *U. talbotii* Wernh.: **C4d**: isorhynchophylline [713], rhynchophylline [713]. *U. thwaitesii* Alston (see *U. elliptica*). *U. tomentosa* (Willd.) DC. (= *U. surinamensis* Miq.): **C3a**: dihydrocorynantheine [739], dihydrocorynantheine *N*-oxide [739], hirsuteine [739], hirsutine [739], hirsutine *N*-oxide [739]; **C4d**: isorhynchophylline [739], isorhynchophylline *N*-oxide [739], isorotundifoline [739], (−)-rhynchophylline (α; [739, 740]), rhynchophylline *N*-oxide [739], rotundifoline [739]; **C5b**: isomitraphylline [739], isomitraphylline *N*-oxide [739], mitraphylline [739]. *U. tonkinensis* Havil. (see *U. homomalla*). *U. toppingii* Merr. (see *U. lanosa* var. *toppingii* forma *toppingii*). *U. trinervis* Havil. (see *U. longiflora* var. *longiflora*). *U. uraiensis* Hayata (see *U. hirsuta*). *U. valetoniana* Merr. et Perry (see *U. nervosa*). *U. velutina* Havil. [= *U. canescens* Korth. ssp. *velutina* (Havil.) Ridsd. ex Phillip. et Heming., *U. clavisepala* Elm.]: **C5b**: isomitraphylline [713], isopteropodine [715], isopteropodine *N*-oxide [715], mitraphylline [713], pteropodine [715], pteropodine *N*-oxide [715], speciophylline [715], speciophyline *N*-oxide [715], uncarine F [715], uncarine F *N*-oxide [715]. *U. wallichii* Korth. (see *U. cordata* var. *ferruginea* forma *insignis*). *U. wangii* How (see *U. scandens*). *U. wrayi* King et Gamble (see *U. callophylla*).

REFERENCES

1. M. V. Kisakürek and M. Hesse in *Indole and Biogenetically Related Alkaloids*, J. D. Phillipson and M. H. Zenk, Eds., Academic Press, London, 1980.
2. H.-P. Ros, R. Kyburz, N. W. Preston, R. T. Gallagher, I. R. C. Bick, and M. Hesse, *Helv. Chim. Acta* **62**, 481 (1979).

3. R. Hegnauer, *Chemotaxonomie der Pflanzen*, Birkhäuser-Verlag, Basel; Vol. 3, 1964, Apocynaceae; Vol. 4, 1966, Loganiaceae; Vol. 6, 1973, Rubiaceae.
4. D. Ganzinger and M. Hesse, *Lloydia* **39,** 326 (1976).
5. J. A. Joule, H. Monteiro, L. J. Durham, B. Gilbert, and C. Djerassi, *J. Chem. Soc.,* 4773 (1965).
6. R. T. Brown and A. A. Charalambides, *Tetrahedron Lett.*, 1649 (1974).
7. R. T. Brown and W. P. Blackstock, *Tetrahedron Lett.*, 3063 (1972).
8. M. Hesse, *Indolalkaloide in Tabellen*, Springer-Verlag, Berlin, 1964.
9. M. Hesse, *Indolalkaloide in Tabellen, Ergänzungswerk*, Springer-Verlag, Berlin, 1968.
10. K. Jankowski, *Experientia* **29,** 519 (1973).
11. J. D. Olbright, J. C. van Meter, and L. Goldman, *Lloydia* **28,** 212 (1965).
12. R. S. Cahn, C. K. Ingold, and V. Prelog, *Angew. Chemie* **78,** 413 (1966).
13. J. Le Men and W. I. Taylor, *Experientia* **21,** 508 (1965).
14. A. J. M. Leeuwenberg, in *Indole and Biogenetically Related Alkaloids*, J. D. Phillipson and M. H. Zenk, Eds., Academic Press, London, 1980.
15. A. R. Battersby, A. R. Burnett, and P. G. Parsons, *J. Chem. Soc. Ser. C,* 1193 (1969).
16. A. R. Battersby, *Specialist Periodical Reports, Alkaloids*, The Chemical Society, 1971, p. 31.
17. G. A. Cordell, *Lloydia* **37,** 219 (1974).
18. I. Kompiš, M. Hesse and H. Schmid, *Lloydia* **34,** 269 (1971).
19. A. I. Scott, *Acc. Chem. Res.* **3,** 151 (1970).
20. A. Husson, Y. Langlois, C. Riche, H.-P. Husson, and P. Potier, *Tetrahedron* **29,** 3095 (1973).
21. M. Andriantsiferana, R. Besselievre, C. Riche, and H.-P. Husson, *Tetrahedron Lett.* 2587 (1977).
22. C. Kan-Fan, G. Massiot, A. Ahond, B. C. Das, H.-P. Husson, P. Potier, A. I. Scott, and C. C. Wei, *J. Chem. Soc., Chem. Commun.*, 164 (1974).
23. L. Douzoua, M. Mansour, M.-M. Debray, L. Le Men-Olivier, and J. Le Men, *Phytochemistry* **13,** 1994 (1974).
24. J. Le Men, private communication (1977).
25. E. Haack, A. Popelak, A. Spingler, and F. Kaiser, *Naturwissenschaften* **41,** 479 (1954).
26. P. Lathuillière, L. Olivier, J. Lévy, and J. Le Men, *Ann. Pharm. Fr.* **24,** 547 (1966).
27. P. N. Edwards and G. F. Smith, *Proc. Chem. Soc.,* 215 (1960).
28. B. L. Moeller, L. Seedorff, and W. Nartey, *Phytochemistry* **11,** 2620 (1972).
29. A. I. Scott and C. L. Yeh, *J. Am. Chem. Soc.* **96,** 2273 (1974).
30. M. Pichon, *Mém. Mus. Nat. Hist. Nat. Nou.* Ser. **24,** 111 (1948), **27,** 153 (1949); *Not. Syst.* Paris **13,** 212, 230 (1948).
31. A. J. M. Leeuwenberg, *Adansonia* **16,** 383 (1976).
32. K. Schumann, *Rubiaceae* in *Die natürlichen Pflanzenfamilien*, Vol. 4, A. Engler and K. Prantl, Eds., W. Engelmann, Leipzig, 1891.
33. C. E. B. Bremekamp, *Acta Botan. Neerl.* **15,** 1 (1966).
34. B. Verdcourt, *Bull. Jard. Botan. Etat Bruxelles* **28,** 209 (1958); in *Flora of Tropical East Africa*, Crown Agents for Oversea Governments and Administration, London.
35. C. E. Ridsdale and R. C. Bakhiuzen van den Brink, *Blumea* **22,** 541 (1975).
36. C. E. Ridsdale, *Blumea* **23,** 177 (1976), **24,** 43, 307 (1978).
37. F. Ehrendofer, Karyologie, Systematik und Evolution der Rubiaceae, lecture in Strassburg, 1977.
38. A. J. M. Leeuwenberg, Ed., *Loganiaceae* in *Die natürlichen Pflanzenfamilien*, Vol. 28dI, Duncker Humblot Berlin 1980.
39. N. G. Bisset and J. D. Phillipson, *Lloydia* **34,** 1 (1971).

40. N. G. Bisset, *Lloydia* **37,** 62 (1974).
41. V. Černy and F. Sorm in *The Alkaloids*, Vol. 9, R. H. F. Manske, Ed. 1967.
42. G. Lewin, N. Kunesch, and J. Poisson, *C. R. Acad. Sci.,* Ser. C **280,** 987 (1975).
43. A. Banerji, M. Chakrabarty, and B. Mukherjee, *Phytochemistry* **11,** 2605 (1972).
44. E. E. Waldner, M. Hesse, W. I. Taylor, and H. Schmid, *Helv. Chim. Acta* **50,** 1926 (1967).
45. F. Mayerl and M. Hesse, *Helv. Chim. Acta* **61,** 337 (1978).
46. M. Hesse, F. Bodmer, C. W. Gemenden, B. S. Joshi, W. I. Taylor, and H. Schmid, *Helv. Chim. Acta* **49,** 1173 (1966).
47. B. Mukherjee, A. B. Ray, A. Chatterjee, and B. C. Das, *Chem. Ind. London*, 1387 (1969).
48. A. Banerji and M. Chakrabarty, *Ind. J. Chem.* **11,** 706 (1973).
49. T. Kishi, M. Hesse, C. W. Gemenden, W. I. Taylor, and H. Schmid, *Helv. Chim. Acta* **48,** 1349 (1965).
50. T. Kishi, M. Hesse, W. Vetter, C. W. Gemenden, W. I. Taylor, and H. Schmid, *Helv. Chim. Acta* **49,** 946 (1966).
51. Z. M. Khan, M. Hesse, and H. Schmid, *Helv. Chim. Acta* **50,** 1002 (1967).
52. D. E. Burke, G. A. Cook, J. M. Cook, K. G. Haller, H. A. Lazar, and P. W. Le Quesne, *Phytochemistry* **12,** 1467 (1973).
53. C. E. Nordman and C. K. Kumra, *J. Am. Chem. Soc.* **87,** 2059 (1965).
54. J. M. Cook and P. W. Le Quesne, *J. Org. Chem.* **36,** 582 (1971).
55. R. C. Elderfield and R. E. Gilman, *Phytochemistry* **11,** 339 (1972).
56. J. M. Cook, P. W. Le Quesne, and R. C. Elderfield, *J. Chem. Soc. Chem. Commun.,* 1306 (1969).
57. J. M. Cook and P. W. Le Quesne, *Phytochemistry* **10,** 437 (1971).
58. J. M. Cook and P. W. Le Quesne, *J. Org. Chem.* **40,** 1367 (1975).
59. S. K. Talapatra and B. Talapatra, *J. Ind. Chem. Soc.* **44,** 639 (1967).
60. S. Mamatas-Kalamaras, T. Sevenet, C. Thal, and P. Potier, *Phytochemistry* **14,** 1849 (1975).
61. W. Boonchuay and W. E. Court, *Planta Med.* **29,** 380 (1976).
62. W. Boonchuay and W. E. Court, *Phytochemistry* **15,** 821 (1976).
63. S. C. Dutta, S. K. Bhattacharya, and A. B. Ray, *Planta Med.* **30,** 86 (1976).
64. R. C. Rastogi, R. S. Kapil, and S. P. Popli, *Experientia* **26,** 1056 (1970).
65. A. Chatterjee, B. Mukherjee, S. Ghosal, and P. K. Banerjee, *J. Ind. Chem. Soc.* **46,** 635 (1969).
66. Y. Morita, M. Hesse, H. Schmid, A. Banerji, J. Banerji, A. Chatterjee, and W. E. Oberhänsli, *Helv. Chim. Acta* **60,** 1419 (1977).
67. T. M. Sharp, *J. Chem. Soc.* 1227 (1934).
68. O. Hesse, *Liebigs Ann. Chem.* **203,** 170 (1880).
69. A. B. Ray and S. C. Dutta, *Experientia* **29,** 1337 (1973).
70. A. Chatterjee and S. Mukhopadhyay, *Ind. J. Chem.* **15B,** 183 (1977).
71. T. R. Govindachari, N. Viswanathan, B. R. Pai, and T. S. Savitri, *Tetrahedron* **21,** 2951 (1965).
72. A. Chatterjee, P. L. Majumder, and B. C. Das, *Chem. Ind. London*, 1388 (1969).
73. A. Chatterjee, P. L. Majumder, and A. B. Ray, *Tetrahedron Lett.*, 159 (1965).
74. H. K. Bojthe, A. Kocsis, I. Mathe, J. Tamas, and O. Clauder, *Acta Pharm. Hung.* **44,** 66 (1974).
75. S.-i. Sakai, *Heterocycles* **4,** 131 (1976).
76. S.-i. Sakai, H. Ohtani, H. Ido, and J. Haginiwa, *Yakugaku Zasshi* **93,** 483 (1973).
77. N. Aimi, Y. Asada, S.-i. Sakai, and J. Haginiwa, *Chem. Pharm. Bull.* **26,** 1182 (1978).
78. J.-M. Panas, A.-M. Morfaux, L. Olivier, and J. Le Men, *Ann. Pharm. Fr.* **30,** 273 (1972).

79. J.-M. Panas, A.-M. Morfaux, L. Le Men-Olivier, and J. Le Men, *Ann. Pharm. Fr.* **30,** 785 (1972).

80. B. Zsadon, M. Szilasi, J. Tamás, and P. Kaposi, *Phytochemistry* **14,** 1438 (1975).

81. M. P. Cava, S. S. Tjoa, Q. A. Ahmed, and A. I. Da Rocha, *J. Org. Chem.* **33,** 1055 (1968).

82. C. Niemann and J. W. Kessel, *J. Org. Chem.* **31,** 2265 (1966).

83. M. Urrea, A. Ahond, H. Jacquemin, S.-K. Kan, C. Poupat, P. Potier, and M.-M. Janot, *C. R. Acad. Sci. Ser. C* **287,** 63 (1978).

84. B. Gilbert, A. P. Duarte, Y. Nakagawa, J. A. Joule, S. E. Flores, J. A. Brissolese, J. Campello, E. P. Carrazzoni, R. J. Owellen, E. C. Blossey, K. S. Brown, and C. Djerassi, *Tetrahedron* **21,** 1141 (1965).

85. B. Gilbert, J. A. Brissolese, N. Finch, W. I. Taylor, H. Budzikiewicz, J. M. Wilson, and C. Djerassi, *J. Am. Chem. Soc.* **85,** 1523 (1963).

86. J. C. Simoẽs, B. Gilbert, W. J. Cretney, M. Hearn, and J. P. Kutney, *Phytochemistry* **15,** 543 (1976).

87. R. H. Burnell and J. D. Medina, *Phytochemistry* **7,** 2045 (1968).

88. R. F. Garcia M. and K. S. Brown, *Phytochemistry* **15,** 1093 (1976).

89. N. J. Dastoor, A. A. Gorman, and H. Schmid, *Helv. Chim. Acta* **50,** 213 (1967).

90. L. D. Antonaccio, N. A. Pereira, B. Gilbert, H. Vorbrueggen, H. Budzikiewicz, J. M. Wilson, L. J. Durham, and C. Djerassi, *J. Am. Chem. Soc.* **84,** 2161 (1962).

91. C. Djerassi, Y. Nakagawa, J. M. Wilson, H. Budzikiewicz, B. Gilbert, and L. D. Antonaccio, *Experientia* **19,** 497 (1963).

92. J. D. Medina and J. A. Hurtado, *Planta Med.* **32,** 130 (1977).

93. P. R. Benoin, R. H. Burnell, and J. D. Medina, *Can. J. Chem.* **45,** 725 (1967).

94. R. H. Burnell and N.-T. Sen, *Phytochemistry* **10,** 895 (1971).

95. M. Pinar and H. Schmid, *Helv. Chim. Acta* **50,** 89 (1967).

96. B. Gilbert, L. D. Antonaccio, and C. Djerassi, *J. Org. Chem.* **27,** 4702 (1962).

97. R. R. Arndt, S. H. Brown, N. C. Ling, P. Roller, C. Djerassi, J. M. Ferreira F., B. Gilbert, E. C. Miranda, S. E. Flores, A. P. Duarte, and E. P. Carrazzoni, *Phytochemistry* **6,** 1653 (1967).

98. D. W. Thomas, H. K. Schnoes, and K. Biemann, *Experientia* **25,** 678 (1969).

99. G. Spiteller and M. Spiteller-Friedmann, *Mh. Chem.* **94,** 779 (1963).

100. G. Spiteller and M. Spiteller-Friedmann, *Mh. Chem.* **93,** 795 (1962).

101. K. H. Palmer, *Can. J. Chem.* **42,** 1760 (1964).

102. J. Schmutz and H. Lehner, *Helv. Chim. Acta* **42,** 874 (1959).

103. F. Fish, M. Qaisuddin, and J. B. Stenlake, *Chem. Ind. London,* 319 (1964).

104. R. R. Arndt and C. Djerassi, *Experientia* **21,** 566 (1965).

105. K. Biemann, M. Friedmann-Spiteller, and G. Spiteller, *Tetrahedron Lett.*, 485 (1961).

106. S. Markey, K. Biemann, and B. Witkop, *Tetrahedron Lett.*, 157 (1967).

107. M. Gorman, A. L. Burlingame, and K. Biemann, *Tetrahedron Lett.*, 39 (1963).

108. H. K. Schnoes, A. L. Burlingame, and K. Biemann, *Tetrahedron Lett.*, 993 (1962).

109. P. Relyveld, Ph.D. thesis, Utrecht, 1966.

110. J. A. Joule, M. Ohashi, B. Gilbert, and C. Djerassi, *Tetrahedron* **21,** 1717 (1965).

111. M. Damak, A. Ahond, P. Potier, and M.-M. Janot, *Tetrahedron Lett.*, 4731 (1976).

112. M. Damak, A. Ahond, H. Doucerain, and C. Riche, *J. Chem. Soc. Chem. Commun.*, 510 (1976).

113. M. Damak, C. Poupat, and A. Ahond, *Tetrahedron Lett.*, 3531 (1976).

114. M. Mansour, L. Le Men-Olivier, J. Lévy, and J. Le Men, *Phytochemistry* **13,** 2861 (1974).

115. F. Titeux, M. Mansour, M.-M. Debray, L. Le Men-Olivier, and J. Le Men, *Phytochemistry* **13,** 1620 (1974).

116. F. Titeux, L. Le Men-Olivier, and J. Le Men, *Phytochemistry* **14,** 565 (1975).

117. L. L. Douzoua, M. Debray, P. Bellet, L. Olivier, and J. Le Men, *Ann. Pharm. Fr.* **30,** 199 (1972).

118. E. Bombardelli, A. Bonati, B. Danieli, B. Gabetta, and G. Mustich, *Fitoterapia* **45,** 183 (1974).

119. F. Titeux, B. Richard, M.-M. Debray, L. Le Men-Olivier, and J. Le Men, *Phytochemistry* **14,** 1648 (1975).

120. V. C. Agwada, J. Naranjo, M. Hesse, H. Schmid, Y. Rolland, N. Kunesch, J. Poisson, and A. Chatterjee, *Helv. Chim. Acta* **60,** 2830 (1977).

121. I. Chardon-Loriaux and H.-P. Husson, *Tetrahedon Lett.,* 1845 (1975).

122. I. Chardon-Loriaux, M.-M. Debray, and H.-P. Husson, *Phytochemistry* **17,** 1605 (1978).

123. M.-M. Janot, J. Le Men, and Y. Gabbai, *Ann. Pharm. Fr.* **15,** 474 (1957).

124. N. R. Farnsworth, W. D. Loub, R. N. Blomster, and M. Gorman, *J. Pharm. Sci.* **53,** 1558 (1964).

125. D. J. Abraham, N. R. Farnsworth, R. N. Blomster, and A. G. Sharkey, *Tetrahedron Lett.*, 317 (1965).

126. W. D. Loub, N. R. Farnsworth, R. N. Blomster, and W. W. Brown, *Lloydia* **27,** 470 (1964).

127. B. Gabetta, E. M. Martinelli, and G. Mustich, *Fitoterapia* **47,** 6 (1976).

128. N. R. Farnsworth, H. H. S. Fong, and R. N. Blomster, *Lloydia* **29,** 343 (1966).

129. D. J. Abraham and N. R. Farnsworth, *J. Pharm. Sci.* **58,** 694 (1969).

130. D. J. Abraham, N. R. Farnsworth, R. N. Blomster, and R. E. Rhodes, *J. Pharm. Sci.* **56,** 401 (1967).

131. P. Rasoanaivo, N. Langlois, and P. Potier, *Phytochemistry* **11,** 2616 (1972).

132. R. Z. Andriamialisoa, N. Langlois, P. Potier, A. Chiaroni, and C. Riche, *Tetrahedron* **34,** 677 (1978).

133. P. Rasoanaivo, N. Langlois, P. Potier, and P. Bladon, *Tetrahedron Lett.*, 1425 (1973).

134. N. Langlois and P. Potier, *Phytochemistry* **11,** 2617 (1972).

135. R. Z. Andriamialisoa, N. Langlois, and P. Potier, *Tetrahedron Lett.*, 2849 (1976).

136. P. Rasoanaivo, A. Ahond, J.-P. Cosson, N. Langlois, P. Potier, J. Guilhem, A. Ducruix, C. Riche, and C. Pascard, *C. R. Acad. Sci. Ser. C.* **279,** 75 (1974).

137. A. Chiaroni, C. Riche, L. Diatta, R. Z. Andriamialisoa, N. Langlois, and P. Potier, *Tetrahedron* **32,** 1899 (1976).

138. R. Z. Andriamialisoa, N. Langlois, and P. Potier, *Tetrahedron Lett.*, 163 (1976).

139. M. Tin-Wa, H. H. S. Fong, R. N. Blomster, and N. R. Farnsworth, *J. Pharm. Sci.* **57,** 2167 (1968).

140. A. Chatterjee, G. K. Biswas, and A. B. Kundu, *Ind. J. Chem.* **11,** 7 (1973).

141. A. I. Scott, P. C. Cherry, and A. A. Qureshi, *J. Am. Chem. Soc.* **91,** 4932 (1969).

142. J. P. Kutney and R. T. Brown, *Tetrahedron* **22,** 321 (1966).

143. K. Stolle and D. Gröger, *Pharm. Zentralhalle* **106,** 285 (1967).

144. W. B. Mors, P. Zaltzman, J. J. Beereboom, S. C. Pakrashi, and C. Djerassi, *Chem. Ind. London*, 173 (1956).

145. G. H. Svoboda, N. Neuss, and M. Gorman, *J. Am. Pharm. Assoc., Sci. Ed.* **48,** 659 (1959).

146. J. P. Kutney, G. Cook, J. Cook. I. Itoh, J. Clardy, J. Fayos, P. Brown, and G. H. Svoboda, *Heterocycles* **2,** 73 (1974).

147. N. Neuss, H. E. Boaz, J. L. Occolowitz, E. Wenkert, F. M. Schell, P. Potier, C. Kan, M. M. Plat, and M. Plat, *Helv. Chim. Acta* **56,** 2660 (1973).

148. N. Neuss, A. J. Barnes, and L. L. Huckstep, *Experientia* **31,** 18 (1975).

149. G. H. Svoboda, M. Gorman, N. Neuss, and A. J. Barnes, *J. Am. Pharm. Assoc., Sci. Ed.* **50,** 409 (1961).

150. N. Neuss, M. Gorman, N. J. Cone, and L. L. Huckstep, *Tetrahedron Lett.*, 783 (1968).
151. S. Tafur, W. E. Jones, D. E. Dorman, E. E. Logsdon, and G. H. Svoboda, *J. Pharm. Sci.* **64,** 1953 (1975).
152. J. W. Moncrief and W. W. Lipscomb, *J. Am. Chem. Soc.* **87,** 4963 (1965).
153. N. Neuss, L. L. Huckstep, and N. J. Cone, *Tetrahedron Lett.*, 811 (1967).
154. G. H. Svoboda, M. Gorman, A. J. Barnes, and A. T. Oliver, *J. Pharm. Sci.* **51,** 518 (1962).
155. S. S. Tafur, J. L. Occolowitz, T. K. Elzey, J. W. Paschal, and D. E. Dorman, *J. Org. Chem.* **41,** 1001 (1976).
156. G. C. Lleander, E. H. Salud, and E. C. Rigor, *Philipp. J. Sci.* **98,** 247 (1972).
157. G. H. Aynilian, S. G. Weiss, G. A. Cordell, D. J. Abraham, F. A. Crane, and N. R. Farnsworth, *J. Pharm. Sci.* **63,** 536 (1974).
158. A. B. Segelman and N. R. Farnsworth, *J. Pharm. Sci.* **63,** 1419 (1974).
159. G. A. Cordell and N. R. Farnsworth, *J. Pharm. Sci.* **65,** 366 (1976).
160. M. B. Patel, L. Thompson, C. Miet, and J. Poisson, *Phytochemistry* **12,** 451 (1973).
161. M. B. Patel, C. Miet, and J. Poisson, *Ann. Pharm. Fr.* **25,** 379 (1967).
162. M. P. Cava, Y. Watanabe, K. Bessho, J. A. Weisbach, and B. Douglas, *J. Org. Chem.* **33,** 3350 (1968).
163. P. A. Crooks and B. Robinson, *J. Pharm. Pharmacol.* **22,** 471 (1970).
164. P. A. Crooks and B. Robinson, *J. Pharm. Pharmacol.* **25,** 820 (1973); *Chem. Abstr.* **80,** 45606v (1974).
165. J. J. Dugan, M. Hesse, U. Renner, and H. Schmid, *Helv. Chim. Acta* **52,** 701 (1969).
166. U. Renner and H. Fritz, *Tetrahedron Lett.*, 283 (1964).
167. U. Renner, D. A. Prins, and W. G. Stoll, *Helv. Chim. Acta* **42,** 1572 (1959).
168. B. C. Das, E. Fellion, and M. Plat, *C. R. Acad. Sci. Ser. C* **264,** 1765 (1967).
169. B. Gabetta, E. M. Martinelli, and G. Mustich, *Fitoterapia* **46,** 195 (1975).
170. E. Bombardelli, A. Bonati, B. Gabetta, E. M. Martinelli, G. Mustich, and B. Danieli, *J. Chem. Soc., Perkin Trans. I,* 1432 (1976).
171. M. Pinar, U. Renner, M. Hesse, and H. Schmid, *Helv. Chim. Acta* **55,** 2972 (1972).
172. D. G. I. Kingston, B. B. Gerhart, F. Ionescu, M. M. Mangino, and S. M. Sami, *J. Pharm. Sci.* **67,** 249 (1978).
173. D. G. I. Kingston, B. B. Gerhart, and F. Ionescu *Tetrahedron Lett.*, 649 (1976).
174. C. Hootele, R. Lévy, M. Kaisin, J. Pecher, and R. H. Martin, *Bull. Soc. Chim. Belg.* **76,** 300 (1967).
175. C. Hootele, J. Pecher, R. H. Martin, G. Spiteller, and M. Spiteller-Friedmann, *Bull. Soc. Chim. Belg.* **73,** 634 (1964).
176. C. Hootele and J. Pecher, *Chimia* **22,** 245 (1968).
177. J. J. Dugan, M. Hesse, U. Renner, and H. Schmid, *Helv. Chim. Acta* **50,** 60 (1967).
178. R. Besselièvre, N. Langlois, and P. Potier, *Bull. Soc. Chim. Fr.*, 1477 (1972).
179. C. Kan-Fan, R. Besselièvre, A. Cavé, B. C. Das, and P. Potier, *C. R. Acad. Sci.* Ser. C **272,** 1431 (1971).
180. C. Kan-Fan, H.-P. Husson, and P. Potier, *Bull. Soc. Chim. Fr.*, 1227 (1976).
181. C. Kan-Fan, B. C. Das, H.-P. Husson, and P. Potier, *Bull. Soc. Chim. Fr.*, 2839 (1974).
182. J. Bruneton, A. Bouquet, and A. Cavé, *Phytochemistry* **12,** 1475 (1973).
183. J. Bruneton, C. Kan-Fan, and A. Cavé, *Phytochemistry* **14,** 569 (1975).
184. J. Bruneton, A. Cavé, and A. Cavé, *Tetrahedron* **29,** 1131 (1973).
185. J. Bruneton, A. Bouquet, and A. Cavé, *Phytochemistry* **13,** 1963 (1974).
186. A. Cavé, J. Bruneton, A. Ahond, A.-M. Bui, H.-P. Husson, C. Kan, G. Lukacs, and P. Potier, *Tetrahedron Lett.*, 5081 (1973).

187. A. Cavé, A. Bouquet, and B. C. Das, *C. R. Acad. Sci.* Ser. C **272,** 1367 (1971).
188. J. P. Kutney and G. B. Fuller, *Heterocycles* **3,** 197 (1975).
189. D. Stauffacher, *Helv. Chim. Acta* **44,** 2006 (1961).
190. B. Talapatra, A. Patra, and S. K. Talapatra, *Phytochemistry* **14,** 1652 (1975).
191. M. Gorman, N. Neuss, N. J. Cone, and J. A. Deyrup, *J. Am. Chem. Soc.* **82,** 1142 (1960).
192. G. Delle Monache, I. L. D'Albuquerque, F. Delle Monache, and G. B. Marini-Bettòlo, *Accad. Nazionale dei Lincei* **52,** 375 (1972).
193. M. Gorman, N. Neuss, N. J. Cone, and J. A. Deyrup, *J. Am. Chem. Soc.* **82,** 1142 (1960).
194. S. M. Kupchan, A. Bright, and E. Macko, *J. Pharm. Sci.* **52,** 598 (1963).
195. H. K. Schnoes, D. W. Thomas, R. Aksornvitaya, W. R. Schleigh, and S. M. Kupchan, *J. Org. Chem.* **33,** 1225 (1968).
196. K. Raj, A. Shoeb, R. S. Kapil, and S. P. Popli, *Phytochemistry* **13,** 1621 (1974).
197. A. C. Santos, G. Aguilar-Santos, and L. L. Tibayan, *Anales Real. Acad. Farm.* **31,** 3 (1965).
198. J. Bruneton, A. Cavé, E. W. Hagaman, N. Kunesch, and E. Wenkert, *Tetrahedron Lett.*, 3567 (1976).
199. J. R. Knox and J. Slobbe, *Austr. J. Chem.* **28,** 1813 (1975).
200. P. Lathuillière, L. Oliver, J. Lévy, and J. Le Men, *Ann. Pharm. Fr.* **28,** 57 (1970).
201. G. Aguilar-Santos, A. C. Santos, and L. M. Joson, *J. Philipp. Pharm. Assoc.* **50,** 321 (1964).
202. V. C. Agwada, Y. Morita, U. Renner, M. Hesse, and H. Schmid, *Helv. Chim. Acta* **58,** 1001 (1975).
203. J. Le Men, P. Potier, L. Le Men-Olivier, J.-M. Panas, B. Richard, and C. Potron, *Bull. Soc. Chim. Fr.*, 1369 (1974).
204. M. P. Cava, S. K. Talapatra, J. A. Weisbach, B. Douglas, R. F. Raffauf, and J. L. Beal, *Tetrahedron Lett.*, 931 (1965).
205. J.-P. Paccioni and H.-P. Husson, *Phytochemistry* **17,** 2146 (1978).
206. R. Paris and M. Pointet, *Ann. Pharm. Fr.* **12,** 547 (1954).
207. R. E. Moore and H. Rapoport, *J. Org. Chem.* **38,** 215 (1973).
208. A. Chiaroni, C. Riche, M. Pais, and R. Goutarel, *Tetrahedron Lett.*, 4729 (1976).
209. H. Rapoport and R. E. Moore, *J. Org. Chem.* **27,** 2981 (1962).
210. H. Rapoport, T. P. Onak, N. A. Hughes, and M. G. Reinecke, *J. Am. Chem. Soc.* **80,** 1601 (1958).
211. R. Kaschnitz and G. Spiteller, *Mh. Chem.* **96,** 909 (1965).
212. P. Rasoanaivo and G. Lukacs, *J. Org. Chem.* **41,** 376 (1976).
213. M. P. Cava, S. K. Talapatra, K. Nomura, J. A. Weisbach, B. Douglas, and E. C. Shoop, *Chem. Ind. London*, 1242 (1963).
214. M. V. Lakshmikantham, M. J. Mitchell, and M. P. Cava, *Heterocycles* **9,** 1009 (1978).
215. A.-M. Bui, M.-M. Debray, P. Boiteau, and P. Potier, *Phytochemistry* **16,** 703 (1977).
216. G. Ferrari, O. Fervidi, and M. Ferrari, *Phytochemistry* **10,** 439 (1971).
217. V. Vecchietti, G. Ferrari, F. Orsini, F. Pelizzoni, and A. Zajotti, *Phytochemistry* **17,** 835 (1978).
218. P. Potier, A.-M. Bui, B. C. Das, J. Le Men, and P. Boiteau, *Ann. Pharm. Fr.*, **26,** 621 (1968).
219. V. Agwada, M. B. Patel, M. Hesse, and H. Schmid, *Helv. Chim. Acta* **53,** 1567 (1970).
220. J. Naranjo, M. Hesse, and H. Schmid, *Helv. Chim. Acta* **53,** 749 (1970).
221. M. F. Bartlett, B. Korzun, R. Sklar, A. F. Smith, and W. I. Taylor, *J. Org. Chem.* **28,** 1445 (1963).
222. A.-M. Morfaux, L. Olivier, J. Lévy, and J. Le Men, *Ann. Pharm. Fr.* **27,** 679 (1969).
223. M. F. Bartlett, R. Sklar, A. F. Smith, and W. I. Taylor, *J. Org. Chem.* **28,** 2197 (1963).

224. R. H. Burnell, A. Chapelle, and M. F. Khalil, *Can. J. Chem.* **52,** 2327 (1974).

225. A.-M. Morfaux, L. Olivier, and J. Le Men, *Bull. Soc. Chim. Fr.*, 3967 (1971).

226. A.-M. Morfaux, L. Le Men-Olivier, J. Lévy, and J. Le Men, *Tetrahedron Lett.*, 1939 (1973).

227. L. Olivier, F. Quirin, B. C. Das, J. Lévy, and J. Le Men, *Ann. Pharm. Fr.*, **26,** 105 (1968).

228. I. Søndergaard and F. Nartey, *Phytochemistry* **15,** 1322 (1976).

229. A.-M. Morfaux, J. Vercauteren, J. Kerharo, L. Le Men-Olivier, and J. Le Men, *Phytochemistry* **17,** 167 (1978).

230. Y. Morita, M. Hesse, and H. Schmid, *Helv. Chim. Acta* **52,** 89 (1969).

231. C. W. L. Bevan, M. B. Patel, A. H. Rees, D. R. Harris, M. L. Marshak, and H. H. Mills, *Chem. Ind. London*, 603 (1965).

232. C. Kump, M. B. Patel, J. M. Rowson, M. Hesse, and H. Schmid, *Pharm. Acta Helv.* **40,** 586 (1965).

233. Y. Morita, M. Hesse, and H. Schmid, *Helv. Chim. Acta* **51,** 1438 (1968).

234. B. W. Bycroft, M. Hesse, and H. Schmid, *Helv. Chim. Acta* **48,** 1598 (1965).

235. C. W. L. Bevan, M. B. Patel, A. H. Rees, and A. G. Loudon, *Tetrahedron* **23,** 3809 (1967).

236. P. L. Majumder, *J. Ind. Chem. Soc.* **45,** 853 (1968).

237. A. A. Kiang and G. F. Smith, *Proc. Chem. Soc.*, 298 (1962).

238. S. Baassou, H. Mehri, and M. Plat, *Phytochemistry* **17,** 1449 (1978).

239. A. Rabaron and M. Plat, *Plant. Med. Phytother.* **7,** 319 (1973); *Chem. Abstr.* **80,** 130499g (1974).

240. H. H. A. Linde, *Helv. Chim. Acta* **48,** 1822 (1965).

241. M. H. Mehri, A. Rabaron, T. Sevenet, and M. M. Plat, *Phytochemistry* **17,** 1451 (1978).

242. M. Damak, A. Ahond, and P. Potier, *Tetrahedron Lett.*, 167 (1976).

243. A. Rabaron, M. H. Mehri, T. Sevenet, and M. M. Plat, *Phytochemistry* **17,** 1452 (1978).

244. A. Rabaron, M. Plat, and P. Potier, *Phytochemistry* **12,** 2537 (1973).

245. H. Mehri, M. Plat, and P. Potier, *Ann. Pharm. Fr.*, **29,** 291 (1971).

246. M. Daudon, M. H. Mehri, M. M. Plat, E. W. Hagaman, and E. Wenkert, *J. Org. Chem.* **41,** 3275 (1976).

247. J. M. Panas, B. Richard, C. Potron, R. S. Razafindrambao, M.-M. Debray, L. Le Men-Olivier, J. Le Men, A. Husson, and H.-P. Husson, *Phytochemistry* **14,** 1120 (1975).

248. F. A. Doy and B. P. Moore, *Austr. J. Chem.* **15,** 548 (1962).

249. N. Peube-Locou, M. Koch, M. Plat, and P. Potier, *Ann. Pharm. Fr.* **30,** 775 (1972); *Chem. Abstr.* **78,** 145190g (1973).

250. N. Peube-Locou, M. Plat, and M. Koch, *Phytochemistry* **12,** 199 (1937).

251. N. Preaux, M. Koch, M. Plat, and T. Sevenet, *Plant. Med. Phytother.* **8,** 250 (1974); *Chem. Abstr.* **83,** 75339q (1975).

252. N. Preaux, M. Koch, and M. Plat, *Phytochemistry* **13,** 2607 (1974).

253. S.-i. Sakai, N. Aimi, K. Takahashi, M. Kitagawa, K. Yamaguchi, and J. Haginawa, *Yakugaku Zasshi* **94,** 1274 (1974).

254. N. Peube-Locou, M. Koch, M. Plat, and P. Potier *Ann. Pharm. Fr.* **30,** 821 (1972).

255. J. Bruneton, T. Sevenet, and A. Cavé, *Phytochemistry* **11,** 3073 (1972).

256. F. A. Doy and B. P. Moore, *Austr. J. Chem.* **15,** 548 (1962).

257. S. R. Johns, J. A. Lamberton, B. P. Moore, and A. A. Sioumis, *Austr. J. Chem.* **28,** 1627 (1975).

258. B. Douglas, J. L. Kirkpatrick, B. P. Moore, and J. A. Weisbach, *Austr. J. Chem.* **17,** 246 (1964).

259. J. Bruneton and A. Cavé, *Ann. Pharm. Fr.* **30,** 629 (1972).

260. J. Bruneton, J.-L. Pousset, and A. Cavé, *C. R. Acad. Sci.* Ser. C **273,** 442 (1971).

261. G. H. Svoboda, G. A. Poore, and M. L. Montfort, *J. Pharm. Sci.* **57,** 1720 (1968).

262. W. Jordan and P. J. Scheuer, *Tetrahedron* **21,** 3731 (1965).

263. P. J. Scheuer and J. T. H. Metzger, *J. Org. Chem.* **26,** 3069 (1961).

264. J. Bruneton and A. Cavé, *Phytochemistry* **11,** 2618 (1972).

265. S. Goodwin, A. F. Smith, and E. C. Horning, *J. Am. Chem. Soc.* **81,** 1903 (1959).

266. C. Kan-Fan, B. C. Das, P. Potier, and M. Schmid, *Phytochemistry* **9,** 1351 (1970).

267. J. P. Cosson and M. Schmid, *Phytochemistry* **9,** 1353 (1970).

268. M. Sainsbury and B. Webb, *Phytochemistry* **11,** 2337 (1972).

269. H.-P. Ros, E. Schöpp, and M. Hesse, *Z. Naturforschung* **33c,** 290 (1978).

270. A. Harmouche, H. Mehri, M. Koch, A. Rabaron, M. Plat, and T. Sevenet, *Ann. Pharm. Fr.* **34,** 31 (1976).

271. W. E. Meyer, J. A. Coppola, and L. Goldman *J. Pharm. Sci.* **62,** 1199 (1973).

272. T. R. Govindachari, B. S. Joshi, A. K. Saksena, S. S. Sathe, and N. Viswanathan, *Tetrahedron Lett.*, 3873 (1965).

273a. A. Harmouche, H. Mehri, M. Koch, A. Rabaron, M. Plat, and T. Sevenet, *Ann. Pharm. Fr.* **34,** 31 (1976).

273b. C. Miet and J. Poisson, *Phytochemistry* **16,** 153 (1977).

274. C. R. Biswas, *Sci. Cult.* **39,** 259 (1973); *Chem. Abstr.* **79,** 113214x (1973).

275. M. Zeches, M.-M. Debray, G. Ledouble, L. Le Men-Olivier, and J. Le Men, *Phytochemistry* **14,** 1122 (1975).

276. J. Le Men, G. Hugel, M. Zeches, M.-J. Hoizey, L. Le Men-Olivier, and J. Lévy, *C. R. Acad. Sc., Ser. C* **283,** 759 (1976).

277. M. J. Hoizey, M.-M. Debray, L. Le Men-Olivier, and J. Le Men, *Phytochemistry* **13,** 1995 (1974).

278. M.-J. Hoizey, L. Olivier, M. Debray, M. Quirin, and J. Le Men, *Ann. Pharm. Fr.* **28,** 127 (1970).

279. F. Quirine, M.-M. Debray, C. Sigaut, P. Thepenier, L. Le Men-Olivier, and J. Le Men, *Phytochemistry* **14,** 812 (1975).

280. F. Picot, F. Lallemand, P. Boiteau, and P. Potier, *Phytochemistry* **13,** 660 (1974).

281. N. Petitfrere, A. M. Morfaux, M. M. Debray, L. Le Men-Olivier, and J. Le Men, *Phytochemistry* **14,** 1648 (1975).

282. M. de Bellefon, M.-M. Debray, L. Le Men-Olivier, and J. Le Men, *Phytochemistry* **14,** 1649 (1975).

283. J. M. Panas, B. Richard, C. Sigaut, M.-M. Debray, L. Le Men-Olivier, and J. Le Men, *Phytochemistry* **13,** 1969 (1974).

284. F. Picot, P. Boiteau, B. C. Das, P. Potier, and M. Andriantsiferana, *Phytochemistry* **12,** 2517 (1973).

285. L. Le Men-Olivier, B. Richard, and J. Le Men, *Phytochemistry* **13,** 280 (1974).

286. M. C. Lévy, M.-M. Debray, L. Le Men-Olivier, and J. Le Men, *Phytochemistry* **14,** 579 (1975).

287. H. Achenbach and E. Schaller, *Chem. Ber.* **109,** 3527 (1976).

288. H. Achenbach and E. Schaller, *Chem. Ber.* **108,** 3842 (1975).

289. H. Achenbach and E. Schaller, *Tetrahedron Lett.*, 351 (1976).

290. F. J. A. Matos, R. Braz F., O. R. Gottlieb, F. W. L. Machado, and M. I. L. M. Madruga, *Phytochemistry* **15,** 551 (1976).

291. J. A. Weisbach, R. F. Raffauf, O. Ribeiro, E. Macko, and B. Douglas, *J. Pharm. Sci.* **52,** 350 (1963).

292. H. Achenbach, *Tetrahedron Lett.*, 4405 (1966).

293. Z. Voticky, L. Jahodář, and M. P. Cava, *Coll. Czech, Chem. Comm.* **42,** 1403 (1977).

294. L. Jahodář, Z. Voticky, and M. P. Cava, *Phytochemistry* **13,** 2880 (1974).

295. B. Hwang, J. A. Weisbach, B. Douglas, R. Raffauf, M. P. Cava, and K. Bessho *J. Org. Chem.* **34,** 412 (1969).

296. R. H. Burnell and J. D. Medina, *Can. J. Chem.* **49,** 307 (1971).

297. P. R. Benoin, R. H. Burnell, and J. D. Medina *Tetrahedron Lett.*, 807 (1968).

298. T. A. Henry, *J. Chem. Soc.*, 2759 (1932).

299. G. Ledouble, L. Olivier, M. Quirin, J. Lévy, J. Le Men, and M.-M. Janot, *Ann. Pharm. Fr.* **22,** 463 (1964).

300. S. Silvers and A. Tulinsky, *Tetrahedron Lett.*, 339 (1962).

301. L. Olivier, J. Lévy, J. Le Men, M.-M. Janot, H. Budzikiewicz, and C. Djerassi, *Ann. Pharm. Fr.* **22,** 35 (1964).

302. L. Olivier, J. Lévy, J. Le Men, M.-M. Janot, H. Budzikiewicz, and C. Djerassi, *Bull. Soc. Chim. Fr.*, 868 (1965).

303. W. I. Taylor, M. F. Bartlett, L. Olivier, J. Lévy and J. Le Men, *Bull. Soc. Chim. Fr.*, 392 (1964).

304. J. Lévy, G. Ledouble, J. Le Men, and M.-M. Janot, *Bull. Soc. Chim. Fr.*, 1917 (1964).

305. G. Büchi, R. E. Manning, and F. A. Hochstein, *J. Am. Chem. Soc.* **84,** 3393 (1962).

306. Z. M. Khan, M. Hesse, and H. Schmid, *Helv. Chim. Acta* **48,** 1957 (1965).

307. W. G. Kump and H. Schmid, *Helv. Chim. Acta* **44,** 1503 (1961).

308. M. Hesse, F. Bodmer, and H. Schmid, *Helv. Chim. Acta* **49,** 964 (1966).

309. A. A. Gorman, N. J. Dastoor, M. Hesse, W. v. Philipsborn, U. Renner, and H. Schmid, *Helv. Chim. Acta* **52,** 33 (1969).

310. M. Pinar, M. Hesse, and H. Schmid, *Helv. Chim. Acta* **56,** 2719 (1973).

311. J. Naranjo, M. Pinar, M. Hesse, and H. Schmid, *Helv. Chim. Acta* **55,** 752 (1972).

312. M. Pinar, M. Hanaoka, M. Hesse, and H. Schmid, *Helv. Chim. Acta* **54,** 15 (1971).

313. Z. M. Khan, M. Hesse, and H. Schmid, *Helv. Chim. Acta* **50,** 625 (1967).

314. A. A. Gorman and H. Schmid, *Mh. Chem.* **98,** 1554 (1967).

315. R. M. Bernal, A. Villegas-Castillo, and O. P. Espejo, *Experientia* **16,** 353 (1960).

316. B. P. Korzun, A. F. S. André, and P. R. Ulshafer, *J. Am. Pharm. Assoc., Sci. Ed.* **46,** 720 (1957).

317. P. J. Madati, M. J. Kayani, H. A. M. Pazi, and A. F. D. Nyamgenda, *Planta Med.* **32,** 258 (1977).

318. M. S. Habib and W. E. Court, *Phytochemistry* **13,** 661 (1974).

319. B. O. G. Schuler and F. L. Warren, *J. Chem. Soc.*, 215 (1956).

320. M. A. Khan and A. M. Ahsan, *Tetrahedron Lett.*, 5137 (1970).

321. M. A. Khan and S. Siddiqui, *Experientia* **28,** 127 (1972).

322. W. Boonchuay and W. E. Court, *Phytochemistry* **29,** 201 (1976).

323. D. A. A. Kidd, *J. Chem. Soc.*, 2432 (1958).

324. C. Miet, G. Croquelois, and J. Poisson, *Phytochemistry* **16,** 803 (1977).

325. B. Danieli, E. Bombardelli, A. Bonati, and B. Gabetta, *Chem. Ind. Milano* **53,** 1042 (1971).

326. E. Bombardelli, A. Bonati, B. Gabetta, and G. Mustich, *Fitoterapia* **43,** 67 (1973); *Chem. Abstr.* **79,** 137335p (1973).

327. B. Danieli, E. Bombardelli, A. Bonati, B. Gabetta, and G. Mustich, *Chem. Ind.* (Milano) **54,** 618 (1972).

328. M. M. Iwu and W. E. Court, *Planta Med.* **34,** 390 (1978).

329. M. M. Iwu and W. E. Court, *Planta Med.* **33,** 360 (1978).

330. M. M. Iwu and W. E. Court, *Phytochemistry* **17,** 1651 (1978).
331. M. M. Iwu and W. E. Court, *Planta Med.* **32,** 158 (1977).
332. M. M. Iwu and W. E. Court, *Experientia* **33,** 1268 (1977).
333. C. K. Atal, *J. Am. Pharm. Assoc., Sci. Ed.* **48,** 37 (1959).
334. M. Gorman, N. Neuss, C. Djerassi, J. P. Kutney, and P. J. Scheuer, *Tetrahedron* **1,** 328 (1957).
335. A. Chatterjee and S. Talapatra, *Naturwissenschaften* **42,** 182 (1955).
336. G. Combes, L. Fonzes, and F. Winternitz, *Phytochemistry* **5,** 1065 (1966).
337. R. Pernet, J. Philippe, and G. Combes, *Ann. Pharm. Fr.* **20,** 527 (1962).
338. G. Combes, L. Fonzes, and F. Winternitz, *Bull. Soc. Chim. Fr.*, 2130 (1966).
339. G. Combes, L. Fonzes, and F. Winternitz, *Phytochemistry* **7,** 477 (1968).
340. G. Combes, L. Fonzes, and F. Winternitz, *Bull. Soc. Chim. Fr.* 761 (1967).
341. J. M. Müller, *Experientia* **13,** 479 (1957).
342. G. Dillemann, R. Paris, and P. Chaumelle, *Ann. Pharm. Fr.* **16,** 504 (1958).
343. P. Timmins and W. E. Court, *Phytochemistry* **13,** 281 (1974).
344. M. B. Patel, J. Poisson, J. L. Pousset, and J. M. Rowson, *J. Pharm. Pharmacol.* **17,** 323 (1965).
345. S. C. Pakrashi, C. Djerassi, R. Wasicky, and N. Neuss, *J. Am. Chem. Soc.* **77,** 6687 (1955).
346. D. S. Rao and S. B. Rao, *J. Am. Pharm. Assoc., Sci. Ed.* **44,** 253 (1955).
347. M. M. Iwu and W. E. Court, *Planta Med.* **33,** 232 (1978).
348. R. Salkin, N. Hosansky, and R. Jaret, *J. Pharm. Sci.* **50,** 1038 (1961).
349. E. Smith, R. S. Jaret, R. J. Shine, and M. Shamma, *J. Am. Chem. Soc.* **89,** 2469 (1967).
350. P. Timmins and W. E. Court, *Planta Med.* **27,** 105 (1975).
351. P. Timmins and W. E. Court, *Planta Med.* **29,** 283 (1976).
352. E. Schlittler, H. Schwarz, and F. Bader, *Helv. Chim. Acta* **35,** 271 (1952).
353. P. Timmins and W. E. Court, *Phytochemistry* **15,** 733 (1976).
354. P. Timmins and W. E. Court, *Planta Med.* **26,** 170 (1974).
355. A. K. Kiang and A. S. C. Wan, *J. Chem. Soc.*, 1394 (1960).
356. A. K. Kiang, S. K. Loh, M. Demanczyk, C. W. Gemenden, G. J. Papariello, and W. I. Taylor, *Tetrahedron* **22,** 3293 (1966).
357. A. Chatterjee, A. K. Ghosh, and M. Chakrabarty, *Experientia* **32,** 1236 (1976).
358. W. J. McAleer, R. G. Weston, and E. E. Howe, *Chem. Ind. London*, 1387 (1956).
359. G. Iacobucci and V. Deulofeu, *J. Org. Chem.* **22,** 94 (1957).
360. O. O. Orazi, R. A. Corral, and M. E. Stoichevich, *Can. J. Chem.* **44,** 1523 (1966).
361. S. C. Pakrashi, C. Djerassi, R. Wasicky, and N. Neuss, *J. Am. Chem. Soc.* **77,** 6687 (1955).
362. A. Chatterjee and A. B. Ray, *J. Sci. Industr. Res., Ser. A* **21,** 515 (1962).
363. D. Banes, A. E. H. Houk, and J. Wolff, *J. Amer. Pharm. Assoc., Sci. Ed.* **47,** 625 (1958).
364. F. E. Bader, D. F. Dickel, C. F. Huebner, R. A. Lucas, and E. Schlittler, *J. Am. Chem. Soc.* **77,** 3547 (1955).
365. P. R. Ulshafer, W. I. Taylor, and R. H. Nugent, *C. R. Acad. Sci.* **244,** 2989 (1957).
366. F. E. Bader, D. F. Dickel, and E. Schlittler, *J. Am. Chem. Soc.* **76,** 1695 (1954).
367. A. Chatterjee and S. Bose, *J. Ind. Chem. Soc.* **38,** 403 (1961).
368. M. Hanaoka, M. Hesse, and H. Schmid, *Helv. Chim. Acta* **52,** 1723 (1970).
369. S. P. Majumdar, P. Potier, and J. Poisson, *Tetrahedron Lett.*, 1563 (1972).
370. C. Djerassi, J. Fishman, M. Gorman, J. P. Kutney, and S. C. Pakrashi, *J. Am. Chem. Soc.* **79,** 1217 (1957).
371. J. Keck, *Naturwissenschaften* **42,** 391 (1955).

372. B. U. Vergara, *J. Am. Chem. Soc.* **77,** 1864 (1955).
373. A. S. Belikov, *Khim. Prir. Soedin.* **64** (1969).
374. A. Stoll and A. Hofmann, *Soc. Biol. Chem. India*, 248 (1955).
375. E. Haack, A Popelak, H. Spingler, and F. Kaiser, *Naturwissenschaften* **41,** 479 (1954).
376. N. Hosansky and E. Smith, *J. Am. Pharm. Assoc., Sci. Ed.* **44,** 639 (1955).
377. M. W. Klohs, F. Keller, R. E. Williams, and G. W. Kusserow, *Chem. Ind. London*, 187 (1956).
378. P. R. Ulshafer, M. L. Pandow, and N. H. Nugent, *J. Org. Chem.* **21,** 923 (1956).
379. A. Hofmann, *Helv. Chim. Acta* **38,** 536 (1955).
380. H. R. Arthur, *Chem. Ind. London*, 85 (1956).
381. H. R. Arthur and S. N. Loo, *Phytochemistry* **5,** 977 (1966).
382. H. R. Arthur, S. R. Johns, J. A. Lamberton, and S. N. Loo, *Austr. J. Chem.* **21,** 1399 (1968).
383. J.-L. Pousset, M. Debray, and J. Poisson, *Phytochemistry* **16,** 153 (1977).
384. N. N. Sabri and W. E. Court, *Phytochemistry* **17,** 2023 (1978).
385. M. B. Patel, J. Poisson, J. L. Pousset, and J. M. Rowson, *J. Pharm. Pharmacol.* **16,** 163T (1964).
386. M. M. Iwu and W. E. Court, *Planta Med.* **32,** 88 (1977).
387. J. Poisson and R. Goutarel, *Bull. Soc. Chim. Fr.,* 1703 (1956).
388. J. Poisson, R. Bergoeing, N. Chauveau, M. Shamma, and R. Goutarel, *Bull. Soc. Chim. Fr.*, 2853 (1964).
389. A. Hofmann and A. J. Frey, *Helv. Chim. Acta* **40,** 1866 (1957).
390. J. Poisson, A. Le Hir, R. Goutarel, and M.-M. Janot, *C. R. Acad. Sci.* **238,** 1607 (1954).
391. J. Poisson, N. Neuss, R. Goutarel, and M.-M. Janot, *Bull. Soc. Chim. Fr.*, 1195 (1958).
392. E. Haack, A. Popelak, and H. Spingler, *Naturwissenschaften* **43,** 328 (1956).
393. J.-L. Pousset and J. Poisson, *Ann. Pharm. Fr.* **23,** 733 (1965).
394. J.-L. Pousset and J. Poisson, *C. R. Acad. Sci., Ser.* 8 **259,** 597 (1964).
395. M. Muquet, J.-L. Pousset, and J. Poisson, *C. R. Acad. Sci., Ser.* C **266,** 1542 (1968).
396. J. Poisson, P. R. Ulshafer, L. E. Paszek, and W. I. Taylor, *Bull. Soc. Chim. Fr.*, 2683 (1964).
397. A. Hofmann and A. J. Frey, *Helv. Chim. Acta* **40,** 1866 (1957).
398. W. I. Taylor, A. J. Frey, and A. Hofmann, *Helv. Chim. Acta* **45,** 611 (1962).
399. G. B. Guise, E. Ritchie, and W. C. Taylor, *Austr. J. Chem.* **18,** 1279 (1965).
400. K. T. D. De Silva, D. King, and G. N. Smith, *J. Chem. Soc. Chem. Commun.*, 908 (1971).
401. G. N. Smith, *J. Chem. Soc. Chem. Commun.,* 912 (1968).
402. A. Chatterjee, A. Banerji, P. Majumder, and R. Majumder, *Bull. Chem. Soc. Jpn.* **49,** 2000 (1976).
403. Y. Ahmad, K. Fatima, A. Rahman, J. L. Occolowitz, B. A. Solheim, J. Clardy, R. L. Garnick, and P. W. Le Quesne, *J. Am. Chem. Soc.* **99,** 1943 (1977).
404. Y. Ahmad, P. W. Le Quesne, and N. Neuss, *J. Chem. Soc. Chem. Commun.*, 538 (1970).
405. J. M. Karle and P. W. Le Quesne, *J. Chem. Soc. Chem. Commun.*, 416 (1972).
406. K. T. D. De Silva, G. N. Smith, and K. E. H. Warren, *J. Chem. Soc. Chem. Commun.*, 905 (1971).
407. U. Renner and P. Kernweisz, *Experientia* **19,** 244 (1963).
408. U. Renner, *Lloydia* **27,** 406 (1964).
409. F. Walls, O. Collera, and A. Sandoval L., *Tetrahedron* **2,** 173 (1958).
410. O. Collera, F. Walls, A. Sandoval, F. García, J. Herrán, and M. C. Perezamador, *Bol. Institut. Quim.* **14,** 3 (1962).
411. A. Henriques, C. Kan-Fan, A. Ahond, C. Riche, and H.-P. Husson, *Tetrahedron Lett.*, 3707 (1978).

412. R. Iglesias and L. Diatta, *Rev. Cenic, Cienc. Fis.* **6,** 141 (1975); *Chem. Abstr.* **84,** 44503e (1976).
413. R. Iglesias and L. Diatta, *Rev. Cenic, Cienc. Fis.* **6,** 135 (1975); *Chem. Abstr.* **84,** 44502d (1976).
414. D. G. I. Kingston, F. Ionescu, and B. T. Li, *Lloydia* **40,** 215 (1977).
415. D. G. I. Kingston, B. T. Li, and F. Ionescu, *J. Pharm. Sci.* **66,** 1135 (1977); *Chem. Abstr.* **87,** 157068x (1977).
416. M. P. Cava, S. K. Mowdood, and J. L. Beal, *Chem. Ind. London,* 2064 (1965).
417. A. Cavé, J. Bruneton, and R. R. Paris, *Plant. Med. Phytother.* **6,** 228 (1972); *Chem. Abstr.* **78,** 26480s (1973).
418. J. C. Gaignault and J. Delourme-Houdé, *Fitoterapia* **48,** 243 (1977).
419. R. Goutarel, J. Poisson, G. Croquelois, Y. Rolland, and C. Miet, *Ann. Pharm. Fr.* **32,** 521 (1974).
420. D. F. Dickel, C. L. Holden, R. C. Maxfield, L. E. Paszek, and W. I. Taylor, *J. Am. Chem. Soc.* **80,** 123 (1958).
421. E. Wenkert and H. E. Gottlieb, *Heterocycles* **7,** 753 (1977).
422. F. Khuong-Huu, M. Cesario, J. Guilhem, and R. Goutarel, *Tetrahedron* **32,** 2539 (1976).
423. A. F. S. André, B. Korzun, and F. Weinfeldt, *J. Org. Chem.* **21,** 480 (1956).
424. S. Goodwin and E. C. Horning, *Chem. Ind. London,* 846 (1956).
425. A. Walser and C. Djerassi, *Helv. Chim. Acta* **48,** 391 (1965).
426. N. C. Ling and C. Djerassi, *J. Am. Chem. Soc.* **92,** 6019 (1970).
427. N. C. Ling, C. Djerassi, and P. G. Simpson, *J. Am. Chem. Soc.* **92,** 222 (1970).
428. J. S. E. Holker, M. Cais, F. A. Hochstein, and C. Djerassi, *J. Org. Chem.* **24,** 314 (1959).
429. S. H. Brown, C. Djerassi, and P. G. Simpson, *J. Am. Chem. Soc.* **90,** 2445 (1968).
430. C. Djerassi, H. J. Monteiro, A. Walser, and L. J. Durham, *J. Am. Chem. Soc.* **88,** 1792 (1966).
431. M. Falco, J. Garnier-Gosset, E. Fellion, and J. Le Men, *Ann. Pharm. Fr.* **22,** 455 (1964).
432. J. Gosset, J. Le Men, and M.-M Janot, *Ann. Pharm. Fr.* **20,** 448 (1962).
433. M.-M. Janot, J. Le Men, and C. Fan, *Ann. Pharm. Fr.* **15,** 513 (1957).
434. B. C. Das, J. Garnier-Gosset, J. Le Men, and M.-M. Janot, *Bull. Soc. Chim. Fr.*, 1903 (1965).
435. J. Gosset-Garnier, J. Le Men, and M.-M. Janot, *Bull. Soc. Chim. Fr.*, 676 (1965).
436. K. Bláha, K. Kavková, Z. Koblicová, and J. Trojánek, *Coll. Czech. Chem. Commun.* **33,** 3833 (1968).
437. J. Bhattacharyya and S. C. Pakrashi, *Tetrahedron Lett.*, 159 (1972).
438. E. Ali, V. S. Giri, and S. C. Pakrashi, *Experientia* **31,** 876 (1975).
439. M. R. Sharipov, M. Khalmirzaev, V. M. Malikov, and S. Y. Yunusov, *Khim. Prir. Soedin.* 413 (1974).
440. V. M. Malikov, M. R. Sharipov, and S. Y. Yunusov, *Khim. Prir. Soedin.*, 760 (1972).
441. N. Abdurakhimova, P. K. Yuldashev, and S. Y. Yunusov, *Khim. Prir. Soedin.*, 310 (1967).
442. S. Kasimirov, P. C. Yuldashev, and S. Y. Yunusov, *Dokl. Adad. Nauk SSR* **162,** 102 (1965).
443. M. R. Yagudaev, V. M. Malikov, and S. Y. Yunusov, *Khim. Prir. Soedin.*, 493 (1974).
444. N. Abdurakhimova, S. Z. Kasymov, and S. Y. Yunusov, *Khim. Prir. Soedin.*, 135 (1968).
445. N. Abdurakhimova, P. K. Yuldashev, and S. Y. Yunusov, *Khim. Prir. Soedin.*, 224 (1965).
446. M. M. Khalmirzaev, V. M. Malikov, and S. Y. Yunusov, *Khim. Prir. Soedin.*, 806 (1973).
447. M. R. Sharipov, M. M. Khalmirzaev, V. M. Malikov, and S. Y. Yunusov, *Khim. Prir. Soedin.*, 401 (1976).
448. M. M. Khalmirzaev, V. M. Malikov, and S. Y. Yunusov, *Khim. Prir. Soedin.*, 411 (1974).
449. M. M. Khalmirzaev, V. M. Malikov, and S. Y. Yunusov, *Khim. Prir. Soedin.*, 681 (1973).
450. P. K. Yuldashev and S. Y. Yunusov, *Khim. Prir. Soedin.*, 110 (1965).

451. D. A. Rakhimov, K. T. Ilyasova, V. M. Malikov, and S. Y. Yunusov, *Khim. Prir. Soedin.*, 521 (1969).

452. M. A. Kuchenkova, P. K. Yuldashev, and S. Y. Yunusov, *Khim. Prir. Soedin.*, 65 (1967).

453. K. T. Ilyasova, V. M. Malikov, and S. Y. Yunusov, *Khim. Prir. Soedin.*, 164 (1971).

454. D. A. Rakhimov, M. R. Sharipov, V. M. Malikov, and S. Y. Yunusov, *Khim. Prir. Soedin.*, 677 (1971).

455. D. A. Rakhimov, V. M. Malikov, M. R. Yagudaev, and S. Y. Yunusov, *Khim. Prir. Soedin.*, 226 (1970).

456. V. M. Malikov, S. Z. Kasymov, and S. Y. Yunusov, *Khim. Prir. Soedin.*, 640 (1970).

457. M. A. Kuchenkova, P. K. Yuldashev, and S. Y. Yunusov, *Dokl. Akad. Nauk Uzb. SSR* **21,** 42 (1964).

458. U. Osmanov, K. N. Aripov, and T. T. Shakirov, *Khim. Prir. Soedin.*, 442 (1973).

459. M. R. Yagudaev, V. M. Malikov, and S. Y. Yunusov, *Khim. Prir. Soedin.*, 260 (1974).

460. D. A. Rakhimov, V. M. Malikov, and S. Y. Yunusov, *Khim. Prir. Soedin.*, 461 (1969).

461. P. K. Yuldashev, U. Ubaev, M. A. Kuchenkova, and S. Y. Yunusov, *Khim. Prir. Soedin.*, 34 (1965).

462. I. Ognyanov, B. Pyuskyulev, B. Bozjanov, and M. Hesse, *Helv. Chim. Acta* **50,** 754 (1967).

463. I. Ognyanov and B. Pyuskyulev, *Chem. Ber.* **99,** 1008 (1966).

464. I. Ognyanov, B. Pyuskyulev, and G. Spiteller, *Mh. Chem.* **97,** 855 (1966).

465. Z. Z. Dzhakeli, *Kim. Prir. Soedin.*, 420 (1978).

466. I. Ognyanov, *Chem. Ber.* **99,** 2052 (1966).

467. G. V. Chkhikvadze, M. M. Khalmirzaev, V. Y. Vachnadze, V. M. Malikov, and S. Y. Yunusov, *Khim. Prir. Soedin.*, 227 (1976).

468. B. Pyuskyulev, I. Ognyanov, and P. Panov, *Tetrahedron Lett.*, 4559 (1967).

469. I. Ognyanov, B. Pyuskyulev, I. Kompis, T. Sticzay, G. Spiteller, M. Shamma, and R. J. Shine, *Z. Naturforschung* **23b,** 282 (1968).

470. V. Y. Vachnadze, V. M. Malikov, K. S. Mudzhiri, and S. Y. Yunusov, *Khim. Prir. Soedin.*, 341 (1972).

471. V. Y. Vachnadze, V. M. Malikov, K. S. Mudzhiri, and S. Y. Yunusov, *Sovbshch. Akad. Nauk Gruz. SSR* **66,** 97 (1972).

472. G. H. Aynilian, N. R. Farnsworth, and J. Trojánek, *Lloydia* **37,** 299 (1974).

473. G. H. Aynilian, C. L. Bell, and N. R. Farnsworth, *J. Pharm. Sci.* **64,** 341 (1975).

474. G. H. Aynilian, C. L. Bell, N. R. Farnsworth, and D. J. Abraham, *Lloydia* **37,** 589 (1974).

475. N. Abdurakhimova, P. K. Yuldashev, and S. Y. Yunusov, *Dokl. Akad. Nauk Uzb. SSR* 33 (1964).

476. M. Plat, R. Lemay, J. Le Men, M.-M. Janot, C. Djerassi, and H. Budzikiewicz, *Bull. Soc. Chim. Fr.*, 2497 (1965).

477. A. Orechoff, H. Gurewitch, and S. Norkina, *Arch. Pharm.* **272,** 70 (1934).

478. J. L. Kaul and J. Trojánek, *Lloydia* **29,** 26 (1966).

479. J. L. Kaul, J. Trojánek, and A. K. Bose, *Chem. Ind. London*, 853 (1966).

480. O. Štrouf and J. Trojánek, *Coll. Czech. Chem. Comm.* **29,** 447 (1964).

481. A. Banerji and M. Chakrabarty, *Phytochemistry* **13,** 2309 (1974).

482. A. Banerji and M. Chakrabarty, *Phytochemistry* **16,** 1124 (1977).

483. P. Potier, R. Beugelmans, J. Le Men, and M.-M. Janot, *Ann. Pharm. Fr.* **23,** 61 (1965).

484. J. L. Kaul, J. Trojánek, and A. K. Bose, *Coll. Czech. Chem. Commun.* **35,** 116 (1970).

485. S. Savaskan, I. Kompiš, M. Hesse, and H. Schmid, *Helv. Chim. Acta* **55,** 2861 (1972).

486. H. K. Schnoes, K. Biemann, J. Mokrý, I. Kompiš, A. Chatterjee, and G. Ganguli *J. Org. Chem.* **31,** 1641 (1966).

487. H. Meisel, W. Döpke, and E. Gründemann, *Tetrahedron Lett.*, 1291 (1971).
488. E. Grossmann, P. Šefčovič, and K. Szász, *Phytochemistry* **12,** 2085 (1973).
489. I. Kompiš and J. Mokrý, *Coll. Czech. Chem. Commun.,* **33,** 4328 (1968).
490. H. Meisel and W. Döpke, *Tetrahedron Lett.*, 1285 (1971).
491. Z. Votický, E. Grossmann, and P. Potier, *Coll. Czech. Chem. Commun.* **42,** 548 (1977).
492. J. Le Men, *Bull. Chim. Therapeut.*, 137 (1971).
493. J. Mokrý and I. Kompiš, *Tetrahedron Lett.*, 1917 (1963).
494. W. Döpke, H. Meisel, E. Gründemann, and G. Spiteller, *Tetrahedron Lett.*, 1805 (1968).
495. J. Mokrý and I. Kompiš, *Lloydia* **27,** 428 (1964).
496. M. Plat, D. D. Manh, J. Le Men, M.-M. Janot, H. Budzikiewicz, J. M. Wilson, L. J. Durham, and C. Djerassi, *Bull. Soc. Chim. Fr.*, 1082 (1962).
497. D. W. Thomas and K. Biemann, *Lloydia* **31,** 1 (1968).
498. G. Büchi, R. E. Manning, and S. A. Monti, *J. Am. Chem. Soc.* **86,** 4631 (1964).
499. F. Puisieux, J.-P. Devissaguet, C. Miet, and J. Poisson, *Bull. Soc. Chim. Fr.*, 251 (1967).
500. D. W. Thomas and K. Biemann, *Tetrahedron* **24,** 4223 (1968).
501. U. Renner, *Experientia* **13,** 468 (1957).
502. U. Renner, *Experientia* **15,** 185 (1959).
503. F. Puisieux, M. B. Patel, J. M. Rowson, and J. Poisson, *Ann. Pharm. Fr.* **23,** 33 (1965).
504. J. Poisson, F. Puisieux, C. Miet, and M. B. Patel, *Bull. Soc. Chim. Fr.*, 3549 (1965).
505. F. Puisieux, J.-P. Devissaguet, C. Miet, and J. Poisson, *Bull. Soc. Chim. Fr.*, 251 (1967).
506. B. Gabetta, E. M. Martinelli, and G. Mustich, *Fitoterapia* **45,** 32 (1974).
507. E. Bombardelli, A. Bonati, B. Gabetta, E. Martinelli, G. Mustich, and B. Danieli, *Phytochemistry* **15,** 2021 (1976).
508. J.-C. Braekman, M. Tirions-Lampe, and J. Pecher, *Bull. Soc. Chim. Belg.* **78,** 523 (1969).
509. N. Defay, M. Kaisin, J. Pecher, and R. H. Martin, *Bull. Soc. Chim. Belg.* **70,** 475 (1961).
510. M. Denayer-Tournay, J. Pecher, R. H. Martin, M. Friedmann-Spiteller, and G. Spiteller, *Bull. Soc. Chim. Belg.* **74,** 170 (1965).
511. G. Tirions, M. Kaisin, J.-C. Braekman, J. Pecher, and R. H. Martin, *Chimia* **22,** 87 (1968).
512. G. Lhoest, R. De Neys, N. Defay, J. Seibl, J. Pecher, and R. H. Martin, *Bull. Soc. Chim. Belg.* **74,** 534 (1965).
513. E. Bombardelli, A. Bonati, B. Gabetta, E. M. Martinelli, G. Mustich, and B. Danieli, *Tetrahedron* **30,** 4141 (1974).
514. E. Bombardelli, A. Bonati, B. Danieli, B. Gabetta, E. M. Martinelli, and G. Mustich, *Experientia* **30,** 979 (1974).
515. M. Quirin, F. Quirin, and J. Le Men, *Ann. Pharm. Fr.* **22,** 361 (1964).
516. E. Bombardelli, A. Bonati, B. Danieli, B. Gabetta, E. M. Martinelli, and G. Mustich, *Experientia* **31,** 139 (1975).
517. P. L. Majumder and B. N. Dinda, *J. Ind. Chem. Soc.* **51,** 370 (1974).
518. P. L. Majumder, B. N. Dinda, and T. K. Chanda, *Ind. J. Chem.* **11,** 1208 (1973).
519. G. B. Guise, M. Rasmussen, E. Ritchie, and W. C. Taylor, *Austr. J. Chem.* **18,** 927 (1965).
520. F. Fish, F. Newcombe, and J. Poisson, *J. Pharm. Pharmacol.* **12,** 41T (1960).
521. F. Fish and F. Newcombe, *J. Pharm. Pharmacol.* **16,** 832 (1964).
522. N. Neuss and N. J. Cone, *Experientia* **15,** 414 (1959).
523. A. Goldblatt, C. Hootele, and J. Pecher, *Phytochemistry* **9,** 1293 (1970).
524. Y. Rolland, G. Croquelois, N. Kunesch, P. Boiteau, M. Debray, J. Pecher, and J. Poisson, *Phytochemistry* **12,** 2039 (1973).
525. B. O. G. Schuler, A. A. Verbeek, and F. L. Warren, *J. Chem. Soc.*, 4776 (1958).

526. Y. Rolland, N. Kunesch, F. Libot, J. Poisson, and H. Budzikiewicz, *Bull. Soc. Chim. Fr.*, 2503 (1975).

527. N. Kunesch, Y. Rolland, J. Poisson, P. L. Majumder, R. Majumder, A. Chatterjee, V. C. Agwada, J. Naranjo, M. Hesse, and H. Schmid, *Helv. Chim. Acta* **60,** 2854 (1977).

528. J. Haginiwa, S.-i. Sakai, A. Kubo, K. Takahasi, and M. Taguchi, *Yakugaku Zasshi* **90,** 219 (1970).

529. S.-i. Sakai, N. Aimi, A. Kubo, M. Kitagawa, M. Hanasawa, K. Katano, K. Yamaguchi, and J. Haginiwa, *Chem. Pharm. Bull.* **23,** 2805 (1975).

530. S. Sakai, N. Aimi, K. Yamaguchi, H. Ohhira, K. Hori, and J. Haginiwa, *Tetrahedron Lett.*, 715 (1975).

531. S. Sakai, N. Aimi, K. Yamaguchi, E. Yamanaka, and J. Haginiwa, *Tetrahedron Lett.*, 719 (1975).

532. N. Aimi, S. Sakai, Y. Iitaka, and A. Itai, *Tetrahedron Lett.*, 2061 (1971).

533. J. Haginiwa, S. Sakai, A. Kubo, and T. Hamamoto, *Yakugaku Zasshi* **87,** 1484 (1967).

534. S. Sakai, A. Kubo, T. Hamamoto, M. Wakabayashi, K. Takahashi, Y. Ohtani, and J. Haginiwa, *Tetrahedron Lett.*, 1489 (1969).

535. S. Sakai, A. Kubo, and J. Haginiwa, *Tetrahedron Lett.*, 1485 (1969).

536. S. Sakai, N. Aimi, A. Kubo, M. Kitagawa, M. Shiratori, and J. Haginiwa, *Tetrahedron Lett.*, 2057 (1971).

537. M.-M. Janot, R. Goutarel, and M. C. P. Barron, *Ann. Pharm. Fr.* **11,** 602 (1953).

538. E. Gellert and H. Schwarz, *Helv. Chim. Acta* **34** 779 (1951).

539. W. Arnold, F. Berlage, K. Bernauer, H. Schmid, and P. Karrer, *Helv. Chim. Acta* **41,** 1505 (1958).

540. A. R. Battersby and D. A. Yeowell, *J. Chem. Soc.*, 4419 (1964).

541. M. Hesse, W. v. Philipsborn, D. Schumann, M. Spiteller-Friedmann, W. I. Taylor, H. Schmid, and P. Karrer, *Helv. Chim. Acta* **47,** 878 (1964).

542. R. Verpoorte, E. W. Kode, H. v. Doorne, and A. B. Svendsen, *Planta Med.* **33,** 237 (1978).

543. C. Galeffi, E. M. Delle Monache, and G. B. Marini-Bettòlo, *Ann. Chim.* (Rome) **63,** 849 (1973).

544. C. G. Casinovi, *Gazz. Chim. Ital.* **87,** 1174 (1957).

545. R. Marini-Bettòlo and F. Delle Monache, *Gazz. Chim. Ital.* **103,** 543 (1973).

546. J. D. Phillipson, S. R. Hemingway, N. G. Bisset, P. J. Houghton, and E. J. Shellard, *Phytochemistry* **13,** 973 (1974).

547. T. Y. Au, H. T. Cheung, and S. Sternhell, *J. Chem. Soc., Perkin Trans.* **I** 13 (1973).

548. I. Iwataki and J. Comin, *Tetrahedron* **27,** 2541 (1971).

549. R. Verpoorte, A. B. Svendsen, and F. Sandberg, *Acta Pharm. Suec.* **12,** 455 (1975); *Chem. Abstr.* **84,** 102333n (1976).

550. R. Verpoorte, *Pharm. Weekblad.* **110,** 447 (1975).

551. M. Koch, J. Garnier, and M. Plat, *Ann. Pharm. Fr.* **30,** 299 (1972).

552. N. G. Bisset and J. D. Phillipson, *Phytochemistry* **13,** 1265 (1974).

553. F. Delle Monache, E. Corio, C. R. Cartoni, A. Carpi, and G. B. Marini-Bettòlo, *Lloydia* **33,** 279 (1970).

554. H. Müller, M. Hesse, P. Waser, H. Schmid, and P. Karrer, *Helv. Chim. Acta* **48,** 320 (1965).

555. W. Rolfsen, L. Bohlin, S. K. Yeboah, M. Geevaratne, and R. Verpoorte, *Planta Med.* **34,** 264 (1978).

556. A. Petitjean, P. Rasoanaivo, and J. M. Razafintsalama, *Phytochemistry* **16,** 154 (1977).

557. H. King, *J. Chem. Soc.*, 955 (1949).

558. G. B. Marini-Bettòlo, M. A. Jorío, A. Pimenta, A. Ducke, and D. Bovet, *Gazz. Chim. Ital.* **84,** 1161 (1954).

559. R. Verpoorte and A. B. Svendsen, *Lloydia* **39,** 357 (1976).
560. R. Verpoorte and A. B. Svendsen, *J. Pharm. Sci.* **67,** 171 (1978).
561. C. Galeffi, A. Lupi, and G. B. Marini-Bettòlo, *Gazz. Chim. Ital.* **106,** 773 (1976).
562. G. B. Marini-Bettòlo, M. A. Ciasca, C. Galeffi, N. G. Bisset, and B.A. Krukoff, *Phytochemistry* **11,** 381 (1972).
563. A. Pimenta, M. A. Jorío, K. Adank, and G. B. Marini-Bettòlo, *Gazz. Chim. Ital.* **84,** 1147 (1954).
564. C. A. Coune and L. J. G. Angenot, *Phytochemistry* **17,** 1447 (1978).
565. C. Coune, *Plant. Med. Phytother.* **12,** 106 (1978).
566. C. Coune and L. Angenot, *Planta Med.* **34,** 53 (1978).
567. M. Koch, E. Fellion, and M. Plat, *Phytochemistry* **15,** 321 (1976).
568. N. G. Bisset, Ph.D. thesis, University of London, 1968.
569. N. G. Bisset, *Chem. Ind. London*, 1036 (1965).
570. K. Biemann, J. S. Grossert, J. M. Hugo, J. Occolowitz, and F. L. Warren, *J. Chem. Soc.*, 2814 (1965).
571. M. Spiteller-Friedmann and G. Spiteller, *Liebigs Ann. Chem.* **712,** 179 (1968).
572. R. Sarfati, M. Paris, and F.-X. Jarreau, *Phytochemistry* **9,** 1107 (1970).
573. N. G. Bisset, J. Bosly, B. C. Das, and G. Spiteller, *Phytochemistry* **14,** 1411 (1975).
574. M. Spiteller-Friedmann and G. Spiteller, *Liebigs Ann. Chem.* **711,** 205 (1968).
575. F. Sandberg, K. Roos, K. J. Ryrberg, and K. Kristiansson, *Tetrahedron Lett.*, 6217 (1968).
576. N. G. Bisset, B. C. Das, and J. Parello, *Tetrahedron* **29,** 4137 (1973).
577. N. G. Bisset and A. A. Khalil, *Phytochemistry* **15,** 1973 (1976).
578. N. G. Bisset, *Tetrahedron Lett.*, 3107 (1968).
579. N. Bisset, *C. R. Acad. Sci. Ser. C.* **261,** 5237 (1965).
580. N. G. Bisset, *Lloydia* **35,** 203 (1972).
581. N. G. Bisset and M. D. Walker, *Phytochemistry* **13,** 525 (1974).
582. N. G. Bisset and J. D. Phillipson, *Lloydia* **39,** 263 (1976).
583. N. G. Bisset, A. K. Choudhury, and M. D. Walker, *Phytochemistry* **13,** 255 (1974).
584. F. Delle Monache, E. Corio, and G. B. Marini-Bettòlo, *Ann. Ist. Super. Sanità* **3,** 564 (1967).
585. N. K. Hart, S. R. Johns, J. A. Lamberton, H. Suares, and R. E. Summons, *Austr. J. Chem.* **24,** 1741 (1971).
586. C. Mathis and P. Duquénois, *Ann. Pharm. Fr.* **21,** 17 (1963).
587. M. A. Iorio, O. Corvillon, H. M. Alves, and G. B. Marini-Bettòlo, *Gazz. Chim. Ital.* **86,** 923 (1956).
588. G. B. Marini-Bettòlo, S. E. Giuffra, C. Galeffi, and E. M. Delle Monache, *Gazz. Chim. Ital.* **103,** 591 (1973).
589. C. Vamvacas, W. v. Philipsborn, E. Schlittler, H. Schmid, and P. Karrer, *Helv. Chim. Acta* **40,** 1793 (1957).
590. E. Schlittler and J. Hohl, *Helv. Chim. Acta* **35,** 29 (1952).
591. E. Bächli, C. Vamvacas, H. Schmid, and P. Karrer, *Helv. Chim. Acta* **40,** 1167 (1957).
592. S. I. Heimberger and A. I. Scott, *J. Chem. Soc. Chem. Commun.*, 217 (1973).
593. A. Guggsiberg, M. Hesse, H. Schmid, and P. Karrer, *Helv. Chim. Acta* **49,** 1 (1966).
594. C. Galeffi, E. M. Delle Monache, and G. B. Marini-Bettòlo, *J. Chromatogr.* **88,** 416 (1974).
595. P. Šefčovič, L. Dúbravková, and F. G. Torto, *Planta Med.* **16,** 143 (1968).
596. N. G. Bisset and A. K. Choudhury, *Phytochemistry* **13,** 265 (1974).
597. K. Warnat, *Helv. Chim. Acta* **14,** 997 (1931).
598. C. Galeffi, M. Nicoletti, I. Messana, and G. B. Marini-Bettòlo, *Tetrahedron* **35,** 2545 (1979).

599. F. Delle Monache, A. G. de Brovetto, E. Cor, and G. B. Marini-Bettòlo, *J. Chromatogr.* **32,** 178 (1968).

600. N. G. Bisset and J. D. Phillipson, *Phytochemistry* **12,** 2049 (1973).

601. H. Singh, V. K. Kapoor, J. D. Phillipson, and N. G. Bisset, *Phytochemistry* **14,** 587 (1975).

602. D. M. Franco, P. T. Aldo, and G. B. Marini-Bettòlo, *Tetrahedron Lett.*, 2009 (1969).

603. M. Koch, M. Plat, and J. Le Men, *Tetrahedron* **25,** 3377 (1969).

604. G. B. Marini-Bettòlo, E. M. Delle Monache, S. E. Giuffra, and C. Galeffi, *Gazz. Chim. Ital.* **101,** 971 (1971).

605. G. B. Marini-Bettòlo, P. de Berredo Carneiro, and G. C. Casinovi, *Gazz. Chim. Ital.* **86,** 1148 (1956).

606. M. Koch, M. Plat, B. C. Das, and J. Le Men, *Bull. Soc. Chim. Fr.*, 3250 (1968).

607. M. Koch, M. Plat, B. C. Das, E. Fellion, and J. Le Men, *Ann. Pharm. Fr.* **27,** 229 (1969).

608. A. Penna, M. A. Iorio, S. Chiavarelli, and G. B. Marini-Bettòlo, *Gazz. Chim. Ital.* **87,** 1163 (1957).

609. C. Galeffi, M. C. Rendina, E. M. Delle Monache, and G. B. Marini-Bettòlo, *Il Farmaco* **26,** 1100 (1971).

610. G. B. Marini-Bettòlo, M. Lederer, M. A. Jorio, and A. Pimenta, *Gazz. Chim. Ital.* **84,** 1155 (1954).

611. A. R. Battersby, R. Binks, H. F. Hodson, and D. A. Yeowell, *J. Chem. Soc.*, 1848 (1960).

612. H. Asmis, H. Schmid, and P. Karrer, *Helv. Chim. Acta* **37,** 1983 (1954).

613. H. Asmis, P. Waser, H. Schmid, and P. Karrer, *Helv. Chim. Acta* **38,** 1661 (1955).

614. A. R. Battersby, H. F. Hodson, G. V. Rao, and D. A. Yeowell, *J. Chem. Soc. C*, 2335 (1967).

615. H. King, *J. Chem. Soc.*, 3263 (1949).

616. K. Adank, D. Bovet, A. Ducke, and G. B. Marini-Bettòlo, *Gazz. Chim. Ital.* **83,** 966 (1953).

617. C. Richard, C. Delaude, L. Le Men-Olivier, and J. Le Men, *Phytochemistry* **27,** 539 (1978).

618. R. Verpoorte, E. W. Kodde, and A. Baerheim-Svendsen, *Planta Med.* **34,** 62 (1978).

619. L. Angenot and N. G. Bisset, *J. Pharm. Belg.* **26,** 585 (1971).

620. L. J. G. Angenot, C. A. Coune, M. J. G. Tits, and K. Yamada, *Phytochemistry* **17,** 1687 (1978).

621. L. Angenot, C. Coune, and M. Tits, *J. Pharm. Belg.* **33,** 11 (1978).

622. L. Angenot, M. Dubois, C. Ginion, W. v. Dorsser, and A. Dresse, *Arch. Int. Pharmacodyn.* **215,** 246 (1975).

623. L. Dupont, J. Lamotte-Brasseur, O. Dideberg, H. Campsteyn, M. Vermeire, and L. Angenot, *Acta Crystal.* **B33,** 1801 (1977).

624. M. Koch, E. Fellion, and M. Plat, *Ann. Pharm. Fr.* **31,** 45 (1973).

625. M. Koch and M. Plat, *C. R. Acad. Sci., Ser.* C **273,** 753 (1971).

626. L. Angenot, O. Dideberg, and L. Dupont, *Tetrahedron Lett.*, 1357 (1975).

627. L. Angenot, C. Coune, and M. Tits, *J. Pharm. Belg.* **33,** 284 (1978).

628. L. Angenot, *Planta Med.* **27,** 24 (1975).

629. L. Angenot, *Plant. Med. Phytother.* **12,** 123 (1978).

630. O. Dideberg, J. Lamotte-Brasseur, L. Dupont, H. Campsteyn, M. Vermeire, and L. Angenot, *Acta Crystal.* **B33,** 1796 (1977).

631. C. Richard, C. Delaude, L. Le Men-Olivier, J. Lévy, and J. Le Men, *Phytochemistry* **15,** 1805 (1976).

632. M. Tits and D. Tavernier, *Plant. Med. Phytother.* **12,** 92 (1978).

633. L. Angenot, N. G. Bisset, and M. Franz, *Phytochemistry* **14,** 2519 (1975).

634. M. J. G. Tits and L. Angenot, *Planta Med.* **34,** 57 (1978).

635. N. G. Bisset and J. D. Phillipson, *J. Pharm. Pharmacol.* **25,** 563 (1973).

636. N. G. Bisset and A. K. Choudhury, *Phytochemistry* **13,** 259 (1974).
637. H.-G. Boit and L. Paul, *Naturwissenschaften* **47,** 136 (1960).
638. R. T. Brown and L. R. Row, *J. Chem. Soc. Chem. Commun.*, 453 (1967).
639. L. Merlini and G. Nasini, *Gazz. Chim. Ital.* **98,** 974 (1968).
640. R. T. Brown, K. V. J. Rao, P. V. S. Rao, and L. R. Row, *J. Chem. Soc. Chem. Commun.*, 350 (1968).
641. R. T. Brown, C. L. Chapple, and G. K. Lee, *J. Chem. Soc. Chem. Comun.*, 1007 (1972).
642. R. T. Brown and A. A. Charalambides, *Phytochemistry* **14,** 2527 (1975).
643. R. T. Brown and A. A. Charalambides, *Tetrahedron Lett.*, 3429 (1974).
644. R. T. Brown and B. F. M. Warambwa, *Phytochemistry* **17,** 1686 (1978).
645. R. T. Brown and A. A. Charalambides, *Experientia* **31,** 505 (1975).
646. W. P. Blackstock, R. T. Brown, and G. K. Lee, *J. Chem. Soc. Chem. Commun.*, 910 (1971).
647. R. T. Brown and A. A. Charalambides, *J. Chem. Soc. Chem. Comm.*, 553 (1974).
648. R. T. Brown and S. B. Fraser, *Tetrahedron Lett.*, 841 (1973).
649. R. T. Brown and A. A. Charalambides, *J. Chem. Soc. Chem. Comm.*, 765 (1973).
650. W. P. Blackstock and R. T. Brown, *Tetrahedron Lett.*, 3727 (1971).
651. W. P. Blackstock, R. T. Brown, C. L. Chapple, and S. B. Fraser, *J. Chem. Soc. Chem. Comm.* 1006 (1972).
652. R. T. Brown and C. L. Chapple, *Tetrahedron Lett.*, 2723 (1976).
653. R. T. Brown and S. B. Fraser, *Tetrahedron Lett.*, 1957 (1974).
654. R. T. Brown and C. L. Chapple, *Tetrahedron lett.*, 1629 (1976).
655. R. T. Brown, S. B. Fraser, and J. Banerji, *Tetrahedron Lett.*, 3335 (1974).
656. S. R. Johns, J. A. Lamberton, and J. L. Occolowitz, *Austr. J. Chem.* **20,** 1463 (1967).
657. J. D. Phillipson and S. R. Hemingway, *Phytochemistry* **13,** 2621 (1974).
658. J. Le Men, C. Kan, P. Potier, and M.-M. Janot, *Ann. Pharm. Fr.* **23,** 691 (1965).
659. T. A. Henry, K. S. Kirby, and G. E. Shaw, *J. Chem. Soc.*, 524 (1945).
660. O. Hesse, *Ber. Dtsch. Chem. Ges.* **10,** 2152 (1877).
661. J. Melchio, A. Bouquet, M. Pais, and R. Goutarel, *Tetrahedron Lett.*, 315 (1977).
662. Raymond-Hamet, *C. R. Acad. Sci.* **212,** 305 (1941).
663. M.-M. Janot and R. Goutarel, *C. R. Acad. Sci.* **218,** 852 (1944).
664. Raymond-Hamet, *C. R. Acad. Sci.* **197,** 860 (1933).
665. M.-M. Janot and R. Goutarel, *C. R. Acad. Sci.* **220,** 617 (1945).
666. N. A. Cu, R. Goutarel, and M.-M. Janot, *Bull. Soc. Chim. Fr.*, 1292 (1957).
667. T. H. van der Meulen and G. J. M. van der Kerk, *Rec. Trav. Chim. Pays-Bas* **83,** 141 (1964).
668. H.-P. Husson, C. Kan-Fan, T. Sévenet, and J.-P. Vidal, *Tetrahedron Lett.*, 1889 (1977).
669. H. Bohrmann, C. Lau-Cam, J. Tashiro, and H. W. Youngken, *Phytochemistry* **8,** 645 (1969).
670. C. A. Lau-Cam and J. Tashiro, *Phytochemistry* **10,** 1655 (1971).
671. E. J. Shellard and K. Sarpong, *J. Pharm. Pharmacol.* **22,** 34 S (1970).
672. A. H. Beckett, E. J. Shellard, and A. N. Tackie, *J. Pharm. Pharmacol.* **15,** 166 T (1963).
673. W. F. Trager, C. M. Lee, J. D. Phillipson, R. E. Haddock, D. B. Dwuma-Badu, and A. H. Beckett, *Tetrahedron* **24,** 523 (1968).
674. J. D. Phillipson, P. Tantivatana, E. Torpo, and E. J. Shellard, *Phytochemistry* **12,** 1507 (1973).
675. E. J. Shellard, A. H. Beckett, P. Tantivatana, J. D. Phillipson, and C. M. Lee, *J. Pharm. Pharmacol.* **18,** 553 (1966).
676. E. J. Shellard and K. Sarpong, *J. Pharm. Pharmacol.* **23,** 559 (1971).
677. E. J. Shellard, J. D. Phillipson, and K. Sarpong, *Phytochemistry* **10,** 2505 (1971).

678. E. J. Shellard, A. H. Beckett, P. Tantivatana, J. D. Phillipson, and C. M. Lee, *Planta Med.* **15,** 245 (1967).

679. E. J. Shellard and K. Sarpong, *Tetrahedron* **27,** 1725 (1971).

680. E. J. Shellard, J. D. Phillipson, and D. Gupta, *Planta Med.* **17,** 146 (1969).

681. E. J. Shellard and P. K. Lala, *Planta Med.* **31,** 395 (1977).

682. E. J. Shellard, *Pharm. Weekblad* **106,** 224 (1971).

683. S. R. Hemingway, P. J. Houghton, J. D. Phillipson, and E. J. Shellard, *Phytochemistry* **14,** 557 (1975).

684. P. J. Houghton and E. J. Shellard, *Planta Med.* **26,** 104 (1974).

685. G. Barger, E. Dyer, and L. J. Sargent, *J. Org. Chem.* **4,** 418 (1939).

686. E. J. Shellard and P. K. Lala, *Planta Med.* **33,** 63 (1978).

687. E. J. Shellard, P. J. Houghton, and P. K. Lala, *Phytochemistry* **16,** 1427 (1977).

688. A. H. Beckett, E. J. Shellard, J. D. Phillipson, and C. M. Lee, *J. Pharm. Pharmacol.* **17,** 753 (1965).

689. E. J. Shellard, P. J. Houghton, and M. Resha, *Planta Med.* **34,** 253 (1978).

690. E. J. Shellard, P. J. Houghton, and M. Resha, *Planta Med.* **34,** 26 (1978).

691. J. B. Hendrickson and J. J. Sims, *Tetrahedron Lett.*, 929 (1963).

692. A. H. Beckett, E. J. Shellard, and A. N. Tackie, *J. Pharm. Pharmacol.* **15,** 158 T (1963).

693. J. D. Phillipson, D. Rungsiyakul, and E. J. Shellard, *Phytochemistry* **12,** 2043 (1973).

694. G. I. Dimitrienko, D. G. Murray, and S. McLean, *Tetrahedron Lett.*, 1961 (1974).

695. S. McLean, G. I. Dmitrienko, and A. Szakolcai, *Can. J. Chem.* **54,** 1262 (1976).

696. G. I. Dmitrienko, A. Szakolcai, and S. McLean, *Tetrahedron Lett.*, 2599 (1974).

697. M. Sainsbury and B. Webb, *Phytochemistry* **14,** 2691 (1974).

698. S. R. Johns, J. A. Lamberton, and A. A. Sioumis, *Austr. J. Chem.* **23,** 1285 (1970).

699. K. L. Stuart and R. B. Woo-Ming, *Tetrahedron Lett.*, 3853 (1974).

700. J.-L. Pousset, A. Bouquet, A. Cavé, A. Cavé, and R.-R. Paris, *C. R. Acad. Sci., Ser. C* **272,** 665 (1971).

701. J. Levesque, J.-L. Pousset, and A. Cavé, *C. R. Acad. Sci., Ser. C* **280,** 593 (1975).

702. J. Levesque, J. L. Pousset, and A. Cavé, *Fitoterapia* **48,** 5 (1977).

703. J. Levesque, J.-L. Pousset, and A. Cavé, *C. R. Acad. Sci., Ser. C* **279,** 1053 (1974).

704. J. Levesque, J.-L. Pousset, A. Cavé, and A. Cavé, *C. R. Acad. Sci., Ser. C* **278,** 959 (1974).

705. J. L. Pousset, J. Levesque, A. Cavé, F. Picot, P. Potier, and R. R. Paris, *Plant. Med. Phytother.* **8,** 51 (1974); *Chem. Abstr.* **81,** 117054j (1974).

706. P. Karrer, R. Schwyzer, and A. Flam, *Helv. Chim. Acta* **35,** 851 (1952).

707. T. H. van der Meulen and G. J. M. van der Kerk, *Rec. Trav. Chim. Pays-Bas* **83,** 148 (1964).

708. R. Goutarel, M.-M. Janot, V. Prelog, and W. I. Taylor, *Helv. Chim. Acta* **33,** 150 (1950).

709. F. Hotellier, P. Delaveau, R. Besselièvre, and J.-L. Pousset, *C. R. Acad. Sci., Ser. C* **282,** 595 (1976).

710. F. Hotellier, P. Delaveau, and J.-L. Pousset, *Phytochemistry* **14,** 1407 (1975).

711. F. Hotellier, P. Delaveau, and J. L. Pousset, *Plant. Med. Phytother.* **11,** 106 (1977); *Chem. Abstr.* **87,** 114684c (1977).

712. R. T. Brown, C. L. Chapple, and A. G. Lashford, *Phytochemistry* **16,** 1619 (1977).

713. J. D. Phillipson, S. R. Hemingway, and C. E. Ridsdale, *Lloydia* **41,** 503 (1978).

714. Raymond-Hamet, *C. R. Acad. Sci.* **259,** 3872 (1964).

715. J. D. Phillipson and S. R. Hemingway, *Phytochemistry* **14,** 1855 (1975).

716. J. D. Phillipson and S. R. Hemingway, *Phytochemistry* **12,** 1481 (1973).

717. A. F. Beecham, N. K. Hart, S. R. Johns, and J. A. Lamberton, *Austr. J. Chem.* **21,** 491 (1968).

718. S. R. Johns and J. A. Lamberton, *Tetrahedron Lett.*, 4883 (1966).
719. J. D. Phillipson and S. R. Hemingway, *J. Pharm. Pharmacol.* **25** (Suppl.), 143 P (1973).
720. C. Cistaro, R. Mondelli, and M. Anteunis, *Helv. Chim. Acta* **59,** 2249 (1976).
721. L. Merlini, G. Nasini, and R. E. Haddock, *Phytochemistry* **11,** 1525 (1972).
722. L. Merlini, R. Mondelli, G. Nasini, and M. Hesse, *Tetrahedron* **26,** 2259 (1970).
723a. L. Merlini, R. Mondelli, G. Nasini, and M. Hesse, *Tetrahedron* **23,** 3129 (1967).
723b. L. Merlini, R. Mondelli, G. Nasini, and M. Hesse, *Tetrahedron Lett.*, 1571 (1967).
724. L. Merlini, G. Nasini, and J. D. Phillipson, *Tetrahedron* **28,** 5971 (1972).
725. C. Cistaro, L. Merlini, R. Mondelli, and G. Nasini, *Gazz. Chim. Ital.* **103,** 153 (1973).
726. W. I. Taylor and Raymond-Hamet, *C. R. Acad. Sci. Ser. D* **262,** 1141 (1966).
727. K. C. Chan, *Tetrahedron Lett.*, 3403 (1968).
728. D. Ponglux, P. Tantivatana, and S. Pummangura, *Planta Med.* **31,** 26 (1977).
729. J. Haginiwa, S.-i. Sakai, K. Takahashi, M. Taguchi, and S. Seo, *Yakugaku Zasshi* **91,** 575 (1971).
730. H. Kondo and T. Ikeda, *Yakugaku Zasshi* **61,** 416, 453 (1941).
731. N. Finch and W. I. Taylor, *Tetrahedron Lett.*, 167 (1963).
732. N. Aimi, E. Yamanaka, J. Endō, S.-i. Sakai, and J. Haginiwa, *Tetrahedron Lett.*, 1081 (1972).
733. J. D. Phillipson and S. R. Hemingway, *Phytochemistry* **12,** 2791 (1973).
734. K. C. Chan, F. Morsingh, and G. B. Yeoh, *J. Chem. Soc. Ser. C*, 2245 (2966).
735. J. D. Phillipson and S. R. Hemingway, *Phytochemistry* **12,** 2795 (1973).
736. J. Haginiwa, S.-i. Sakai, N. Aimi, E. Yamanaka, and N. Shinma, *Yakugaku Zasshi* **93,** 448 (1973).
737. N. Aimi, E. Yamanaska, N. Shinma, M. Fujiu, J. Kurita, S.-i. Sakai, and J. Haginiwa, *Chem. Pharm. Bull. Jpn.* **25,** 2067 (1977).
738. T. Nozoye, Y. Shibanuma, and A. Shigehisa, *Yakugaku Zasshi* **95,** 758 (1975).
739. S. R. Hemingway and J. D. Phillipson, *J. Pharm. Pharmacol.* **26** (Suppl.), 113 P (1974).
740. Raymond-Hamet, *C. R. Acad. Sci.* **235,** 547 (1952).
741. J. A. Goodson, *J. Chem. Soc.*, 2626 (1932).
742. G. Croquelois, N. Kunesch, M. Debray, and J. Poisson, *Plant. Med. Phytother.* **6**, 122 (1972); *Chem. Abstr.*, **77,** 98778x (1972).
743. M. Kučera, V. O. Marquis, and H. Kučerova, *Planta Med.* **21,** 343 (1972).
744. A. Chatterjee and S. Ghosal, *Naturwissenschaften* **48,** 219 (1961).
745. T. M. Sharp, *J. Chem. Soc.*, 287 (1934).
746. G. H. Svoboda, *J. Am. Pharm. Assoc., Sci. Ed.* **46,** 508 (1957).
747. R. C. Elderfield and S. L. Wythe, *J. Org. Chem.* **19,** 683 (1954).
748. W. D. Crow and Y. M. Greet, *Austr. J. Chem.* **8,** 461 (1955).
749. W. D. Crow, N. C. Hancox, S. R. Johns, and J. A. Lamberton, *Austr. J. Chem.* **23,** 2489 (1970).
750. S. Mamatas-Kalamaras, T. Sévenet, C. Thal, and P. Potier, *Phytochemistry* **14,** 1637 (1975).
751. B. C. Das, J.-P. Cosson, and G. Lukacs, *J. Org. Chem.* **42,** 2785 (1977).
752. B. C. Das, J. P. Cosson, G. Lukacs, and P. Potier, *Tetrahedron Lett.*, 4299 (1974).
753. N. K. Hart, S. R. Johns, and J. A. Lamberton, *Austr. J. Chem.* **25,** 2739 (1972).
754. G. Lewin, N. Kunesch, A. Cavé, T. Sevenet, and J. Poisson, *Phytochemistry* **14,** 2067 (1975).

Subject Index

Organism Index

All binomials and genera appear in *italics.*